AF595362

Weed Ecology and New Approaches for Management

Weed Ecology and New Approaches for Management

Editors

Anna Kocira
Mariola Staniak

MDPI • Basel • Beijing • Wuhan • Barcelona • Belgrade • Manchester • Tokyo • Cluj • Tianjin

Editors
Anna Kocira
Institute of Agricultural Sciences
State School of Higher Education in Chełm
Chełm
Poland

Mariola Staniak
Department of Forage Crop Production
Institute of Soil Science and Plant Cultivation - State Research Institute
Puławy
Poland

Editorial Office
MDPI
St. Alban-Anlage 66
4052 Basel, Switzerland

This is a reprint of articles from the Special Issue published online in the open access journal *Agriculture* (ISSN 2077-0472) (available at: www.mdpi.com/journal/agriculture/special_issues/Weed_Ecology_Approaches).

For citation purposes, cite each article independently as indicated on the article page online and as indicated below:

LastName, A.A.; LastName, B.B.; LastName, C.C. Article Title. *Journal Name* **Year**, *Volume Number*, Page Range.

ISBN 978-3-0365-1512-0 (Hbk)
ISBN 978-3-0365-1511-3 (PDF)

Contents

MDPI

Editorial
Weed Ecology and New Approaches for Management

Anna Kocira [1,*,†] and Mariola Staniak [2,*,†]

1 Institute of Agricultural Sciences, State School of Higher Education in Chełm, Pocztowa 54, 22-100 Chełm, Poland
2 Department of Forage Crop Production, Institute of Soil Science and Plant Cultivation—State Research Institute, Czartoryskich 8, 24-100 Puławy, Poland
* Correspondence: akocira@pwsz.chelm.pl (A.K.); staniakm@iung.pulawy.pl (M.S.)
† These authors contributed equally to this work.

Citation: Kocira, A.; Staniak, M. Weed Ecology and New Approaches for Management. *Agriculture* **2021**, *11*, 262. https://doi.org/10.3390/agriculture11030262

Received: 1 March 2021
Accepted: 15 March 2021
Published: 19 March 2021

1. Introduction

The rich biodiversity of agricultural fields and their surroundings enhances natural ecosystems and has a positive impact on their productivity and resistance, e.g., by maintaining a balance between crop pathogens and pests, ensuring pollination and nutrient cycles [1]. Segetal flora, commonly known as weeds, are present in all types of crops, and their composition, density and biomass depend on many factors, including the species of crop, habitat conditions, the performance of crop management procedures, and the time of year [2]. From the point of view of human management, weeds are undesirable species in fields, because their presence reduces the yield of crops and contributes to the deterioration of the crop quality. Some weeds are also harmful because they can be hosts of pests or crop pathogens, parasites or semiparasites, or have poisonous properties towards animals and humans. However, apart from the negative aspects, weeds also have many positive functions in agrocenoses, such as increasing the number of pollinators, providing habitats and food for beneficial insects, and some secrete substances that inhibit the development of pests [3].

Weed control is important in growing systems, requiring the integration of different plant protection strategies and methods. The Special Issue "*Weed Ecology and New Approaches for Management*" contains 14 original research articles and 1 review article covering topics related to the biology and damage of weeds, especially related to the health and yielding of crops and the biodiversity of segetal weeds, as well as integrated methods of weed control and herbicide resistance. It includes articles related to the effects of tillage and management intensity on species diversity and weed abundance in winter cereals [4,5] and legumes e.g., soybean [6], pea [7] and lupine [7,8]. Several articles have shown that crop species and habitat conditions significantly affect the abundance and botanical composition of the associated flora (segetal weeds) [4,5,9]. Moreover, it was found that the effectiveness of the use of plant protection products in wheat cultivation depended on the cultivar [10]. Increasing the intensity of the production also affects the yields of lupine [8] and spelt wheat [10], as well as the health of soybean [6] and chamomile [11]. In perennial crops (willow), segetal weeds are the dominant flora, although with the age of the plantation their number decreases in favor of apophytes [12]. Dangerous to the native flora are invasive species that require monitoring, such as exotic *Tamarix* species, introduced to South Africa as ornamental plants [13]. Two articles on potato cultivation stated that good results in weed control can be achieved by combining mechanical and chemical methods [14] or by the pre-emergence application of the mixture of herbicides with an adjuvant [15]. In turn, la Cruz et al. [16] identified glyphosate resistance in a new species of *Amaranthus viridis* in citrus orchards. Furthermore, it was confirmed that the level of damage after herbicide application and the regeneration rate of camelina plants depended on the cultivar [17]. The review article presents the ecosystem advantages of legume cover crops, including their effect on weed control [18].

2. Biodiversity of Weeds and Tillage Systems

During the period of the intensive development of conventional agriculture, human activity was directed mainly at the elimination of undesirable flora in crops, therefore many weed species are now threatened with extinction. Studies carried out in eight European countries on the effects of pesticides on the biodiversity of segetal flora showed a negative relationship between grain yield and weed diversity. Changes in agricultural production occurring in the last 50 years resulted in a threefold reduction in the diversity of weeds in wheat cultivation [19]. The studies by Chamorro et al. [20] carried out in Catalonia showed not only a reduction in total weed infestation (69%), but also the frequency of weeds (58%) and species richness (47%) over a period of about 50 years, with the most dramatic decreases in the groups of segetal (75%) and rare species (87%).

An opportunity for the maintenance of the population of segetal flora is the promotion of the idea of sustainable and organic agriculture, in which the aim is to reduce the number of species accompanying cultivated plants to such a level that they do not significantly reduce crop yields [21]. On organic farms, weed infestation is limited without the use of herbicides, e.g., through crop management practices. Farmers on organic farms are also allowed to use biostimulants, organic foliar fertilizers which can increase the resistance of cultivated plants to stressful conditions, improve the health of plants and even reduce weed infestation [22], as can be seen from the use of Herbagreen Basic and Bio-algeen, which reduced the infestation chamomile by pathogens and reduced the number and increased biodiversity of weeds [11]. In the organic cultivation of soft-stemmed species, it is good practice to use low support plants that do not compete for with the main crop for light, protect it against lodging and, by taking up additional space, increase its competitiveness, and effectively reduce weed and fungal pathogen pressure [23].

On a correctly managed organic farm, the occurrence of weeds is effectively reduced but, in principle and in practice, they are not completely eliminated. It is considered that the greater the diversity in the species composition of a weed community, the less harmful it is to the crop [24,25]. Berbeć et al. [4] showed higher species diversity and weed abundance in the ecological system than in the conventional low-input system, but in both farming systems the Shannon–Wiener diversity index had high values, while the Simpson dominance index had low values, indicating their high importance for biodiversity.

The germination rate of weed seeds and their subsequent growth and competitiveness against the crop depend on the tillage [26]. In intensive ploughing systems, weed seeds are evenly distributed in the arable layer, while in reduced tillage systems, a large number of them are concentrated in the surface layer of the soil [27]. Feledyn-Szewczyk et al. [5] showed that weed infestation and soil weed seed bank in winter wheat were considerably higher in simplified systems (reduced tillage, direct sowing) than in the ploughing system, although not in all years of the study. Bojarszczuk and Podleśny [7] showed that simplified tillage significantly increased the weed infestation in legume. The high importance of the tillage system on the biodiversity of the segetal flora was confirmed by the biodiversity indices. Gawęda et al. [6] found that the abundance and dry weight of the weeds in the soybean cultivated in the no-till system was higher than in the conventional plough system. Tillage simplification also contributes to changes in the species composition of weeds [28], and to an increase in plant infection by fungal pathogens in soybean [6].

3. Biodiversity of Weeds in Annual Crops

There is a conflict between the productivity of the crop and the weeds, which compete with the crop for limited environmental resources. Oerke [29] found that potential yield losses due to competition between weeds and crops can amount to around 34%. The most unfavorable situation occurs when there is dominance of one or several weed species, e.g., in fields treated with herbicides. Weed compensation may consequently lead to a significant decrease in the yield of the cultivated crop [30]. Faligowska et al. [8] showed that the seed yield and its quality (protein content) of lupine was determined mainly by the intensity of farming. The yield and quality of narrow-leaved lupine seeds in the conventional system was higher than in a low-input or a medium-input system. The low-input technology provided the lowest seed production cost and the highest income, despite the fact that the weed infestation in lupines in this cropping system was higher than in conventional cropping [8,31].

The large percentage of cereals in the structure of crops in Poland facilitates the growth of segetal plants in arable fields, but the abundance and botanical composition of the associated flora depend primarily on the crop species and habitat conditions [4,5]. Assessment of the diversity of the segetal flora carried out by Sawicka et al. [9] showed a higher abundance of weeds in spring wheat and triticale than in winter wheat. In turn, the highest species diversity was observed in triticale, and the lowest in winter wheat cultivation.

Haliniarz et al. [10] showed that the effectiveness of the use of plant protection products in the cultivation of spelt wheat depended on the cultivar. Under intensive cultivation, lower weed infestation and higher yields were obtained for the cultivar Rokosz compared to Schwabenspelz. Further, Andruszczak [32] showed that the chemical protection of spelt wheat improved yield and quality of grains, but this depended significantly on the wheat cultivar.

4. Biodiversity of Weeds in Permanent Crops

A relatively new agricultural practice is the cultivation of perennial species on arable land for energy purposes. However, replacing annual with perennial crops has an impact on biodiversity in terms of the genetic, species and habitat levels [33]. Feledyn-Szewczyk et al. [34] conducted a study assessing the impact of perennial energy crops on the diversity of associated flora, and showed changes in the flora communities, which consisted in an increase in the proportion of perennial species at the expense of annual species typical for agricultural crops, and an increase in the abundance of species characteristic for ruderal habitats, meadows and forests.

In perennial plantations, the accompanying flora have different development conditions than in annual crops due to different cultivation technologies. In the case of trees and bushes intended for energy purposes, the long period between harvests (2–3 years) influences the habitat conditions related to temperature, humidity and limited light, which determine the species composition and abundance of weeds. The age of the crop and soil conditions are also important [35]. The study of Janicka et al. [12] showed that a significant proportion of the flora in cultivation *Salix viminalis* constituted segetal weeds occurring commonly in cereals and root crops. With an increase in the age of the plantations, the number of segetal species decreased in favor of apophytes. Salix plantation included several medicinal and melliferous species important for humans and biodiversity, but also invasive species dangerous for native flora that need monitoring. Examples of plants with high invasive potential are the exotic species *Tamarix*, which were introduced to South Africa as ornamental plants and are now threatening riparian ecosystems over 10 million ha of land [13]. Setshedi et al. [13] state that in order to protect local biodiversity, the continuous monitoring of these species and the constant and consistent removal of dangerous plants are needed to restore native ecosystems.

5. Chemical and Nonchemical Methods of Plant Protection

To successfully manage weeds it is important to understand the interactions between weed control methods and the weed spectrum, and to properly manage the cultivation system to prevent weed emergence and keep the weed seed bank in the soil low [36]. Proper cultivation and agricultural practices, including a well-planned crop rotation, sowing cover crops, early sowing dates and optimal row spacing are important aspects of weed management in all farming systems [9].

Chemical weed control, on the other hand, should be based on a detailed knowledge of, among other things, the environmental factors influencing the effectiveness of herbicides, both during and after their application [37]. The factors determining the absorption, translocation and degradation of active substances in the plant, like air temperature and humidity are also very important [38]. Pszczółkowski et al. [14] confirmed that the variable course of the atmospheric conditions in the years of the study affected the effectiveness of the herbicides, which differentiated both the abundance of monocotyledonous and dicotyledonous weeds, their dry and fresh weight and the species composition of segetal weeds.

The selection of herbicides, or combinations of herbicides, doses, and timing of applications should be adapted to the degree of the weed infestation, with the aim of controlling a wide range of weeds [37]. In the study of Pszczółkowski et al. [14], good effects in reducing weed infestation in potato cultivation were obtained by combining mechanical and chemical methods (linuron and clomazone). In turn, Barbaś et al. [37] showed that the pre-emergence application of herbicides (metribuzin and rimsulfuron) with an added adjuvant (SN oil) can contribute to better weed control, reduce herbicide dosage, environmental impact and costs and above all, provide a broader spectrum of the active substance action as compared to mechanical weed control.

At the same time, the use of covers (polyethylene film or polypropylene agrotextile) in very early potato cultivation caused an increase in the number and weight of weeds and enriched the species composition of segetal weeds, probably due to higher air and soil temperature [14].

During the action of herbicides on weeds, a selection process occurs by which the least sensitive plants survive, but each weed population is more or less uniform and the process of resistance development can vary in intensity and speed [39]. La Cruz et al. [16] identified glyphosate resistance in a new species, *Amaranthus viridis*. Based on the enzymatic activity tests, it was found that at least one target site-type mechanism was involved in the resistance mechanism. Therefore, a number of procedures should be used to prevent or significantly delay weed resistance to herbicides [15].

The toxicity and rate of degradation of biocides depend primarily on the rate and the structure of the active substance, which is the primary plant protection product [40]. In addition, determining the effects of herbicides on crop development is also important in the context of different cultivars, which is related to their intraspecies variability, which can be observed in the example of *Camelina sativa* [17]. Sobiech et al. [17] showed that camelina cultivar Przybrodzka showed the lowest level of damage after herbicide application, and that plant damage after p-ethyl quizalofop and propakisafop completely disappeared (after 42 days). Herbicides can directly reduce photosynthetic activity or indirectly damage plants, which in turn reduces photosynthetic efficiency [41]. Sobiech et al. [39] found that picloram contributed the most damage to camelina, which also had the biggest effect on PSII function. However, the level of chlorophyll fluorescence parameter values of the plants indicated little damage to PSII for all substances and the possibility of subsequent regeneration of the plant.

The benefits of agricultural conservation practices have increased farmers' interest in cropping legume cover crops (CCs) [42]. Legume CCs are considered a systemic approach to weed control. They reduce weeds and provide other benefits to farming systems, such as improving soil quality [43]. In particular, the wider use of legume CCs is justified in cropping systems with the limited use of mineral fertilizers and pesticides [18].

6. Conclusions

It can be expected that further popularization of the idea of sustainable and organic agriculture, in which the abundance of segetal weeds in agricultural crops is kept at a sufficiently low level, will bring measurable effects by obtaining yields at an acceptable level and by maintaining the biodiversity of arable fields. Attention should also be paid to solutions that allow for a better understanding of the interactions between weed control methods and the extent of their occurrence, which is helpful for effective weed management in plant crops. In chemical protection, weed resistance is an important problem that has increased in recent years and is directly linked to the effectiveness of the crop protection, so research should be intensified to address this issue.

Author Contributions: Conceptualization, A.K. and M.S.; writing—original draft preparation, A.K. and M.S.; and writing—review and editing, A.K. and M.S. All authors have read and agreed to the published version of the manuscript.

Conflicts of Interest: The authors declare no conflict of interest.

References

1. Garbach, K.; Milder, J.C.; Montenegro, M.; Karp, D.S. Biodiversity and ecosystem services in agroecosystems. *Encycl. Agric. Food Syst.* **2014**, *2*. [CrossRef]
2. Petit, S.; Boursault, A.; Le Guilloux, M.; Munier-Jolain, N.; Reboud, X. Weeds in agricultural landscape. A review. *Agron. Sustain. Dev.* **2011**, *31*, 309–317. [CrossRef]
3. Hole, D.G.; Perkins, A.J.; Wilson, J.D.; Alexander, I.H.; Grice, P.V.; Evans, A.D. Does organic farming benefit biodiversity? *Biol. Conserv.* **2005**, *122*, 113–130. [CrossRef]
4. Berbeć, A.K.; Staniak, M.; Feledyn-Szewczyk, B.; Kocira, A.; Stalenga, J. Organic but also low-input conventional farming systems support high biodiversity of weed species in winter cereals. *Agriculture* **2020**, *10*, 413. [CrossRef]
5. Feledyn-Szewczyk, B.; Smagacz, J.; Kwiatkowski, C.A.; Harasim, E.; Woźniak, A. Weed flora and soil seed bank composition as affected by tillage system in three-year crop rotation. *Agriculture* **2020**, *10*, 186. [CrossRef]
6. Gawęda, D.; Haliniarz, M.; Bronowicka-Mielniczuk, U.; Łukasz, J. Weed infestation and health of the soybean crop depending on cropping system and tillage system. *Agriculture* **2020**, *10*, 208. [CrossRef]
7. Bojarszczuk, J.; Podleśny, J. Segetal diversity in selected legume crops depending on soil tillage. *Agriculture* **2020**, *10*, 635. [CrossRef]
8. Faligowska, A.; Panasiewicz, K.; Szymańska, G.; Ratajczak, K.; Sulewska, H.; Pszczółkowska, A.; Kocira, A. Influence of farming system on weed infestation and on productivity of narrow-leaved lupin (*Lupinus angustifolius* L.). *Agriculture* **2020**, *10*, 459. [CrossRef]
9. Sawicka, B.; Krochmal-Marczak, B.; Barbaś, P.; Pszczółkowski, P.; Ćwintal, M. Biodiversity of weeds in fields of grain in South-Eastern Poland. *Agriculture* **2020**, *10*, 589. [CrossRef]
10. Haliniarz, M.; Gawęda, D.; Nowakowicz-Dębek, B.; Najda, A.; Chojnacka, S.; Łukasz, J.; Wlazło, Ł.; Różańska-Boczula, M. Evaluation of the weed infestation, grain health, and productivity parameters of two spelt wheat cultivars depending on crop protection intensification and seeding densities. *Agriculture* **2020**, *10*, 229. [CrossRef]
11. Kwiatkowski, C.A.; Haliniarz, M.; Harasim, E. Weed infestation and health of organically grown chamomile (*Chamomilla recutita* (L.) Rausch.) depending on selected foliar sprays and row spacing. *Agriculture* **2020**, *10*, 168. [CrossRef]
12. Janicka, M.; Kutkowska, A.; Paderewski, J. Diversity of segetal flora in *Salix viminalis* L. crops established on former arable and fallow lands in Central Poland. *Agriculture* **2020**, *11*, 25. [CrossRef]
13. Setshedi, K.T.A.; Newete, S.W. The impact of exotic Tamarix species on riparian plant biodiversity. *Agriculture* **2020**, *10*, 395. [CrossRef]
14. Pszczółkowski, P.; Barbaś, P.; Sawicka, B.; Krochmal-Marczak, B. Biological and agrotechnical aspects of weed control in the cultivation of early potato cultivars under cover. *Agriculture* **2020**, *10*, 373. [CrossRef]
15. Barbaś, P.; Sawicka, B.; Krochmal Marczak, B.; Pszczółkowski, P. Effect of mechanical and herbicide treatments on weed densities and biomass in two potato cultivars. *Agriculture* **2020**, *10*, 455. [CrossRef]
16. Alcántara-de la Cruz, R.; da Silva Amaral, G.; de Oliveira, G.M.; Rufino, L.R.; de Azevedo, F.A.; de Carvalho, L.B.; das Graças Fernandes da Silva, M.F. Glyphosate resistance in Amaranthus viridis in Brazilian citrus orchards. *Agriculture* **2020**, *10*, 304. [CrossRef]
17. Sobiech, Ł.; Grzanka, M.; Kurasiak-Popowska, D.; Radzikowska, D. Phytotoxic effect of herbicides on various camelina [*Camelina sativa* (L.) Crantz] genotypes and plant chlorophyll fluorescence. *Agriculture* **2020**, *10*, 185. [CrossRef]
18. Kocira, A.; Staniak, M.; Tomaszewska, M.; Kornas, R.; Cymerman, J.; Panasiewicz, K.; Lipińska, H. Legume cover crops as one of the elements of strategic weed management and soil quality improvement. A Review. *Agriculture* **2020**, *10*, 394. [CrossRef]
19. Arslan, Z.F. Decrease in biodiversity in wheat fields due to changing agricultural practices in five decades. *Biodivers. Conserv.* **2018**, *27*, 3267–3286. [CrossRef]
20. Chamorro, L.; Masalles, R.M.; Sans, F.X. Arable weed decline in Northeast Spain: Does organic farming recover functional biodiversity? *Agric. Ecosyst. Environ.* **2016**, *223*, 1–9. [CrossRef]
21. Ponce, C.; Bravo, C.; Garcia de Leon, D.; Magana, M.; Alonso, J.C. Effects of organic farming on plant and arthropod communities: A case study in Mediterranean dryland cereals. *Agric. Ecosyst. Environ.* **2011**, *141*, 193–201. [CrossRef]
22. Kwiatkowski, C.A.; Haliniarz, M.; Harasim, E.; Kołodziej, B.; Yakimovich, A. Foliar applied biopreparations as a natural method to increase the productivity of garden thyme (*Thymus vulgaris* L.) and to improve the quality of herbal raw material. *Acta Sci. Pol. Hortorum Cultus* **2020**, *19*, 107–118. [CrossRef]

23. Kraska, P.; Andruszczak, S.; Kwiecińska-Poppe, E.; Staniak, M.; Różyło, K.; Rusecki, R. Supporting crop and different row spacing as factors influencing weed infestation in lentil crop and seed yield under organic farming conditions. *Agronomy* **2020**, *10*, 9. [CrossRef]
24. Baessler, C.; Klotz, S. Effects of changes in agricultural land-use on landscape structure and arable weed vegetation over the last 50 years. *Agric. Ecosyst. Environ.* **2006**, *115*, 3–50. [CrossRef]
25. Berbeć, A.K.; Feledyn-Szewczyk, B. Biodiversity of weeds and soil seed bank in organic and conventional farming systems. *Res. Rural Dev.* **2018**, *2*, 12–19. [CrossRef]
26. Sekutowski, T.; Smagacz, J. Share of anthropophytes in the crop sequence: Winter wheat—Maize—Spring wheat depending on tillage system. *Acta Agrobot.* **2014**, *67*, 117–122. [CrossRef]
27. Tørresen, K.S.; Skuterud, R.; Tandsaether, H.J.; Hagemo, M.B. Long-term experiments with reduced tillage in spring cereals. I. Effects on weed flora, weed seedbank and grain yield. *Crop Protect.* **2003**, *22*, 185–200. [CrossRef]
28. Weber, R.; Sekutowski, T.; Owsiak, Z. Variability of weed infestation of winter wheat cultivars in relation to tillage systems. *Prog. Plant Prot.* **2014**, *54*, 178–184. [CrossRef]
29. Oerke, E.C. Crop losses to pests. *J. Agric. Sci.* **2006**, *144*, 31–43. [CrossRef]
30. Harker, N.K.N.; O'Donovan, J.T. Recent weed control, weed management, and integrated weed management. *Weed Technol.* **2013**, *27*, 1–11. [CrossRef]
31. Faligowska, A.; Panasiewicz, K.; Szymańska, G.; Szukala, J.; Koziara, W.; Pszczółkowska, A. Productivity of white lupin (*Lupinus albus* L.) as an effect of diversified farming systems. *Legum. Res. Int. J.* **2017**, *40*, 872–877. [CrossRef]
32. Andruszczak, S. Reaction of winter spelt cultivars to reduced tillage system and chemical plant protection. *Zemdirb. Agric.* **2017**, *104*, 15–22. [CrossRef]
33. Börjesson, P. Environmental effects of energy crop cultivation in Sweden—I: Identification and quantification. *Biomass Bioenergy* **1999**, *16*, 137–154. [CrossRef]
34. Feledyn-Szewczyk, B.; Matyka, M.; Staniak, M. Comparison of the effect of perennial energy crops and agricultural crops on weed flora diversity. *Agronomy* **2019**, *9*, 695. [CrossRef]
35. Janicka, M.; Kutkowska, A.; Paderewski, J. Diversity of vascular flora accompanying *Salix viminalis* L. crops depending on soil conditions. *Glob. Ecol. Conserv.* **2020**, *23*, e01068. [CrossRef]
36. Boydston, R.A. Managing weeds in potato rotations without herbicides. *Am. J. Potato Res.* **2010**, *87*, 420–427. [CrossRef]
37. Barbaś, P.; Sawicka, B. Dependence of potato crop on weed infestation. *Agronom. Res.* **2020**, *8*, 346–359. [CrossRef]
38. Mayerová, M.; Madaras, M.; Soukup, J. Effect of chemical weed control on crop yields in different crop rotations in a long-term field trial. *Crop. Prot.* **2018**, *114*, 215–222. [CrossRef]
39. Scalla, R. *Les Herbicides. Mode D'action et Principes D'utilisation*; INRA: Paris, France, 1990.
40. Awasthi, R.; Gaur, P.M.; Turner, N.C.; Vadez, V.; Siddique, K.H.M.; Nayyar, H. Effects of individual and combined heat and drought stress during seed filling on the oxidative metabolism and yield of chickpea (*Cicer arietinum*) genotypes differing in heat and drought tolerance. *Crop. Pasture Sci.* **2017**, *68*, 823. [CrossRef]
41. Kaiser, Y.I.; Menegat, A.; Gerhards, R. Chlorophyll fluorescence imaging: A new method for rapid detection of herbicide resistance in *Alopecurus myosuroides*. *Weed Res.* **2013**, *53*, 399–406. [CrossRef]
42. Oliveira, M.C.; Butts, L.; Werle, R. Assessment of cover crop management strategies in Nebraska, US. *Agriculture* **2019**, *9*, 124. [CrossRef]
43. Lüscher, A.; Mueller-Harvey, I.; Soussana, J.F.; Rees, R.M.; Peyraud, J.L. Potential of legume-based grassland-livestock systems in Europe: A review. *Grass Forage Sci.* **2014**, *69*, 206–228. [CrossRef] [PubMed]

Article

Organic but Also Low-Input Conventional Farming Systems Support High Biodiversity of Weed Species in Winter Cereals

Adam Kleofas Berbeć [1,*], Mariola Staniak [2], Beata Feledyn-Szewczyk [1], Anna Kocira [3] and Jarosław Stalenga [1]

[1] Department of Systems and Economics of Crop Production, Institute of Soil Science and Plant Cultivation-State Research Institute, Czartoryskich 8, 24-100 Puławy, Poland; bszewczyk@iung.pulawy.pl (B.F.-S.); stalenga@iung.pulawy.pl (J.S.)

[2] Department of Forage Crop Production, Institute of Soil Science and Plant Cultivation-State Research Institute, Czartoryskich 8, 24-100 Puławy, Poland; staniakm@iung.pulawy.pl

[3] State School of Higher Education in Chełm, Institute of Agricultural Sciences, Pocztowa 54, 22-100 Chełm, Poland; akocira@pwszchelm.edu.pl

* Correspondence: aberbec@iung.pulawy.pl; Tel.: +48-81-4786-824

Received: 10 August 2020; Accepted: 16 September 2020; Published: 18 September 2020

Abstract: In recent years, the European Union has been paying particular attention to the problem of biodiversity loss. The possibilities of its assessment and conservation are included in the latest European Union (EU) policies and reflected in the European Biodiversity Strategy. The biodiversity of weeds in winter cereals in organic and conventional low-input farms in Eastern Poland was investigated during a 3-year period. Significantly more species and larger abundance were found in organic than in conventional farming systems. The biodiversity of these communities was described by Shannon's diversity and Simpson's dominance indices, which showed diversity to be well maintained in both farming systems; however, significantly higher Shannon's index and significantly lower Simpson's index values were observed in organic farms. Both farming systems were the mainstay of endangered and rare species, as well as some invasive weed species. Weed communities of organic farms were dominated mostly by *Setaria pumila* and *Elymus repens*, while conventional farms were dominated by *Juncus bufonius* and *Setaria pumila*. The study showed the importance of organic farming systems for biodiversity conservation. It was also shown that low-input (traditional) conventional farms are also beneficial for biodiversity conservation.

Keywords: biodiversity; weed; organic farming; low-input conventional farming; Shannon's index; Simpson's index

1. Introduction

An agri-ecosystem is an exceptional living environment for wild species of plants (weeds) but also other organisms, such as microorganisms, invertebrates and other higher organisms, which all, including the crop diversity, build up the biodiversity of agricultural lands. Biodiversity is inextricably linked with the provision of ecosystem services. High biodiversity of microorganisms and predatory invertebrates can, for example, bring positive results in the form of biological pest control, organic matter decomposition rate, carbon cycle enhancement, etc. [1]. Higher biodiversity means greater stability of the ecosystem and hence better delivery of ecosystem services. This has a direct impact on the results of agricultural production, especially in farming systems based on these services (e.g., organic farming) [2–5]. The European Union has implemented the European Biodiversity Strategy as one of the major objectives of the EU policies. The main goals of the strategy are conservation of biodiversity, ecosystem services and loss prevention in all ecosystems and also in agricultural areas [6]. Lately, the intensification of agricultural production, excessive water consumption and environmental pollution (mostly with pesticides and nutrient contamination) caused the loss of plant, invertebrate and vertebrate species on arable fields [7–9]. The main aim of the strategy is to prevent further species loss, conservation of their natural habitats and sustainable use of biodiversity of species. Plants occupy the lowest level in the trophic chain; therefore, conservation of their biodiversity will enhance species richness of organisms at higher trophic levels (e.g., animals) [10–12].

Weed infestation is currently one of the most important factors limiting agricultural production, especially in organic farming. Weeds cause loss of both yield quality and quantity. According to Oerke [13], the potential crop losses due to weed–crop competition can amount to around 34% and can be higher than losses caused by pests (18%) and pathogens (16%). On the other hand, weeds are an important, integral part of the agro-ecosystem as they are the basis of the food chain [12]. Some authors reported that the limited abundance of weeds in agricultural production (by even up to 64%) is mostly due to use of herbicides and fertilizers but also due to agricultural practices that promote crop competitiveness against weeds (indirect weed control practices like new varieties of crops) [14–16]. Currently, a few species that were able to adapt to the intensive production conditions of conventional agriculture are dominating the weed species community. According to Arslan [17], the changes in agricultural production in the last 50 years resulted in a threefold reduction in the biodiversity of weeds in wheat production. The abundance of some species, common on arable fields in the past, is currently dramatically low. Moreover, species that used to represent the greatest threat to crops and crop yield are currently disappearing, which was confirmed by Chamorro et al. [18]. Those authors also showed the loss of biodiversity of weed species that are particularly important for birds, pollinators and other animals by nearly 50%. The investigation into weed functional groups showed a reduction in the number of segetal and rare species by 75% and 87%, respectively. Interestingly, the biodiversity of weeds in organic farming systems was twice as high for segetal and species important for fauna and four times as high for rare species compared to the conventional farming system [18]. Travlos et al. [19] reported that, in many cases, organic farming causes on average 30% higher species richness than conventional farming systems and also favors the existence of habitats for rare weed species. Successful agricultural production under organic farming conditions requires a high provision of ecosystem services to ensure efficient functioning of key processes such as biological pest control, cycling of carbon and nutrients, soil fertility building and proper water management [20].

Organic farming is an environmentally friendly alternative to conventional farming. However, the agriculture of Poland is strongly polarized, with high-input agriculture in the northern–western regions of the country and low-input, traditional family farms in the east. This polarization of the county is clearly visible in the numbers shown by Kopiński and Matyka [21], e.g., the average farm size in western regions is more than twice as high (14–18 ha) as that in eastern regions (6 ha), the average nitrogen, phosphorus and potassium fertilizers (NPK) usage is also almost twice as high in western regions (139–194 kg ha^{-1}) than in eastern regions (113 kg ha^{-1}), and annual work unit (AWU) per hectare of agricultural land is also more than three times lower in western regions (8–11 AWU ha^{-1} AL) than in eastern regions (29 AWU ha^{-1} AL). The differences are caused by many natural, organizational and production conditions. It would be interesting to evaluate how these agricultural conditions are reflected in the biodiversity of wild flora, especially when compared to organic farming systems, in which biodiversity of weeds should be well-maintained. The aim of the study was to compare species diversity in two farming systems—organic and low-input conventional—in Eastern Poland.

2. Materials and Methods

2.1. Site Characteristics and Experimental Design

The study on the biodiversity of weeds in winter cereals was carried out in 2012–2014 within the project on the protection of species diversity of valuable natural habitats on agricultural lands on

Natura 2000 areas in the Lublin Voivodeship (KIK/25). Biodiversity monitoring took place annually, between 10 June and 5 July. Fourteen pairs of study surfaces (study squares of a surface of 9 ha), with a predominant share of conventional farming system or a predominant share of certified organic farming system, were selected. Each square had at least a 50% share of organic or conventional agricultural area. Pairs of organic–conventional squares were located as close to each other as possible, to ensure similar soil and climate conditions. Moreover, study squares were at least 500 m away from forests and other shelterbelts to minimize the impact of non-agricultural ecosystems. The fields with winter cereals—wheat (*Triticum* L.), rye (*Secale cereale* L.), triticale (×*Triticosecale* Wittm. ex A. Camus), barley (*Hordeum* L.) or cereal mixtures—were selected annually within 14 pairs of study squares. In the first year of the study (2012), each study square was included in the study, as there was at least one field with winter cereals. In the following years (2013 and 2014), the number of tested pairs of organic–conventional fields was lower due to lack of appropriate crops (winter cereals) within the given squares. The total sample size (2012–2014) involved 38 conventional and 38 organic winter cereal fields (14 pairs in 2012, 12 pairs in 2013 and 12 pairs in 2014). All fields were located in the valleys of Wieprz, Tysmienica and Bug rivers, in the vicinity of Natura 2000 areas in one of the easternmost regions of Poland—the region of Lublin. Light, sandy soils dominated on tested fields. The locations of Lublin region in Poland and fourteen study squares in the Lublin region are shown in Figure 1.

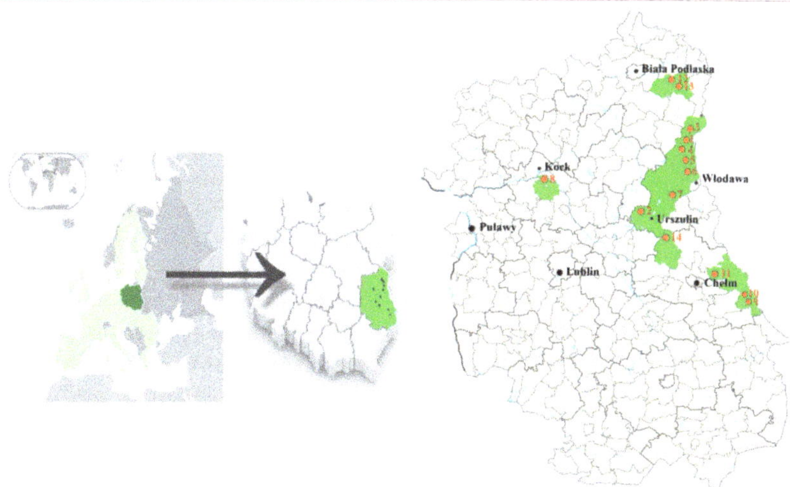

Figure 1. The locations of fourteen pairs of study squares.

2.2. Weed Infestation Analyses

All plant species present in the winter crop field that were not winter cereals were considered as weeds (wild flora). Weed species and their abundance were counted on each field once during the growing season in late June or early July. Five surfaces (replications) of 1 m^2 each were investigated on all selected fields. The surfaces were lined up in a straight line with 10 m spacing between them. Species that were not recorded within the surface, but present in its direct vicinity (2 m^2), were also recorded and added to species list (with the minimal abundance of 1 plant m^{-2}). The biodiversity of weeds was described with the total number of species and their abundance, Shannon's diversity index (H') [22] and Simpson's dominance index (SI) [23]. Shannon's diversity index (H') and Simpson's dominance index (SI) were calculated according to the following equations:

$$H' = -\sum_{i=1}^{s} p_i \ln p_i \tag{1}$$

$$SI = \sum p_i^2 \quad (2)$$

where p—the proportion (n/N) of individuals of one particular species found (n) divided by the total number of individuals found (N), ln—the natural log, s—the number of species.

Shannon's index (H') depends on the number of species and their mutual quantitative proportions. The higher the value of the index, the more diversified the community is. Simpson's dominance index (SI) can range from 0 to 1; values close to 1 indicate a clear dominance of one or several species.

2.3. Statistical Analyses

Non-parametric Mood's test (median-based) was used to determine the significance of differences as most of the biodiversity parameters deviated from the normal distribution. The significance level of the test was set at $\alpha = 0.05$. The significance of differences of diversity indices was calculated using t-test for diversity indices also set at $\alpha = 0.05$. The differences between years of the study was estimated on the basis of single factor ANOVA and pairwise Mann–Whitney post-hoc test ($\alpha = 0.05$). All results of statistical analysis with p-values lower than or equal to 0.05 were considered statistically significant. The statistical analysis was done with Real Statistic add-in for Microsoft Excel.

2.4. Weather Conditions

The average temperature and the sum of precipitations on test sites was similar in all 3 years of the study, with a few exceptions. The year of 2013 had almost twice as much precipitation in May as years 2012 and 2013 (Figure 2). In 2012, the average temperature in the first decade of February was twice as low as in 2013 and 2014 (Figure 3).

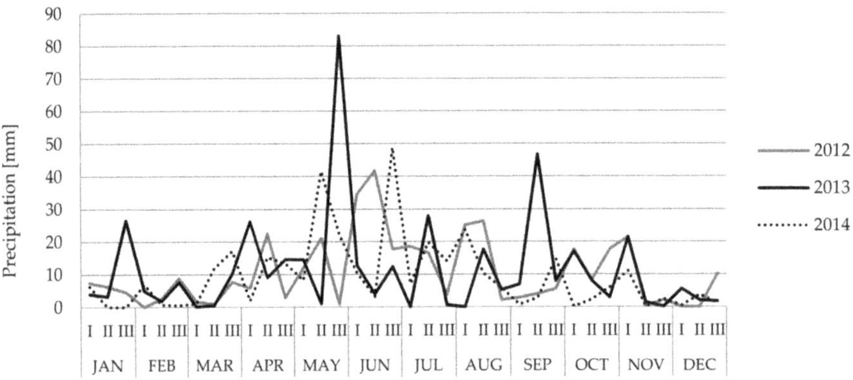

Figure 2. The 10-day sum of precipitations in following months in the years of investigation. I, II, II—the first, the second and the third ten days of month.

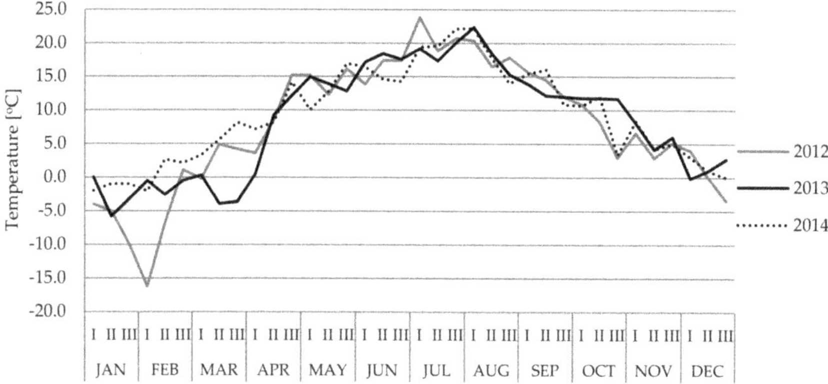

Figure 3. The 10-day average temperature in following months in the years of investigation. I, II, II—the first, the second and the third ten days of month.

2.5. *Organic and Conventional Farms*

A survey with farm owners of winter cereal fields was conducted annually in November (after harvesting of crops) to gather the details of agricultural practices on the fields included in the study. Selected features of tested organic and conventional farms are given in Table 1. Organic and conventional farms had similar agricultural areas, of 28.3 and 24.5 ha, respectively. Organic farms included in the study were slightly larger than the regional average (average area of organic farm in the region of Lublin—15.6 ha) [24]. This was most probably due to the study squares (preselected 9 ha study surface) selection criteria (fields at least 500 m away from forest and possible biodiversity reservoirs), which promoted fields of larger area. Farms of both farming systems had similar quality of soils (mostly sandy soils of IV and V class) (Table 1). The use of some certified mineral fertilizers of natural origin is allowed in organic farming systems in Poland. However, the share of fields with natural mineral fertilization was so low (5%) that the median of the sum of mineral NPK for organic farms was 0 kg ha^{-1} (Table 1). The main difference between organic and conventional farms was the fertilization strategy. In total, 92.1% of conventional and only 42.1% of organic fields were fertilized. More fields were organically fertilized in organic than in conventional farms (36.8% and 13.2%, respectively). There were no organic fields with mixed organic and mineral fertilizers, while 21.0% of conventional fields were treated with both organic and mineral fertilizers. The amount of used mineral fertilization was significantly higher in conventional farms (Table 1). Plant protection products (PPP) were used only in conventional farming systems. Ninety-five percent of fields were sprayed with PPP, 74% of fields were sprayed with only herbicides, while 21% of fields were sprayed with insecticides or fungicides. The average yield of winter cereals was significantly higher in conventional farming systems than in organic farming systems (3.1 and 2.0 Mg ha^{-1}, respectively).

Table 1. The main features (medians) of the tested organic (ORG) and conventional (CONV) farms.

Feature	Median	
	ORG	CONV
Soil class (I—best; VI—worse)	IV and V	IV and V
Agricultural area of the farm (ha)	28.3 a	24.5 a
Sum of mineral NPK (kg ha^{-1})	0 a	121 b
Number of mechanical weed control treatments	1 a	1 a
Number of plant protection product treatments	0 a	1 b
Average yields of cereals (Mg ha^{-1})	2.0 a	3.1 b
Share (%) of fields with:		
Single harrowing	44.7	50.0
Double harrowing	36.8	26.3
More than two harrowings	10.5	5.3
No mechanical weed control	7.9	18.4
Only mineral fertilization	5.3	57.9
Only organic fertilization	36.8	13.2
Mineral and organic fertilization	0.0	21.0
No fertilization	57.9	7.9
Herbicide use	0	74.0
Use of PPP other than herbicides	0	21.0

PPP—plant protection products; different lowercase letters indicate significant differences according to Mood's test ($p < 0.05$).

2.6. Correlation Analysis

Spearman's correlation coefficients were calculated to determine the influence and strength of the relationship between various habitat and agrotechnical factors on the diversity of segetal flora. For the analysis of the correlation, some survey data characterizing the habitat and management strategy were selected, which could affect the biodiversity of flora and soil seed bank, such as area of the tested winter cereal field, number of commercial crops cultivated (complexity of crop rotation), amounts of nitrogen (N), phosphorus (P) and potassium (K) brought in along with mineral and natural fertilization, number of mechanical and chemical treatments of weed infestation, grain yield and share of fields in farms that are covered with vegetation during winter period (share of "green" fields). The correlation was considered significant if correlation coefficient was equal or lower than 0.05 ($\varrho \leq 0.05$).

3. Results and Discussion

3.1. Number of Species and Their Abundance

The results showed significant differences in the number and the abundance of weed species between the years of the study (Figure 4). This was most probably due to the differences in weather conditions in subsequent years of research. In 2013, March was colder than in 2012 and 2014, with snow cover lasting even until mid-April, which caused a delay in cereals development. The sum of rainfall in 2013, which was almost twice as high as in 2012 and 2014, could also have been particularly important as it made it impossible for some farmers to perform weed control at the optimal time (local inundations of fields). In connection with the above, further statistical analysis was conducted separately for each year of the study.

In total, there were 149 weed species found in both organic and conventional farming systems. No statistically significant differences between the number of species in organic (133) and conventional farming systems (123) were found (Table 2). In total, 107 species were found in fields of both farming systems (72%). Significantly more (26) weed species occurred only in organic farming systems than only in conventional winter cereals (16).

Most of the unique species for both farming systems were found only in one year and on one field. *Trifolium repens*, *Bromus hordeaceus*, *Persicaria amphibia*, *Rhinanthus serotinus* and *Scutellaria galericulata* were found in organic winter cereals in more than one year and in more than one place, while *Lamium purpureum* and *Hypericum perforatum* were found in more than one year and more than one place in the conventional farming system (these were the most widespread and the most constant over time, unique species for the given farming system). A list of species that were unique for organic and conventional farming systems is given in Table 3. In the previous study (unpublished data) on the biodiversity of weeds in spring cereals (the same study area), more unique weed species were found in organic than in conventional farming systems. Moreover, *Rhinanthus serotinus* was also found to be one of the species unique to the organic farming system (it was found in two years of the study in one field of spring cereals). The average (median) number of species was significantly higher in all years of the study in the organic farming system. The same refers to the abundance of weeds (Table 2). According to Kleijn et al. [25], there are many weed species adapted to extensive agriculture, but it is hard to choose species that are typical for the intensive conventional farming conditions of Western Europe. There is a common trend that weed abundance and richness are positively affected by organic farming [26]. Furthermore, diversity of weed species seems to be enhanced under low-input farming system conditions, while low N fertilization enhanced the effective control of weeds [18]. Moreover, a low frequency of PPP application supported the biodiversity of weed flora.

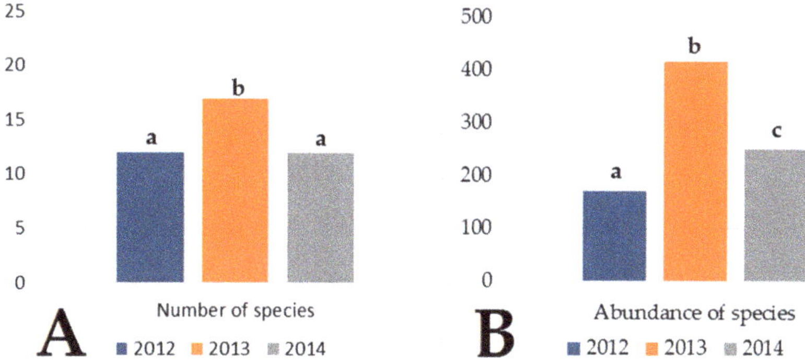

Figure 4. Comparison of median numbers of species (**A**) and their abundance (**B**) (plants m^{-2}) in years of investigation. Different letters indicate significant differences according to single factor ANOVA and pairwise Mann–Whitney post-hoc test ($p < 0.05$).

Table 2. Number and abundance of weed species (plants m^{-2}) in organic (ORG) and conventional (CONV) farming systems in 2012–2014.

Parameters	ORG	CONV
Total number of species	149	
Total number of species in the farming system	133 a	123 a
Species unique for system	26 a	16 b
Species common for both farming systems	107	
Median of Number of Weed Species per Field		
2012	13.0 a	9.0 b
2013	19.0 a	14.0 b
2014	14.5 a	9.0 b
Median of Abundance of Weed Flora (Plants m^{-2})		
2012	225 a	94 b
2013	470 a	319 b
2014	339 a	128 b

Different lowercase letters indicate significant differences according to Mood's test ($p < 0.05$).

Table 3. List of species unique to organic (ORG) and conventional (CONV) farming systems with number of years of study when given species were found and number of places (study squares) where species were found.

Species	Number of Years in Which Species Was Found	Number of Study Squares Where Species Was Found
	ORG	
Trifolium repens L.	2	4
Bromus hordeaceus L.	2	2
Persicaria amphibia (L.) Delarbre	2	2
Rhinanthus serotinus (Schönh.) Oborny	2	2
Scutellaria galericulata L.	2	2
Carduus crispus L.	2	1
Vicia villosa Roth	1	4
Trifolium hybridum L.	1	2
Matricaria chamomilla L.	1	2
Rumex obtusifolius L.	1	2
Phragmites australis (Cav.) Trin. ex Steud	1	1
Campanula persicifolia L.	1	1
Peucedanum palustre (L.) Moench	1	1
Pisum sativum L.	1	1
Brassica juncea (L.) Czern.	1	1
Lotus corniculatus L	1	1
Valeriana officinalis L.	1	1
Linaria vulgaris Mill.	1	1
Lupinus albus L.	1	1
Lupinus angustifolius L.	1	1
Gypsophila paniculata L.	1	1
Papaver argemone L.	1	1
Potentilla anserina L.	1	1
Erysimum cheiranthoides L.	1	1
Rorippa palustris L. Besser	1	1
Lysimachia vulgaris L.	1	1
	CONV	
Lamium purpureum L.	2	4
Hypericum perforatum L.	2	2
Agrostis gigantea Roth	1	3
Cardaminopsis arenosa (L.) Hayek	1	2
Festuca rubra L.	1	1
Helictotrichon pubescens (Huds.) Besser ex Schult. and Schult. f.	1	1
Geranium sanguineum L.	1	1
Geranium dissectum L.	1	1
Campanula rapunculoides L.	1	1
Lathyrus tuberosus L.	1	1
Myosurus minimus L.	1	1
Matricaria discoidea DC.	1	1
Arabidopsis thaliana (L.) Heynh.	1	1
Amaranthus retroflexus L.	1	1
Thlaspi arvense L.	1	1
Bidens frondosa L.	1	1

3.2. Shannon's Diversity and Simpson's Dominance Indices

The values of Shannon's diversity (H') index were higher for weed communities present in organic than for conventional farming systems in 2012 and 2014. In 2013, the opposite relation was visible (higher Shannon's diversity index values in conventional farming system). This could be due to the almost twice as high rainfall in May of 2013 as in 2012 and 2014, which resulted in local floods of fields. This made it impossible to carry out weed management practices on time, which resulted in increased numbers of weed species and abundance of weeds. The effects became particularly visible in the conventional farming system, as it is, to a greater extent, dependent on the timeliness of weed control treatments, especially spraying with herbicides. The increased weed infestation resulted in increased Shannon's diversity index values, which, in 2013, was significantly higher than in organic farms. Moreover, biodiversity of weeds described by Simpson's dominance (SI) index performed better (lower values of index) in organic than in conventional farms in 2012 and 2013, with no differences in 2014 (Table 4). The lack of differences in Simpson's dominance index in 2014 might be the result of higher weed infestation in 2013. More weeds were able to germinate, bloom and produce seeds that made greater inflow of seeds into soil seed bank. This might result in higher weed infestation in following years, which made the differences between the two systems a little blurred, in this case, causing no differences in Simpson's dominance index. Both Shannon's and Simpson's biodiversity indices showed that weed communities are significantly better maintained in organic farming systems, but relatively high values of Shannon's index (H′ = 3.9–4.8) and low values of Simpson's index (SI = 0.03–0.07) prove that biodiversity of weeds in conventional farms was also high. This was probably due to the fact that conventional farms were mostly of extensive character (low PPP use, low mineral fertilization) (Table 1). This proves the importance of low-input conventional farming for biodiversity conservation. This type of farming is common especially in Eastern and South-Eastern Poland, where economic and organizational conditions (small farms, fragmentation of agricultural land and low profitability) resulted in widespread low-input farming that, in these specific regions, can be considered traditional (low fertilizer input, especially mineral fertilizers, very low PPP consumption). The Lubelskie region, where the presented study was located, is one of the regions of Poland with the lowest intensity of agricultural production [21]. The high value of biodiversity of weeds of both organic and low-input conventional farming systems was confirmed by Berbeć and Feledyn-Szewczyk [27] for spring cereals and soil seed banks as well as by Jastrzębska et al. [28] for, among others, winter triticale. The results showed that the biodiversity of more intensive farming systems (high-input conventional, monoculture) is lower than in more sustainable farming systems (integrated and organic farming systems), which was confirmed by the presented study [29,30]. Armengot et al. [31] found that values of Shannon's biodiversity index in organic farming systems can be almost twice as high as for conventional farming systems (H′ = 2.5 and H′ = 1.5 respectively). It was shown that the biodiversity of organic farms and low-input conventional farms can be much more similar, but still some significant differences can be found. The results revealed very low values of Simpson's dominance index for both farming systems, which indicates that weed communities were not dominated by a single species. Despite the low values of Simpson's index, the differences between the tested systems were still significant. Feledyn Szewczyk and Duer [29] also found that conventional cereals were more dominated by single species than organic ones. This was also confirmed by the presented study.

Table 4. Values of Shannon's diversity (H′) and Simpson's dominance (SI) indices for weed communities in organic (ORG) and conventional (CONV) farming systems.

Year	Index	ORG	CONV
	Biodiversity Indices		
2012	Shannon (H′)	4.152 b	3.860 a
	Simpson (SI)	0.031 a	0.045 b
2013	Shannon (H′)	4.518 a	4.777 b
	Simpson (SI)	0.021 a	0.071 b
2014	Shannon (H′)	4.305 b	4.138 a
	Simpson (SI)	0.028 a	0.025 a

3.3. Dominant Species

Weed species with a total share in community exceeding 5% are given in Tables 5 and 6 (organic and conventional farming system, respectively). Winter cereals were dominated by *Poaceae* weeds (*Setaria pumila*, *Elymus repens*, *Apera spica-venti*) and *Juncus bufonius*. These species are common in cereals of different farming systems all over Europe [32]. As for all annual weed species, their survival in agricultural (arable) ecosystems depends on the success of seed germination in subsequent growing seasons [33–35]. *Setaria pumila* and *Elymus repens* seemed to be the most dominant species of the organic farming system (27–42% of weed community annually). A large share of *Elymus repens* in the organic weed community may indicate the rather poor quality of agricultural treatments (especially weed control), which indirectly contributed to rather low yields of cereals. This may also be related to high share of cereals in crop rotation. *Elymus repens* was also dominant in conventional farming systems, which shows that weed control was an issue for farmers in both farming systems.

Table 5. Share (%) of dominant (more than 5% of the community) species in organic winter cereals.

Species	2012	2013	2014	2012–2014
Setaria pumila (Poir.) Roem. and Schult	32.3	23.9	14.8	23.4
Elymus repens (L.) Gould	9.2	9.8	11.8	10.3
Apera spica-venti (L.) P. Beauv	8.5	3.0	6.2	5.4
Poa bulbosa L.	0.0	5.1	10.2	5.3
Rumex acetosella L.	10.0	3.4	1.8	4.7
Polygonum lapathifolium L. subsp. lapathifolium	6.1	5.3	2.4	4.6
Scleranthus annuus L.	4.5	3.1	5.9	4.3
Echinochloa crus-galli (L.) P. Beauv.	0.0	7.6	1.5	3.7
Anthemis arvensis L.	0.0	3.0	5.2	2.9
Number of dominant species (>5% of weed community)	5	5	6	4

Table 6. Share (%) of dominant (more than 5% of the community) weed species in conventional farming system.

Species	2012	2013	2014	2012–2014
Juncus bufonius L.	6.0	31.9	3.4	21.3
Setaria pumila (Poir.) Roem. and Schult	25.2	12.6	4.9	13.4
Echinochloa crus-galli (L.) P. Beauv.	16.0	4.6	7.7	7.3
Apera spica-venti (L.) P. Beauv	2.7	6.1	15.4	7.3
Poa bulbosa L.	0.0	10.0	1.8	6.5
Viola arvensis Murr.	4.9	3.6	12.0	5.5
Elymus repens (L.) Gould	6.0	2.7	8.0	4.4
Centaurea cyanus L.	0.6	0.6	8.3	2.2
Spergula arvensis L.	6.8	0.3	0.1	1.4
Anthemis arvensis L.	0.0	0.2	5.8	1.3
Scleranthus annuus L.	0.0	0.0	5.5	1.2
Rumex acetosella L.	5.8	0.1	0.1	1.1
Number of dominant species (>5% of weed community)	6	4	7	6

Echinochloa crus-galli and *Setaria pumila* used to be considered in the past as weeds typical to root crops and maize, with rather occasional occurrence in cereals [36]. The presented study showed a large share of *Echinochloa crus-galli* and *Setaria pumila* in the cereals in both farming systems. Its dominance in the weed community of cereals of the Lubelskie region was confirmed by Ziemińska-Smyk [37] and the previous study of the current authors [25].

3.4. Endangered Species

Endangered weed species have been selected on the basis of the Polish Red List of Plant and Fungi [38]. In total, five endangered weed species were found in winter cereals (Table 7). In the organic farming system, three species marked as vulnerable in the Polish Red List of Plant and Fungi were found (*Bromus secalinus, Ranunculus arvensis, Anagallis foemina*) and one species which is subject to partial legal protection in Poland (*Helichrysum arenarium*). Additionally, a few specimens of *Myosurus minimus* (vulnerable species according to Polish Red List of Plant and Fungi) were found in conventional farming system. Moreover, other weed species which are currently disappearing under conventional, intensive agriculture conditions were found in both farming systems. Those species included *Agrostema githago* L., *Arnoseris minima* L., *Consolida regalis* S.F. Gray, *Euphorbia exigua* L., *Geranium sanguineum* L., *Lathyrus tuberosus* L. and *Veronica dillenii* Crantz. Spring cereals seem to be slightly more important to rare and endangered weed species as one more endangered weed species and a few more rare species were found by the authors in their previous study in spring cereals [27] than in the presented study in the same study area. The presence of endangered and rare weed species proves the high environmental value of both organic and low-input conventional farming systems (which are still common in Eastern Poland) and their importance in biodiversity conservation.

Table 7. Average abundance (plants m^{-2}) of endangered weed species in organic (ORG) and conventional (CONV) winter cereals.

Species and Endangerment Category	Average Abundance (Plants m^2)	
	ORG	CONV
Bromus secalinus L. [V]	0.1	0.5
Ranunculus arvensis L. [V]	0.3	0.1
Anagallis foemina Mill. [V]	0.3	0.2
Myosurus minimus L. [V]	–	<0.1
Helichrysum arenarium (L.) Moench [P]	0.8	0.3

[V]—vulnerable according to Polish Red List of Plant and Fungi [31]; [P]—subject to partial legal protection in Poland.

3.5. Invasive Weed Species

Invasive weed species are currently observed in all types of ecosystems in Poland [39,40]. According to Tokarska-Guzik et al. [40], 84 species of plants are considered as invasive plant species in Poland. Ten of these species are typical segetal plants (weeds) that can be found on arable fields. These species are *Alopecurus myosuroides* Huds., *Amaranthus retroflexus* L., *Anthoxanthum aristatum* Boiss., *Avena fatua* L., *Echinochloa crus-galli* (L.) Beauv, *Galinsoga ciliata* Ruiz and Pav., *Galinsoga parviflora* Cav., *Setaria pumila* (Poir.) Roem. and Schult., *Setaria viridis* (L.) P. Beauv. and *Veronica persica* Poir. Most of those species were found in winter cereals in organic and conventional farming systems in the presented study (Table 8). Rural areas are made of both agricultural land which is devoted to the cultivation of plants and livestock raising and other habitats which are not being used in agricultural production (fallow lands, field boundaries, roads, ponds, in- and mid-fields shrubs, etc.), which can be a habitat for invasive plants. Such habitats, subjected to human activity, under favorable conditions, are the mainstay from which invasive plant species can penetrate into adjacent ecosystems [41], but they are also the sources of agricultural biodiversity [42].

In the presented study, there were in total 10 invasive weed species observed both in organic and conventional farming systems (Table 8). Nine species were observed in both faming systems; one extra species (*Amaranthus retroflexus*) was observed only in the conventional farming system. Most of the observed invasive weed species were classified as weeds of the first (the lowest) class of invasiveness according to Tokarska-Guzik et al. [36]. Only one species of the second class (*Erigeron annuus*) was observed (in both organic and conventional farming systems). There were no invasive weed species of the third and fourth (the highest) invasiveness classes. The most common invasive weed species in both farming systems were *Setaria pumila* and *Echinochloa crus-galli*. Unfortunately, these species were also one of the most dominant species in the investigated weed community of organic and conventional farming systems (Tables 5 and 6). Other authors found *Conyza canadensis*, which was found in the presented study in both farming systems with rather low average abundance of 2 plants per m^{-2} (organic farming system) and less than 1 plant per m^{-2} (conventional farming system). This invasive plant species was found by other authors to be one of the most abundant invasive weed species in both agricultural and other ecosystems [43,44].

Table 8. Average abundance of invasive weed species in organic (ORG) and conventional (CONV) winter cereals (average for 2012–2014).

Species	Invasiveness Class *	Average Abundance (Plants m^{-2})	
		ORG	CONV
Setaria pumila (Poir.) Roem. and Schult.	I	88.7	37.0
Echinochloa crus-galli L.	I	14.2	20.3
Setaria viridis (L.) P. Beauv.	I	5.1	1.2
Avena fatua L.	I	<0.1	0.7
Conyza Canadensis (L.) Cronquist	I	2.3	0.7
Veronica persica Poir.	I	0.7	1.3
Galinsoga parviflora Cav.	I	0.1	<0.1
Oxalis stricta L.	I	0.5	0.9
Amaranthus retroflexus L.	I	0	0.3
Erigeron annuus (L.) Pers	II	0.3	1.1

* Invasiveness class from I (the lowest) to IV (the highest) according to Tokarska Guzik et al. [36].

3.6. Correlation Analysis

The main cause of biodiversity loss in conventional farms is the intensification of agricultural production linked with high consumption of PPPs, including herbicides [45–48]. The results of the correlation analysis of the presented study are shown in Table 9. In most cases, tested biodiversity parameters in conventional farms were significantly negatively correlated with the number of chemical weed control treatments and field area (higher biodiversity on smaller fields). In conventional farms, Shannon's diversity index was negatively correlated with the sum of mineral NPK fertilization. This relationship was confirmed also by other authors [49,50]. The loss of biodiversity on fields with higher NPK might be caused by higher intensity of production (e.g., use of herbicides), but also higher mineral NPK fertilization can cause a drop in soil pH, which changes the conditions of the habitat and makes fewer species from soil seed banks able to grow in more acidic habitats. Confirmation of this may be the fact that organic fertilization had no effect on biodiversity on conventional fields. Fried et al. (2020) [51] found increased fertilization to be one of the most important factors that caused changes in weed population (mainly due to the occurrence of more nutrient-demanding weed species).

Shannon's diversity index in conventional farms was also positively correlated with percentage of fields in farms that were covered with vegetation during wintertime (share of "green fields"). A literature study shows that conventional farms that cultivated aftercrops as green manure can have an increased Shannon diversity index of weeds [52,53].

Armengot et al. (2013) [27] found that mechanical weed management had no negative effect on the biodiversity of weeds. The presented study confirmed this. Moreover, the tested biodiversity parameters (number of weed species and their abundance, Shannon's diversity and Simpson's dominance indices) in the tested organic farming system were not correlated with any of the tested field or farm parameters (farm or field area, number of cultivated crop species, percentage of fields that were covered with vegetation during wintertime, soil class, sum of NPK fertilization, number of mechanical weed management treatments, crop species or yield). This shows that biodiversity in the tested organic farms was less sensitive to disturbances caused by management strategy or organizational conditions than biodiversity in conventional farms. Pressure from external factors caused conventional farms to be less stable in terms of biodiversity.

Table 9. Spearman's correlation coefficient matrix of selected variables for organic and conventional farming system.

	NS	Ab	H′	SI	NC	SGF	FA	Min NPK	Org NPK	NH	PPP	Yield
NS		0.186	0.707 *	−0.608 *	0.117	0.200	−0.150	0.143	0.145	0.173	0	0.130
Ab	0.678 *		0.008	−0.013	0.171	0.154	−0.124	0.142	0.055	0.037	0	0.024
H′	0.529 *	0.283		−0.971 *	−0.104	−0.059	0.048	0.037	0.097	0.264	0	0.114
SI	−0.388 *	−0.202	−0.943 *		0.062	0.092	−0.100	−0.067	−0.054	−0.207	0	−0.151
NC	−0.184	−0.260	−0.113	0.129		0.029	−0.210	0.279	0.124	−0.013	0	0.077
SGF	0.221	0.161	0.322 *	−0.194	−0.108		−0.372 *	0.254	0.223	0.215	0	0.031
FA	−0.517 *	−0.400 *	−0.390 *	0.351 *	0.029	−0.158		0.008	−0.537 *	−0.031	0	−0.169
Min NPK	−0.145	−0.095	−0.378 *	0.286	0.143	−0.465 *	0.301		−0.121	−0.121	0	0.069
Org NPK	0.165	0.076	0.238	−0.248	−0.395 *	0.095	−0.251	−0.265		0.306	0	0.260
NH	0.027	0.107	0.203	−0.216	0.044	0.043	−0.217	−0.176	−0.152		0	−0.079
PPP	−0.430 *	−0.430 *	−0.548 *	0.448 *	−0.059	−0.137	0.402 *	0.356 *	0.044	−0.481 *		0.000
Yield	−0.232	−0.395 *	−0.147	−0.003	−0.324 *	−0.091	0.138	0.309	−0.027	−0.212	0.503 *	

Significant Spearman's correlation coefficient at * $p < 0.05$. Correlations for organic farming system are in upper right part of the matrix. Correlations for conventional farming systems are in lower left part of the matrix. NS—number of species; Ab—abundance; H′—Shannon's index; SI—Simpson's index; NC—number of crops; SGF—share of "green" fields; FA—field area; Min NPK—mineral NPK fertilization; Org NPK—organic NPK fertilization; NH—number of harrowings; PPP—number of PPP treatments.

4. Conclusions

The current research has shown significantly higher species diversity and abundance of weeds in the organic than in the conventional farming system. Shannon's diversity index had high values and Simpson's dominance index had low values in both farming systems, which showed their high biodiversity importance. The presence of the rare and endangered species showed that both organic and low-input conventional farming systems are the mainstays of valuable weed species. Winter cereal fields were dominated by some *Poacea* species, which might be due to less intense weed management. Correlation analysis showed weed community diversity of organic fields of winter cereals to be more resistant to external disturbances linked with management strategy than biodiversity of conventional fields.

Author Contributions: Conceptualization, J.S.; methodology, J.S., B.F.-S., M.S.; investigation, B.F.-S., M.S., A.K.B.; statistical analysis, A.K.B.; references, A.K.B., M.S., B.F.-S., A.K.; writing—original draft preparation, A.K.B.; writing—review and editing, M.S.; visualization, M.S., B.F.-S.; supervision, J.S.; project administration and technical preparation, A.K. All authors have read and agreed to the published version of the manuscript.

Funding: The research was funded by "Protection of species diversity of valuable natural habitats on agricultural lands on Natura 2000 areas in the Lublin Voivodeship (KIK/25)" project, a part of Polish–Swiss cooperation programme. The APC was funded by the Polish Ministry of Science and Higher Education within the statutory activity of the Institute of Soil Science and Plant Cultivation–State Research Institute.

Acknowledgments: The study was conducted within the project, Protection of species diversity of valuable natural habitats on agricultural lands on Natura 2000 areas in the Lublin Voivodeship (KIK/25), as part of Polish–Swiss cooperation.

Conflicts of Interest: The authors declare no conflict of interest. The funders had no role in the design of the study; in the collection, analyses, or interpretation of data; in the writing of the manuscript, or in the decision to publish the results.

References

1. Garbach, K.; Milder, J.C.; Montenegro, M.; Karp, D.S. Biodiversity and Ecosystem Services in Agroecosystems. *Encycl. Agric. Food Syst.* **2014**, 2. [CrossRef]
2. Loreau, M.; Naeem, S.; Inchausti, P.; Bengtsson, J.; Grime, J.P.; Hector, A.; Hooper, D.U.; Huston, M.A.; Raffaelli, D.; Schmid, B.; et al. Biodiversity and ecosystem functioning: Current knowledge and future challenges. *Science* **2001**, *294*, 804–808. [CrossRef]
3. Hooper, D.U.; Chapin Iii, F.S.; Ewel, J.J.; Hector, A.; Inchausti, P.; Lavorel, S.; Lawton, J.H.; Lodge, D.M.; Loreau, M.; Naeem, S.; et al. Effects of biodiversity on ecosystem functioning: A consensus of current knowledge. *Ecol. Monogr.* **2005**, *75*, 3–35. [CrossRef]

4. Balvanera, P.; Pfisterer, A.B.; Buchmann, N.; He, J.S.; Nakashizuka, T.; Raffaelli, D.; Schmid, B. Quantifying the evidence for biodiversity effects on ecosystem functioning and services. *Ecol. Lett.* **2006**, *9*, 1146–1156. [CrossRef]
5. Cardinale, B.J.; Matulich, K.L.; Hooper, D.U.; Byrnes, J.E.; Duffy, E.; Gamfeldt, L.; Balvanera, P.; O'Connor, M.I.; Gonzalez, A. The functional role of producer diversity in ecosystems. *Am. J. Bot.* **2011**, *98*, 572–592. [CrossRef]
6. European Commission. Communication from the Commission to the European Parliament, the Council, the Economic and Social Committee and the Committee of the Regions Our Life Insurance, Our Natural Capital: An Eu Biodiversity Strategy to 2020. COM/2011/0244 Final. Available online: www.eur-lex.europa.eu/legal-content/EN/TXT/?uri=CELEX%3A52011DC0244 (accessed on 4 August 2020).
7. Wagner, M.; Bullock, J.M.; Hulmes, L.; Hulmes, S.; Pywell, R. Cereal density and N-fertilizer effects on the flora and biodiversity value of arable headlands. *Biodivers. Conserv.* **2017**, *26*, 85–102. [CrossRef]
8. Meyer, S.; Wesche, K.; Krause, B.; Leuschner, C. Dramatic losses of specialist arable plants in Central Germany since the 1950s/60s–A cross-regional analysis. *Divers. Distrib.* **2013**, *19*, 1175–1187. [CrossRef]
9. Hallmann, C.A.; Sorg, M.; Jongejans, E.; Siepel, H.; Hofland, N.; Schwan, H.; Stenmans, W.; Müller, A.; Sumser, H.; Hörren, T.; et al. More than 75 percent decline over 27 years in total flying insect biomass in protected areas. *PLoS ONE* **2017**, *12*, e0185809. [CrossRef]
10. Schumacher, M.; Ohnmacht, S.; Rosenstein, R.; Gerhards, R. How management factors influence weed communities of cereals, their diversity and endangered weed species in Central Europe. *Agriculture* **2018**, *8*, 172. [CrossRef]
11. Scherber, C.; Eisenhauer, N.; Weisser, W.W.; Schmid, B.; Voigt, W.; Fischer, M.; Schulze, E.-D.; Roscher, C.; Weigelt, A.; Allan, E.; et al. Bottom-up effects of plant diversity on multitrophic interactions in a biodiversity experiment. *Nature* **2010**, *468*, 553–556. [CrossRef]
12. Marshall, E.J.P.; Brown, V.K.; Boatman, N.D.; Lutman, P.J.W.; Square, G.R.; Ward, L.K. The role of weeds in supporting biological diversity within crop fields. *Weed Res.* **2003**, *43*, 77–89. [CrossRef]
13. Oerke, E.-C. Crop losses to pests. *J. Agric. Sci.* **2006**, *144*, 31–43. [CrossRef]
14. Gerhards, R.; Dieterich, M.; Schumacher, M. Rückgang von ackerunkräutern in baden-württemberg—Ein vergleich von vegetationskundlichen erhebungen in den jahren 1948/49, 1975–1978 und 2011 im raum mehrstetten—Empfehlungen für landwirtschaft und naturschutz. *Gesunde Pflanz.* **2013**, *65*, 151–160. [CrossRef]
15. Urban, M.; Adamczewski, K.; Dobrzański, A. Ecological niche of weeds in spring cereals cultivars grown after winter oilseed rape. *Prog. Plant Prot.* **2013**, *53*, 105–109. (In Polish). Available online: https://pdfs.semanticscholar.org/c950/1dabafdfba2d60e8f1ca16ae81739e92ee67.pdf (accessed on 18 September 2020).
16. Feledyn-Szewczyk, B.; Nakielska, M.; Jończyk, K.; Berbeć, A.K.; Kopiński, J. Assessment of the suitability of 10 winter triticale cultivars (x Triticosecale Wittm. ex A. Camus) for organic agriculture: Polish case study. *Agronomy* **2020**, *10*, 1144. [CrossRef]
17. Arslan, Z.F. Decrease in biodiversity in wheat fields due to changing agricultural practices in five decades. *Biodivers. Conserv.* **2018**, *27*, 3267–3286. [CrossRef]
18. Chamorro, L.; Masalles, R.M.; Sans, F.X. Arable weed decline in Northeast Spain: Does organic farming recover functional biodiversity? *Agric. Ecosyst. Environ.* **2016**, *223*, 1–9. [CrossRef]
19. Travlos, I.S.; Chelmona, N.; Roussis, I.; Bilalls, D.J. Weed-species abundance and diversity indicates in relation to tillage systems and fertilization. *Front. Environ. Sci.* **2018**, *6*, 11. [CrossRef]
20. Clergue, B.; Amiaud, B.; Pervanchon, F.; Lasserre-Joulin, F.; Plantureux, S. Biodiversity: Function and assessment in agricultural areas. A review. *Agron. Sustain. Dev.* **2005**, *25*, 309–327. [CrossRef]
21. Kopiński, J.; Matyka, M. Ocena regionalnego zróżnicowania współzależności czynników przyrodniczych i organizacyjno-produkcyjnych w polskim rolnictwie. *Zagadnienia Ekon. Rolnej* **2016**, *1*, 57–79. [CrossRef]
22. Shannon, C.E.A. Mathematical theory of communications. *Bell Syst. Tech.* **1948**, *27*, 379–423. [CrossRef]
23. Simpson, E.H. Measurement of diversity. *Nature* **1949**, *163*, 688. [CrossRef]
24. Central Statistical Office. *Statistical Yearbook*; Central Statistical Office: Warsaw, Poland, 2019; p. 94. (In Polish)
25. Kleijn, D.; Rundlöf, M.; Scheper, J.; Smith, H.G.; Tscharntke, T. Does conservation on farmland contribute to halting the biodiversity decline? *Trends Ecol. Evol.* **2011**, *26*, 474–481. [CrossRef] [PubMed]
26. Armengot, L.; Sans, F.X.; Fischer, C.; Flohre, A.; José-Maria, L.; Tscharntke, T.; Thies, C. The β-diversity of arable weed communities on organic and conventional cereal farms in two contrasting regions. *Appl. Veg. Sci.* **2012**, *15*, 571–579. [CrossRef]

27. Berbeć, A.K.; Feledyn-Szewczyk, B. Biodiversity of weeds and soil seed bank in organic and conventional farming systems. *Res. Rural Dev.* **2018**, *2*, 12–19. [CrossRef]
28. Jastrzębska, M.; Jastrzębski, W.P.; Hołdyński, C.; Kostrzewska, M.K. Weed species diversity in organic and integrated farming systems. *Acta Agrobot.* **2013**, *66*, 113–124. [CrossRef]
29. Feledyn-Szewczyk, B.; Duer, I. Weed infestation in spring wheat cultivated in organic crop production system in comparison with other crop production systems. *J. Res. Appl. Agric. Eng.* **2007**, *52*, 40–44. (In Polish)
30. Feledyn-Szewczyk, B.; Matyka, M.; Staniak, M. Comparison of the effect of perennial energy crops and agricultural crops on weed flora diversity. *Agronomy* **2019**, *9*, 695. [CrossRef]
31. Armengot, L.; José-María, L.; Chamorro, L.; Sans, F.X. Weed harrowing in organically grown cereal crops avoids yield losses without reducing weed diversity. *Agron. Sustain. Dev.* **2013**, *33*, 405–411. [CrossRef]
32. Glemnitz, M.; Radics, L.; Hoffmann, J.; Czimber, G. Weed species richness ad species composition of different arable field types—A comparative analysis along a climate gradient from South to North Europe. *J. Plant Dis. Protect.* **2006**, *20*, 577–586.
33. Froud-Williams, R.J.; Chancellor, R.J.; Drennan, D.S.H. Influence of cultivation regime upon buried weed seeds in arable cropping systems. *J. Appl. Ecol.* **1983**, *20*, 199–208. [CrossRef]
34. Kremer, R.J. Management of weed seed banks with microorganisms. *Ecol. Appl.* **1993**, *3*, 42–52. [CrossRef] [PubMed]
35. Graziani, F.; Onofri, A.; Pannacci, E.; Tei, F.; Guiducci, M. Size and composition of weed seedbank in long-term organic and conventional low-input cropping systems. *Eur. J. Agron.* **2012**, *39*, 52–61. [CrossRef]
36. Wesołowski, M. Występowanie wybranych gatunków chwastów w uprawach rolniczych. In *Makroregion Południowo-Wschodni*; IUNG: Pulawy, Poland, 1988; R220/8; pp. 1–43. (In Polish)
37. Ziemińska-Smyk, M. Zmiany w zachwaszczeniu upraw zbóż na Zamojszczyźnie gatunkami z rodziny traw (*Poaceae*). *Zesz. Nauk. UP Wrocław Rol. C* **2012**, *584*, 159–163. (In Polish)
38. Mirek, Z. *Red List of Plants and Fungi in Poland*; Szafer Institute of Botany, Polish Academy of Sciences: Krakow, Poland, 2006.
39. Tokarska-Guzik, B.; Bzdęga, K.; Nowak, T.; Lewandowska, A.; Gancarek, M.; Frelich, M. Alien plants in Poland: Research directions and putting the results into practice. *Biodiv. Res. Conserv.* **2014**, *35*, 57–74. [CrossRef]
40. Tokarska-Guzik, B.; Dajdok, Z.; Zając, M.; Zając, A.; Urbisz, A.; Danielewicz, W.; Hołdyński, C. *Rośliny Obcego Pochodzenia w Polsce ze Szczególnym Uwzględnieniem Gatunków Inwazyjnych*; Generalna Dyrekcja Ochrony Środowiska: Warsaw, Poland, 2012; pp. 35–69. (In Polish)
41. Jackowiak, B. Modele ekspansji roślin synantropijnych i transgenicznych. *Phytocoen. 11 Semin. Geobot.* **1999**, *6*, 4–16.
42. Kujawa, K.; Bernacki, Z.; Arczyńska-Chudy, A.; Janku, K.; Karg, J.; Kowalska, J.; Oleszczuk, M.; Sienkiewicz, P.; Sobczyk, D.; Weyssenhoff, H. Flower strips as rarely used in Poland tool for enhancement of Integrated Pest Management in cultivated fields, and for enlargement of biodiversity in agricultural areas. *Prog. Plant Prot.* **2018**, *58*, 115–128. (In Polish) [CrossRef]
43. Bomanowska, A.; Ferchmin, M.; Kirpluk, I.; Otręba, A. Inwazyjne gatunki roślin we florze Puszczy Kampinoskiej. In *Inwazyjne Gatunki Roślin w Kampinoskim Parku Narodowym i w Jego Sąsiedztwie*; Kampinoski Park Narodowy: Izabelin, Poland, 2014; pp. 25–35. (In Polish)
44. Chmielecki, B.; Kucharski, L. Gatunki inwazyjne w dolinie środkowej Warty. *Biul. Uniejowski* **2018**, *7*, 77–95. (In Polish) [CrossRef]
45. Hyvönen, T.; Salonen, J. Weed species diversity and community composition in cropping practices at two intensity levels—A six-year experiment. *Plant Ecol.* **2002**, *154*, 73–81. [CrossRef]
46. Szeleźniak, E.; Grabiński, J.; Nieróbca, P. Wpływ intensywności technologii produkcji wybranych gatunków zbóż na ich zachwaszczenie. *Acta Sci. Pol. Agric.* **2007**, *6*, 83–90. (In Polish)
47. Koocheki, A.; Nassiri, M.; Alimoradi, L.; Ghorbani, R. Effect of cropping systems and crop rotations on weeds. *Agron. Sustain. Dev.* **2009**, *29*, 401–408. [CrossRef]
48. Berbeć, A.K. Weed Diversity in Spring Cereals and Weed Seed Bank in Conventional and Organic Farms in the Province of Lublin. Ph.D. Thesis, Institute of Soil Science and Plant Cultivation State Research Institute, Puławy, Poland, 2016.
49. Tang, L.; Cheng, C.; Wan, K.; Li, R.; Wang, D.; Tao, Y.; Pan, J.; Xie, J.; Chen, F. Impact of fertilizing pattern on the biodiversity of a weed community and wheat growth. *PLoS ONE* **2014**, *9*, e84370. [CrossRef] [PubMed]

50. Haliniarz, M.; Gawęda, D.; Kwiatkowski, C.; Frant, M.; Różańska-Boczula, M. Weed biodiversity in field Pea under reduced tillage and different mineral fertilization conditions. *Bulg. J. Agric. Sci.* **2014**, *20*, 1340–1348.
51. Fried, G.; Chauvel, B.; Munoz, F.; Reboud, X. Which traits make weeds more successful in maize crops? Insights from a three-decade monitoring in France. *Plants* **2020**, *9*, 40. [CrossRef] [PubMed]
52. Majchrzak, L.; Piechota, T. Wpływ technologii uprawy na zachwaszczenie pszenicy jarej. *Fragm. Agron.* **2013**, *31*, 94–101.
53. Wortman, S.E.; Lindquist, J.L.; Haar, M.J.; Francis, C.A. Increased weed diversity, density and above-ground biomass in long-term organic crop rotations. *Renew. Agric. Food Syst.* **2010**, *25*, 281–295. [CrossRef]

Review

Legume Cover Crops as One of the Elements of Strategic Weed Management and Soil Quality Improvement. A Review

Anna Kocira [1,*], Mariola Staniak [2], Marzena Tomaszewska [1], Rafał Kornas [1], Jacek Cymerman [1], Katarzyna Panasiewicz [3] and Halina Lipińska [4]

1 Institute of Agricultural Sciences, State School of Higher Education in Chełm, Pocztowa 54, 22-100 Chełm, Poland; mtomaszewska@pwsz.chelm.pl (M.T.); rkornas@pwsz.chelm.pl (R.K.); jcymerman@pwsz.chelm.pl (J.C.)

2 Department of Forage Crop Production, Institute of Soil Science and Plant Cultivation—State Research Institute, Czartoryskich 8, 24-100 Puławy, Poland; mstaniak@iung.pulawy.pl

3 Agronomy Department, Poznań University of Life Sciences, Dojazd 11, 60-632 Poznań, Poland; katarzyna.panasiewicz@up.poznan.pl

4 Department of Grassland and Landscape Forming, University of Life Sciences in Lublin, Akademicka 13, 20-950 Lublin, Poland; halina.lipinska@up.lublin.pl

* Correspondence: akocira@pwsz.chelm.pl; Tel.: +48-82-565-8895

Received: 19 August 2020; Accepted: 3 September 2020; Published: 5 September 2020

Abstract: The benefits of conservation practices increased the interest of farmers in the cultivation of cover crops (CCs). This review aims to present and analyze the state of the art on the cultivation of legume CCs, including their importance in protecting crops against weeds, as well as their effects on organic matter and nitrogen content in the soil, physical and biological properties of the soil, and its erosion. The multi-purpose character of legume CCs is visible in their positive effect on reducing weed infestation, but also on the soil: reducing its compaction and erosion, improving its structural and hydraulic properties, increasing the content of organic matter and activity of soil microorganisms, or increasing its nitrogen content due to symbiotic N_2 fixing. This review demonstrates that a wider use of legume CCs in organic farming is needed. The benefits of legume CCs for successive crops in these cultivation conditions, both in terms of inhibiting weed populations and improving fertility and soil properties, also need to be identified. Further research is also needed to determine the potential impact of legume CCs on the improvement of the quality of degraded soils, or those with less favorable physicochemical properties.

Keywords: legume cover crop; weed control; organic matter; nitrogen; soil physical and biological properties; soil erosion

---

1. Introduction

Currently, more and more attention is paid to production systems based on managing natural resources in such a way as to ensure that the needs of contemporary and future generations are met. This trend fits well with conservation agriculture, which aims to protect, improve and more efficiently use natural resources through the integrated management of soil, water, and biological resources, in combination with external inputs. This approach contributes to environmental protection, but also improves agricultural production and maintains it at a high level [1].

All the benefits of conservation practices increased farmers' interest in the cultivation of cover crops (CCs) [2]. CCs are defined as non-cash crops that can be grown before or together with the main crops to keep the soil covered with vegetation for as long as possible—even all year round [3]. CCs have a positive effect on agroecosystems by reducing soil erosion and nitrate leaching, increasing water infiltration and maintaining soil moisture. They also suppress weeds and reduce the occurrence of pests, nematodes and various soil pathogens, and improve soil quality, e.g., by increasing the content of organic matter and the availability of nutrients [4–13]. Species used as CCs plants should produce a lot of biomass, which is important for uniform coverage of the soil surface. Moreover, their C:N ratio should be balanced, and they should be resistant to rapid decomposition, thus protecting the soil, even from the early stages of growth and development of the main crop [14,15]. The soil exposure promotes weed infestation, increases susceptibility to erosion, while high C:N can extract nitrogen from the system, reducing its availability to successive plants [16]. Cover crops can be introduced into plant production systems by: (i) cultivation during the off-season and destruction before main crop cultivation—a common practice in annual growing systems; (ii) cultivation with the main crop for a part, or all of its growing season as living mulch—a common practice in perennial growing systems [17].

The CCs that are most commonly used in farming belong to families *Fabaceae*, *Brassicaceae*, and *Poaceae*. Due to the varied growth rate, the amount of biomass produced, the uniformity of the soil cover and the C:N ratio, they play a different role in the main crop. Numerous studies confirm that legume CCs improve soil quality and thus provide more favorable conditions for the growth, development and yielding of main crops, while playing a significant role in reducing weed infestation [12,18–25]. In addition to ecosystem services, legumes are also a rich source of protein in human and animal nutrition. However, the use of these plants without prior thermal treatment may be limited by the presence of anti-nutritional and toxic factors such as proteolytic inhibitors, phytohemagglutinins, lathyrogens, cyanogenetic compounds, compounds causing favism, and factors affecting digestibility and saponins [26].

This review aims to present and analyze the state of the art on the cultivation of legume Ccs, including their importance in protecting crops against weeds, their effect on organic matter and nitrogen content in the soil, as well as physical and biological properties of the soil, and its erosion.

2. Effect of Legume CCs on Weed Control

CCs are an essential tool in the integrated management of weeds, including those resistant to herbicides. They provide a competitive advantage, contributing to the good condition of the soil and inhibiting weed infestation [17,27–29]. Oerke [30] reports that weeds are one of the most important factors contributing to the reduction of the yield, by up to 34%. The estimated yield loss caused by weeds is up to two times higher than the loss caused by other pests (insects, pathogens), which amount to approximately 8% to 10% [27]. The large difference between the potential and actual estimates of crop loss due to weeds indicates that current weed control is relatively effective [31].

The weed management system is based on information on interactions between plants, including the competitive ability of the main crops on various stages of their development to inhibit weed growth and expansion [32]. CCs introduced into the cultivation of the main plant contribute to the reduction of weed infestation by preventing seed germination and emergence of weed seedlings. Thus, they limit the growth and development of weeds and reduce the number of seeds in the weed seed bank in the soil (by increasing seed predation, limiting seed recruitment) [33,34].

Appropriate selection of crop species, and even varieties, often has a decisive impact on the level of weed infestation control [35,36]. Characteristic features such as growth rates [37], rapid germination, early above-ground growth and vigor [38], shading ability [37], crop height, tillering capacity [38,39], rapid leaf area and canopy establishment, large leaf area development and duration [38,40], long stem, high biomass upright growth [15,41,42], which are also related to early light interception [43] and allelopathy [40,44,45] affect the interactions between the main crop and the weeds. These features are leveraged by CCs to reduce weed infestation in the main crops. The effectiveness of CCs in this area often depends on the CC species, the composition of weed population, the area covered, persistence of crop residues, physical impedance, reduction in light transmittance to the soil, and decreased daily soil temperature fluctuations [7,19,46–50] as well as management practices, e.g., cutting or mulching [20,51,52]. However, the effectiveness of weed control by legume CCs depends both on their morphological, phenological and physiological characteristics [19,35,41] (Table 1), as well as on the growing conditions [53,54].

Table 1. Effect of different species of legume CCs on the reduction of weed infestation.

Cover Crops	Crop	Dominant Weed Species	Weed Control	References
Medicago lupulina L.	*Triticum aestivum* L.	-	reduced weed DM (50%)	[55]
M. lupulina L.; *Medicago sativa* L.; *Trifolium pratense* L.; *Trifolium repens* L.	*T. aestivum L.*	-	reduced weed density (40%–57%); reduced aerial DM of *M. lupulina* (about 35 kg DM ha^{-1}) and *M. sativa* (about 16 kg DM ha^{-1})	[13]
M. sativa L.; *T. pratense* L.	*T. aestivum* L. × *Triticosecale* Wittmack	-	reduced weed DM (68%—*M. sativa*; 38%—*T. pratense*); reduced of weed density (65%)	[54]
M. sativa L.; *Lupinus albus* L.	*T. aestivum* L. × *Secale cereale* L.	*Chenopodium album* L., *Poa annua* L., *Stellaria media* (L.) Vill.	reduced weed biomass (54%—*M. sativa*; 42%—*L. albus*)	[43]
M. sativa L.; *T. pratense* L.; *Pisum sativum* L.	*T. aestivum* L.	*Descurainia sophia* (L.) Webb. ex Prantl; *Sonchus oleraceus* L.; *Kochia scoparia* (L.) Schrad.	reduced weed DM (45%—*M. sativa*; 63%—*P. sativum*); increased of weed DM (11%—*T. pratense*)	[53]
Medicago polymorpha L.; *Medicago truncatula* Gaertn.; *Trifolium alexandrinum* L.; *T. pratense* L.	*T. aestivum* L. and *Zea mays* L. rotation system	*Capsella bursa- pastoris* (L.) Medik.; *S. media* (L.) Vill.; *Thlaspi arvense* L.; *T. aestivum* L. (volunteer wheat)	reduced density (41–78%) and dry weight (26%–80%) of winter annual weeds; reduced dry weights of summer annual (70%–*Medicago* spp.) and perenial weeds (35%–75% *Medicago* spp., *T. alexandrinum*)	[56]
M. lupulina L. (United Kingdom); *T. repens* L. (Norway, Germany, Sweden); *Trifolium subterraneum* L. (Germany, Switzerland); mixture of *T. repens* and *Lolium perenne* L. (Sweden); mixture of *M. lupulina*, *Sinapis alba* L., *Brassica napus* L. and *Raphanus sativus* L. (United Kingdom)	*T. aestivum* L. (first year); *Hordeum vulgare* L. in United Kingdom and Norway, and *Z. mays* L. at the other sites (second year)	*S. media* (L.) Vill.; *C. album* L.; *Rumex* spp.; *Tripleurospermum inodorum* (L.) Sch.Bip.; *Elymus repens* (L.) Gould	reduced weed cover throughout the intercrop period (55% to 1% depending on site); no reduced weed biomass or density	[57]
M. lupulina L.; *T. pratense* L.; *T. repens* L.; *T. incarnatum* L.; *Trifolium resupinatum* L.; *M. alba* Medik.; *V. sativa* L.; mixture of *M. lupulina* L. and *L. multiflorum* Lam.	*H. vulgare* L.; *T. aestivum* L.	*Galeopsis* L. spp.; *Myosotis arvensis* (L.) Hill; *S. media* (L.) Vill.; *Viola arvensis* Murr.; *Taraxacum officinale* Weber in Wiggers; *T. inodorum* (L.) Sch.Bip.; *Cirsium arvense* (L.) Scop.); *P. annua* L.	reduced weed density and biomass in *T. aestivum* above 50% (in *H. vulgare*—no effect)	[58]
Medicago scutellata Mill.; *Vicia villosa* Roth.; *T. subterraneum* L.	*Solanum tuberosum* L.	*Lolium temulentum* L.; *S. media* (L.) Vill.	reduced weed biomass (22%–57%)	[59]

Table 1. *Cont.*

Cover Crops	Crop	Dominant Weed Species	Weed Control	References
M. lupulina L.; mixture of *M. lupulina* L. + *Lolium multiflorum* Lam. var. *westerwoldicum* Mansh.	*Beta vulgaris* L.	*Agropyron repens* (L.) P. Beauv.; *C. album* L.; *Echinochloa crus-galli* (L.) Beauv.; *Galium aparine* L.; *V. arvensis* Murr.; *Amaranthus retroflexus* L.; *Solanum nigrum* L.; *S. media* (L.) Vill.	reduced weed number (25%–38%—*M. lupulina*; 44%–55%—mixture of CCs) and air-dry weight of weeds (21%–44%—*M. lupulina*; 45%–51%—mixture of CCs)	[60]
M. lupulina L.; *V. villosa* Roth.; *T. subterraneum* L.; *T. pratense* L.; *T. repens* L.; *Trifolium incarnatum* L.	-	*Chamomilla suaveolens* (Pursh) Rydb.; *Matricaria perforata* Merat; *P. annua* L.	reduced weed dry weight (*V. villosa* reduced 95% compared to *T. repens*)	[61]
monoculture or mixture of *Trifolium hybridum* L., *Lotus corniculatus* L., *M. lupulina* L., *T. incarnatum* L.; *Lolium multiflorum* Lam.; *Lotus pedunculatus* Cav.; *M. sativa* L., *Festuca pratensis* Huds.; *Lathyrus pratensis* L.; *L. perenne* L.; *T. pratense* L.; *Onobrychis viciifolia* Scop.; *Phleum pratense* L.; *T. repens* L.; *Melilotus alba* Medik.; *V. sativa* L.	-	*S. media* (L.) Vill.; *Sonchus arvensis* L.; *Veronica persica* Poiret; *Persicaria maculosa* L.; *Ranunculus repens* L.; *V. arvensis* Murray	reduced weed aboveground biomass (44%–92%—legumes, 72%–90%—*T. pratense*)	[62]
monoculture or mixture of *Trifolium hybridum* L. (AC) and *M. lupulina* L. (BM); AC:BM ratios (100:0, 67:33, 50:50, 33:67, 0:100)	-	-	reduced weed aboveground biomass (39%–96%, depending on the harvest date)	[20]
T. pratense L.	*T. aestivum* L. × *Triticosecale* Wittmack	-	reduced weed density (38%)	[63]
T. pratense L.	*T. aestivum* L.	*Ambrosia artemisiifolia* L.	reduced weed biomass (28%–43%)	[64]
T. pratense L.; *T. repens* L.; mixture of *T. pratense* L. and *Phleum pratense* L.; mixture of *T. pratense* L. and *Lolium* L.	*T. aestivum* L.; *Avena sativa* L.	*Spergula arvensis* L.; *S. media* (L.) Vill.; *V. arvensis* Murray; *C. album* L.; *Erodium cicutarium* (L.) L'Herit; *C. arvense* (L.) Scop.	reduced weed biomass (74%)—mixture of *T. pratense* and *Lolium*; increased seed bank and density of emerged weed (4.5 and 10 times in cloves)	[65]
T. incarnatum L.; *T. subterraneum* L.	*Z. mays* L.	*S. nigrum* L.; *C. album* L.; *A. retroflexus* L.; *Ammi majus* L.; *Cynodon dactylon* (L.) Pers.; *Geranium dissectum* L.; *Polygonum aviculare* L.; *V. persica* Poiret; *Xanthium strumarium* L.; *E. crus-galli* (L.) Beauv.	reduced weed biomass (22%–46%—*T. incarnatum*; 21%–67%—*T. subterraneum*)	[7]
T. pratense L.; *V. villosa* Roth.	*Z. mays* L.	*A. retroflexeus* L.; *Convolvulus arvensis* L.; *Acroptilon repens* (L.) DC.; *Cuscuta* sp.	reduced weed biomass (77%—*V. villosa*)	[66]
T. pratense L.; *V. sativa* L.	-	*Lamium aplexicaule* L.; *Papaver rhoeas* L.; *Sinapis arvensis* L.; *Chamomilla recutita* L.; *Phalaris minor* Retz.	reduced weed number (34%–68%—*V. sativa*; 19%–48%—*T. pratense*) and dry weight (58%–78%—*V. sativa*; 29%–44%—*T. pratense*)	[35]

Table 1. *Cont.*

Cover Crops	Crop	Dominant Weed Species	Weed Control	References
T. repens L.; *T. pratense* L.; *V. villosa* Roth; *Vicia benghalensis* L.; *Trifolium resupinatum* L.	-	*A. retroflexus* L.; *C. album* L.; *Portulaca oleracea* L.; *Persicaria longiseta* (De Bruyn) Kitag.; *S. nigrum* L.	reduced weed dry weight (29%–53%—*V. villosa*)	[19]
T. hybridum L.; *Trifolium michelianum* Savi var. *balansae* (Boiss.) Azn.; *T. alexandrinum* L.; *T. incarnatum* L.; *Trifolium resupinatum* L.; *T. pratense* L.; *T. repens* L.	-	*Brassica juncea* (L.)	reduced mustard biomass (29%–57%—without mowing)	[41]
mixture of *Vicia faba* L. (35%); *Vicia dasycarpa* Ten. (15%); *V. benghalensis* L. (15%); *P. sativum* L. (25%); *A. sativa* L. (10%)	-	*Malva parviflora* L.; *C. bursa-pastoris* L.; *S. media* L.; *Lamium amplexicaule* L.; *Urtica urens* L.; *Sonchus* spp.; *P. annua* L.	reduced weed DM linearly with increasing seeding rate (82%–100%)	[67]
V. dasycarpa Ten.; *V. faba* L.; *Lupinus angustifolius* L.	-	*Bromus carthartictus* Vahl; *C. bursa-pastoris* (L.) Medik.; *C. album* L.; *M. parviflora* L.; *S. media* L.	reduced weed density (23%–80%) and dry weight (30%–80%)	[68]
P. sativum L. and *V. sativa* L.—LSL; *T. incarnatum* L. and *Trifolium squarrosum* L.—SSL; *A. sativa* L. and *H. vulgare* L.—POA; *R. sativus* L. and *Brassica nigra* L.—BRS; mixtures of SSL, LSL + POA, SSL + BRS, LSL + POA + BRS, SSL + POA + BRS, LSL + SSL + POA + BRS	-	*Senecio vulgaris* L.; *Helmintotheca echioides* L.; *Alopecurus myosuroides* L.; *R. repens* L.; *Juncus tenageja* Ehrh.; *Lolium multiflorum* Lam.	reduced weed biomass (93%—mixture of LSL + POA compared to monoculture of *P. sativum*; 54%—mixture of SSL + POA compared to monoculture of *T. incarnatum*)	[69]
V. villosa Rotch.	*Apium graveolens* L.	*S. media* (L.) Vill.; *Amaranthus blitoides* S. Wats; *Cyperus esculentus* L.; *C. bursa-pastoris* (L.) Medik.; *P. oleracea* L.	reduced weed biomass (70%)	[70]
V. villosa Rotch.	*Solanum lycopersicum* L.	*A. retroflexus* L.; *Digitaria sanguinalis* (L.) Scop.; *P. oleracea* L.	reduced weed density (72%–79%) and aboveground biomass (40%)	[71]
V. sativa L.	*Z. mays* L.	*Ipomoea grandifolia* (Dammer) O'Donell; *Euphorbia heterophylla* L.; *D. sanguinalis* (L.) Scop.; *Cyperus rotundus* L.	reduced weed DM (76%) and numer (58%)	[14]
V. villosa Roth.; mixture of *V. villosa* and *S. cereale* L.	*Z. mays* L.	*L. amplexicaule* L.; *S. media* (L.) Vill.; *P. annua* L.	decreased weed biomass (92%—*V. villosa*; 97%—mixture of cover crops)	[72]
V. villosa Roth.	*Glycine max* (L.) Merr.	*Amaranthus rudis* Sauer; *Setaria faberi* Herrm.	decreased weed biomass (26%, in rolled system compared to the burndown system)	[51]

Table 1. *Cont.*

Cover Crops	Crop	Dominant Weed Species	Weed Control	References
mixture of *V. villosa* Roth. and *S. cereale* L.	*G. max* (L.) Merr.	*A. retroflexus* L.; *A. artemisiifolia* L.; *C. album* L.; *Polygonum convolvulus* L.; *Panicum dichotomiflorum* Michx.; *Setaria faberi* Herrm.; *Setaria glauca* L.; *C. esculentus* L.; *T. officinale* Weber in Wiggers	reduced weed density (67%–85%—*C. album A. retroflexus*, *Setaria* spp.), without *C. esculentus* L.	[73]
V. villosa Roth.; *P. sativum* L.	*Brassica oleracea* L. var. *acephala*	*E. cruss galli* (L.) P.B.; *C. dactylon* (L.) Pers.; *C. arvensis* L.; *C. album* L.; *P. oleracea* L.; *A. retroflexus* L.; *C. arvense* (L.) Scop.	reduced weed dry biomass (81%—*V. villosa*, 48%—*P. sativum*) and density (66%—*V. villosa*, 15%—*P. sativum*)	[42]
mixture of *V. villosa* Rotch. and *S. cereale* L.	*Brassica oleracea* var. *capitata* f. rubra	*T. arvense* L.; *C. bursa-pastoris* (L.) Medik.; *Galinsoga parviflora* Cav.; *L. amplexicaule* L.	reduced weed number (25%) and fresh biomass (50%)	[74]
mixture of *V. villosa* Rotch. and *S. cereale* L.	*Capsicum annuum* L.; *B. oleracea* var. *capitata* f. rubra	*C. album* L.; *C. bursa-pastoris* L.; *S. vulgaris* L.; *Matricaria inodora* L.; *L. amplexicaule* L.; *G. parviflora* Cav.; *E. crus-galli* (L.) Beauv.; *U. urens* L.; *Fallopia convolvulus* (L.) Á. Löve; *Polygonum persicaria* L.; *A. retroflexus* L.; *T. arvense* L.; *S. media* (L.) Vill.; *E. cicutarium* (L.) L'Herit	reduced weed number and biomass (39%–58%—cover crops mulching, 10%–45%—cover crops incorporated into soil)	[75]
V. villosa Rotch.; mixture of *V. villosa* Rotch. and *S. cereale* L.	*Solanum lycopersicum* L.; *Cucurbita pepo* L.; *C. annuum* L.	*C. bursa-pastoris* (L.) Medik.; *Setaria* spp.; *C. album* L.; *A. retroflexus* L.	reduced weed density (96%—mixture of cover crops, 80%—*V. villosa*)	[76]
Crotalaria juncea L.	*S. lycopersicum* L.	*Digitaria horizontalis* Willd.; *Gnaphalium spicatum* Lam.; *Cyperus* sp.; *G. parviflora* Cav.; *Amaranthus* sp.	reduced weed DM (97%)	[77]

CCs can be introduced before the main crop, i.e., after crushing, treating with a herbicide, plowing, or left on the surface of the soil as dead mulch (the successive crop is directly drilled through the mulch) or grown together with the main crop as living mulch [20]. In both cultivation systems, CCs, and especially legume CCs, can be successfully used in limiting weed infestations by using two mechanisms: direct and indirect [17,18,20]. Adapting the phenology of legume CCs to the cultivation of the main crop uses its niche pre-emption, which can be a direct, successful competition for weeds in terms of habitat and resources, e.g., light, soil, water, and nutrients [43,78–82]. An indirect effect of CCs legume on weeds is also observable, manifested by physical [83,84] or chemical suppression [52,85–87]. The weed suppression mechanism based on the competing abilities of legume CCs is visible when used as a smother crop or the living mulch.

Bilalis et al. [35] demonstrated that *V. sativa* caused a reduction in the light available for weeds, which led to an observable reduction in the number of weeds and their dry weight. The competitive abilities of common vetch were related to its morphological features, including high overall leaf area, greater height and length of stems, number of shoots per plant, biomass and upright growth. A clear reduction in the emergence of weeds was also observed in the cultivation of *V. villosa* [42,61]. Additionally, in the case of *M. lupulina*, both the morphological features and the rapid growth rate of the plant determined its ability to suppress weeds. In the case of legume CCs characterized by a slower growth rate (*T. hybridum*), their allelopathic properties were decisive for the effectiveness of weed control at an early stage [20]. In turn, Ross et al. [41] stated that the competitiveness of *Trifolium* sp. is determined not only by its morphological features, but also by the method of cultivation. A higher level of weed suppression has been reported on the low-productivity site than on the high-productivity site. Greater competitiveness of *T. alexandrinum* was determined by the following features: upright growth habit, long stems, high biomass production, and late flowering, and of *T. hybridum*: upright growth and long stems. However, in the case of *T. squarrosum*, its high and long-term stability of the weed suppression capacity was decided by its mat-like prostrate growth during early developmental stages [69].

Numerous reports confirmed that annual weeds were suppressed to a greater extent by legume CCs residues, with the end result being more pronounced for small-seeded summer annual weeds than for large-seeded annuals [46,84,88]. Brandsæter and Netland [61] have shown that *T. incarnatum* limited weed infestation to a greater extent than *T. subterraneum*. This proves that the lower weed suppression effectiveness of *T. incarnatum* occurs when perennial and monocotyledonous weeds dominate the flora. Leaving the remains of legume CCs has a lower suppression effect on perennial weeds [73]. Other authors have also reported a lower impact of legume CCs on the reduction of perennial weed infestation (especially winter-seeded species kept only for a single growing season) [56,89–91]. Sjursen et al. [65] also found the lack of the suppressive effect of undersowing clover (*T. pratense, T. repens*) on annual weeds with regard to both short-term (weed emergence) and long-term (seed bank) effects, possibly due to the successive fertilization effect with legume CCs.

CCs, including legume CCs, can suppress not only weeds that are already resistant to herbicides, but also a significant number and biomass of herbicide-susceptible weeds, thus reducing the intensity of their selection in terms of future resistance [92,93]. Reducing weed abundance, biomass or seed production in legume CCs, and suppressing seedling emergence from the weed seed bank reduces the selection pressure at the end of cover crop [92,94], particularly in no-till systems, where herbicides are used to terminate cover crop.

Legume CCs can also compete for resources necessary for growth and development, also with the main crops [95]. It follows that legume CCs not only compete with weeds by inhibiting their emergence and growth, but can also suppress the main crops [78]. Therefore, when selecting CCs species, attention should be paid to their faster growth and shorter vegetation period compared to the main plants, in order to minimize competition, e.g., for the light. It is also necessary to provide the soil with water in order to minimize competition between the crop and CCs [12,96]. The competition of CCs for water, especially in early spring, may be visible in the cultivation of the main crop in arid and semi-arid regions, where soil moisture is a factor limiting plant production [97]. On the other hand, it was found that legume CCs compete to a lesser extent for N from soil with the main crops [5]. In addition, chemical and physical factors that affect weeds can also affect crops and the CCs themselves. For example, the main crops may be inhibited by allelochemicals released by some legume CCs. Moreover, in the absence of rotation in the cultivation of CCs, accumulation of allelochemicals and even pest (weeds, pests, pathogens) populations may occur. This will negatively affect not only the main crops, but also the CCs themselves [98].

Legumes CCs (*Mucuna deeringiana* (Bort) Merr., *Canavalia ensiformis* (L.) DC., *Leucaena leucocephala* (Lam.) de Wit, *Lysiloma latisiliquum* (L.) Benth.) used as both living CCs and dead mulches (incorporated on soil surface) have reduced the weed biomass. The greatest reduction in weed biomass (68%) was reported for *M. deeringiana* as living CC in corn cultivation. The inhibitory effect of these legumes on the growth and development of weeds is related to their allelopathic properties. The aqueous leachates of all four legumes showed strong phytotoxic effect on the rooth growth of *E. crus-galli* and *Amaranthus hypochondriacus* (L.) [52]. Additionally, aqueous leachates from fresh leaves and volatiles of *Tephrosia vogelii* Hook. inhibit the seed germination and seedling growth of *Festuca arundinacea* Schreb., *Cynodon dactylon* (L.) Pers. and *Digitaria sanguinalis* (L.) Scop. Additionally, mulching with this legume CC resulted in reduction of the weed biomass (15.8%) in corn cultivation [99]. The allelopathic effect in legumes is varietal dependent and might have a genetic basis. The inhibitory effect of *M. sativa* cultivars on weeds was proportional to the number and quantity of growth inhibitors (phenolic compounds), which showed strong allelopathic activity. As a result, the weed suppression of legume CC might be proportional to its allelopathic magnitude [100]. When *M. sativa* was incorporated into soil (as mulch) for weed-control purposes, phenolic acids detected in the soil reached maximum concentrations after 10 to 15 days and were efficacious until 20 to 25 days. Chemicals released from allelopathic plants incorporated into soil are toxic and cause inhibition of certain species and could be applied as a biological tool for weed management [101].

However, depending on the desired purpose of the main plant cultivation, the cultivation of legume CCs can be regulated in terms of sowing time, growth rhythm, retention of N or the biological N-fixing potential. Therefore, they are less competitive with the main crop than the weeds [78].

It has been demonstrated that a higher density of legume CCs also increases competitive interactions and resource consumption between the main crop species. This in turn provides an effective weed reduction [20,102]. With greater coverage of the soil surface with CCs at an early stage of their growth a more effective suppression of weeds was observed, especially at a high sowing density of species characterized by high seed weight. Increasing the coverage with the CC from the early growth stage, even if the plant is low, it can inhibit weed biomass production. Moreover, the cultivation of low CCs as a mixed cropping or intercropping system is beneficial for the main crop, especially for the higher species, as there is little competition for light between the main crop and the cover crop [19]. Bilalis et al. [35] found that *V. sativa* was more effective in reducing weed infestation than *T. pratense*, which was characterized by higher competitiveness in relation to weeds due to a higher number of developed shoots and greater dry matter. This is confirmed by Uchino et al. [19], in whose research *V. villosa* proved more effective in supressing the weed biomass, as compared to *T. repens*, *T. pratense*, *T. resupinatum*, and *V. benghalensis*. Additionally, *M. sativa* is more competitive with weeds than *T. pratense* under semi-arid conditions [53,103,104]. Although some studies have confirmed the positive correlation between CCs biomass and weed suppression [42,105–107], Elsalahy et al. [20] showed that in the case of legumes, this relationship may be weak and inconsistent. The mechanisms that determine the impact of legume CCs biomass on weed biomass are the contrasting characteristics of legume species in the mix (growth rate, response to environmental conditions), which allow the legume CCs to mutually complement throughout the cultivation period [20,69], as well as their allelopathic properties. Under conditions where higher biomass production is limited by environmental conditions, the allelopathic abilities of legume CCs may play an important role in reducing weed infestation [20]. The chemical weed suppression is seen in alfalfa (*M. sativa*), which gives off allelochemicals (phenolic compounds) harmful to weeds, e.g., *Doparium junceum* Hamilt., *Lindernia pyxidaria* L., *Elatine trianda* Schk. var. *pedicellata* Krylov, *Eleocharis acicularis* (L.) Roem. et Schult., *C. album* L., *E. crus galli* (L.) Beauv. [67,108–111]. *V. villosa* [112–114], *T. pratense* [115,116], *T. incarnatum* [112], and *T. hybridum* [111] also show allelopathic effect against weeds. Kruidhof et al. [43] demonstrated that the suppression of spring weed emergence by plant residues CCs (*M. lupulina*—54% on average) may be related to the release of allelochemicals (saponins, flavonoids and phenolic acids) into the soil. Therefore, in terms of weed control, successive crop cultivation should take into account the physical properties of plants introduced into the soil, determined by the pre-treatment of the residues and the nature of the CCs [20,117].

The indirect effect of legume CCs on weeds is creating a physical barrier (as incorporated or soil surface-placed cover crop residues), which can prevent the newly formed seeds from reaching the soil surface and contribute to depletion of the weed seed bank in the soil, and as a result, inhibit or delay germination and emergence of weeds [115]. In this case, a more effective weed suppression is seen in CCs legume producing more biomass, especially at the beginning of the growing season. Osipitan et al. [118] found that early weed control with CCs is even compared to chemical and mechanical weed control methods in crop systems. In addition, an earlier sowing date, higher seeding density, and a delayed termination date of CCs favors more biomass, and is therefore more effective in suppressing weed infestation, especially for annual summer weeds [35,73,119]. In addition, cultivation practices such as delaying planting and termination dates of CCs may selectively affect weed populations by changing their structure into less competitive species [73]. It is one of the potential mechanisms contributing to the change in the crops' tolerance of weed biomass [120].

Legume CCs can also initiate the germination of weed seeds, leading to faster depletion of the soil weed seed bank, especially in long-term cultivation [91]. Moonen and Bàrberi [91] found a 22% reduction in the soil weed seed bank, even seven years after introducing CC legumes (*T. subterraneum*) into the no-till corn cultivation system. In turn, Sjursen et al. [65] observed that the annual seeding of clovers (*T. pratense* and *T. repens*) favored the increase of weed biomass and the size of the seed bank. This is most likely due to the ability of legumes to fix atmospheric nitrogen, which also becomes available to weeds.

Legume CCs (e.g., *M. sativa*) can also increase the activity of *V. arvensis*, *A. myosuroides* seed predators (both vertebrates and invertebrates) by changing the quality of their habitats [121]. Seed losses reaching 25 to 50% caused by seed predators are sufficient to significantly slow down the growth of the weed population [122]. Youngerman et al. [123] showed that seeds of the main plants can also be damaged by invertebrate seed predators. However, they less prefer CCs seeds with hard seed coats, for example *V. villosa*. Blubaugh et al. [124] found that CCs increased the frequency of weed seed consumption by omnivorous predators by 73%. As a result the cover crops improve biocontrol, not only by promoting increased activity of omnivores, but also by facilitating their function as seed predators on an individual-level.

The presence of dominant weed species in plant crops can be explained by their divergent characteristics, which have allowed them to react differently to environmental fluctuations and management practices, in accordance with a process called "the storage effect". According to the theory of the storage effect, for the species in the weed community to survive, there must be species-specific responses to changing environmental conditions. The responses should then lead to differentiated competition (covariance between the environment and competition), and the durability of the seed bank should buffer the growth of weed populations in years with unfavorable environmental conditions [125]. However, CCs can sometimes have a positive effect on the germination or growth of weeds [117]. Legume CCs play a part in the fixing of atmospheric nitrogen and therefore they can indirectly create more favorable conditions for weed growth than the other CCs [59].

Any undesirable plant in the main crop cultivation, despite its ecosystem advantages, is considered a weed. Therefore, the self-seeding of legume CCs in succeeding crops can be a problem, especially in the case of species that are highly competitive with those of main crops. Their control may be difficult, especially in the case of perennial species characterized by a rapid growth rate, producing a large number of diasporas and that are resistant to abiotic stress conditions.

3. Effect of Legume CCs Cultivation Systems on Weed Control

As mentioned before, CCs can be introduced into plant production systems as living mulch by sowing into the main crop, or as dead mulch, i.e., sowing the CCs after the main crop harvest and terminating their growth before cultivation of the secondary main crop, by plowing under, crushing or treating with glyphosate. In this case the successive plant is sown directly into the mulch [57]. The use of CCs for weed control is particularly important in no-till systems [119].

Living mulches based on CCs have a positive effect on reducing weed infestation as well as on the health of the crop by reducing pest infection or infestation [126–129]. However, some authors report that when using mulch, excess moisture in the soil may be unfavorable for the soil, increasing the infestation of pathogens and in some cases, pests (snails, rodents) [130,131].

It has been shown that legumes used as live mulch are able to effectively control weeds, but at high density [20]. The effectiveness of legume CCs in suppressing weeds is mainly the result of their higher competitive abilities (e.g., for light and habitat) compared to weeds [20,102,132]. Teasdale and Mohler [47] demonstrated that weed suppression was more effective in the case of *V. villosa* mulch, in early growing season, during the spring biomass accumulation. However, as the season progressed and the mulch degraded, this effect wore down. A notable reduction in weed biomass was observed after cutting legume CCs due to the reduction of weed growth and direct mortality, especially when the cutting height is less than 7 cm. In addition, dead mulch from legume CCs may promote capturing seeds of some wind-dispersed weed species [127]. The competitive equilibrium between legume CCs and weeds is changed also due to the different regrowth of cut weeds, especially annual weeds [20]. However, emergence of weeds can be stimulated by introducing cut mulch into the soil [53,133]. This is confirmed by the reports of Golian et al. [74] who incorporated *S. cereale* and *V. villosa* mixes into the soil. This not only did not reduce weed infestation, but even increased the amount of weeds by 15%.

Legume CCs sown in late summer, for example in cereal crops, provide a winter cover and are terminated before sowing the main crops [51]. Terminating the growth of CCs before the main crop is essential for their usefulness, as regrowing CCs can interfere with the growth and development of the main crop, which causes a decline in the yield [134]. Moreover, the method of CC termination may influence their weed suppression potential in successive crops [50]. In no-till cropping systems, a herbicide may be used to complete the cultivation of CCs, including legume CCs, but physical methods are also available [135,136]. Mowing is one of the physical methods of terminating the cultivation of CCs without disturbing the soil, but in this method regrowth of CCs and an aggregated spatial distribution of crop residues occur [135]. The cover-crop roller-crimper [137] can also be used to terminate the growth of CCs.

It has been found that plant residue management can regulate the diversity and composition of the weed population by introducing qualitative and quantitative changes [138,139]. Residual CCs in the form of dead mulch change soil properties, which can impact weed control by affecting the survival, dormancy, predation, and long-term viability of weed seeds [140,141]. As dead mulch, legume CCs shade the surface of the soil and thus capture solar radiation. This reduces not only light permeability, but also the daily amplitude of soil temperature. Undoubtedly, this inhibits weed seed germination, because the dormancy of seeds of many species of annuals is controlled by light and temperature [47,73,142]. This is confirmed by studies in which the remains of legume CCs (*V. villosa*) inhibit the emergence of weeds due to the reduced access of light to the soil surface [47,143]. Additionally, in the case of mulch produced from a mix of CCs (*O. viciifolia* and *H. vulgare*) these mechanisms contributed to the inhibition of weed infestation [87]. It follows that dead mulch suppresses weeds by creating a physical and chemical barrier [140,144].

The effectiveness of CCs as dead mulch is determined by the ability of CCs to suppress weeds during its active growth and the residual effect of CC mulch after senescence [7]. In addition, the effectiveness of weed control is also influenced by the even distribution of residual CCs and the way they are placed in the main crop. Residues of CCs on the soil surface decompose more slowly and gradually release agrochemicals, as compared to the residues mixed with the soil [117,145]. Along with an increase in the amount of mulch there was also a decrease in the occurrence of small-seeded annual weeds [46,84]. Teasdale and Mohler [84] reported that weed suppression was impacted to a greater degree by the amount of residual CCs than by their type, however, in the case of perennial weeds, increasing the amount of residue CCs had no damping effect.

Decaying residue of CCs may create a more favorable environment for weed germination and development [144]. Leaving residue on the soil surface may result in better availability of water, which supports weed development [47,146]. It was also shown that a small amount of plant residue of *V. villosa* promotes light transmission and triggers a phytochrome reaction, occasionally stimulating the emergence of weeds, especially *Amaranthus hybridus* [147]. This is also confirmed by the studies by Mischler et al. [90], who demonstrated that *V. villosa* stimulates the germination and growth of weeds—especially in conditions of increased temperature and soil moisture [15]. This is due to the relatively rapid decomposition of *V. villosa* biomass, even during its growth. CCs plant residue may also interfere with sowing or mechanical weed control, which in turn may have a negative effect on successive tillage [148].

CCs used as intercrops can potentially reduce weed infestation, which positively impacts the crop yield. Intercrops maintain high competition with weeds, irrespective of the species and productivity of the weeds, the biomass of the main crop or the availability of nitrogen in the soil [149]. Weed infestation is inhibited by reducing the space available for weeds in the habitat, and by reducing their resources, e.g., water, light, and nutrients. Nitrogen and light play an important role in reducing weed infestation due to the complementary abilities of intercrop species, their use of nitrogen (mineral nitrogen in soil and atmospheric nitrogen N_2), as well as light capture and soil surface coverage [149,150]. These mechanisms are often interrelated and depend on the temporal and spatial growth dynamics of the above-ground and underground parts of the plant [151]. Therefore, this may explain the increase in intercrop yields and the inhibition of weed infestation [152,153]. Intercrop species with different characteristics in terms of growth rate and response to environmental conditions favor the inhibition of weed infestation. For example, species with varying growth rates can cause a temporary asynchronicity in species growth dynamics; a fast-growing species competes with weeds early in the growing season and a slow-growing species competes with later-emerging weeds. This enables effective weed management throughout the growing season, while also reducing direct competition for resources between species in the intercrop [20]. High and constant inhibition of weed infestation in the legume/cereal intercrop was demonstrated even with a low share of the cereal plant in the intercrop biomass, while for a sole legume crop, a decrease and high variability in weed suppression were noted [149,154]. For example, in the sole crop, *P. sativum* is a weak competitor in relation to weeds [155], therefore, with greater pressure from weeds, this species reacts with a limited use of N and a decrease in the seed yield [156,157], which is also confirmed by studies on other legume species [158,159].

Numerous studies have also confirmed the effectiveness of the relay intercropping of a forage legume (RIL) in controlling the main crop weeds [18,53–55,63,64,160]. In RIL, CCs are left in the field until the crop is harvested, which enables weed control before undersowing [65]. This allows avoiding the bare soil period, which favors the emergence of weeds between the cash crop harvest and the full establishment of a CC, when it is sown between two cash crops [18]. To achieve a significant reduction of weed biomass, it is necessary to maintain the CCs legume until late autumn [18,53]. According to Blaser et al. [54,63] and Amossé et al. [18] the presence of legume CCs effectively reduced weed density even during the harvest of the crop. Legume CCs were found effective in reducing the number of weeds emerging in spring, as well as in increasing the mortality of weed seedlings. These mechanisms are also associated with limited resources (e.g., light, water), which negatively affects the growth and development of weed seedlings [43]. Leaving CCs until autumn effectively suppresses the emergence of weeds, which leads to reducing the weed seed bank in the soil [161]. Amossé et al. [18] concluded that both legume CCs and cereal crops are necessary to reduce weed abundance and biomass during RIL. The reduction of weed infestation in RIL is thus managed by the cumulative effect of the crop and the legume CCs. The combination of winter crops and RIL decreases weed infestation in spring, reducing the number of annuals. Therefore, in crop rotations with a high proportion of spring crops and with weed populations adapted to this crop cycle, the use of RIL is very effective in reducing the number of spring weeds [18].

Elsalahy et al. [20] reported higher capability of some small-seeded legume mixes to suppress weeds compared to the capabilities of monocultures of these species. These authors believe that this is most likely due to complementary growth features of the plants over time, i.e., a faster growing species (*M. lupulin*) suppresses weeds at the beginning of cultivation, and a slower growing species (*T. hybridum*) inhibits the later growth of weeds [20]. A better reduction of weed infestation was also found in mixes of leguminous plants with other species, as compared to the monocultures of these species [62,69,86,162]. These mixes allow obtaining a higher biomass of CCs, as well as retarding the decomposition of plant residues, e.g., legumes. This, in turn, affects their greater ability to reduce weed infestation [86]. The effectiveness of the mixes in reducing weed infestation is determined not only by the produced biomass, but also by other plant properties, such as allelopathic activity or canopy architecture, and the related soil shading surface [86,87]. Ranaldo et al. [69] confirmed that biodiversity in CC mixes has a more beneficial effect on the suppression of weeds, e.g., by producing more CC biomass, also under difficult field conditions. The mechanism explaining the effect of species diversity in CC mixes on weed control is the complementarity of resource use. Smith et al. [163] formulated the Resource Pool Diversity Hypothesis according to which, with a more diverse pool of soil resources, a lower abundance of weeds is expected. The increase in the ability to suppress weeds is influenced by the differentiation of the functional traits of the CCs (e.g., the growth form, root system, nutrient use strategy), i.e., the synergy between complementary species traits in the CC mix [69]. This will ensure a more complementary and consistent production of CC biomass throughout the growing season, even under changing environmental conditions [105,106,164]. However, Elsalahy et al. [20] confirmed that the ability to suppress weeds was influenced to a greater extent by the species identity in legume mix, than by the species diversity. In addition, the changing proportion of species in the legume mix has no significant impact on the suppression of weeds. However, a mix of legumes with a single dominant species had better effects in suppressing weeds compared to a mix comprising a balanced share of individual species [20]. In turn, Suter et al. [165] confirmed that mixes of species help control weeds, especially when individual species are functionally diverse. Due to the functional complementarity of the individual species, a higher sowing rate can be applied in multi-species mixes, as compared to the sowing standard for species grown in a monoculture. Moreover, obtaining a higher plant density thanks to the use of many species in the mix is more effective in controlling weeds [62,166]; this is a practice used in organic farming conditions. However, Smith et al. [27] argue that CCs mixes do not suppress weeds any more than the top CCs with the highest weed suppression capacity, grown as a monoculture. Moreover, according to the authors, a more diverse mix of CCs (14 species) does not outperform a less diverse mix. Research by Panasiewicz et al. [167] with narrow-leaved lupine in crop rotations with 75% cereal composition also indicate that the method of tillage has a significant influence on weed infestation and consequently, the seed yield. Dry weigt of weeds under conventional tillage was significantly lower than in reduced tillage and no-tillage. There was also a significant difference between dry weight of weeds in reduced tillage and no-tillage. The weight of weeds in no-tillage was approximately two times greater than in reduced tillage.

In the organic system undoubted benefits can be generated by introduction of legume CCs. In chisel-plow based organic system (OR) with CCs (crimson clover before corn, rye before soybean) and manure applied for nutrients and postplanting cultivation, higher soil combustible C and N concentrations at all depth intervals to 30 cm compared to other cultivation systems were reported for weed control. In addition, corn grown under organic system had more N available compared to conventional no-tillage. Systems that incorporate high amounts of organic inputs from manure and

CCs can improve soil more than conventional no-tillage systems despite reliance on a minimum level of tillage. Authors suggest that if adequate weed control could be achieved in reduced tillage organic systems, higher soil quality could be achieved with yield-enhancing benefits compared to conventional no-tillage systems [89]. Additionally cover crop-based, organic rotational no-till crop production (mechanical termination of CCs with a roller-crimper and no-till planting crop into CC mulches) becomes a viable strategy for reduction of tillage in organic annual grain systems. Authors suggest that N-release synchrony with corn demand and improvement of weed suppression can be improved with grass-legume mixture. Integration of high-residue inter-row cultivation improves weed control consistency and may reduce reliance on optimization of CC biomass accumulation for weed suppression. However, breeding efforts are required to improve CC germplasm and develop regionally-adapted varieties [15].

Other ecosystem functions are also supported by legume CC, as conservative biological control with legume and non-legume CCs is an interesting alternative to the chemical control of many insects [168,169]. It was reported that cover crops regulate the tetranychidae mite populations in citrus orchards [168,170] as well as reduce thrips infestation in cotton cultivation [169]. However, legume cover crops may not affect the population of the root weevil in citrus groves or its feeding damage [171].

Cover crops provide a constant and abundant supply of food sources for natural enemies (predators and parasitoids) during the flowering period, allowing natural enemies to build up in the system, thereby keeping pest populations at acceptable levels. [172]. Altieri et al. [173] confirmed that correct agroecosystem diversification strategies such as CCs usually regulate pests by restoring the natural control of insect pests. Legume CCs (such as *Neonotonia wightii*) act as a reservoir for phytoseiid mites, thus contributing to the biological control of phytophagous mites in citrus orchards [174]. However, increase of the floral diversity in sown CCs could constitute a complementary method in management programs, by providing a greater amount of alternative food resources and alternative hosts to enhance the conservation and biological control of natural enemy populations [175]. Different CC species, with different blooming phenologies, provide habitat and resources for potential wild pollinators, particularly native bees. As a result, flowering CCs can be used for pollinator conservation purposes [176,177].

4. Effect of Legume CCs on the Soil Environment

4.1. *Effect of Legume CCs on the Content of Organic Matter and Nitrogen in Soil*

The main factors causing the degradation of agricultural soils are: intensive agrotechnical treatments, excessive chemical treatment, cultivation in a monoculture and inappropriate melioration. In consequence, soil organic matter is reduced, which in turn is associated with the deterioration of the physicochemical properties, sorption capacity, and biological activity of the soil. The high content of organic matter stabilizes the soil structure as well as reducing its susceptibility o compaction and erosion. It is also an indicator of soil fertility [178]. An important component of soil are microorganisms, which play a key role in the process of biomass decomposition and humus formation, nutrient circulation and increasing soil resistance to harmful factors [23,179]. Diversity of soil microorganisms is a prerequisite for maintaining soil fertility and has a positive effect on plant health. It conditions their proper growth and development, and, consequently, yielding [180].

One of the most important factors influencing the maintenance of soil fertility and productivity is a rich crop rotation, which should include CCs [22,181]. It increases the productivity of crops at a reduced workload, and at the same time cuts down the negative impact on the environment. In addition to enriching the crop rotation by increasing biodiversity, crop rotation brings additional benefits such as: increasing the content of organic matter in the soil, reducing weed pressure, reducing the spread of pathogens, and protecting the soil in periods between main crops [182,183]. Keeping CCs during the autumn and winter period is especially important as it limits nutrient losses and erosion, especially in light soils. The lack of plant cover in this period causes disturbances in the soil structure, reduces the infiltration of water into the substrate, and at the same time increases the leaching or blowing away of the arable layer [184].

Fabaceae are excellent ground cover crops; growing them brings a number of benefits for improving soil quality. Due to the biological ability to fix atmospheric nitrogen, their biomass is used entirely as green fertilizer or crop residues, which enrich the soil with nutrients, including organic carbon and nitrogen. This is beneficial from the point of view of the successive crop (Table 2) [185–187].

Table 2. Effect of legume CCs on the content of organic carbon and nitrogen in soil.

Legume Cover Crops	Crop/Tillage	Soil Texture	Potentially Mineralizable Carbon (kg·ha^{-1})	Permanganate Oxidizable Carbon (kg·ha^{-1})	Potentially Mineralizable Nitrogen (mg·kg^{-1})	Soil Inorganic N	Soil Organic C	References
no legume CCs *P. sativum* L. mixture of *P. sativum* L. and *A. sativa* L. mixture of *P. sativum* L. and *B. napus* L. mixture of *P. sativum* L., *A. sativa* L. and *B. napus* L. mixture of *P. sativum* L., *A. sativa* L., *B. napus* L., *V. villosa* Rotch, *R. sativus* L. and *H. vulgare* L.	*T. aestivum* L./no-till	clay loam	133 202 192 210 158 190	795 769 831 776 810 850				[188]
no legume CCs mixture of *S. cereale* L. and *V. villosa* Roth. mixture of *S. cereale* L. and *T. incarnatum* L. mixture of *S. cereale* L., *A. sativa* L., *R. sativus* var. niger J. Kern., *Brassica campestris* L. and *T. incarnatum* L.	*Z. mays* L. *G. max* (L.) Merr./no-till	silt loam			28.4 34.7 32.6 35.8	**(mg·kg^{-1})** 15.5 18.5 20.6 19.4		[189]
no legume CCs *M. sativa* L. *T. pratense* L. *P. sativum* L.	*T. aestivum* L./ conventional tillage, fall planted	sandy clay loam				**(kg·ha^{-1})** 19.6 43.8 34.6 30.0		[53]
no legume CCs *M. sativa* L. *T. pratense* L. *P. sativum* L.	*T. aestivum* L./conventional tillage, spring planted	sandy clay loam				**(kg·ha^{-1})** 20.8 27.1 18.5 24.2		[53]
no legume CCs *P. sativum* L. *V. villosa* Rotch/	*Z. mays* L./conventional tillage	silt loam				**(g·kg^{-1})** 1.22 1.26 1.28	**(g·kg^{-1})** 15.7 16.0 15.8	[190]
no legume CCs *T. incarnatum* L. *V. villosa* Rotch.	*Sorghum bicolor* (L.) Moench/ conventional tillage	sandy clay loam				**(g·kg^{-1})** 1.0–1.3 1.3–1.5 1.3–1.5	**(g·kg^{-1})** 8.5–10.1 10.6–12.8 10.2–11.8	[191]
no legume CCs mixture of *S. cereale* L., *P, sativum* L. and *T. incarnatum* L.	*Z. mays* L. *G. max* (L.) Merr./conventional tillage	silt loam			5.78 7.62	**(weight %)** 0.11 0.10		[192]

Table 2. *Cont.*

Legume Cover Crops	Crop/Tillage	Soil Texture	Potentially Mineralizable Carbon (kg·ha^{-1})	Permanganate Oxidizable Carbon (kg·ha^{-1})	Potentially Mineralizable Nitrogen (mg·kg^{-1})	Soil Inorganic N	Soil Organic C	References
no legume CCs *S. cereale* L. (in *Z. mays*) and *V. villosa* Rotch. (in *G. max*) *S. cereale* L. (in *Z. mays*) and mixture of *S. cereale* L. and *V. villosa* Rotch. (in *G. max*)	*Z. mays* L. *Glycine max* (L.) Merr./no-till	silt loam				**(mg·kg^{-1})** 0.26 0.29 0.21		[193]
no legume CCs *Mucuna pruriens* (L.) DC. no legume CCs *M. pruriens* (L.) DC.	*Z. mays* L./ no-till	sandy loam					**(Mg·ha^{-1})** 6.76/0–5 9.42/0–5 19.2/0–20 25.1/0–20	[194]

The amount of organic carbon and nitrogen introduced into the soil depends on many factors such as: quantity, quality, and management of biomass or crop residues, as well as frequency of cultivation and climatic conditions [195]. Amossé et al. [18] showed that legume ground cover plants such as *M. lupulina*, *M. sativa*, *T. pratense* and *T. repens* significantly enriched the soil with symbiotic nitrogen, increasing the yield of Z. mays as a successor plant by 31%. However, they did not significantly affect the yield of *T. aestivum* (winter wheat) into which they were sown. Nitrogen content in biomass of ground cover plants, analyzed in late autumn, after *T. aesivum* harvest, showed that from 71% to 96% of N came from symbiotically fixed nitrogen, which enriched the soil by 38 to 67 kg N·ha^{-1}. Legume CCs also usually have a C:N ratio of under 20:1, which promotes faster nitrogen mineralization. This reduces the amount of nitrogen available to the main crop [21]. This is especially important in management systems where there are nitrogen deficiencies in the soil, such as organic cereal cultivation, where the use of nitrogen in mineral fertilizers is unacceptable [149,196–200].

In addition, in cereal/legume intercrops grown in arid, semi-arid, tropical, and temperate climates the intercropping may be advantageous when nitrogen availability corresponding to soil nitrogen plus N-fertilizer is below a determined threshold (12 g N m^{-2}) due to a high degree of complementary nitrogen use between the two species for low N levels [201–203].

According to Abdalla et al. [166], CCs which included legumes increased the yield of the successive crop by an average of 13%, in contrast to CCs with no legumes, which reduced the yield of the main crop by an average of 4%. According to the authors of the presented meta-analysis of 106 studies in 372 locations, including different countries and climatic zones, the advantages of CCs (legumes or legume mixes), justify their widespread adoption as good agricultural practice in sustainable management. However, the authors believe that in order to increase the efficiency of CCs, field management technology should be optimized to local climatic conditions, water abundance, soil type, and the cultivation system.

Biomass and post-harvest residues of CCs positively affect the quality of physical and chemical parameters of the soil, e.g., by increasing the content of organic matter, the amount of stable soil aggregates, and the content of available nitrogen [22,24]. The beneficial effects of legumes increase when the crop rotation is more diversified. Studies have shown that the use of three or more plant species and, additionally, a legume plant as intercrop allows for a greater accumulation of soil organic matter than in the case of two plants in a crop rotation, along with a legume intercrop [22]. On the other hand, García-González et al. [204] showed that the use of a mix of *H. vulgare* with *V. sativa* as a CC for several years increased the soil organic matter and field water capacity, as well as reduced leaching of nitrogen compounds from the soil. Greater stabilization of the soil structure, the formation of a lumpy structure and an increase in the amount of waterproof soil aggregates were also observed.

The literature on the subject reports that non-legume CCs significantly reduce the leaching of nitrates to the groundwater by incorporating nitrogen into the biomass [205,206], while opinions regarding legumes vary. According to Abdalla et al. [181], all analyzed cover plants (legumes, non-legumes, and legume mixes) significantly reduced nitrogen leaching. Justes et al. [207] reported that the efficiency of *Fabaceae* as CCs in reducing nitrates was half that of *Poaceae* and *Brassicaceae*. In turn, Valkama et al. [208] demonstrated that legumes were not effective at capturing nitrogen, contrary to *Poaceae*. More efficient soil nitrogen capturing by *Poaceae* and *Brassicaceae* than *Fabaceae* species is most likely influenced by their more developed root system and greater resistance to lower temperatures, which favors the extension of their growing season [59,133].

Apart from reduction of N leaching to ground water, CCs contribute to increase SOC sequestration (soil carbon sequestration) without having significant effects on direct N_2O emissions [181]. Lal [209] suggest that the SOC can be enriched by use of apposite crop rotations. Crop rotation can improve biomass production and thereafter the soil C sequestration, principally the rotations of legumes with non-legumes. The legume-based rotations are more efficient in converting biomass C into SOC in compression to the grass-based rotation. Hajduk et al. [210] suggest that the inclusion of legumes in rotation has the potential to guarantee the in-situ availability of N, which in turn played a vital role in generating higher biomass C. It also promotes the release of C via root exudation in to the rhizospheric zone. N fixed by the root nodules of legumes also accelerates the C sequestration potential of succeeding crop in the rotation, by improved microbial functionaries and biomass production by successive crop. When provided with the legumes, they enhance the nitrogen use efficiency and produce more root biomass and thus C inputs in soil [211]. In legumes based rotation endorsed the accumulation of liable C pool in soil ecosystem, which is beneficial for successive plants [209]. Legume and non-legume CCs in mixtures also have a better effect on the main crops, preventing a reduction in yield or even increasing it (by 13%) without significantly affecting N in grain. CCs can also mitigate the net greenhouse gas balance (NGHBG), contributing to the resilience of farming systems to environmental changes, such as climate change, through increased fertility, productivity and better water quality, which undoubtedly affects succeeding crops [181].

4.2. *Effect of Legume CCs on Soil Fauna and Microflora*

Due to the large amount of post-harvest residues and root secretions, *Fabaceae* CCs increase the amount of carbon and nitrogen compounds introduced into the soil, which is the main source of energy for soil microorganisms [179]. Peralta et al. [180] showed a significant increase in the functional groups of microorganisms responsible for the suppressive properties of the soil with a diversified crop rotation including *Z. mays*—*G. max*—*T. aesivum* plus two CCs—*T. pratense* and *S. cereale*, as compared to a *Z. mays* monoculture. In soils where rotation was applied, the pool of the gene responsible for the production of the bacterial antibiotic pyrrolithrin was increased compared to the soil under *Z. mays* monoculture cultivation. This proves an intensive development of microorganisms and plant protection properties. A more simplified crop rotation, which included only two plant species, such as: *T. aesivum* and *B. napus*, or *Z. mays* and *G. max* had no significant effect on soil microbial activity [180]. Using *T. repens* as a CC increased the microbial biomass and the population of earthworms (*Lumbricidae*) in the cultivation of *T. aesivum* [212], while in *Z. mays* cultivation, populations of all soil microorganisms, i.e., bacteria, fungi, nematodes, and mites, were increased [213]. According to Bardgett and van der Putten [214], increasing the diversity and number of species of cultivated plants has a positive effect on soil fertility, due to diversification of the available food sources for microorganisms and, consequently, for plants. The biomass of microorganisms also increases, and the structure of the soil microbiome is exposed to change, which has a huge impact on their functionality as well as soil fertility. Crop rotation also significantly impacts the increasing biodiversity in the soil environment [182]. In turn, Somenahally et al. [23] observed an increase in the biomass of mycorrhizal fungi in *T. aesivum* cultivation with the use of legumes as CCs.

4.3. *Effect of Legume CCs on the Physical Properties of Soil*

The physical properties of soil are related to the size and distribution of soil particles, as well as the movement of liquids and gases in the soil, which affect the appearance and structure of the soil. The main physical properties of soil are texture, structure, density, porosity, soil water, soil temperature, and color. The infiltration properties of soil are determined by its physical properties, such as bulk density, porosity, sorption, and aggregation. CCs may change the physical and structural properties of soil, both directly through the influence of roots on the formation of pores and aggregates, and indirectly, by introducing plant residues into the soil, and their decomposition [215].

It has been shown that leaving the CCs biomass as a dead mulch on the soil surface, with no tillage, potentially contributes to the reduction of soil bulk density, increasing soil porosity, sorption and aggregation, and thus increases its infiltration properties. Reduced bulk density, as well as increased porosity and macro-aggregation of the soil also reduce the potential for runoff, erosion and evaporation by increasing the potential for faster water capture and greater water availability to plants. However, the positive effect of dead mulch on the physical properties of the soil, in particular on increasing infiltration and reducing evaporation, depends on the climatic conditions. In the case of prolonged drought, no differences were found to bare soil [216].

4.3.1. Effect of Legume CCs on Soil Structure and Aggregation

The structure of the soil is the shape, size, and arrangement of primary and secondary particles in a specific structural pattern. Residues of CC plants, including legume CCs, improve soil structure by influencing soil strength, porosity and hydraulic properties. They also increase the diffusion capacity of soil air, as well as the formation of soil aggregates (basic units of soil structure) [217–219] (Table 3).

Table 3. Effect of legume CCs on the physical properties of soil.

Legume Cover Crops	Crop/Tillage	Soil Texture	Bulk Density (g cm^{-3})	Water-Stable Agregates (%)	Mean Weight Diameter (mm)	Total Aggregate-Associated Carbon (dag kg^{-1})	Soil Particle Size Fractions (>0.002 mm)	Soil Total Porosity (%)	Soil Capillary Porosity (%)	Soil Penetration Resistance (MPa)	References
no legume CCs *V. villosa* Rotch.	*Z. mays* L./no-till	fine sandy loam	1.44 1.47								[220]
no legume CCs *T. incarnatum* L. *V. villosa* Rotch.	*Z. mays* L./no-till	clay loam		56.3 55.0 58.2							[191]
no legume CCs *T. incarnatum* L. *V. villosa* Rotch.	*S. bicolor* (L.) Moench/ conventional tillage	sandy clay loam		28.9 37.9 36.7							[191]
no legume CCs *V. villosa* Rotch. mixture of *S. cereale* L. and *V. villosa* Rotch.	*Z. mays* L. *G. max* (L.) Merr./no-till	silt loam	1.32 1.23 1.23	38.0 43.0 44.0							[24]
no legume CCs *G. max* (L.) Merr	*T. aestivum* L. *S. bicolor* (L.) Moench /no-till	silt loam	1.21 1.24		0.42 0.76						[221]
mixture of *Phaseolus vulgaris* L. and *Pennisetum glaucum* L. mixture of *P. vulgaris* L. and *P. glaucum* L. mixture of *P. vulgaris* L. and *P. glaucum* L.	*Z. mays* L./no-till minimum tillage conventional tillage	Typic Hapludult	1.58/0–10 1.71/10–20 1.47/0–10 1.48/10–20 1.64/0–10 1.77/10–20		1.79/0–5 1.47/5–10 1.05/10–20 1.79/0–5 1.16/5–10 1.05/10–20 1.53/0–5 1.26/5–10 1.42/10–20	0.96/0–5 0.57/5–10 0.36/10–20 1.08/0–5 0.66/5–10 0.54/10–20 0.51/0–5 0.39/5–10 0.54/10–20				1.94/0–10 4.07/10–20 1.64, 0–10 6.55/10–20 2.06/0–10 5.13/10–20	[217]
no legume CCs mixture of *V. faba* L. and *V. villosa* Rotch no legume CCs mixture of *V. faba* L. and *V. villosa* Rotch	*T. aestivum* L./plough or conservation tillage	silt loam	1.56/5–10 1.50/5–10 1.62/15–20 1.53/15–20				8 10 (plough tillage) 5 7 (conservation tillage)	38.5/5–10 40.7/5–10 36.4/15–20 39.1/15–20	34.4/5–10 35.0/5–10 32.4/15–20 33.0/15–20		[218]

Table 3. *Cont.*

Legume Cover Crops	Crop/Tillage	Soil Texture	Bulk Density (g cm^{-3})	Water–Stable Agregates (%)	Mean Weight Diameter (mm)	Total Aggregate-Associated Carbon (dag kg^{-1})	Soil Particle Size Fractions (>0.002 mm)	Soil Total Porosity (%)	Soil Capillary Porosity (%)	Soil Penetration Resistance (MPa)	References
no legume CCs *S. cereale* L. (in *Z. mays*) and *V. villosa* Rotch. (in *G. max*) *S. cereale* L. (in *Z. mays*) and mixture of *S. cereale* L. and *V. villosa* Rotch. (in *G. max*)	*Z. mays* L. *G. max* (L.) Merr./no-till	silt loam	1.41 1.37 1.37	38.6 42.1 44.1						1.41 1.49 1.46	[193]
no legume CCs *T. incarnatum* L. *V. villosa* Rotch	*Solanum lycopersicum* L. *Solanum melongena* L./ conventional tillage	sandy loam			1.70 1.66 1.71						[222]
no legume CCs mixture of *S. cereale* L., *P. sativum* L., and *T. incarnatum* L.			1.61 1.65								[192]

It has been shown that supplying plant residues, e.g., CCs, improves soil structure [219]. It depends not only on the total organic C content, but is a function of many factors, including the chemical composition of the organic matter and the applied management system [223].

An important physical property of soil is aggregation. It affects the soil's bulk density, porosity, followed by water infiltration, water consumption efficiency, wind and water erosion and, consequently, the yield. Many factors affect soil aggregation, especially organic matter content and soil texture. The mechanism of binding soil particles into stable aggregates varies depending on factors related to soil parent material, climate, vegetation, and management practices [224]. It was observed that earlier removal of crop residues and the reduction of organic matter content in the soil results in a decrease in the stability of aggregates [225]. The tillage system also affects aggregation. It was found that in conventional soil cultivation, the stability of aggregates and decomposition of plant residues decreased. This, in turn, led to the degradation of the soil structure. It was found that no-till soil not only increases the stability of the aggregates, but also improves the content of organic matter within the aggregates. It follows that post-harvest CC residues increase soil aggregation and its structural stability [191,226–228]. The use of CCs in cultivation increased the stability of aggregates in the soil (expressed as mean weight diameter) and the content of diluted acid-extractable polysaccharides. The diluted acid-extractable polysaccharides fraction has been shown to represent active binding agents under short-term cover crops [228]. Application *V. villosa* as a CC also contributed to the improvement of water-aggregate stability by 9% to 17% compared to cultivation without CCs [24]. Other authors also confirmed the beneficial effects of legume CCs (*T. incarnatum*, *V. villosa*) to increase water-stable aggregates, even by 1.2 to 2 times compared to cultivation without CCs [25,191]. Improving the stability of soil aggregates by introducing CCs into the cultivation also affects other physical properties of the soil, leading to improved water storage, soil macroporosity, as well as increased content of organic carbon and nutrients. As a result, this supports the growth of the plant root system, reduces soil erosion (protecting it against raindrops) and improves microbial activity [12]. The beneficial effect of roots on soil structure is seen in reduced soil volume density in the topsoil or in changed pore size distribution, but without increasing the total porosity [229,230]. This is confirmed by Carof et al. [231], who observed larger functional pores and an increased number of tubules in the soil under a no-till legume CC system, attributing these changes to the activity of the roots. The impact of CCs on increasing soil porosity and reducing the amount of occluded pores was also reported by Villamil et al. [24].

Numerous reports have confirmed that the stability of soil aggregates is associated with an increased concentration of organic carbon [232–234]. However, the improvement in water-stable aggregates may depend on the soil texture class, as legume CCs (*T. incarnatum*, *V. villosa*) improved this feature in sandy clay loam, but not in clay loam [191]. In addition, the use of a legume CC (*V. villosa*) also decreased penetration resistance and bulk density of the soil due to plant residues and increased soil organic matter compared to cultivation without CCs [24].

4.3.2. Effect of Legume CCs on Soil Water Management

By increasing soil aggregation, CCs improve its hydraulic properties, i.e., water infiltration, water retention capacity and saturated hydraulic conductivity (Table 4). It was found that the long-term effects of legume CCs, e.g., *V. villosa* and *T. incarnatum*, contributed to increasing soil porosity, saturated hydraulic conductivity, and water retention capacity [235], thus protecting its top layer. In turn, increasing water infiltration improves the ability to capture precipitation and store water [12].

Table 4. Effect of legume CCs on soil water management.

Legume Cover Crops	Crop/Tillage	Soil Texture	Water Infiltration (cm)	Soil Water Content /Depth (cm)	Soil Moisture	Water Stability Index (%)	References
V. villosa Rotch.	*Z. mays* L./no-till	silt loam	0.52				[236]
no legume CCs *G. max* (L.) Merr	*T. aestivum* L. *S. bicolor* (L.) Moench /no-till	silt loam	5.7 11.4				[221]
no legume CCs mixture of *A. sativa* L. and *V. villosa* Rotch.	*S. lycopersicum* L./ conventional tillage	loam	19.0 20.3				[237]
no legume CCs *Trifolium fragiferum* L.	orchard/conventional tillage	sandy loam	3.3 6.3				[237]
			cumulative (mm)	($m^3 \cdot m^{-3}$)			
no legume CCs	-	sandy loam	40	0.56			[238]
V. dasycarpa L.			41	0.72			
L. angustifolius L.			43	0.68			
no legume CCs			28	0.74			
mixture of *A. sativa* L.(90%) and *V. dasycarpa* L. (10%)			41	0.77			
mixture of *A. sativa* L.(70%) and *V. dasycarpa* L. (30%)			44	0.77			
mixture of *A. sativa* L.(50%) and *V. dasycarpa* L. (50%)			57	0.80			
				($g \cdot kg^{-1}$)			
no legume CCs	*T. aestivum* L./ conventional tillage, fall planted	sandy clay loam		192/90			[52]
M. sativa L.				171/90			
T. pratense L.				190/90			
P. sativum L.				188/90			
				($g \cdot kg^{-1}$)			
no legume CCs	*T. aestivum* L./ conventional tillage, spring planted	sandy clay loam		188/90			[52]
M. sativa L.				189/90			
T. pratense L.				189/90			
P. sativum L.				185/90			
				(%)			
no legume CCs	*T. aestivum* L./no-till	clay loam		15.0			[188]
P. sativum L.				11.5			
mixture of *P. sativum* L. and *A. sativa* L.				11.0			
mixture of *P. sativum* L. and *B. napus* L.				11.5			
mixture of *P. sativum* L., *A. sativa* L. and *B. napus* L.				10.7			
mixture of *P. sativum* L., *A. sativa* L., *B. napus* L., *V. villosa* Rotch, *R. sativus* L. and *H. vulgare* L.				11.0			

Table 4. *Cont.*

Legume Cover Crops	Crop/Tillage	Soil Texture	Water Infiltration (cm)	Soil Water Content /Depth (cm)	Soil Moisture	Water Stability Index (%)	References
no legume CCs *T. pratense* L. no legume CCs *T. pratense* L.	*T. aestivum* L. *Z. mays* L. *G. max* (L.) Merr./no-till	clay loam		(%) 27.1/0–10 27.0/0–10 31.2/0–30 30.8/0–30			[239]
no legume CCs *T. pratense* L. no legume CCs *T. pratense* L.	*T. aestivum* L. *Z. mays* L. *G. max* (L.) Merr./conventional tillage	clay loam		(%) 24.1/0–10 23.4/0–10 27.9/0–30 29.7/0–30			[239]
no legume CCs mixture of *S. cereale* L. and *V. villosa* Roth. mixture of *S. cereale* L. and *T. incarnatum* L. mixture of *S. cereale* L., *A. sativa* L., *Raphanus sativus* var. niger J. Kern., *B. campestris* L. and *T. incarnatum* L.	*Z. mays* L. *G. max* (L.) Merr./no-till	silt loam			(mm) 17 20 19 21		[189]
no legume CCs mixture of *V. faba* L. and *V. villosa* Rotch no legume CCs mixture of *V. faba* L. and *V. villosa* Rotch	*T. aestivum* L.	silt loam			(%) 16.7/5–10 17.6/5–10 (%) 16.9/15–20 18.2/15–20	40.4 70.2	[218]
mixture of *P.vulgaris* L. and *P. glaucum* L. mixture of *P. vulgaris* L. and *P. glaucum* L. mixture of *P. vulgaris* L. and *P. glaucum* L.	*Z. mays* L./no-till minimum tillage conventional tillage	Typic Hapludult			(g g^{-1}) 0.06/0–10 0.03/10–20 0.05/0–10 0.03/10–20 0.10/0–10 0.06/10–20		[217]
no legume CCs mixture of *P. sativum* L. ssp. *arvense* and *S. cereale* L. mixture of *V. villosa* Rotch and *S. cereale* L.	*Z. mays* L./no-till	loamy sand clay loam S		(%) volumetric 0.182/10 0.178/10 0.198/10			[86]
no legume CCs mixture of *P. sativum* L. ssp. *arvense* and *S. cereale* L. mixture of *V. villosa* Rotch and *S. cereale* L.	*Z. mays* L./no-till	clay loam S		(%) volumetric 0.251/10 0.257/10 0.261/10			[86]

The use of legume CCs as dead mulch contributes to reducing water loss by evapotranspiration. This improves water infiltration and moisture retention, inhibiting surface runoff, reducing soil degradation, slow nutrient release and delivering them to the rhizosphere of the plant, which ultimately improves resource efficiency [218,238,240–243]. Many authors have reported the beneficial effect of legume CCs on increasing soil moisture [195,244,245]. This effect is likely to result from improving the soil's structural features, which increase the infiltration rate and reduce soil water evaporation [246,247]. Singh et al. [216] confirmed that infiltration and water permeability is greater under no-tillage than under plow tillage because of the greater number of macrospores and increased microbial activity.

Increasing soil moisture is more favorably influenced by a greater variation of CCs in the mix [187,189]. CCs characterized by a deep root system (e.g., *L. angustifolius*), can also be particularly effective in increasing the effective root depth and subsoil water storage capacity [21,248]. However, Ghimire et al. [188] found that legume CCs can reduce soil water content by 2% to 3% compared to leaving the soil fallow (at the end of CC cultivation). However, leaving residual CCs between main crops increased the water content as well by 2% to 4% compared to fallow soil. Thorsted et al. [148] confirmed that in supporting the cultivation of the main crop with a legume intercrop, the water consumption in the soil increases due to the increased canopy transpiration from these plant species. Some authors found that a CC intercrop stimulated higher soil moisture only in the top layer (0–10 cm), while their effect on moisture in deeper layers turned out to be insignificant [249,250]. On the other hand, Harasim et al. [218] proved an increase in soil moisture at a depth of 15 to 20 cm with legume CCs cultivated as intercrops. RIL directly affects the availability of water for cultivated plants during the growing season, especially at a depth of 90 cm [251]. However, a lower legume CC used as intercrop has a lesser impact on the limited water availability than the main crop. Amossé et al. [251] found that the availability of water for the main crop is primarily determined by the development and size of the legume CCs root system, which the authors prove on the example of *M. sativa*. The legume is characterized by a strongly developed root system and, despite the lower above-ground part, it had a greater impact on reducing soil moisture than *T. pratense*.

It was found that in RIL cultivation, the time interval between sowing the main crop and the legume CCs allowed for sufficient development of the root system of the crop (before sowing the legume CCs) so that the plant could take up water from deeper soil layers. On the other hand, competition for soil water between the main crop and the legume CCs could have led to its roots growing to a greater depth. Therefore, despite the drying out of the top layer of soil, which is additionally increased by the sown legumes, the legume may not disturb the water uptake of the main crop if its root system is well developed and would not be limited by excessive soil compaction [251].

The negative influence of legume CCs as living mulch in the cultivation of the main crops is reported by Thorsted et al. [148] and Martens et al. [252] according to whom the competition between the legume CCs (*T. repens*, *T. pratense*) and a crop (*T. aestivum*) is decisive in terms of water availability in soil, apart from environmental conditions. Nielsen et al. [245] demonstrated that in a semi-arid environment, CCs use limited water reserves in the soil, which may also affect the availability of water for the successive crop cultivation and, consequently, its yield. Under these environmental conditions, the cultivation of mixed-species CCs is also unjustified as they use a similar amount of water as single-species CC cultivations.

On the other hand, in semi-arid Texas, cover crop as mulch (incorporated on soil surface) increased available soil water by 73% and more than doubled grain sorghum yields [253]. The availability of soil water at CC planting and depletion during growth are always a concern in semi-arid and arid regions. Therefore the potential benefits must be balanced against possible negative effects on the cash crop. Additionally, conservation tillage increases storage of precipitation in the soil through increased infiltration and reduced evaporation. This additional water supplements growing season precipitation and irrigation to meet crop water needs on the semi-arid regions [254].

4.3.3. Effect of Legume CCs on Soil Temperature and Light Availability

CCs help to mitigate soil temperature fluctuations by capturing solar radiation and isolating the soil, which prevents it from overheating. Numerous authors have confirmed that legume CCs used as mulch increases light reflection due to the lighter color of the soil surface. This in turn lowers the soil temperature and ultimately improves resource efficiency [218,240–243]. It was shown that residues of legume CCs reduce daily fluctuations in soil temperature in no-till systems, lowering its maximum and increasing the minimum, compared to conventional tillage systems. However, the effectiveness of CCs on soil temperature is determined by both the size of the CCs canopy cover and the amount of plant residues [21]. Cultivating CCs as living mulch allowed decreasing the maximum temperature by approximately 5 °C at 5 cm soil depth, compared to dead mulch left on the soil surface [255]. This is also confirmed by the studies of other authors, who additionally found an increase in the minimum soil temperature by 1 °C after application of legume CCs [46,221]. In turn, leaving the legume CC (*T. alexandrium*) after harvesting the main plant contributed to a delayed and shallower (by 0.2 m) soil freezing in late autumn, and earlier thawing and heating of the soil in spring (by approximately 3 °C) compared to the soil not covered with CCs [256]. Harvest residues of legume CCs regulate soil temperature, contributing to warming in winter and cooling in spring and summer [216]. Decreasing the maximum soil temperature during the summer period reduces evaporation from the soil surface and also increases water retention in the soil. On the other hand, the impact of CCs on the soil temperature in spring depends on the climatic conditions: in warm climates, slowing down the heating of the soil is beneficial for plants, while in cooler climates it may delay the germination and emergence of seeds, which adversely affects sowing [12]. Drury et al. [239] confirmed that decreasing the daily maximum temperature by using legume CCs as mulch can be particularly harmful in cool regions. In the case of plants grown in warmer regions, the use of legume CCs as mulch decreased the maximum soil temperature, favoring the growth and development of plants [257]. On the other hand, the darker color of the *V. villosa* mulch only slightly increased the soil temperature at a depth of 5 cm (by 0.5 °C) compared to the lighter mulch of *T. aestivum* [258].

However, according to Dabney et al. [21] soil temperature is more impacted by CCs used as living mulch or mulch left on the soil surface than CC harvest residues. Moreover, the effectiveness of soil temperature regulation by residual CCs is determined by the rate of their decomposition, and legume CCs tend to decompose as residue CCs faster than other CCs [46]. Hence, it is commonly believed that legume CCs have a lower soil temperature regulating effect than cereal crops. Moreover, according to Martens et al. [252], legume CCs also reduce temperature fluctuations at 5 cm above the soil surface.

Carof et al. [79] have shown that legume CCs, especially in the early stages of crop growth, can shade and compete with them for light. Therefore, when selecting legume Ccs, the species that should be taken into account are those with active growth and a relatively short growing season, so that they compete as little as possible with the crop for light. In the RIL system, the shading effect of the main crop (*T. aestivum*) occurred before its harvest, but only in the highest species of legumes (*M. sativa*) [251]. However, legume CCs can also compete for light with weeds whose seeds require light to germinate (e.g., small-seeded annual weeds), limiting their growth and development [136]. Plant residues of CCs can act as a physical barrier to emerging weeds by blocking their access to light, or by creating conditions similar to those in deeper soil layers (no light, lower daily temperature amplitude) [78]. However, living CCs legumes have a greater suppressive effect on weeds at all stages of their development than their crop residues. This is due to the fact that live CC legumes absorb red light to a greater extent and reduce the far-red ratio sufficient to inhibit phytochrome-mediated seed germination of weeds. In contrast, harvest residues of legume CCs have little effect on this ratio [126].

Low light levels hinder not only growth but also fixing N_2 in legumes, which need light to maintain symbiosis with *Rhizobium* and assimilation of N. This is more pronounced in the species *T. repens* than in *T. pratense* sown in the main crop [160].

4.4. Effect of Legume CCs on Wind and Water Soil Erosion

One of the benefits of growing CCs is to prevent soil degradation as a result of wind and water erosion. Water erosion affects both soil and water quality, as floodwater flows through various water bodies (including irrigation ditches, water reservoirs, drinking water). On the other hand, wind erosion mainly affects soil quality as it reduces its fertility [259].

Legume CCs effectively reduce the decline of the topsoil due to water erosion, as shown in the example *V. villosa*, or *V. villosa/T. aestivum* mix. In cultivation of *Gossypium hirsutum* L., the latter caused an 82% lower soil loss (conventional till) or 73% (no-till), compared to cultivation without CCs [260]. In turn, a decrease in soil loss by 39%, 68%, and 92%, respectively, was found in *Lens culinaris* Medik. (winter lentil) and *P. sativum* (spring pea) as CCs (*T. aestivum* as the main crop) [95] and a mix of *S. cereale* and *V. villosa* as CCs (with *Z. mays* as the main crop) [261]. On the other hand, runoff reduction was decreased by 13% (with a mix of *S. cereale* and *V. villosa* as CCs, *Z. mays* as the main crop) [261], 42% (with *L. culinaris* as CC, *T. aestivum* as the main crop), and up to 73% (with *P. sativum* as CCs, *T. aestivum* as the main crop) [96].

CCs are an effective technique to protect soil from erosion, especially when grown between primary crops, when they extend the period of soil coverage with plants [215]. However, at the beginning of winter, when CCs freeze, their aboveground biomass is less effective at protecting soil from water erosion, although their root system may still play an important role in improving soil resistance. Thick root cover plants are less effective in preventing soil loss by concentrated flow erosion than the finely branched root cover plants [262].

The influence of cover crops on erosive processes depends on how much they reduce the soil detachment and transporting forces. CCs reduce inter-row erosion, primarily by increasing the amount of soil covering with live plants or their remains, and its duration. Inter-row erosion is caused by the detachment of soil particles by the impacting raindrops. In contrast, live CCs or their debris intercept raindrops and dissipate the impact energy, contributing to the reduction of inter-row erosion [215].

CCs also reduce the loss of sediment and the dissolved nutrients (total P and NO_3—N), thus influencing the improvement of water quality, soil fertility, and plant yield [215]. CCs protect soil by absorbing the energy of raindrops, reducing detachment of aggregates and promoting formation of stable aggregates in the soil, increasing topsoil roughness, delaying the onset of runoff and intercepting it, reducing runoff speed, and increasing water infiltration [221]. However, the effectiveness of CCs in reducing soil water erosion is determined by the speed of plant growth and the production of biomass. Moreover, mixes of legume CCs with other species (e.g., grasses) are more effective in increasing the uniform surface cover than the monoculture cultivation, which ultimately helps to reduce soil erosion to a greater extent [263].

CCs also help to reduce wind erosion, which adversely affects the quality of the soil, especially in a semi-arid environment in the period when strong winds occur in the absence of plant cover (late winter and early spring). The use of winter or spring CCs that grow well in these conditions helps to reduce wind erosion. Blanco-Canqui et al. [78] observed a reduction in susceptibility to soil wind erosion by growing CCs (e.g., legumes and grasses) during fallow in wheat crop rotation. This reduced the fraction of soil susceptible to erosion (<0.84 mm in diameter) by 80% and increased the size of dry aggregates by 60%. Among the legume CCs, spring lentil contributed most to the increase in soil aggregate size distribution and the 1.6-fold reduction of the soil fraction susceptible to wind erosion. However, due to the faster decomposition of plant residues and the lower plant residue height, legume CCs are less effective at reducing wind erosion compared to grasses. Reducing the risk of wind erosion by CC cultivation is possible due to the improvement of the physical properties of the soil by improving its structure, as well as by increasing the organic C content of the soil and the physical retention of soil aggregates by the CCs' root system. Moreover, increasing the content of organic C in soil has an anti-erosion effect. It increases the stability of aggregates and reduces the fraction of soil susceptible to wind erosion, as organic C can physically, chemically and biologically fix soil particles to form stable macro-aggregates [264].

5. Conclusions

Legume CCs seem to be a useful agronomic solution in crops, especially when the aim is to protect natural resources, apart from improving or maintaining agricultural production at a high level. However, it is difficult to propose technical cultivation strategies that are applicable to all conditions. The reason for this is that legume CCs species have different agrotechnical requirements as well as morphological, phenological, and physiological characteristics that can affect the productivity of the main crop. Therefore, some authors argue that using multi-species CCs mixes that include legumes may be a good solution as the characteristics of individual species would complement each other or work synergistically.

In addition, legume CCs are seen as a systemic approach to weed control. Apart from reducing weed infestation, they provide many other agro-system benefits, including soil improvement. Living legume CCs inhibit the development of weed populations through niche pre-emption, while their crop residues inhibit or delay weed emergence and growth by creating a physical and chemical barrier (allelopathic effect). The multi-purpose character of legume CCs is visible in reducing compaction and erosion of the soil, improving its structural and hydraulic properties, increasing the content of organic matter and activity of soil microorganisms, or increasing its nitrogen content due to symbiotic N_2 fixing. This review shows that a wider use of legume CCs is needed in crops, especially in those with limited use of pesticides and mineral fertilizers (organic). It is also necessary to determine the benefits of legume CCs for successive crops under these growing conditions, both in terms of inhibiting weed population and improving soil fertility and its properties. Further research is also needed to determine the potential impact of legume CCs on the improvement of the quality of degraded soils, or those with less favorable physicochemical properties.

Author Contributions: Conceptualization, A.K., M.S., K.P. and H.L.; writing—original draft preparation, A.K., M.S., M.T., R.K., J.C., K.P., H.L.; writing—review and editing, A.K., M.S., K.P., H.L.; supervision, A.K. All authors have read and agreed to the published version of the manuscript.

Funding: This research received no external funding.

Conflicts of Interest: The authors declare no conflict of interest.

References

1. Meena, M.S.; Singh, K.M. Conservation Agriculture: Innovations, Constraints and Strategies for Adoption. *Munich Pers. RePEc Arch.* **2013**, 49380. Available online: https://mpra.ub.uni-muenchen.de/49380/ (accessed on 31 July 2020). [CrossRef]
2. Oliveira, M.C.; Butts, L.; Werle, R. Assessment of Cover Crop Management Strategies in Nebraska, US. *Agriculture* **2019**, *9*, 124. [CrossRef]
3. Melander, B.; Rasmussen, I.A.; Bàrberi, P. Integrating physical and cultural methods of weed control-examples from European research. *Weed Sci.* **2005**, *53*, 369–381. [CrossRef]
4. Hartwig, N.L.; Ammon, H.U. Cover crops and living mulches. *Weed Sci.* **2002**, *50*, 688–699. [CrossRef]
5. Sarrantonio, M.; Gallandt, E. The role of cover crops in North American cropping systems. *J. Crop Prod.* **2003**, *8*, 53–74. [CrossRef]
6. Blackshaw, R.E.; Moyer, J.R.; Huang, H.C. Beneficial effects of cover crops on soil health and crop management. *Rec. Res. Dev. Soil Sci.* **2005**, *1*, 15–35.
7. Bàrberi, P.; Mazzoncini, M. Changes in weed community composition as influenced by cover crop and management system in continuous corn. *Weed Sci.* **2001**, *49*, 491–499. [CrossRef]
8. Shennan, C. Cover crops, nitrogen cycling and soil properties in semi-irrigated vegetable production systems. *HortScience* **1992**, *27*, 749–754. [CrossRef]
9. Wyland, L.J.; Jackson, L.E.; Chaney, W.E.; Klonsky, K.; Koike, S.T.; Kimple, B. Winter cover crops in a vegetable cropping system: Impacts on nitrate leaching, soil water, crop yield, pests and management. *Agric. Ecosyst. Environ.* **1996**, *59*, 1–17. [CrossRef]
10. Ritter, W.F.; Scarborough, R.W.; Chirniside, A.E.M. Winter cover crops as a best management practice for reducing nitrogen leaching. *J. Contam. Hydrol.* **1998**, *34*, 1–15. [CrossRef]

11. Shrestha, A.; Knezevic, S.Z.; Roy, R.C.; Ball-Coelho, B.R.; Swanton, C.J. Effect of tillage, cover crop and crop rotation on the composition of weed flora in a sandy soil. *Weed Res.* **2002**, *42*, 76–87. [CrossRef]
12. Blanco-Canqui, H.; Shaver, T.M.; Lindquist, J.L.; Shapiro, C.A.; Elmore, R.W.; Francis, C.A.; Hergert, G.W. Cover Crops and Ecosystem Services: Insights from Studies in Temperate Soils. *Agron. J.* **2015**, *107*, 2449–2474. [CrossRef]
13. Jabran, K.; Tursum, N.; Isik, D.; Demir, Z. Use of Living, Mowed, and Soil-Incorporated Cover Crops for Weed Control in Apricot Orchards. *Agronomy* **2018**, *8*, 150. [CrossRef]
14. Cutti, L.; Lamego, F.P.; de Aguiar, A.D.M.; Kaspary, T.E.; Gonsiorkiewicz-Rigon, C.A. Winter cover crops on weed infestation and maize yield. *Rev. Caatinga* **2016**, *29*, 885–891. [CrossRef]
15. Wallace, J.M.; Williams, A.; Liebert, J.A.; Ackroyd, V.J.; Vann, R.A.; Curran, W.S.; Keene, C.L.; VanGessel, M.J.; Ryan, M.R.; Mirsky, S.B. Cover Crop-Based, Organic Rotational No-Till Corn and Soybean Production Systems in the Mid-Atlantic United States. *Agriculture* **2017**, *7*, 34. [CrossRef]
16. Rizzardi, M.A.; Silva, L.F. Influence of Black Oats and Rape as Cover Crops on Chemical Weed Control Timing in No-till Corn. *Planta Daninha* **2006**, *24*, 669–675. [CrossRef]
17. Lemessa, F.; Wakjira, M. Cover Crops as a Means of Ecological Weed Management in Agroecosystems. *J. Crop Sci. Biotechnol.* **2015**, *18*, 133–145. [CrossRef]
18. Amossé, C.; Jeuffroyb, M.H.; Celettea, F.; Davida, C. Relay-intercropped forage legumes help to control weeds in organic grain production. *Eur. J. Agron.* **2013**, *49*, 158–167. [CrossRef]
19. Uchino, H.; Iwama, K.; Jitsuyama, Y.; Ichiyama, K.; Sugiura, E.; Yudate, T. Stable characteristics of cover crops for weed suppression in organic farming systems. *Plant Prod. Sci.* **2011**, *14*, 75–85. [CrossRef]
20. Elsalahy, H.; Döring, T.; Bellingrath-Kimura, S.; Arends, D. Weed Suppression in Only-Legume Cover Crop Mixtures. *Agronomy* **2019**, *9*, 648. [CrossRef]
21. Dabney, S.M.; Delgado, J.A.; Reeves, D.W. Using winter cover crops to improve soil and water quality. *Commun. Soil Sci. Plan.* **2001**, *32*, 1221–1250. [CrossRef]
22. Tiemann, L.K.; Grandy, A.S.; Atkinson, E.E.; Marin-Spiotta, E.; McDaniel, M.D. Crop rotational diversity enhances belowground communities and functions in an agroecosystem. *Ecol. Lett.* **2015**, *18*, 761–771. [CrossRef] [PubMed]
23. Somenahally, A.; DuPont, J.I.; Brady, J.; McLawrence, J.; Northup, B.; Gowda, P. Microbial communities in soil profile are more responsive to legacy effects ofwheat-cover crop rotations than tillage systems. *Soil Biol. Biochem.* **2018**, *123*, 126–135. [CrossRef]
24. Villamil, M.B.; Bollero, G.A.; Darmody, R.G.; Simmons, F.W.; Bullock, D.G. No-till corn/soybean systems including winter cover crops: Effects on soil properties. *Soil Sci. Soc. Am. J.* **2006**, *70*, 1936–1944. [CrossRef]
25. Hubbard, R.K.; Strickland, T.C.; Phatak, S. Effects of cover crop systems on soil physical properties and carbon/nitrogen relationships in the Coastal Plain of southeastern USA. *Soil Tillage Res.* **2013**, *126*, 276–283. [CrossRef]
26. Gupta, Y.P. Anti-nutritional and toxic factors in food legumes: A review. *Plant Food Hum. Nutr.* **1987**, *37*, 201–228. [CrossRef] [PubMed]
27. Balbinot, A.A., Jr.; Fleck, N.G. Weed management in the corn crop through plant spatial arrangement and characteristics of genotypes. *Cienc. Rural* **2005**, *35*, 245–252. [CrossRef]
28. Smith, R.G.; Warren, N.D.; Cordeau, S. Are cover crop mixtures better at suppressing weeds than cover crop monocultures? *Weed Sci.* **2020**, *68*, 186–194. [CrossRef]
29. Wiggins, M.; McClure, M.; Hayes, R.; Steckel, L. Integrating Cover Crops and POST Herbicides for Glyphosate-Resistant Palmer Amaranth (*Amaranthus palmeri*) Control in Corn. *Weed Technol.* **2015**, *29*, 412–418. [CrossRef]
30. Oerke, E.C. Crop losses to pests. *J. Agric. Sci.* **2006**, *144*, 31–43. [CrossRef]
31. den Hollander, N.G. Growth Characteristics of Several Clover Species and Their Suitability for Weed Suppression in a Mixed Cropping Design. Ph.D Thesis, Wageningen University, Wageningen, The Netherlands, 2012.
32. Tollenaar, M.; Dibo, A.A.; Aqilera, A.; Weise, S.F.; Swanton, C.J. Effect of crop density on weed interference in maize. *Agron. J.* **1994**, *86*, 591–595. [CrossRef]
33. Phatak, S.C. An integrated sustainable vegetable production system. *HortScience* **1992**, *27*, 738–741. [CrossRef]

34. Bastiaans, L.; Paolini, R.; Baumann, D.T. Integrated Crop Management: Opportunities and Limitations for Prevention of Weed Problems. In Proceedings of the 12th EWRS (European Weed Research Society) Symposium 2002, Wageningen, The Netherlands, 24–27 June 2002; pp. 8–9.
35. Bilalis, D.; Karkanis, A.; Efthimiadou, A. Effects of two legume crops, for organic green manure, on weed flora, under Mediterranean conditions: Competitive ability of five winter season weed species. *Afr. J. Agric. Res.* **2009**, *4*, 1431–1441.
36. Efthimiadou, A.P.; Karkanis, A.C.; Bilalis, D.J.; Efthimiadis, P. Review: The phenomenon of crop-weed competition; a problem or a key for sustainable weed management. *J. Food Agric. Environ.* **2009**, *7*, 861–868.
37. Lemerle, D.; Verbeek, B.; Orchard, B. Ranking the ability of wheat varietes to compete with *Lolium rigidum*. *Weed Res.* **2001**, *41*, 197–209. [CrossRef]
38. Pester, T.A.; Burnside, O.C.; Orf, J.H. Increasing crop competitiveness to weeds through crop breeding. *J. Crop Prod.* **2011**, *2*, 59–76. [CrossRef]
39. Korres, N.E.; Froud-Williams, R.J. Effects of winter wheat cultivars and seed rate on the biological characteristics of naturally occurring weed flora. *Weed Res.* **2002**, *42*, 417–428. [CrossRef]
40. Seavers, G.P.; Wright, K.J. Crop canopy development and structure influence weed suppression. *Weed Res.* **2002**, *39*, 319–328. [CrossRef]
41. Ross, S.M.; King, J.R.; Izaurralde, R.C.; O'Donovan, J.T. Weed suppression by seven clover species. *Agron. J.* **2001**, *93*, 820–827. [CrossRef]
42. Mennan, H.; Ngouajio, M.; Kaya, E.; Isık, D. Weed Management in Organically Grown Kale Using Alternative Cover Cropping Systems. *Weed Technol.* **2009**, *23*, 81–88. [CrossRef]
43. Kruidhof, H.M.; Bastiaans, L.; Kropff, M.J. Ecological weed management by cover cropping: Effects on weed growth in autumn and weed establishment in spring. *Weed Res.* **2008**, *48*, 492–502. [CrossRef]
44. Khanh, T.D.; Chung, M.I.; Xuan, T.D.; Tawata, S. The exploitation of crop allelopathy in sustainable Agricultural production. *J. Agron. Crop Sci.* **2005**, *191*, 172–184. [CrossRef]
45. Weston, L.A. Utilization of allelopathy for weed management in agroecosystems. *Agron. J.* **1996**, *88*, 860–866. [CrossRef]
46. Mohler, C.L.; Teasdale, J.R. Response of weed emergence to rate of *Vicia villosa* Roth and *Secale cereale* L. residue. *Weed Res.* **1993**, *33*, 487–499. [CrossRef]
47. Teasdale, J.R.; Mohler, C.L. Light transmittance, soil temperature, and soil moisture under residue of hairy vetch and rye. *Agron. J.* **1993**, *85*, 673–680. [CrossRef]
48. Teasdale, J.R. Contribution of cover crops to weed management in sustainable agricultural systems. *J. Prod. Agric.* **1996**, *9*, 475–479. [CrossRef]
49. Blanco-Canqui, H. Crop residue removal for bioenergy reducessoil carbon pools: How can we offset carbon losses? *Bioenergy Res.* **2013**, *6*, 358–371. [CrossRef]
50. Campiglia, E.; Radicetti, E.; Brunetti, P.; Mancinelli, R. Do cover crop species and residue management play a leading role in pepper productivity? *Sci. Hortic.* **2014**, *166*, 97–104. [CrossRef]
51. Davis, A.S. Cover-Crop Roller—Crimper Contributes to Weed Management in No-Till Soybean. *Weed Sci.* **2010**, *58*, 300–309. [CrossRef]
52. Caamal-Maldonado, J.A.; Jiménez-Osornio, J.J.; Torres-Barragán, A.; Anaya, A.L. The use of allelopathic legume cover and mulch species for weed control in cropping systems. *Agron. J.* **2001**, *93*, 27–36. [CrossRef]
53. Blackshaw, R.E.; Molnar, L.J.; Moyer, J.R. Suitability of legume cover crop winter wheat intercrops on the semi-arid Canadian prairies. *Can. J. Plant Sci.* **2010**, *90*, 479–488. [CrossRef]
54. Blaser, B.C.; Singer, J.W.; Gibson, L.R. Winter cereal canopy effect on cereal and interseeded legume productivity. *Agron. J.* **2011**, *103*, 1180–1185. [CrossRef]
55. Hartl, W. Influence of undersown clovers on weeds and on the yield of winter wheat in organic farming. *Agric. Ecosyst. Environ.* **1989**, *27*, 389–396. [CrossRef]
56. Fisk, J.W.; Hesterman, O.B.; Shrestha, A.; Kells, J.J.; Harwood, R.R.; Squire, J.M.; Sheaffer, C.C. Weed Suppression by Annual Legume Cover Crops in No-Tillage Corn. *Agron. J.* **2001**, *93*, 319–325. [CrossRef]
57. Reimer, M.; Ringselle, B.; Bergkvist, G.; Westaway, S.; Wittwer, R.; Baresel, J.P.; van der Heijden, M.G.A.; Mangerud, K.; Finckh, M.R.; Brandsæter, L.O. Interactive Effects of Subsidiary Crops and Weed Pressure in the Transition Period to Non-Inversion Tillage, A Case Study of Six Sites Across Northernand Central Europe. *Agronomy* **2019**, *9*, 495. [CrossRef]

58. Salonen, J.; Ketoja, E. Undersown cover crops have limited weed suppression potential when reducing tillage intensity in organically grown cereals. *Org. Agric.* **2019**. [CrossRef]
59. Campiglia, E.R.; Paolini, G.C.; Mancunelli, R. The effects of cover cropping on yield and weed control of potato in a transitional system. *Field Crop. Res.* **2009**, *112*, 16–23. [CrossRef]
60. Buraczyńska, D.; Ceglarek, C. The role of green manures, in form of undersown cover crops, and straw in sugar beet cultivation Part I. Sugar beet plantations infestation with weeds. *Biul. IHAR* **2004**, *234*, 171–180.
61. Brandsæter, L.O.; Netland, J. Winter Annual Legumes for Use as Cover Crops in Row Crops in Northern Regions: I. Field Experiments. *Crop Sci.* **1999**, *39*, 1369–1379. [CrossRef]
62. Döring, T.F.; Storkey, J.; Baddeley, J.A.; Collins, R.P.; Crowley, O.; Howlett, S.A.; Jones, H.E.; McCalman, H.; Measures, M.; Pearce, H.; et al. Weeds in organic fertility-building leys: Aspects of species richness and weed management. *Org. Farming* **2017**, *3*, 51–65. [CrossRef]
63. Blaser, B.C.; Gibson, L.R.; Singer, J.W.; Jannink, J.L. Optimizing seeding rates for winter cereal grains and frost-seeded red clover intercrops. *Agron. J.* **2006**, *98*, 1041–1049. [CrossRef]
64. Mutch, D.R.; Martin, T.E.; Kosola, K.R. Red clover (*Trifolium pratense*) suppression of common ragweed (*Ambrosia artemisiifolia*) in winter wheat (*Triticum aestivum*). *Weed Technol.* **2003**, *17*, 181–185. [CrossRef]
65. Sjursen, H.; Brandsæter, L.O.; Netland, J. Effects of repeated clover undersowing, green manure ley and weed harrowing on weeds and yields in organic cereals. *Acta Agric. Scand. Sect. B Soil Plant Sci.* **2012**, *62*, 138–150. [CrossRef]
66. Yeganehpoor, F.; Salmasi, S.Z.; Abedi, G.; Samadiyan, F.; Beyginiya, V. Effects of cover crops and weed management on corn yield. *J. Saudi Soc. Agric. Sci.* **2015**, *14*, 178–181. [CrossRef]
67. Brennan, E.B.; Boyd, N.S.; Smith, R.F.; Foster, P. Seeding Rate and Planting Arrangement Effects on Growth and Weed Suppression of a Legume-Oat Cover Crop for Organic Vegetable Systems. *Agron. J.* **2009**, *101*, 979–988. [CrossRef]
68. Murungu, F.S.; Chiduza, C.; Muchaonyerwa, P. Biomass accumulation, weed dynamics and nitrogen uptake by winter cover crops in a warm-temperate region of South Africa. *Afr. J. Agric. Res.* **2010**, *5*, 1632–1642.
69. Ranaldo, M.; Carlesi, S.; Costanzo, A.; Barberi, P. Functional diversity of cover crop mixtures enhances biomass yield and weed suppression in a Mediterranean agroecosystem. *Weed Res.* **2019**, *60*, 96–108. [CrossRef]
70. Charles, K.S.; Ngouajio, M.; Warncke, D.D.; Poff, K.L.; Hausbeck, M.K. Integration of cover crops and fertilizer rates for weed management in celery. *Weed Sci.* **2006**, *54*, 326–334. [CrossRef]
71. Campiglia, E.; Caporali, F.; Radicetti, E.; Mancinelli, R. Hairy vetch (*Vicia villosa* Roth.) cover crop residue management for improving weed control and yield in no-tillage tomato (*Lycopersicon esculentum* Mill.) production. *Eur. J. Agron.* **2010**, *33*, 94–102. [CrossRef]
72. Seman-Varner, R.; Varco, J.J.; O'Rourke, M.E. Winter Cover Crop and Fall-Applied Poultry Litter Effects on Winter Cover and Soil Nitrogen. *Agron. J.* **2019**, *111*, 3301–3309. [CrossRef]
73. Mirsky, S.B.; Curran, W.S.; Mortensen, D.M.; Ryan, M.R.; Shumway, D.L. Timing of Cover-Crop Management Effects on Weed Suppression in No-Till Planted Soybean using a Roller-Crimper. *Weed Sci.* **2011**, *59*, 380–389. [CrossRef]
74. Golian, J.; Anyszka, Z.; Kosson, R.; Grzegorzewska, M. Effectiveness of selected methods of weed management and their effect on nutrition value and storage ability of red head cabbage. *J. Res. Appl. Agric. Eng.* **2016**, *61*, 144–150.
75. Kohut, M.; Anyszka, Z.; Golian, J. Changes in infestation and yielding of selected vegetable species depending on weed management method. *J. Res. Appl. Agric. Eng.* **2013**, *58*, 255–260.
76. Leavitt, M.J.; Sheaffer, C.C.; Wyse, D.L.; Allan, D.L. Rolled Winter Rye and Hairy Vetch Cover Crops Lower Weed Density but Reduce Vegetable Yields in No-tillage Organic Production. *HortScience* **2011**, *46*, 387–395. [CrossRef]
77. Da Silva, A.C.; Hirata, E.K.; Monquero, P.A. Straw yield and cover crop weed suppression in a no tillage system for processing tomato. *Pesq. Agropec. Bras.* **2009**, *44*, 22–28.
78. Teasdale, J.R.; Brandsæter, L.O.; Calegari, A.; Skora Neto, F. Cover Crops and Weed Management. In *Non-Chemical Weed Management: Principles, Concepts and Technology*; Upadhyaya, M.K., Blackshaw, R.E., Eds.; CABI: Wallingford, UK, 2007; pp. 49–64.

79. Carof, M.; de Tourdonnet, S.; Saulas, P.; le Floch, D.; Roger-Estrade, J. Undersowing wheat with different living mulches in a no-till system. II. Competition for light and nitrogen. *Agron. Sustain. Dev.* **2007**, *27*, 357–365. [CrossRef]
80. Fradgley, N.S.; Creissen, H.E.; Pearce, H.; Howlett, S.A.; Pearce, B.D.; Döring, T.F.; Girling, R.D. Weed Suppression and Tolerance in Winter Oats. *Weed Technol.* **2017**, *31*, 740–751. [CrossRef]
81. Weidlich, E.W.A.; von Gillhaussen, P.; Delory, B.M.; Blossfeld, S.; Poorter, H.; Temperton, V.M. The Importance of Being First: Exploring Priority and Diversity Effects in a Grassland Field Experiment. *Front. Plant Sci.* **2017**, *7*, 1–12. [CrossRef]
82. Mediene, S.; Valantin-Morison, M.; Sarthou, J.P.; Tourdonnet, S.; Gosme, M.; Bertrand, M.; Roger-Estrade, J.; Aubertot, J.N.; Rusch, A.; Motisi, N.; et al. Agroecosystem management and biotic interactions: A review. *Agron. Sustain. Dev.* **2011**, *31*, 491–514. [CrossRef]
83. Teasdale, J.R.; Beste, C.E.; Potts, W.E. Response of weeds to tillage and cover crop residue. *Weed Sci.* **1991**, *39*, 195–199. [CrossRef]
84. Teasdale, J.R.; Mohler, C.L. The quantitative relationship between weed emergence and the physical properties of mulches. *Weed Sci.* **2000**, *48*, 385–392. [CrossRef]
85. Weston, L.A.; Duke, S.O. Weed and crop allelopathy. *Crit. Rev. Plant Sci.* **2003**, *22*, 367–389. [CrossRef]
86. Wells, M.S.; Reberg-Horton, S.C.; Mirsky, S.B. Planting Date Impacts on Soil Water Management, Plant Growth, and Weeds in Cover-Crop-Based No-Till Corn Production. *Agron. J.* **2016**, *108*, 162–170. [CrossRef]
87. Diyanat, M. Weed Management in Organic Horticulture by Cover Crop in Iran. *Int. J. Adv. Biol. Biom. Res.* **2015**, *3*, 153–162.
88. Bhowmik, P.C. Challenges and opportunities in implementing allelopathy for natural weed management. *Crop Prot.* **2003**, *22*, 661–671. [CrossRef]
89. Teasdale, J.R.; Coffman, C.B.; Mangum, R.W. Potential long-term benefits of no-tillage and organic cropping systems for grain production and soil improvement. *Agron. J.* **2007**, *99*, 1297–1305. [CrossRef]
90. Mischler, R.; Duiker, S.; Curran, W.S.; Wilson, D. Hairy Vetch Management for No-Till Organic Corn Production. *Agron. J.* **2010**, *120*, 355–362. [CrossRef]
91. Moonen, A.C.; Bàrberi, P. Size and composition of the weed seedbank after 7 years of different cover crop-maize management systems. *Weed Res.* **2004**, *44*, 163–177. [CrossRef]
92. Wallace, J.M.; Curran, W.S.; Mortensen, D.A. Cover crop effects on horseweed (*Erigeron canadensis*) density and size inequality at the time of herbicide exposure. *Weed Sci.* **2019**, *67*, 327–338. [CrossRef]
93. Wiggins, M.S.; Hayes, R.M.; Steckel, L.E. Evaluating cover crops and herbicides for glyphosate-resistant Palmer amaranth (*Amaranthus palmeri*) control in cotton. *Weed Technol.* **2016**, *30*, 415–422. [CrossRef]
94. Brainard, D.C.; Bellinder, R.R.; Kumar, V. Grass-legume mixtures and soil fertility affect cover crop performance and weed seed production. *Weed Technol.* **2011**, *35*, 473–479. [CrossRef]
95. Lu, Y.; Watkins, K.; Teasdale, J.R.; Abdul-Baki, A. Cover crops in sustainable food production. *Food Rev. Int.* **2000**, *16*, 121–157. [CrossRef]
96. Blanco-Canqui, H.; Holman, J.D.; Schlegel, A.J.; Tatarko, J.; Shaver, T. Replacing fallow with cover crops in a semiarid soil: Effects on soil properties. *Soil Sci. Soc. Am. J.* **2013**, *77*, 1026–1034. [CrossRef]
97. Teasdale, J.R. Principles and Practices for Using Cover Crops in Weed Management Systems. In *Weed Management for Developing Countries*; Add. 1, FAO Paper 120; Labrada, R., Ed.; Fao: Rome, Italy, 2003.
98. Kalinova, J. Allelopathy and Organic Farming. In *Sociology, Organic Farming, Climate Change and Soil Science, Sustainable Agriculture Reviews*; Lichtfouse, E., Ed.; Springer Science+Business Media BV.: Dordrecht, The Netherlands, 2010; Volume 3, pp. 379–418.
99. Wang, R.; Yang, X.; Song, Y.; Zhang, M.; Hu, L.; Su, Y.; Zeng, R. Allelopathic potential of *Tephrosia vogelii* Hook. f.: Laboratory and field evaluation. *Allelopathy J.* **2011**, *28*, 53–62.
100. Xuan, T.D.; Shinkichi, T.; Khanh, T.D.; Min, C.I. Biological control of weeds and plant pathogens in paddy rice by exploiting plant allelopathy: An overview. *Crop Prot.* **2005**, *24*, 197–206. [CrossRef]
101. Xuan, T.D.; Tawata, S.; Khanh, T.D.; Chung, I.M. Decomposition of allelopathic plants in soil. *J. Agron. Crop Sci.* **2005**, *191*, 162–171. [CrossRef]
102. Hauggaard-Nielsen, H.; Andersen, M.K.; Jørnsgaard, B.; Jensen, E.S. Density and relative frequency effects on competitive interactions and resource use in pea-barley intercrops. *Field Crop. Res.* **2006**, *95*, 256–267. [CrossRef]

103. Exner, D.N.; Cruse, R.M. Interseeded forage legume potential as winter ground cover, nitrogen source, and competitor. *J. Prod. Agric.* **1993**, *6*, 226–231. [CrossRef]
104. Schlegel, A.J.; Havlin, J.L. Green fallow for the central great plains. *Agron. J.* **1997**, *89*, 762–767. [CrossRef]
105. Finney, D.M.; White, C.M.; Kaye, J.P. Biomass production and carbon/nitrogen ratio influence ecosystem services from cover crop mixtures. *Agron. J.* **2016**, *108*, 39–52. [CrossRef]
106. Bybee-Finley, K.; Mirsky, S.B.; Ryan, M.R. Crop Biomass Not Species Richness Drives Weed Suppression in Warm-Season Annual Grass–Legume Intercrops in the Northeast. *Weed Sci.* **2017**, *65*, 669–680. [CrossRef]
107. Hiltbrunner, J.; Liedgens, M.; Bloch, L.; Stamp, P.; Streit, B. Legume cover crops as living mulches for winter wheat: Components of biomass and the control of weeds. *Eur. J. Agron.* **2007**, *26*, 21–29. [CrossRef]
108. Weston, L.; Inderjit, S. Allelopathy: A Potential Tool in the Development of Strategies for Biorational Weed Management. In *Non-Chemical Weed Management: Principles, Concepts and Technology*; Upadhyaya, M.K., Blackshaw, R.E., Eds.; CABI: Wallingford, UK, 2007; pp. 65–76.
109. Xuan, T.D.; Tsuzuki, E.; Uematsu, H.; Terao, H. Weed control with alfalfa pellets in transplanting rice. *Weed Biol. Manag.* **2001**, *1*, 231–235. [CrossRef]
110. Xuan, T.D.; Tsuzuki, E.; Terao, H.; Matsuo, M.; Khanh, T.D. Correlation between growth inhibitory exhibition and suspected allelochemicals (phenolic compounds) in the extract of alfalfa (*Medicago sativa* L.). *Plant Prod. Sci.* **2003**, *6*, 165–171. [CrossRef]
111. Elsalahy, H.; Bellingrath-Kimura, S.; Döring, T. Allelopathic effects in species mixtures of legumes. In *Improving Sown Grasslands through Breeding and Management, Proceedings of the Joint 20th Symposium of the European Grassland Federation and the 33rd Meeting of the EUCARPIA, Zurich, Switzerland, 24–27 June 2019*; Huguenin-Elie, O., Studer, B., Kölliker, R., Reheul, D., Probo, M., Barre, P., Feuerstein, U., Roldán-Ruiz, I., Mariotte, P., Hopkins, A., Eds.; Organising Committee: Zurich, Switzerland, 2019; p. 128.
112. White, R.H.; Worsham, A.D.; Blum, U. Allelopathic potential of legume debris and aqueous extracts. *Weed Sci.* **1989**, *37*, 674–679. [CrossRef]
113. Kamo, T.; Hiradate, S.; Fujii, Y. First isolation of natural cyanamide as a possible allelochemical from hairy vetch *Vicia Villosa*. *J. Chem. Ecol.* **2003**, *29*, 275–283. [CrossRef]
114. Bradow, J.M.; Connick, W.J., Jr. Volatile Seed Germination Inhibitors from Plant Residues. *J. Chem. Ecol.* **1990**, *16*, 645–666. [CrossRef]
115. Ohno, T.; Doolan, K.; Zibilske, L.M.; Liebman, M.; Gallandt, E.R.; Berube, C. Phytotoxic effects of red clover amended soils on wild mustard seedling growth. *Agric. Ecosyst. Environ.* **2000**, *78*, 187–192. [CrossRef]
116. Ohno, T.; Doolan, K.L. Effects of red clover decomposition on phytotoxicity to wild mustard seedling growth. *Appl. Soil Ecol.* **2001**, *16*, 187–192. [CrossRef]
117. Kruidhof, H.M.; Bastiaans, L.; Kropff, M.J. Cover crop residue management for optimizing weed control. *Plant Soil* **2009**, *318*, 169–184. [CrossRef]
118. Osipitan, O.A.; Dille, J.A.; Assefa, Y.; Knezevic, S.Z. Cover Crop for Early Season Weed Suppression in Crops: Systematic Review and Meta-Analysis. *Agron. J.* **2018**, *110*, 2211–2221. [CrossRef]
119. Mennan, H.; Jabran, K.; Zandstra, B.H.; Pala, F. Non-Chemical Weed Management in Vegetables by Using Cover Crops: A Review. *Agronomy* **2020**, *10*, 257. [CrossRef]
120. Ryan, M.R.; Mortensen, D.A.; Bastiaans, L.; Teasdale, J.R.; Mirsky, S.B.; Curran, W.S.; Seidel, R.; Wilson, D.O.; Hepperly, P.R. Elucidating the apparent maize tolerance to weed competition in long-term organically managed systems. *Weed Res.* **2010**, *50*, 25–36. [CrossRef]
121. Meiss, H.; le Lagadec, L.; Munier-Jolain, N.; Waldhardt, R.; Petit, S. Weed seed predation increases with vegetation cover in perennial forage crops. *Agric. Ecosyst. Environ.* **2010**, *138*, 10–16. [CrossRef]
122. Medd, R.W.; Ridings, H.I. Relevance of Seed Kill for the Control of Annual Grass Weeds in Crops. In Proceedings of the VII International Symposium on the Biological Control of Weeds, Rome, Italy, 6–11 March 1989; pp. 645–650.
123. Youngerman, C.Z.; di Tommaso, A.; Losey, J.E.; Ryan, M.R. Cover crop seed preference of four common weed seed predators. *Renew. Agric. Food Syst.* **2019**, 1–11. [CrossRef]
124. Blubaugh, C.K.; Hagler, J.R.; Machtley, S.A.; Kaplan, I. Cover crops increase foraging activity of omnivorous predators in seed patches and facilitate weed biological control. *Agric. Ecosyst. Environ.* **2016**, *231*, 264–270. [CrossRef]
125. Teasdale, J.R.; Mirsky, S.B.; Cavigelli, M.A. Weed species and traits associated with organic grain crop rotations in the mid-Atlantic region. *Weed Sci.* **2019**, *67*, 595–604. [CrossRef]

126. Teasdale, J.R.; Daughtry, C.S.T. Weed Suppression by Live and Dessicated Hairy Vetch. *Weed Sci.* **1993**, *41*, 207–212. [CrossRef]
127. Masiunas, J.B.; Eastburn, D.M.; Mwaja, V.N.; Estman, C.E. The Impact of Living and Cover Crop Mulch System on Pests and Yield of Snap Beans and Cabbage. *J. Sustain. Agric.* **1997**, *9*, 61–89. [CrossRef]
128. Teasdale, J.R.; Mangum, R.W.; Radhakrishnan, J.; Cavigelli, M.A. Weed seedbank dynamics in three organic farming crop rotations. *Agron. J.* **2004**, *96*, 1429–1435. [CrossRef]
129. Hoffman, M.L.; Regnier, E.E. Contributions to Weed Suppression from Cover Crops. In *Sustainable Weed Management*; Singh, P., Batish, D.R., Kohli, R.K., Eds.; Food Products Press: New York, NY, USA, 2006; pp. 51–75.
130. Blackshaw, R.E.; Semach, G.; Li, X.; O'donovan, J.T.; Neil Harker, K. An integrated weed management approach to managing foxtail barley (*Hordeum jubatum*) in conservation tillage systems. *Weed Technol.* **1999**, *13*, 347–353. [CrossRef]
131. Streit, B.; Rieger, S.B.; Stamp, P.; Richner, W. The effect of tillage intensity and time of herbicide application on weed communities and populations in maize in Central Europe. *Agric. Ecosyst. Environ.* **2002**, *92*, 211–224. [CrossRef]
132. Sanderson, M.A.; Brink, G.; Stout, R.; Ruth, L. Grass-Legume proportions in forage seed mixtures and effects on herbage yield and weed abundance. *Agron. J.* **2013**, *105*, 1289–1297. [CrossRef]
133. Blum, U.; King, L.D.; Gerig, T.M.; Lehman, M.E.; Worsham, A.D. Effects of clover and small grain cover crops and tillage techniques on seedling emergence of some dicotyledonous weed species. *Am. J. Altern. Agric.* **1997**, *4*, 146–161. [CrossRef]
134. Singer, J.W.; Kohler, K.A.; McDonald, P.B. Self-seeding winter cereal cover crops in soybean. *Agron. J.* **2007**, *99*, 73–79. [CrossRef]
135. Creamer, N.G.; Dabney, S.M. Killing cover crops mechanically: Review of recent literature and assessment of new research results. *Am. J. Alt. Agric.* **2002**, *17*, 32–40. [CrossRef]
136. Teasdale, J.R.; Rosecrance, R.C. Mechanical versus herbicidal strategies for killing a hairy vetch cover crop and controlling weeds in minimum-tillage corn production. *Am. J. Altern. Agric.* **2003**, *18*, 95–102. [CrossRef]
137. Ashford, D.L.; Reeves, D.W. Use of a mechanical roller–crimper as an alternative kill method for cover crops. *Am. J. Altern. Agric.* **2003**, *18*, 37–45. [CrossRef]
138. Judice, W.E.; Griffin, J.L.; Etheredge, L.M.; Jones, C.A. Effects of Crop Residue Management and Tillage on Weed Control and Sugarcane Production. *Weed Technol.* **2007**, *21*, 606–611. [CrossRef]
139. Mahmood, A.; Ihsan, M.Z.; Khaliq, A.; Hussain, S.; Cheema, Z.A.; Naeem, M.; Daur, I.; Hussain, H.A.; Alghabari, F. Crop Residues Mulch as Organic Weed Management Strategy in Maize. *Clean Soil Air Water* **2015**, *43*, 1–8. [CrossRef]
140. Khaliq, A.; Matloob, A.; Hussain, A.; Hussain, S.; Aslam, F.; Zamir, S.I.; Chattha, M.U. Wheat Residue Management Options Affect Productivity, Weed Growth and Soil Properties in Direct-Seeded Fine Aromatic Rice. *Clean Soil Air Water* **2015**, *43*, 1259–1265. [CrossRef]
141. Gallandt, E.R.; Molloy, T.; Lynch, R.P.; Drummond, F.A. Effect of cover-cropping systems on invertebrate seed predation. *Weed Sci.* **2005**, *53*, 69–76. [CrossRef]
142. Fahad, S.; Hussain, S.; Chauhan, B.S.; Saud, S.; Wu, C.; Hassan, S.; Tanveer, M.; Jan, A.; Huang, J. Weed Growth and Crop Yield Loss in Wheat as Influenced by Row Spacing and Weed Emergence Times. *Crop Prot.* **2015**, *71*, 101–108. [CrossRef]
143. Ranaivoson, L.; Naudin, K.; Ripoche, A.; Affholder, F.; Rabeharisoa, L.; Corbeels, M. Agro-Ecological functions of crop residues under conservation agriculture. A review. *Agron. Sustain. Dev.* **2017**, *37*, 1–17. [CrossRef]
144. Reddy, K.N. Effect of Cereal and Legume Cover Crop Residues on Weeds, Yield and Net Return in Soybean (*Glycine max* L.). *Weed Technol.* **2001**, *15*, 660–668. [CrossRef]
145. Sims, B.; Corsi, S.; Gbehounou, G.; Kienzle, J.; Taguchi, M.; Friedrich, T. Sustainable Weed Management for Conservation Agriculture: Options for Smallholder Farmers. *Agriculture* **2018**, *8*, 118. [CrossRef]
146. Liebl, R.; Simmons, F.W.; Wax, L.M.; Stoller, E.W. Effect of rye (*Secale cereale*) mulch on weed control and soil moisture in soybean (*Glycine max*). *Weed Technol.* **1992**, *6*, 838–846. [CrossRef]
147. Teasdale, J.R.; Pillai, P. Contribution of ammonium to stimulation of smooth pigweed (*Amaranthus hybridus* L.) germination by extracts of hairy vetch (*Vicia villosa* Roth) residue. *Weed Biol. Manag.* **2005**, *5*, 19–25. [CrossRef]

148. Thorsted, M.D.; Weiner, J.; Olesen, J.E. Above- and below-ground competition between intercropped winter wheat *Triticum aestivum* and white clover *Trifolium repens*. *J. Appl. Ecol.* **2006**, *43*, 237–245. [CrossRef]
149. Bedoussac, L.; Journet, E.P.; Hauggaard-Nielsen, H.; Naudin, C.; Corre-Hellou, G.; Jensen, E.S.; Prieur, L.; Justes, E. Ecological principles underlying the increase of productivity achieved by cereal-grain legume intercrops in organic farming. A review. *Agron. Sustain. Dev.* **2015**, *35*, 911–935. [CrossRef]
150. Bedoussac, L.; Justes, E. Dynamic analysis of competition and complementarity for light and N use to understand the yield and the protein concentration of a durum wheat-winter pea intercrop. *Plant Soil* **2010**, *330*, 37–54. [CrossRef]
151. Dreccer, M.; Schapendonk, A.; Slafer, G.; Rabbinge, R. Comparative response of wheat and oilseed rape to nitrogen supply: Absorption and utilisation efficiency of radiation and nitrogen during the reproductive stages determining yield. *Plant Soil* **2000**, *220*, 189–205. [CrossRef]
152. Poggio, S.L. Structure of weed communities occurring in monoculture and intercropping of field pea and barley. *Agric. Ecosyst. Environ.* **2005**, *109*, 48–58. [CrossRef]
153. Banik, P.; Midya, A.; Sarkar, B.K.; Ghose, S.S. Wheat and chickpea intercropping systems in an additive experiment. Advantages and weed smothering. *Eur. J. Agron.* **2006**, *24*, 325–332. [CrossRef]
154. Corre-Hellou, G.; Dibet, A.; Hauggaard-Nielsen, H.; Crozat, Y.; Gooding, M.; Ambus, P.; Dahlmann, C.; von Fragstein, P.; Pristeri, A.; Monti, M.; et al. The competitive ability of pea-barley intercrops against weeds and the interactions with crop productivity and soil N availability. *Field Crop. Res.* **2011**, *122*, 264–272. [CrossRef]
155. McDonald, G.K. Competitiveness against grass weeds in field pea genotypes. *Weed Res.* **2003**, *43*, 48–58. [CrossRef]
156. Corre-Hellou, G.; Crozat, Y. N2 fixation and N supply in organic pea (*Pisum sativum* L.) cropping systems as affected by weeds and peaweevil (*Sitona lineatus* L.). *Eur. J. Agron.* **2005**, *22*, 449–458. [CrossRef]
157. Bojarszczuk, J.; Księżak, J.; Staniak, M. Evaluation of weed infestation of triticale and pea mixtures grown for fodder seeds. *J. Res. Appl. Agric. Eng.* **2017**, *62*, 42–48.
158. Staniak, M.; Bojarszczuk, J.; Księżak, J. Undersown serradella (*Ornithopus sativus* L.) as an element of weed control in triticale crops. *J. Res. Appl. Agric. Eng.* **2017**, *62*, 144–148.
159. Księżak, J.; Staniak, M.; Bojarszczuk, J. Evaluation of mixtures of yellow lupine (*Lupinus luteus* L.) with spring cereals grown for seeds. *Appl. Ecol. Environ. Res.* **2018**, *16*, 1683–1696. [CrossRef]
160. Bergkvist, G.; Stenberg, M.; Wetterlind, J.; Båth, B.; Elfstrand, S. Clover cover crops under-sown in winter wheat increase yield of subsequent spring barley—Effect of N dose and companion grass. *Field Crop. Res.* **2011**, *120*, 292–298. [CrossRef]
161. Mirsky, S.B.; Gallandt, E.R.; Mortensen, D.A.; Curran, W.S.; Shumway, D.L. Reducing the germinable weed seedbank with soil disturbance and cover crops. *Weed Res.* **2010**, *50*, 341–352. [CrossRef]
162. Saucke, H.; Ackermann, K. Weed suppression in mixed cropped grain peas and false flax (*Camelina sativa*). *Weed Res.* **2006**, *46*, 453–461. [CrossRef]
163. Smith, R.G.; Mortensen, D.A.; Ryan, M.R. A new hypothesis for the functional role of diversity in mediating resource pools and weed-crop competition in agroecosystems. *Weed Res.* **2010**, *50*, 37–48. [CrossRef]
164. Döring, T.F.; Baddeley, J.A.; Brown, R.; Collins, R.; Crowley, O.; Cuttle, S.; Howlett, S.A.; Jones, H.E.; Mccalman, H.; Measures, M.; et al. *Using Legume-Based Mixtures to Enhance the Nitrogen Use Effciency and Economic Viability of Cropping Systems: Final Report (LK09106/HGCA3447)*; Project Report 513; HGCA/Agriculture and Horticulture Development Board: Kenilworth, UK, 2013. Available online: https://orgprints.org/24662/1/PR513.pdf (accessed on 2 August 2020).
165. Suter, M.; Hofer, D.; Luscher, A. Weed suppression enhanced by increasing functional trait dispersion and resource capture in forage ley mixtures. *Agric. Ecosyst. Environ.* **2017**, *240*, 329–339. [CrossRef]
166. Weiner, J.; Andersen, S.B.; Wille, W.K.M.; Griepentrog, H.W.; Olsen, J.M. Evolutionary Agroecology: The potential for cooperative, high density, weed-suppressing cereals. *Evol. Appl.* **2010**, *3*, 473–479. [CrossRef] [PubMed]
167. Panasiewicz, K.; Faligowska, A.; Szymańska, G.; Szukała, J.; Ratajczak, K.; Sulewska, H. The Effect of Various Tillage Systems on Productivity of Narrow-Leaved Lupin-Winter Wheat-Winter Triticale-Winter Barley Rotation. *Agronomy* **2020**, *10*, 304. [CrossRef]

168. Aguilar-Fenollosa, E.; Ibáñez-Gual, M.V.; Pascual-Ruiz, S.; Hurtado, M.; Jacas, J.A. Effect of ground-cover management on spider mites and their phytoseiid natural enemies in clementine mandarin orchards (I): Bottom-up regulation. *Biol. Control* **2011**, *59*, 158–170. [CrossRef]
169. Bowers, C.; Toews, M.; Liu, Y.; Schmidt, J.M. Cover crops improve early season natural enemy recruitment and pest management in cotton production. *Biol. Control* **2020**, *141*, 104–149. [CrossRef]
170. Aguilar-Fenollosa, E.; Pascual-Ruiz, S.; Hurtado, M.A.; Jacas, J.A. Efficacy and economics of ground cover management as a conservation biological control strategy against *Tetranychus urticae* in clementine mandarin orchards. *Crop Prot.* **2011**, *30*, 1328–1333. [CrossRef]
171. Lapointe, S.L. Leguminous cover crops and their interactions with citrus and *Diaprepes abbreviatus* (Coleoptera: Curculionidae). *Fla. Entomol.* **2003**, *86*, 80–85. [CrossRef]
172. Altieri, M.A.; Nicholls, C.I.; Ponti, L. Crop Diversification Strategies for Pest Regulation in IPM Systems. In *Integrated Pest Management*; Radcliffe, E.B., Hutchinson, W.D., Cancelado, R.E., Eds.; Cambridge University Press: Cambridge, UK, 2009; pp. 116–130.
173. Altieri, M.A.; Nicholls, C.I. Vegetational Designs to Enhance Biological Control of Insect Pests in Agroecosystems. In *Natural Enemies of Insect Pests in Neotropical Agroecosystems*; Souza, B., Vázquez, L., Marucci, R., Eds.; Springer Nature: Cham, Switzerland, 2019; pp. 3–13. [CrossRef]
174. Mailloux, J.; le Bellec, F.; Kreiter, S.; Tixier, M.S.; Dubois, P. Influence of ground cover management on diversity and density of phytoseiid mites (Acari: Phytoseiidae) in Guadeloupean citrus orchards. *Exp. Appl. Acarol.* **2010**, *52*, 275–290. [CrossRef] [PubMed]
175. Damien, M.; le Lann, C.; Desneux, N.; Alford, L.; al Hassan, D.; Georges, R.; van Baaren, J. Flowering cover crops in winter increase pest control but not trophic link diversity. *Agric. Ecosyst. Environ.* **2017**, *247*, 418–425. [CrossRef]
176. Ellis, K.E.; Barbercheck, M.E. Management of Overwintering Cover Crops Influences Floral Resources and Visitation by Native Bees. *Environ. Entomol.* **2015**, *44*, 999–1010. [CrossRef] [PubMed]
177. Saunders, M.E.; Luck, G.W.; Mayfield, M.M. Almond orchards with living ground cover host more wild insect pollinators. *J. Insect Conserv.* **2013**, *17*, 1011–1025. [CrossRef]
178. Lal, R. Restoring soil quality to mitigate soil degradation. *Sustainability* **2015**, *7*, 5875–5895. [CrossRef]
179. Brookes, P.C.; Cayuela, M.L.; Contin, M.; de Nobili, M.; Kemmitt, S.J.; Mondini, C. The mineralization of fresh and humified soil organic matter by the soil microbial biomass. *Waste Manag.* **2008**, *28*, 716–722. [CrossRef]
180. Peralta, A.L.; Sun, Y.; McDaniel, M.D.; Lennon, J.T. Crop rotational diversity increases disease suppressive capacity of soil microbiomes. *Ecosphere* **2018**, *9*, e02235. [CrossRef]
181. Abdalla, M.; Hastings, A.; Cheng, K.; Yue, Q.; Chadwick, D.; Espenberg, M.; Truu, J.; Rees, R.M.; Smith, P. A critical review of the impacts of cover crops on nitrogen leaching, net greenhouse gas balance and crop productivity. *Glob. Chang. Biol.* **2019**, *25*, 2530–2543. [CrossRef]
182. Venter, Z.S.; Jacobs, C.; Hawkins, H.J. The impact of crop rotation on soil microbial diversity: A meta-analysis. *Pedobiol. J. Soil Ecol.* **2016**, *59*, 215–223. [CrossRef]
183. D'Acunto, L.; Andrade, J.F.; Poggio, S.L.; Semmartin, M. Diversifying crop rotation increased metabolic soil diversity and activity of the microbial community. *Agric. Ecosyst. Environ.* **2018**, *257*, 159–164. [CrossRef]
184. Battany, M.; Grismer, M.E. Rainfall runoff and erosion in Napa Valley vineyards: Effects of slope, cover and surface roughness. *Hydrol. Process.* **2000**, *14*, 1289–1304. [CrossRef]
185. Lüscher, A.; Mueller-Harvey, I.; Soussana, J.F.; Rees, R.M.; Peyraud, J.L. Potential of legume-based grassland-livestock systems in Europe: A review. *Grass Forage Sci.* **2014**, *69*, 206–228. [CrossRef] [PubMed]
186. Olson, K.; Ebelhar, S.A.; Lang, J.M. Long-term effects of cover crops on crop yields, soil organic carbon stocks and sequestration. *Open J. Soil Sci.* **2014**, *4*, 284–292. [CrossRef]
187. Wortman, S.E.; Francis, C.A.; Bernards, M.L.; Drijber, R.A.; Lindquist, J.L. Optimizing cover crop benefits with diverse mixtures and an alternative termination method. *Agron. J.* **2012**, *104*, 1425–1435. [CrossRef]
188. Ghimire, R.; Ghimire, B.; Mesbah, A.O.; Sainju, U.M.; Idowu, O.J. Soil Health Response of Cover Crops in Winter Wheat—Fallow System. *Agron. J.* **2019**, *111*, 2108–2115. [CrossRef]
189. Chu, M.; Jagadamma, S.; Walker, F.R.; Eash, N.S.; Buschermohle, M.J.; Duncan, L.A. Effect of Multispecies Cover Crop Mixture on Soil Properties and Crop Yield. *Agric. Environ. Lett.* **2017**, *2*, 170030. [CrossRef]
190. Kuo, S.; Jellum, E.J.; Sanju, U. The Effect of Winter Cover Cropping on Soil and Water Quality. In Proceedings of the Western Nutrient Management Conference, Salt Lake City, UT, USA, 9–10 March 1995; pp. 56–64.

191. McVay, K.A.; Radcliffe, D.E.; Hargrove, W.L. Winter legume effects on soil properties and nitrogen fertilizer requirements. *Soil Sci. Soc. Am. J.* **1989**, *53*, 1856–1862. [CrossRef]
192. Sarr, S.; Gebremedhin, M.; Coyne, M.; Topè, A.; Sistani, K.; Lucas, S. Do Conservation Practices Bring Quick Changes to Key Soil Properties for Resource-Limited Farmers? *J. Ky. Acad. Sci.* **2019**, *80*, 6–16. [CrossRef]
193. Villamil, M.B.; Miguez, F.E.; Bollero, G.A. Multivariate Analysis and Visualization of Soil Quality Data for No-Till Systems. *J. Environ. Qual.* **2008**, *37*, 2063–2069. [CrossRef]
194. Amado, T.J.C.; Bayer, C.; Conceiqdo, P.C.; Spagnollo, E.; de Campos, B.H.C.; da Veiga, M. Potential of Carbon Accumulation in No-Till Soils with Intensive Use and Cover Crops in Southern Brazil. *J. Environ. Qual.* **2006**, *35*, 1599–1607. [CrossRef] [PubMed]
195. Smith, P.; Martino, D.; Cai, Z.; Gwary, D.; Janzen, H.; Kumar, P.; McCarl, B.; Ogle, S.; O'Mara, F.; Rice, C.; et al. Greenhouse gas mitigation in agriculture. *Philos. Trans. R. Soc. B Biol. Sci.* **2008**, *363*, 789–813. [CrossRef] [PubMed]
196. Carlsson, G.; Huss-Danell, K. Nitrogen fixation in perennial forage legumes in the field. *Plant Soil* **2003**, *253*, 353–372. [CrossRef]
197. Casagrande, M.; David, C.; Valantin-Morison, M.; Makowski, D.; Jeuffroy, M.H. Factors limiting the grain protein content of organic winter wheat in south-eastern France: A mixed-model approach. *Agro. Sustain. Dev.* **2009**, *29*, 565–574. [CrossRef]
198. Berry, P.M.; Sylvester-Bradley, R.; Philipps, L.; Hatch, D.J.; Cuttle, S.P.; Rayns, F.W.; Gosling, P. Is the productivity of organic farms restricted by the supply of available nitrogen? *Soil Use Manag.* **2002**, *18*, 248–255. [CrossRef]
199. David, C.; Jeuffroy, M.H.; Henning, J.; Meynard, J.M. Yield variation in organic winter wheat: A diagnostic study in the Southeast of France. *Agron. Sustain. Dev.* **2005**, *25*, 213–223. [CrossRef]
200. Mazzoncini, M.; Canali, S.; Giovannetti, M.; Castagnoli, M.; Tittarelli, F.; Antichi, D.; Nannelli, R.; Cristani, C.; Bárberi, P. Comparison of organic and conventional stockless arable systems: A multidisciplinary approach to soil quality evaluation. *Appl. Soil Ecol.* **2010**, *44*, 124–132. [CrossRef]
201. Fujita, K.; Ofosubudu, K.G.; Ogata, S. Biological nitrogen fixation in mixed legume-cereal cropping systems. *Plant Soil* **1992**, *141*, 155–175. [CrossRef]
202. Jensen, E.S. Grain yield, symbiotic N2 fixation and interspecific competition for inorganic N in pea-barley intercrops. *Plant Soil* **1996**, *182*, 25–38. [CrossRef]
203. Naudin, C.; Corre-Hellou, G.; Pineau, S.; Crozat, Y.; Jeuffroy, M.H. The effect of various dynamics of N availability on winter pea-wheat intercrops: Crop growth, N partitioning and symbiotic N2 fixation. *Field Crop. Res.* **2010**, *119*, 2–11. [CrossRef]
204. García-González, I.; Hontoria, C.; Gabriel, J.L.; Alonso-Ayuso, M.; Qemada, M. Cover crops to mitigate soil degradation and enhance soil functionality in irrigated land. *Geoderma* **2018**, *322*, 81–88. [CrossRef]
205. Basche, A.D.; Miguez, F.E.; Kaspar, T.C.; Castellano, M.J. Do cover crops increase or decrease nitrous oxide emissions? *A meta-analysis. J. Soil Water Conserv.* **2014**, *69*, 471–482. [CrossRef]
206. Quemada, M.; Baranski, M.; Nobel-de Lange, M.N.J.; Vallejo, A.; Cooper, J.M. Meta-analysis of strategies to control nitrate leaching in irrigated agricultural systems and their effects on crop yield. *Agric. Ecosys. Environ.* **2013**, *174*, 1–10. [CrossRef]
207. Justes, E.; Beaudoin, N.; Bertuzzi, P.; Charles, R.; Constantin, J.; Dürr, C.; Réchauchère, O. *The Use of Cover Crops in the Reduction of Nitrate Leaching: Impact on the Water and Nitrogen Balance and Other Ecosystem Services*; Summary of the study report; INRA: Paris, France, 2012.
208. Valkama, E.; Lemola, R.; Känkänen, H.; Turtola, E. Meta-analysis of the effects of under-sown catch crops on nitrogen leaching loss and grain yields in the Nordic countries. *Agric. Ecosys. Environ.* **2015**, *203*, 93–101. [CrossRef]
209. Lal, R. Enhancing eco-efficiency in agro-ecosystems through soil C sequestration. *Crop Sci.* **2010**, *50* (Suppl. 1), S-120. [CrossRef]
210. Hajduk, E.; Właśniewski, S.; Szpunar-Krok, E. Influence of legume crops on content of organic C in sandy soil. *Soil Sci. Annu.* **2015**, *66*, 52–56. [CrossRef]
211. Meena, R.S.; Kumar, S.; Yadav, G.S. Soil Carbon Sequestration in Crop Production. In *Nutrient Dynamics for Sustainable Crop Production*; Meena, R.S., Ed.; Springer Nature Singapore Pte Ltd.: Singapore, 2020; pp. 1–39. [CrossRef]

212. Schmidt, O.; Curry, J.P.; Hackett, R.A.; Purvis, G.; Clements, R.O. Earthworm communities in conventional wheat monocropping and lowinput wheat-clover intercropping systems. *Ann. Appl. Biol.* **2001**, *138*, 377–388. [CrossRef]
213. Nakamoto, T.; Tsukamoto, M. Abundance and activity of soil organisms in fields of maize grown with a white clover living mulch. *Agric. Ecosys. Environ.* **2006**, *115*, 34–42. [CrossRef]
214. Bardgett, R.D.; van der Putten, W.H. Belowground biodiversity and ecosystem functioning. *Nature* **2014**, *515*, 505–511. [CrossRef]
215. Kaspar, T.C.; Singer, J.W. The Use of Cover Crops to Manage Soil. *USDAARS UNL Fac.* **2011**, 1382. Available online: https://digitalcommons.unl.edu/usdaarsfacpub/1382 (accessed on 2 August 2020).
216. Singh, R.; Serawat, M.; Singh, A. Effect of Tillage and Crop Residue Management on Soil Physical Properties. *J. Soil Salin. Water Qual.* **2018**, *10*, 200–206.
217. Oliveira, F.C.C.; Ferreira, G.W.D.; Souza, J.L.S.; Vieira, M.E.O.; Pedrotti, A. Soil physical properties and soil organic carbon content in northeast Brazil: Long-term tillage systems effects. *Sci. Agric.* **2020**, *77*, e20180166. [CrossRef]
218. Harasim, E.; Antonkiewicz, J.; Kwiatkowski, C.A. The Effects of Catch Crops and Tillage Systems on Selected Physical Properties and Enzymatic Activity of Loess Soil in a Spring Wheat Monoculture. *Agronomy* **2020**, *10*, 334. [CrossRef]
219. Lynch, J.M.; Bragg, E. Microorganisms and soil aggregate stability. *Adv. Soil Sci.* **1985**, *5*, 133–171. [CrossRef]
220. Wagger, M.G.; Denton, H.P. Influence of cover crop and wheel traffic on soil physical properties in continuous notill corn. *Soil Sci. Soc. Am. J.* **1989**, *53*, 1206–1210. [CrossRef]
221. Blanco-Canqui, H.; Mikha, M.M.; Presley, D.R.; Claassen, M.M. Addition of cover crops enhances no-till potential for improving soil physical properties. *Soil Sci. Soc. Am. J.* **2011**, *75*, 1471–1482. [CrossRef]
222. Sainju, U.M.; Whitehead, W.F.; Singh, B.P. Cover crops and nitrogen fertilization effects on soil aggregation and carbon and nitrogen pools. *Can. J. Soil Sci.* **2003**, *83*, 155–165. [CrossRef]
223. Dormaar, J.F. Chemical properties of soil and waterstable aggregates after sixty-seven years of cropping to spring wheat. *Plant Soil* **1983**, *75*, 51–61. [CrossRef]
224. Kay, B.D. Rates of change of soil structure under different cropping systems. *Adv. Soil Sci.* **1990**, *12*, 1–52. [CrossRef]
225. Blanco-Canqui, H.; Lal, R. Corn stover removal for expanded uses reduces soil fertility and structural stability. *Soil Sci. Soc. Am. J.* **2009**, *73*, 418–426. [CrossRef]
226. Cannell, R.Q.; Hawes, J.D. Trends in tillage practices in relation to sustainable crop production with special reference to temperate climates. *Soil Tillage Res.* **1994**, *30*, 245–282. [CrossRef]
227. Singh, B.; Chanasyk, D.S.; McGill, W.B.; Nyborg, M.P.K. Residue and tillage management effects on soil properties of a typic cryoboroll under continuous barley. *Soil Tillage Res.* **1994**, *32*, 117–133. [CrossRef]
228. Liu, A.G.; Ma, B.L.; Bomke, A.A. Effects of cover crops on soil aggregate stability, total organic carbon, and polysaccharides. *Soil Sci. Soc. Am. J.* **2005**, *69*, 2041–2048. [CrossRef]
229. Henderson, C.W.L. Lupin as a biological plough: Evidence for, and effects on wheat growth and yield. *Aust. J. Exp. Agric.* **1989**, *29*, 99–102. [CrossRef]
230. Rosolem, C.A.; Foloni, J.S.S.; Tiritan, C.S. Root growth and nutrient accumulation in cover crops as affected by soil compaction. *Soil Tillage Res.* **2002**, *65*, 109–115. [CrossRef]
231. Carof, M.; de Tourdonnet, S.; Coquet, Y.; Hallaire, V.; Roger-Estrade, J. Hydraulic conductivity and porosity under conventional and no-tillage and the effect of three species of cover crop in northern France. *Soil Use Manag.* **2007**, *23*, 230–237. [CrossRef]
232. Blanco-Canqui, H.; Shapiro, C.A.; Wortmann, C.S.; Drijber, R.A.; Mamo, M.; Shaver, T.M.; Ferguson, R.B. Soil organic carbon: The value to soil properties. *J. Soil Water Conserv.* **2013**, *68*, 129A–134A. [CrossRef]
233. Chenu, C.; le Bissonnais, Y.; Arrouays, D. Organic matter influence on clay wettability and soil aggregate stability. *Soil Sci. Soc. Am. J.* **2000**, *64*, 1479–1486. [CrossRef]
234. Pagliai, M.; Vignozzi, N.; Pellegrini, S. Soil structure and the effect of management practices. *Soil Tillage Res.* **2004**, *79*, 131–143. [CrossRef]
235. Keisling, T.C.; Scott, H.D.; Waddle, B.A.; Williams, W.; Frans, R.E. Winter cover crops influence on cotton yield and selected soil properties. *Commun. Soil Sci. Plant Anal.* **1994**, *25*, 3087–3100. [CrossRef]
236. Steele, M.K.; Coale, F.J.; Hill, R.L. Winter annual cover crop impacts on no-till soil physical properties and organic matter. *Soil Sci. Soc. Am. J.* **2012**, *76*, 2164–2173. [CrossRef]

237. Folorunso, O.A.; Rolston, D.E.; Prichard, T.; Louie, D.T. Soil surface strength and infiltration rate as affected by winter cover crops. *Soil Technol.* **1992**, *5*, 189–197. [CrossRef]
238. Mupambwa, H.A.; Wakindiki, I.I.C. Winter cover crops effects on soil strength, infiltration and water retention in a sandy loam Oakleaf soil in Eastern Cape, South Africa. *S. Afr. J. Plant Soil* **2012**, *29*, 121–126. [CrossRef]
239. Drury, C.F.; Tan, C.S.; Welacky, T.W.; Oloya, T.O.; Hamill, A.S.; Weaver, S.E. Red clover and tillage influence on soil temperature, water content, and corn emergence. *Agron. J.* **1999**, *91*, 101–108. [CrossRef]
240. Malhi, S.; Lemke, R. Tillage, Crop Residue and N Fertilizer Effects on Crop Yield, Nutrient Uptake, Soil Quality and Nitrous Oxide Gas Emissions in a Second 4-yr Rotation Cycle. *Soil Tillage Res.* **2007**, *96*, 269–283. [CrossRef]
241. Bajgai, Y.; Kristiansen, P.; Hulugalle, N.; McHenry, M. Comparison of Organic and Conventional Managements on Yields, Nutrients and Weeds in a Corn-Cabbage Rotation. *Renew. Agric. Food Syst.* **2015**, *30*, 132–142. [CrossRef]
242. Khaliq, A.; Shakeel, M.; Matloob, A.; Hussain, S.; Tanveer, A.; Murtaza, G. Influence of Tillage and Weed Control Practices on Growth and Yield of Wheat. *Phillips. J. Crop Sci.* **2013**, *38*, 54–62.
243. Mupangwa, W.; Twomlow, S.; Walker, S. Reduced Tillage, Mulching and Rotational Effects on Maize (*Zea mays* L.), Cowpea (*Vigna unguiculata* (Walp) L.) and Sorghum (*Sorghum bicolor* L. (Moench)) Yields under Semi-Arid Conditions. *Field Crop. Res.* **2012**, *132*, 139–148. [CrossRef]
244. Blanco-Canqui, H.; Claassen, M.; Presley, D. Summer cover crops fix nitrogen, increase crop yield, and improve soil–crop relationships. *Agron. J.* **2012**, *104*, 137–147. [CrossRef]
245. Nielsen, D.C.; Lyon, D.J.; Hergert, G.W.; Higgins, R.K.; Calderón, F.J.; Vigil, M.F. Cover Crop Mixtures Do Not Use Water Differently than Single-Species Plantings. *Agron. J.* **2015**, *107*, 1025–1038. [CrossRef]
246. Colla, G.; Mitchell, J.P.; Joyce, B.A.; Huyet, L.M.; Wallender, W.W.; Temple, S.R. Soil physical properties and tomato yield and quality in alternative cropping systems. *Agron. J.* **2000**, *92*, 924–932. [CrossRef]
247. Lotter, D.W.; Seidel, R.; Liebhardt, W. The performance of organic and conventional cropping systems in an extreme climate year. *Am. J. Altern. Agric.* **2003**, *18*, 146–154. [CrossRef]
248. Wang, Z.; Zhao, X.; Wu, P.; Chen, X. Effects of water limitation on yield advantage and water use in wheat (*Triticum aestivum* L.)/maize (*Zea mays* L.) strip intercropping. *Eur. J. Agron.* **2015**, *71*, 149–159. [CrossRef]
249. Irmak, S.; Sharma, V.; Mohammed, A.T.; Djaman, K. Impacts of cover crops on soil physical properties: Field capacity, permanent wilting point, soil-water holding capacity, bulk density, hydraulic conductivity, and infiltration. *Am. Soc. Agric. Biol. Eng.* **2018**, *61*, 1307–1321. [CrossRef]
250. Licht, M.A.; Al-Kaisi, M. Strip-tillage effect on seedbed soil temperature and other soil physical properties. *Soil Tillage Res.* **2005**, *80*, 233–249. [CrossRef]
251. Amossé, C.; Jeuffroy, M.H.; David, C. Relay intercropping of legume cover crops in organic winter wheat: Effects on performance and resource availability. *Field Crop. Res.* **2013**, *145*, 78–87. [CrossRef]
252. Martens, J.R.T.; Hoeppner, J.F.; Entz, M.H. Legume Cover Crops with Winter Cereals in Southern Manitoba: Establishment, Productivity, and Microclimate Effects. *Agron. J.* **2001**, *93*, 1086–1096. [CrossRef]
253. Unger, P.W.; Vigil, M.F. Cover crops effects on soil water relationships. *J. Soil Water Cons.* **1998**, *53*, 241–244.
254. Clark, A. *Managing Cover Crops Profitably*, 3rd ed.; Handbook Series Book 9; Sustainable Agriculture Research & Education (SARE) Program, University of Maryland: College Park, MD, USA, 2012.
255. Zibilskie, L.M.; Makus, D.J. Black oat cover crop management effects on soil temperature and biological properties on a Mollisol in Texas, USA. *Geoderma* **2009**, *149*, 379–385. [CrossRef]
256. Kahimba, F.C.; Sri Ranjan, R.; Froese, J.; Entz, M.; Nason, R. Cover crop effects on infiltration, soil temperature, and soil moisture distribution in the Canadian prairies. *Appl. Eng. Agric.* **2008**, *24*, 321–333. [CrossRef]
257. Thiagalingam, K.; Dalgliesh, N.P.; Gould, N.S.; McCown, R.L.; Cogle, A.L.; Chapman, A.L. Comparison of no-tillage and conventional tillage in the development of sustainable farming systems in the semi-arid tropics. *Aust. J. Exp. Ag.* **1996**, *36*, 995–1002. [CrossRef]
258. Sharratt, B.S.; Flerchinger, G.N. Straw color for altering soil temperature and heat flux in the subarctic. *Agron. J.* **1995**, *87*, 814–819. [CrossRef]
259. Hansen, L.; Ribaudo, M. *Economic Measures of Soil Conservation Benefits: Regional Values for Policy Assessment*; USDA Technical Bulletins 1922; 2008. Available online: https://www.ers.usda.gov/webdocs/publications/47548/11517_tb1922_reportsummary.pdf?v=0 (accessed on 10 June 2020).

260. Langdale, G.W.; Blevins, R.L.; Karlen, D.L.; McCool, D.K.; Nearing, M.A.; Skidmore, E.L.; Thomas, A.W.; Tyler, D.D.; Williams, J.R. Cover Crop Effects on Soil Erosion by Wind and Water. In *Cover Crops for Clean Water, Proceedings of an International Conference, Jackson, TN, USA, 9–11 April 1991*; Hargrove, W.L., Ed.; Soil and Water Conservation Society: Ankeny, IA, USA, 1999; pp. 15–22.
261. Martin, C.K.; Cassel, D.K. Soil loss and silage yield for three tillage management systems. *J. Prod. Agric.* **1992**, *5*, 581–586. [CrossRef]
262. De Baets, S.; Poesen, J.; Meersmans, J.; Serlet, L. Cover Crops and Their Erosion-Reducing Effects during Concentrated Flow Erosion. *Catena* **2011**, *85*, 237–244. [CrossRef]
263. Wortman, S.E.; Francis, C.A.; Lindquist, J.L. Cover crop mixtures for the western Corn Belt: Opportunities for increased productivity and stability. *Agron. J.* **2012**, *104*, 699–705. [CrossRef]
264. Colazo, J.C.; Buschiazzo, D.E. Soil dry aggregate stability and wind erodible fraction in a semiarid environment of Argentina. *Geoderma* **2010**, *159*, 228–236. [CrossRef]

Article

Weed Infestation and Health of the Soybean Crop Depending on Cropping System and Tillage System

Dorota Gawęda [1], Małgorzata Haliniarz [1,*], Urszula Bronowicka-Mielniczuk [2] and Justyna Łukasz [1]

[1] Department of Herbology and Plant Cultivation Techniques, University of Life Sciences in Lublin, Akademicka 13, 20-950 Lublin, Poland; dorota.gaweda@up.lublin.pl (D.G.); justynalukasz1992@gmail.com (J.Ł.)

[2] Department of Applied Mathematics and Computer Science, University of Life Sciences in Lublin, Głęboka 28, 20-612 Lublin, Poland; urszula.bronowicka@up.lublin.pl

* Correspondence: malgorzata.haliniarz@up.lublin.pl; Tel.: +48-81-4456654

Received: 16 April 2020; Accepted: 5 June 2020; Published: 7 June 2020

Abstract: This study evaluated weed infestation and health of the soybean crop grown in crop rotation (CR) and monoculture (CM) under conventional tillage (CT) and no-tillage (NT) conditions. The research proved that growing soybean in monoculture and under no-tillage conditions increases weed infestation and infection of soybean with fungal diseases. In these treatments, increased numbers of most of the dominant species were also found. A significantly higher percentage of monocotyledonous species and a much lower percentage of dicotyledonous ones in total weed dry weight were shown in the CR treatment relative to CM and in the NT system compared to CT. The biodiversity of the weed community was similar in monoculture and crop rotation, and slightly greater in the NT system in comparison with CT conditions. In both tillage systems, *Amaranthus retroflexus* was the weed species that most infested the soybean crop. In soybean grown after itself, *Amaranthus retroflexus* was the weed that occurred in the greatest numbers, while, in crop rotation, this was *Echinochloa crus-galli*. In all years of the study, soybean was infected with *Septoria glycines* to the highest degree, which was followed by *Cercospora sojina*, whereas infection with *Ascochyta* sp. was the lowest. Weather conditions in individual years of the study were proven to affect weed infestation and infection of soybean with fungal diseases. The study results prove that cropping systems and tillage systems significantly affect weed infestation and health of the soybean crop.

Keywords: *Glycine max* (L.) Merr.; cropping system; tillage system; weed infestation; biological diversity; fungal diseases

---

1. Introduction

Soybean is one of the world's leading crops and, at the same time, the most important *Fabaceae* species. It is characterized by a very valuable chemical seed composition because soybeans contain 18–22% of oil with a high content of unsaturated fatty acids and 33–45% protein with an excellent amino acid composition [1,2]. Due to this, the soybean finds a very wide application in human and animal nutrition, and it is a raw material for many branches of industry. Since soybeans can be grown with minimal use of chemical protection (primarily herbicides) and with reduced nitrogen fertilization, under Polish conditions, soybean cultivation is environmentally-friendly and largely compatible with integrated crop protection [2].

In Poland, the general principles of integrated crop protection are governed by the provisions of the Act of 8 March 2013 on crop protection products (Dz. U. (Journal of Laws) item 455) and the Regulation of the Minister of Agriculture and Rural Development of 18 April 2013 regarding requirements for integrated crop protection (Dz. U. (Journal of Laws) item 505) [3]. In accordance with the currently applicable principles of integrated crop protection, a great role is ascribed to natural fertility of soil and its biological activity. Therefore, the number of ploughings and their depth should be reduced and other implements that deeply loosen the soil without turning it over should be used instead of the plough [2]. The destructive impact of conventional plough tillage on the structure of the topsoil layer and reduced soil biological diversity, among others, make an argument for its abandonment [4,5]. When selecting a tillage system, the specific conditions of a particular farm should be taken into consideration in order to ensure optimal plant growth and developmental conditions [6,7]. Cultural practices affect the amount of crop residue remaining on the soil surface, which, in consequence, determines the growth of pests and weeds as well as the development of diseases [8,9]. Exact mixing of crop residue on which crop pathogens could have survived with soil is considered to be one of the tasks of conventional tillage [10]. On the other hand, an improvement in soil quality, the level of microbial activity, nutrient cycling, and microbial diversity that can be achieved by using no-tillage systems can significantly determine natural resistance of plants to diseases [11].

It is particularly justified to use no-tillage in soybean under rainfall deficit conditions during a growing season. The yield potential of this legume crop is substantially limited by climatic conditions, predominant temperature, and the amount and distribution of rainfall [12,13]. Therefore, in areas with low rainfall levels, it is important to use cultural practices that will ensure retention of the greatest possible amount of rainwater. Agronomic practices that involve replacing the plough with implements and do not turn the soil over offer such a possibility. Thus, no-tillage performs best in a dry climate or in a climate with an uneven distribution of rainfall where crop yields are frequently equal to or higher than in the case of conventional tillage [14].

A correctly designed crop rotation is also the basis for the integrated crop protection system. Monoculture impoverishes both organic matter and soil microbial life [15,16], which leads to increased weed infestation, and deteriorates the phytosanitary condition of a crop [17–19]. Due to the fact that, under Polish conditions, soybeans are not severely attacked by dangerous diseases, an attempt was made to investigate its response when it is grown a in short-term monoculture.

The aim of the present study was to evaluate the effect of the cropping system and tillage system on weed infestation and health of the soybean crop.

2. Materials and Methods

2.1. Location of the Experiment and Soil and Climatic Conditions

This study was conducted at the Uhrusk Experimental Farm (51°18′11′′ N, 23°36′50′′ E) over the period of 2014 to 2017. A field experiment was set up on Rendzic Phaeozem [20] with the grain-size distribution of sandy loam. The soil on which the experiment was carried out was characterized by alkaline pH (in 1 M KCl = 7.7), very high phosphorus availability (229.8 mg P kg^{-1} soil), high potassium availability (150.2 mg K kg^{-1} soil), and very low magnesium availability (16 mg Mg kg^{-1} soil). The humus content was 1.5%, whereas the percentage of fine particles (<0.02 mm) in the 0–30 cm layer amounted to 20.7%.

Throughout the duration of the experiment, all growing seasons of soybeans were characterized by a higher mean air temperature and rainfall total than the long-term mean (Figures 1 and 2).

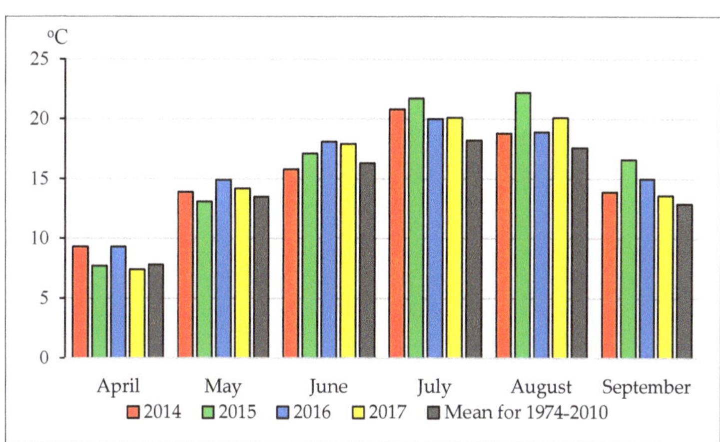

Figure 1. Mean monthly air temperature (°C) at the Bezek Meteorological Station from 2014 to 2017.

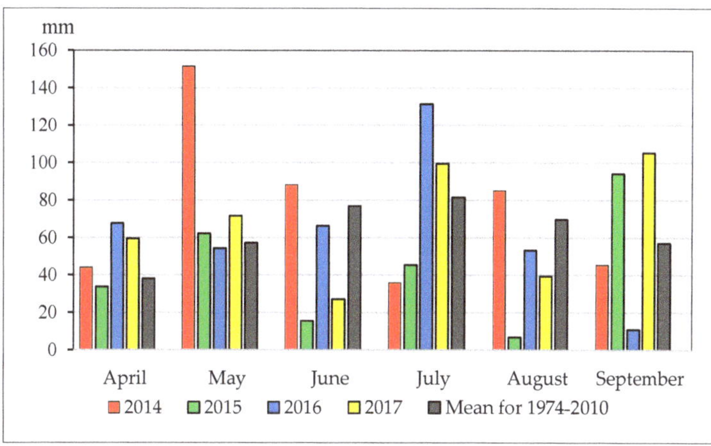

Figure 2. Total rainfall and rainfall distribution (mm) at the Bezek Meteorological Station from 2014 to 2017.

2.2. Experimental Design and Agronomic Practices

The soybean cultivar 'Merlin' was grown in an experimental area of 768 m^2 (area of all crops in crop rotation) and 192 m^2 (area of soybean in monoculture). The total area of the soybean in each year of the experiment was 384 m^2 (soybeans in crop rotation had a total area of 192 m^2 and soybeans in a monoculture had a total area of 192 m^2), while the single plot area was 32 m^2. A field experiment was set up as a split-block design in triplicate. Two experimental factors were included in this study.

I. CS—cropping system: CR—soybean grown in crop rotation after winter wheat (crop rotation: soybean—winter wheat—winter oilseed rape—winter wheat), CM—soybean monoculture cropping.
II. TS—tillage system: CT—conventional tillage, NT—no-tillage.

The experiment diagram shows the distribution of the cropping system and tillage systems and the place of soybeans in crop rotation during individual years of the experiment (Table 1). In each year of the research, soybean was sown in a different crop rotation field after winter wheat. Levels of one factor (cropping system involving crop rotation and monoculture) are arranged in rows, and levels of the second factor (tillage system involving conventional tillage and no-tillage) are arranged in columns. The levels of both factors arranged perpendicularly are crossed to form a combination of these factors. Random levels of one factor determined sub-blocks for levels of the other factor and vice versa. Therefore, the accuracy of assessing the effects of both factors is the same. In all years of research, conventional tillage (CT) and no-tillage (NT) was used for all plants sown in crop rotation (CR) and the soybean monoculture (CM).

Table 1. Experiment scheme. *Soybean location in the experiment. CR—crop rotation, CM—monoculture, CT—plough tillage, NT—no-tillage.

	CT	NT	CT	NT	CT	NT	
CM	CM CT	CM NT	CM CT	CM NT	CM CT	CM NT	*soybean 2014–2017
CR	CR CT	CR NT	CR CT	CR NT	CR CT	CR NT	*soybean 2015
	CR CT	CR NT	CR CT	CR NT	CR CT	CR NT	*soybean 2016
	CR CT	CR NT	CR CT	CR NT	CR CT	CR NT	*soybean 2017
	CR CT	CR NT	CR CT	CR NT	CR CT	CR NT	*soybean 2014
	replicates I		replicates II		replicates III		

crop rotation: soybean—winter wheat—winter oilseed rape—winter wheat.

Under the conventional tillage system, the following tillage operations were carried out: skimming + harrowing, harrowing, autumn ploughing, spring: harrowing, NPK fertilizers application, cultivating with harrowing, seeding, and harrowing. Under the no-tillage system, the following tillage operations were done: stubble cultivator: grubber + cage roller instead of skimming and cultivating instead of autumn ploughing, and spring: the same tillage operations as under the conventional tillage system.

Mineral NPK fertilization was applied before sowing soybean and fertilizer rates were determined by taking into account the crop's nutritional requirements and soil nutrient availability. Fertilization was applied at the following rates: N—30 kg ha^{-1} (34.5% ammonium nitrate), P—40 kg ha^{-1} (40% superphosphate), and K—80 kg ha^{-1} (60% potassium salt).

In the years 2014 and 2016, the sowing time was in the last 10 days of April, whereas, in 2015 and 2017, the sowing time was in the first 10 days of May. Each time before sowing, seeds were dressed with *Bradyrhizobium japonicum* bacteria and the seed dressing Vitavax 200 FS (carboxin, thiuram) at a rate of 80 g and 80 g (active ingredients) per 100 kg of seeds. Seeds were sown at a depth of 3 cm, a row spacing of 20 cm, and with a planned density of 100 plants per 1 m^2.

Weed management in the soybean crop involved mechanical and chemical weed control. Immediately after seeding, a mixture of the following herbicides was applied: Afalon Dyspersyjny 450 SC (linuron) + Dual Gold 960 EC (metolachlor-S) at a rate of 450 g + 1728 g ha^{-1} (active ingredients). Mechanical weed control included harrowing 3–4 days after seeding and double harrowing after emergence of soybeans (at the first trifoliate leaf stage—BBCH 12, and at the third trifoliate leaf stage—BBCH 13).

Soybeans were harvested at full maturity of soybeans (BBCH 89) in the second 10 days of September in 2014 and 2016, while, in 2015 and 2017, soybeans were harvested at full maturity in the third 10 days of September.

2.3. Scope of Study and Statistical Analysis

The evaluation of the species' composition of the weed community as well as of weed density and dry weight, which was performed by the dry-weight-rank method at the pod and seed ripening stage (BBCH 81/82). The sampling area was delineated by a quadrat frame with the dimensions of 1 m × 0.5 m in two randomly selected points of each plot. Weed species nomenclature followed Mirek et al. [21].

The Shannon-Wiener diversity index (H') and the Simpson dominance index (SI) were calculated according to the following formula [22,23].

$$H' = -\Sigma \text{ Pi ln Pi} \quad (1)$$

$$SI = \Sigma \text{ Pi } 2 \quad (2)$$

where:

Pi—probability of species occurrence in a sample,

Pi = n/N,

n—number of individuals in species,

N—total number of individuals in the sampling area,

ln—natural logarithm.

The higher the value of the Shannon-Wiener index, the greater the diversity of a community. The range of Simpson index values is from 0 to 1 with values close to 1 indicating a clear dominance of one or several species.

The evaluation of soybean health was carried out at the first pod set stage (BBCH 70) on 50 plants from each plot. The determination was made using a five-point scale (0–4) expressing increasing plant infection, where 0 denotes leaves without disease symptoms, 1 denotes up to 10% of leaf surface infected, 2 represents up to 50%, 3 represents up to 75%, and 4 represents 100% of leaf surface infected. Having established the degree of plant infection, the number of plants within a specific degree of infection was determined for all replicates. Next, the disease index was calculated according to McKinney's formula [24].

$$Wch = (\sum \text{a})/(\sum \text{b}) \times 100 \quad (3)$$

where:

Wch—disease index;

Σa—sum of the numeric indices in the scale multiplied by the number of plants corresponding to a given index;

Σb—total number of examined plants multiplied by the highest index in the scale.

The study results collected over the period 2014–2017 were analyzed using two-way analysis of variance with repeated measurements over the four growing periods. The significance of differences was estimated by Tukey's test at a significance level of $\alpha = 0.05$. ANALWAR–5.3.FR statistical software was used for calculations. The distribution conformity with normal distributions was verified with the Shapiro-Wilk test, while homogeneity of the variance was tested with the Bartlett test. When necessary to homogenize the variance, the data were subjected to angular transformation before the variance analysis [25]. Before performing the statistical calculations, the percentage values were transformed using the following equation.

$$y = arcsin\sqrt{x} \quad (4)$$

3. Results

3.1. Weed Infestation of Soybean Crops

Cropping systems and tillage systems caused significant differences in weed infestation of the soybean crop (Figures 3 and 4). On average over the study period, in the soybean monoculture, the number and dry weight of weeds was, respectively, higher by 69.4% and 28.6% than in crop rotation. The significant interactions between the cropping system and the tillage system were not found (Figures 5 and 6). The weed of dry weight did not differ between both tillage systems only in the first year of the experiment. The highest weed of dry weight was found in 2014 in the soybean monoculture both under plough tillage (CT) and no-tillage (NT) conditions. Weed infestation of the soybean crop expressed as dry weight of weeds was highest in the first year of the experiment (2014), which was characterized by a higher amount of precipitation than the other years of the experiment and the long-term mean, especially in May and June. The lowest number and dry weight of weeds were found in 2015 and 2016 (Figures 3 and 4).

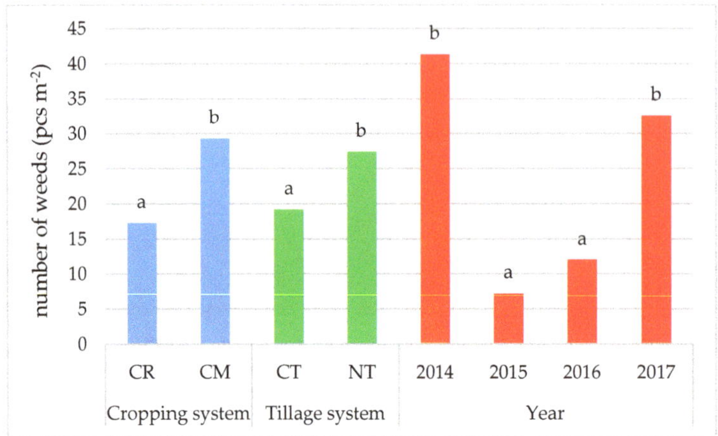

Figure 3. Mean number of weeds in soybean crop depending on cropping system, tillage system, and years of research. CR—crop rotation, CM—monoculture, CT—plough tillage, and NT—no-tillage. Different letters denote significant differences ($p \leq 0.05$) among different cropping systems (CR and CM), tillage systems (CT and NT), and years of research. The same letter means it is not significantly different.

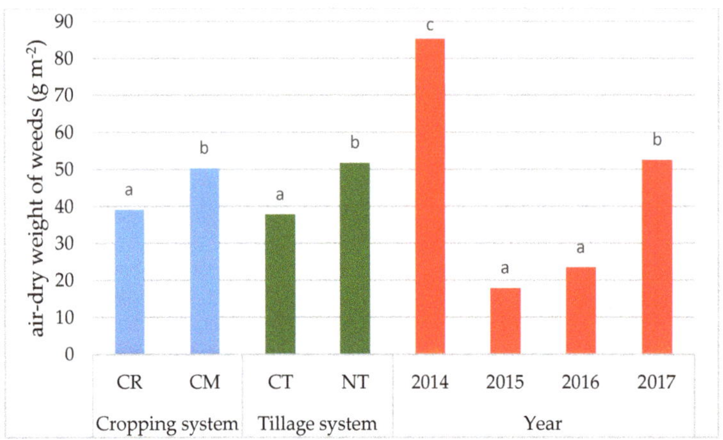

Figure 4. Air-dry weight of weeds in soybean crop depending on cropping system, tillage system, and years of research. CR—crop rotation, CM—monoculture, CT—plough tillage, NT—no-tillage. Different letters denote significant differences ($p \leq 0.05$) among different cropping systems (CR and CM), tillage systems (CT and NT), and years of research. The same letter means it is not significantly different.

The statistical analysis revealed that the studied factors significantly modified the percentage of monocotyledonous and dicotyledonous species in total weed dry weight (Figure 7). In crop rotation (CR), the percentage of monocotyledonous species was significantly higher, whereas the percentage of dicotyledonous ones was much lower than in the soybean monoculture (CM). The NT system significantly increased the percentage of monocotyledonous species and decreased the percentage of dicotyledonous ones in total weed dry weight in comparison with the CT system. As far as the interactions between the experimental factors are concerned, it was proven that the percentage of monocotyledonous species in total weed dry weight was lower in the monoculture under conventional tillage conditions, while the percentage of dicotyledonous ones was higher relative to the other experimental treatments (Figure 8).

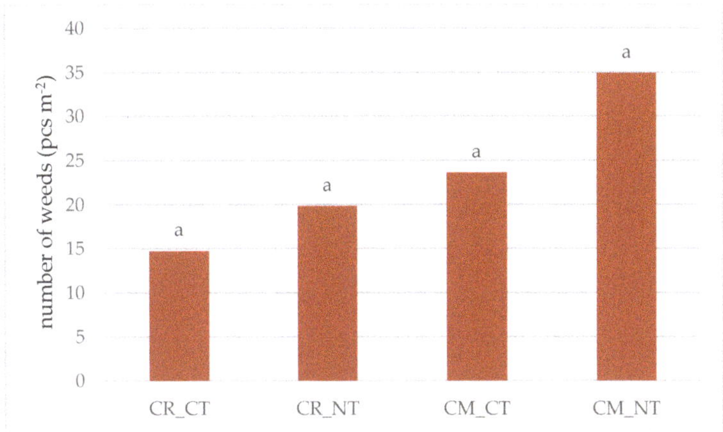

Figure 5. Interactive dependencies of the cropping system and tillage system in shaping the number of weeds in soybean crop (mean for 2014–2017). CR_CT—crop rotation and plough tillage, CR_NT—crop rotation and no-tillage, CM_CT—monoculture and plough tillage, CM_NT—monoculture and no-tillage. The same letter means it does not have a significantly different ($p \leq 0.05$) interaction among crop rotation (CR), monoculture (CM), plough tillage (CT), and no-tillage (NT).

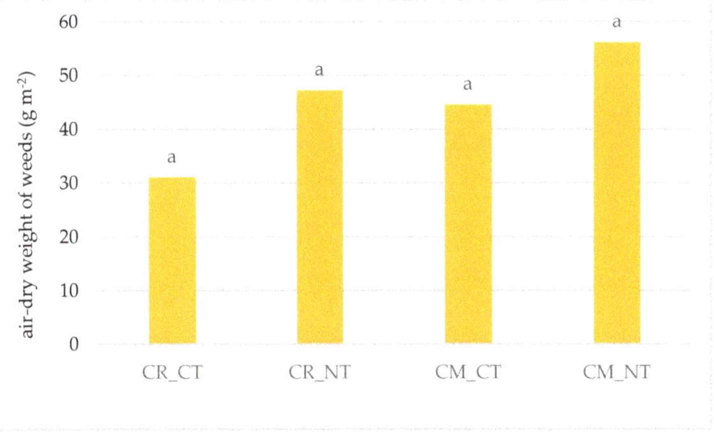

Figure 6. Interactive dependencies of the cropping system and tillage system in shaping air-dry weight of weeds in soybean crop (mean for 2014–2017). CR_CT—crop rotation and plough tillage, CR_NT—crop rotation and no-tillage, CM_CT—monoculture and plough tillage, CM_NT—monoculture and no-tillage. The same letter means it is not significantly different ($p \leq 0.05$) in terms of the interaction among crop rotation (CR), monoculture (CM), plough tillage (CT), and no-tillage (NT).

The agrophytocenosis of the soybean crop grown after winter wheat (CR) included 22 weed species out of which 18 belonged to annual weeds (81.8%) (Table 2). Four perennial weed species and 17 annual ones, which accounted for 81.0% of the total number of species, were observed in a soybean monoculture (CM). In crop rotation, *Echinochloa crus-galli* was the most numerous taxon, followed by *Amaranthus retroflexus*, whereas, in a soybean monoculture, *Amaranthus retroflexus* was predominant.

The number of *Amaranthus retroflexus*, *Chenopodium album*, *Avena fatua*, and *Solanum nigrum* was significantly higher in monoculture (CM) than in CR. Soybean cultivation in crop rotation (CR) also significantly reduced the numbers of the perennial species *Elymus repens*. In CR, *Anchusa arvensis*, *Veronica persica*, *Viola arvensis*, and *Convolvulus arvensis* were found to be present, but they did not inhabit soybeans grown after itself (CM). *Euphorbia helioscopia*, *Thlaspi arvense*, and *Plantago major* infested only a soybean monoculture.

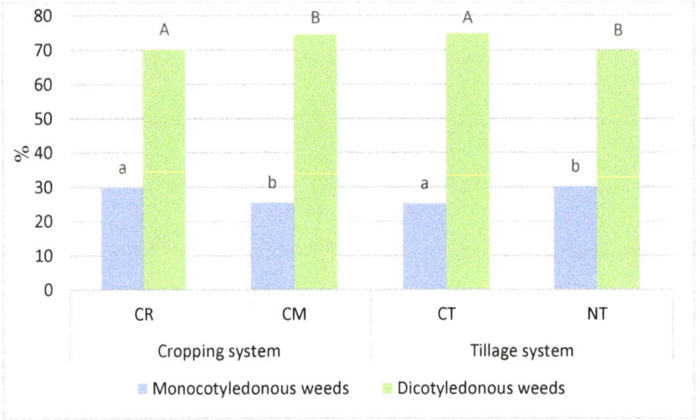

Figure 7. Percentage content of weight of monocotyledonous and dicotyledonous weeds in total weed dry weight depending on the cropping system and tillage system (mean for 2014–2017). CR—crop rotation, CM—monoculture, CT—plough tillage, NT—no-tillage. Different lowercase letters denote significant differences ($p \leq 0.05$) of monocotyledonous weeds among different cropping systems (CR and CM) and tillage systems (CT and NT). Different capital letters denote significant differences ($p \leq 0.05$) of dicotyledonous weeds among different cropping systems (CR and CM) and tillage systems (CT and NT).

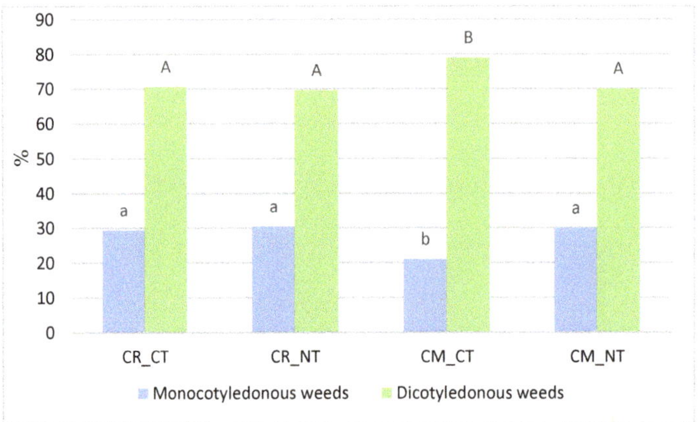

Figure 8. Percentage content of monocotyledonous and dicotyledonous weeds weight in total weed dry weight depending on the interaction of the cropping system and tillage system (mean for 2014–2017). CR_CT—crop rotation and plough tillage, CR_NT—crop rotation and no-tillage, CM_CT—monoculture and plough tillage, and CM_NT—monoculture and no-tillage. Different lowercase letters denote significant differences ($p \leq 0.05$) of monocotyledonous weeds for interaction among crop rotation (CR), monoculture (CM), plough tillage (CT), and no-tillage (NT). Different capital letters denote significant differences ($p \leq 0.05$) of dicotyledonous weeds for interaction among crop rotation (CR), monoculture (CM), plough tillage (CT), and no-tillage (NT).

Table 2. Species composition of weeds per 1 m^2 in soybean crop depending on the cropping system and tillage system (mean for 2014–2017).

Species	Cropping System		Tillage System	
	CR	CM	CT	NT
I. Short-lived				
Amaranthus retroflexus L.	3.4a	7.3b	4.5a	6.1a
Anagallis arvensis L.	0.1a	0.1a	0.1a	0.1a
Anchusa arvensis (L.) M. Bieb.	0.3a	-	-	0.3a
Avena fatua L.	1.5a	2.8b	1.4a	2.9b
Capsella bursa-pastoris (L.) Medik.	0.8a	1.2a	1.2a	0.9a
Chenopodium album L.	0.8a	3.6b	2.2a	2.2a
Echinochloa crus-galli (L.) P. Beauv.	4.2a	2.5a	3.1a	3.7a
Euphorbia helioscopia L.	-	0.2a	0.1a	0.2a
Fallopia convolvulus (L.) Á. Löve	0.6a	0.3a	0.4a	0.6a
Galium aparine L.	0.8a	0.2a	0.4a	0.6a
Lamium amplexicaule L.	0.1a	0.2a	0.2a	0.1a
Matricaria maritima ssp. inodora (L.) Dostál	0.5a	1.2a	0.6a	1.2a
Melandrium album (Mill.) Garcke	0.2a	0.2a	0.1a	0.2a
Polygonum aviculare L.	1.6a	1.8a	1.9a	1.4a
Solanum nigrum L. Emend. Mill.	0.4a	2.0b	0.5a	1.9b
Sonchus asper (L.) Hill.	0.6a	0.5a	0.4a	0.6a
Stellaria media (L.) Vill.	0.3a	0.4a	0.3a	0.4a
Thlaspi arvense L.	-	0.1a	0.1a	-
Veronica persica Poir.	0.1a	-	-	0.1a
Viola arvensis Murr.	0.3a	-	0.2a	0.1a
II. Perennial				
Cirsium arvense (L.) Scop.	0.3a	1.0a	0.5a	0.7a
Convolvulus arvensis L.	0.1a	-	-	0.1a
Elymus repens (L.) Gould	0.1a	3.4b	0.8a	2.7a
Plantago major L.	-	0.2a	0.1a	0.1a
Sonchus arvensis L.	0.2a	0.1a	0.1a	0.2a
Number of weed species	22a	21a	22a	24a

CR—crop rotation, CM—monoculture, CT—plough tillage, NT—no-tillage. Different letters denote significant differences ($p \leq 0.05$) of the number of weed species among different cropping systems (CR and CM) and tillage systems (CT and NT). The same letter means it is not significantly different.

A total of 22 weed species occurred in the soybean crop grown under conventional tillage (CT), whereas, in the no-tillage (NT) treatment, 24 species occurred (Table 2). *Anchusa arvensis, Veronica persica*, and the perennial species *Convolvulus arvensis* were found in the NT treatment, but they did not infest soybeans grown under the conventional system. In the CT treatment, on the other hand, the species *Thlaspi arvense* was observed, which was not found in the NT treatment. *Amaranthus retroflexus* occurred in the greatest numbers in both tillage systems. In the NT treatment, the number of *Avena fatua* and *Solanum nigrum* significantly increased compared to CT.

In crop rotation, 22 weed species were found under no-tillage conditions, while, in the conventional tillage treatment, there were six fewer species (Table 3). The following annual weeds: *Anchusa arvensis, Lamium amplexicaule, Veronica persica*, and perennial weeds: *Convolvulus arvensis, Elymus repens, Sonchus arvensis*, did not occur in soybeans grown in crop rotation under CT, but they were found in the non-tilled plots. In the soybean monoculture, greater weed species' richness (21 species) was observed in the CT treatment where *Lamium amplexicaule, Thlaspi arvense*, and *Sonchus arvensis* were found to be present, which are species that did not occur in the non-tilled plots (NT). Both in crop rotation and soybean monoculture, conventional tillage slightly reduced the number of the most numerous weed species. In crop rotation under CT conditions, the number of individuals of the dominant taxon, *Echinochloa crus-galli*, was 25.0% lower than under no-tillage (NT). However, these differences were not statistically significant. In the soybean crop grown in the monoculture, conventional tillage (CT) slightly reduced the number of individuals of the most numerous species, *Amaranthus retroflexus*.

Table 3. Interactive dependencies of the cropping system and tillage system in shaping weed species' composition per 1 m^2 in soybean crops (mean for 2014–2017).

Species	CR_CT	CR_NT	CM_CT	CM_NT
I. Short-lived				
Amaranthus retroflexus L.	3.2a	3.6a	5.9a	8.7a
Anagallis arvensis L.	0.1a	0.1a	0.1a	0.1a
Anchusa arvensis (L.) M. Bieb.	-	0.7a	-	-
Avena fatua L.	1.2a	1.8a	1.5a	4.1a
Capsella bursa-pastoris (L.) Medik.	1.0a	0.6a	1.3a	1.2a
Chenopodium album L.	0.6a	1.0a	3.8a	3.3a
Echinochloa crus-galli (L.) P. Beauv.	3.6a	4.8a	2.5a	2.5a
Euphorbia helioscopia L.	-	-	0.1a	0.3a
Fallopia convolvulus (L.) Á. Löve	0.5a	0.7a	0.2a	0.4a
Galium aparine L.	0.5a	1.1a	0.2a	0.2a
Lamium amplexicaule L.	-	0.2a	0.5a	-
Matricaria maritima ssp. inodora (L.) Dostál	0.4a	0.6a	0.8a	1.7a
Melandrium album (Mill.) Garcke	0.1a	0.2a	0.2a	0.2a
Polygonum aviculare L.	1.8a	1.3a	2.0a	1.5a
Solanum nigrum L. Emend. Mill.	0.5a	0.4a	0.5a	3.4b
Sonchus asper (L.) Hill.	0.4a	0.7a	0.5a	0.6a
Stellaria media (L.) Vill.	0.3a	0.4a	0.4a	0.4a
Thlaspi arvense L.	-	-	0.2a	-
Veronica persica Poir.	-	0.2a	-	-
Viola arvensis Murr.	0.4a	0.2a	-	-
II. Perennial				
Cirsium arvense (L.) Scop.	0.1a	0.5a	1.0a	1.0a
Convolvulus arvensis L.	-	0.2a	-	-
Elymus repens (L.) Gould	-	0.2a	1.7a	5.1a
Plantago major L.	-	-	0.1a	0.2a
Sonchus arvensis L.	-	0.3a	0.1a	-

CR_CT—crop rotation and plough tillage, CR_NT—crop rotation and no-tillage, CM_CT—monoculture and plough tillage, CM_NT—monoculture and no-tillage. Different letters denote significant differences ($p \leq 0.05$) of the number of weed species for an interaction among crop rotation (CR), monoculture (CM), plough tillage (CT), and no-tillage (NT).

The evaluation of the biological diversity of the segetal flora showed that the Shannon-Wiener index (H′) and the Simpson index (SI) did not differ significantly in crop rotation (CR) and a soybean monoculture (CM). The soybean crop grown in the no-tillage (NT) system was characterized by slightly greater diversity of the weed community than under CT. This is indicated by the higher value of the Shannon diversity index and a slightly lower value of the Simpson dominance index (Figure 9).

The interaction of the cropping system and tillage system did not cause significant variations in the Shannon-Wiener diversity index and the Simpson dominance index (Figure 10). The agrophytocenosis of the soybean crop grown in crop rotation in the NT treatment was characterized by the greatest diversity, while the lowest one was found in the case of crop rotation and CT conditions. Nonetheless, the difference in the values of the diversity and dominance indices between the individual treatments was not significant.

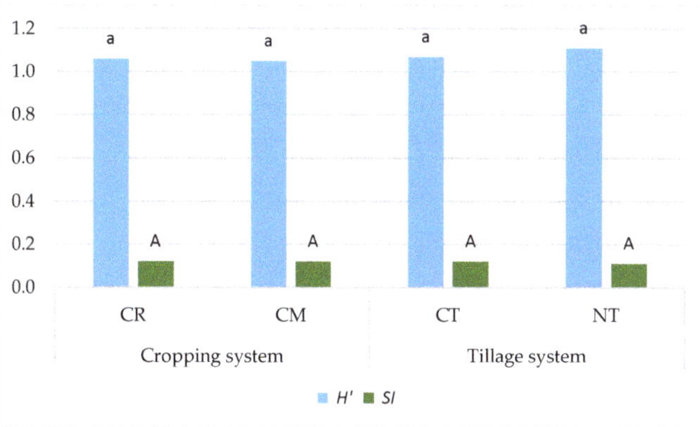

Figure 9. Shannon-Wiener's diversity index (*H*′) and Simpson's dominance index (*SI*) of the weed community in soybean crop depending on the cropping system and tillage system (mean for 2014–2017). CR—crop rotation, CM—monoculture, CT—plough tillage, and NT—no-tillage. Different lowercase letters denote significant differences ($p \leq 0.05$) of Shannon-Wiener's diversity index (*H*′) among different cropping systems (CR and CM) and tillage systems (CT and NT). Different capital letters denote significant differences ($p \leq 0.05$) of Simpson's dominance index *(SI)* among different cropping systems (CR and CM) and tillage systems (CT and NT).

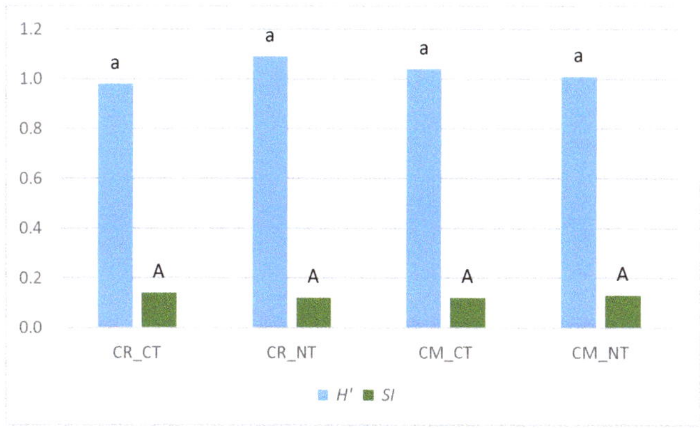

Figure 10. Interactive dependencies of the cropping system and tillage system in shaping Shannon-Wiener's diversity index (*H*′) and Simpson's dominance index (*SI*) of the weed community in soybean crop (mean for 2014–2017). CR_CT—crop rotation and plough tillage, CR_NT—crop rotation and no-tillage, CM_CT—monoculture and plough tillage, CM_NT—monoculture and no-tillage. Different lowercase letters denote significant differences ($p \leq 0.05$) of Shannon-Wiener's diversity index *(H*′*)* for interaction among crop rotation, monoculture, plough tillage (CT), and no-tillage (NT). Different capital letters denote significant differences ($p \leq 0.05$) of Simpson's dominance index *(SI)* for an interaction among crop rotation (CR), monoculture (CM), plough tillage (CT), and no-tillage (NT).

3.2. *Infection of Soybean Plants with Diseases*

In every year of the study, soybeans were infected with Ascochyta blight (*Ascochyta* sp.), Septoria leaf spot (*Septoria glycines*), and frogeye leaf spot (*Cercospora sojina*). The severity of diseases observed was dependent on the factors studied, but it was generally low in the 2015 growing season characterized by the lowest rainfall level. The growing seasons in 2014 and 2017, during which the highest amount of rainfall was recorded, promoted the growth of fungal pathogens. During these seasons, infection of soybean with Ascochyta blight, Septoria leaf spot, and frogeye leaf spot was found to be the highest. In 2017, the severity of symptoms of all diseases observed was significantly higher than in 2015 and 2016. Soybean plants were infected with the pathogen *Septoria glycines* to the greatest degree, which was followed by *Cercospora sojina*, while infection with *Ascochyta* sp. Was the lowest (Figures 11–13).

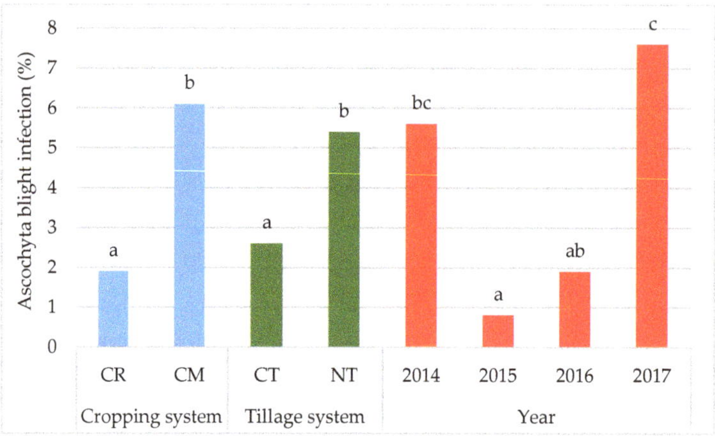

Figure 11. Mean infection of soybean with Ascochyta blight (*Ascochyta* sp.) depending on the cropping system, tillage system, and years of research. CR—crop rotation, CM—monoculture, CT—plough tillage, NT—no-tillage. Different letters denote significant differences ($p \leq 0.05$) among different cropping systems (CR and CM), tillage systems (CT and NT), and years of research. The same letter means it is not significantly different.

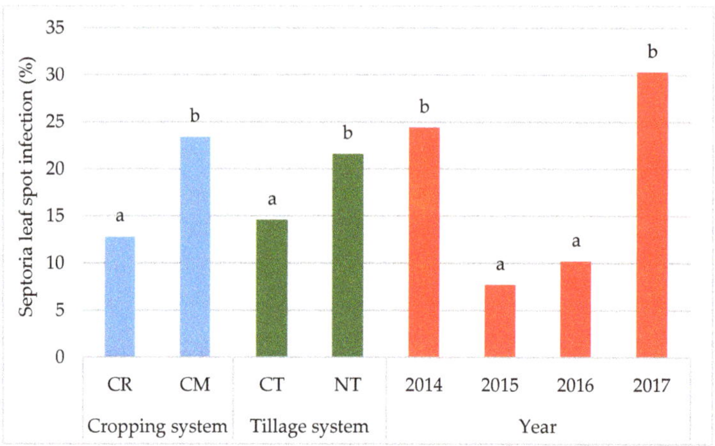

Figure 12. Mean infection of soybeans with Septoria leaf spot (*Septoria glicynes*) depending on the cropping system, tillage system, and years of research. CR—crop rotation, CM—monoculture, CT—plough tillage, NT—no-tillage. Different letters denote significant differences ($p \leq 0.05$) among different cropping systems (CR and CM), tillage systems (CT and NT), and years of research. The same letter means it is not significantly different.

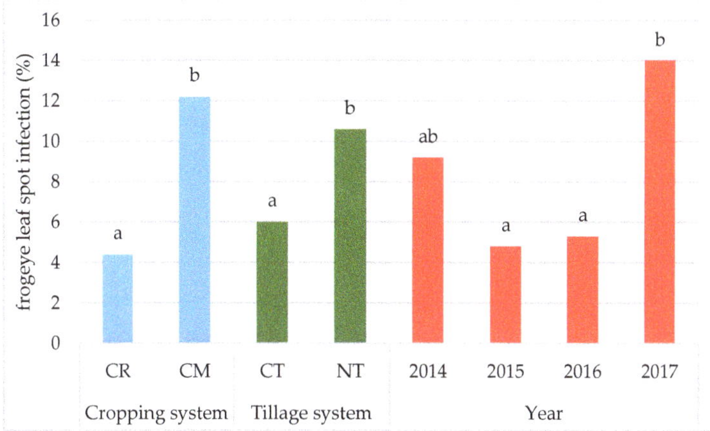

Figure 13. Mean infection of soybeans with frogeye leaf spot (*Cercospora sojina*) depending on the cropping system, tillage system, and years of study. CR—crop rotation, CM—monoculture, CT—plough tillage, NT—no-tillage. Different letters denote significant differences ($p \leq 0.05$) among different cropping systems (CR and CM), tillage systems (CT and NT), and years of research. The same letter means being not significantly different.

On average, over the four-year study period, the experimental factors were proven to influence the health of soybean plants (Figures 11–13). In the monoculture, the severity of infection with the pathogens causing Ascochyta blight, Septoria leaf spot, and frogeye leaf spot was higher by, respectively, more than three times, almost two times, and nearly three times when compared to crop rotation. The no-tillage system promoted the severity of all soybean diseases observed in this experiment. It was proven that, under the CT system, plant infection by the pathogens *Ascochyta* sp., *Septoria glycines*, and *Cercospora sojina* was lower by 51.9%, 32.4%, and 43.4%, respectively, than under no-tillage (NT).

4. Discussion

Weeds are strong competitors for soybeans. With heavy weed infestation, plants exhibit nutrient deficiency symptoms and are pale green, while seed yield is lower and of worse quality. Therefore, the profitability of cultivation of this legume crop is greatly dependent on effective elimination of weed infestation [2]. Weed infestation of crops can be effectively reduced, among others, by selecting an appropriate position of a given crop in crop rotation [26]. The study presented in this paper proved that growing soybeans after winter wheat as a previous crop reduces weed infestation when compared to what was determined in the monoculture. The results of a study by Cardina et al. [27] reveal that monoculture cropping promotes the accumulation of weed seeds in the soil. In the study of these authors, the number of weed seeds was also affected by the interaction between tillage and crop rotation since the highest seed density was shown in the monoculture under NT conditions. Our research did not prove the interaction between the cropping system and tillage system impacted weed infestation of the soybean crop, but a slightly higher number of weeds was observed in the soybean monoculture under NT conditions when compared to that found in the other experimental treatments.

The results of studies regarding the effect of tillage systems on weed infestation of crops are not unequivocal. Santín-Montanyá et al. [28] did not prove conventional tillage and no-tillage systems to have a significant impact on the number of weeds in the winter wheat crop. However, the results of these authors' research, similarly to those presented in our paper, show that weed community diversity, as expressed by the Shannon index, increased under the NT system. Sebayang and Rifai [18], on the other hand, found that the dry weed weight in the soybean crop did not differ significantly between conventional tillage, reduced tillage, and no-tillage. Nonetheless, many authors point out that the no-tillage system increases weed infestation of crops [29–32]. Our experiment also demonstrated a significant increase in the number and weight of weeds under the NT system in comparison with CT. Such an increase may be due to the accumulation of freshly shed weed seeds in the topsoil layer where they germinate and sprout in great numbers in the succeeding crop [33–35]. As a matter of fact, the tillage system does not affect the seed bank size but changes both the composition and distribution of diaspores in the soil profile [36]. The study by Cardina et al. [27] revealed that the number of weed seeds in the soil was highest under NT and generally decreased with increasing tillage intensity. These authors also proved that, under the NT system, weed seeds accumulated near the soil surface (from 0 to 5 cm) whereas, under the other tillage systems, they were evenly distributed across the entire tilled layer. In the opinion of Małecka et al. [37], increased weed infestation under the NT system can generally be observed in the first crop rotation cycle, whereas, during the subsequent period, weed infestation stabilizes and even some positive aspects of no-till field preparation are noticeable.

In the opinion of Chovancova et al. [38], Małecka-Jankowiak et al. [39], and Mancinelli et al. [40], the tillage system modifies soil properties, and, thus, affects plant growth and development. Its effect can be a change in the species' composition of weeds growing in a crop, as shown, among others, by the research presented in this paper. Velykis and Satkus [41] also proved that no-tillage caused a change in the weed species' composition in comparison with the conventional system. In particular, it increased the presence of *Galium aparine* and *Chenopodium album* in the cultivation of a legume crop (pea). According to Campiglia et al. [42], no-tillage can lead to increased occurrence of perennial weeds, which are difficult to control. This is confirmed by our study, which recorded increased numbers of the species *Elymus repens* under the NT system when compared to CT.

From the beginnings of agriculture development, weed control has been carried out using various practices such as soil tillage or crop rotation, while, in recent years, crop protection products have also been used. Widespread application of herbicides has deeply modified agro-ecosystems, negatively affecting their species' diversity and causing herbicide resistance to occur, which has resulted in the appearance of the superweeds, that is, weeds resistant to glyphosate [43,44]. This threatens many functions of an ecosystem, which is of key importance for food production [43] because biodiversity in crop fields and their surroundings performs a number of biological functions in nutrient cycling and utilization as well as in the maintenance of balance among pathogens attacking crops [45,46]. Therefore, the aim of international policy and conducted research has become biodiversity conservation, including the species diversity of agroecosystems [47,48]. According to Clements et al. [49], if diversity increases and the number of ecological interactions also increases, weed species should be viewed as an interactive community rather than an unrelated set of targets for control. The study presented in our paper demonstrates that the NT positively influences the weed community diversity in the soybean crop relative to the CT system. However, no changes were found in the diversity of segetal flora depending on the cropping system. On the other hand, a study by Suwary et al. [50] proved that monoculture cropping contributes to decreased weed species' diversity. Under conditions of winter wheat monoculture cropping, Sekutowski and Domaradzki [51] found a greater dominance of weeds and, at the same time, lower weed species' biodiversity as a result of using tillage reductions.

Weeds occurring in a crop contribute to a change in thermal and moisture conditions in it, which, thus, creates favorable conditions for developing many fungal pathogens [52]. Unwanted vegetation can be an indirect host for pathogenic fungi and contribute to the spread of diseases and greater infection of the crop [53–55]. Our research is also a confirmation of these theories because, in the present study, an increased incidence of fungal diseases was found in the treatments with the highest number and dry weight of weeds. Given the above, it is very important to timely and effectively eliminate weeds as a potential reservoir of fungal diseases [56]. Some biologically active substances used to control weed infestation can also contribute to reducing fungal diseases of soybeans. Plenty of attention has been devoted to glyphosate, which is a commonly used active substance. It results in reduced germination of spores of the fungus Phakopsora pachyrhizi, which causes soybean rust and leads to a lower incidence of this disease in plants [57,58]. Application of lactofen and sulfentrazone promotes increased production of phytoalexin in leaves, which is a substance that inhibits the occurrence of Sclerotinia sclerotiorum in soybeans [59,60]. It should, however, be indicated that lactofen can cause minor transient damage to the crop plant [61] through which pathogens can easily penetrate into the plant [62]. In the literature, reports can be found that, among others, glyphosate application leads to increased numbers of fungi and disturbances in soil biological balance [63]. Crops are then more susceptible to infection by diverse fungal pathogens [64].

Symptoms of many diseases caused by fungi can be observed in soybeans, but, under Polish conditions, these diseases have not been economically important so far. It is also rarely necessary to use fungicides [2]. In our study, soybeans were infected with Septoria leaf spot (*Septoria glycines*) and frogeye leaf spot (*Cercospora sojina*). Soybeans were infected to the lowest degree with Ascochyta blight (*Ascochyta* sp.). Nonetheless, Amin and Melkamu [65], Bretag et al. [66], Davidson and Kimber [67] as well as Sudheesh et al. [68] have shown that the fungus *Ascochyta* sp. can pose a major threat to legume crops and cause a significant loss in yields.

The severity of infection of plants with fungal diseases is largely associated with weather conditions in a particular growing season. Both our research and the studies of other authors reveal that years characterized by a high amount of rainfall most favor the development of fungal diseases [69,70]. In our research in thermal terms, the years 2015 and 2017 proved to be unfavorable for soybean development (Figures 1 and 2). In 2015, in the sowing month and during the initial growth of soybeans (May), when its thermal requirements are very high, a lower temperature was recorded when compared to the long-term mean. The lowest rainfall level was noted in the second growing season (2015), which significantly contributed to reduced weed infestation and decreased infection of soybeans with fungal diseases. The highest rainfall was observed in 2014 and 2017. The year 2017 was unfavorable for soybean development, predominantly in terms of rainfall distribution. There was a rainfall deficit in June (flowering stage), while, in July (flowering/seed filling) and in September, excessive rainfall was recorded. Such a distribution of precipitation favored the development of fungal diseases. In 2017, a lower temperature was observed in April than in the other years of the study when compared to the long-term mean. Due to this and because of torrential rains, sowing of soybeans was done in the first 10 days of May.

Numerous studies, including the study presented in this paper, have proven that monoculture cropping contributes to increased infection of plants with fungal diseases [17,70]. Strom et al. [71] demonstrated that long-term corn and soybean monoculture was associated with substantial changes in fungal communities and their distribution, as the number of host-specific fungal pathogens increased.

In the studies by Kraska et al. [72] and Suproniene et al. [73], tillage systems did not cause differences in infection of wheat by a complex of fungal diseases. The results of our experiment, however, show that the no-tillage system is conducive to the increased prevalence of fungal diseases of soybeans. Vrandečić et al. [74] also proved that fungal infection of soybeans is much higher under reduced tillage and no-tillage conditions compared to that found under conventional tillage. According to these authors, the amount and intensity of rainfall and temperature were the most important factors influencing the level of infection.

The research presented in this paper demonstrates that the no-tillage system and monoculture cropping promoted weed infestation and infection of soybeans with fungal diseases when compared to the conventional tillage system and crop rotation. The results on soybean yield reported in a paper by Gawęda et al. [75] proved that the seed yields in the experiment in question were much lower in the no-tillage treatment. A decreasing trend in yield in a soybean monoculture was observed as late as the fourth year of the experiment [75]. It was also found that infection of soybeans with fungal diseases was greatest in 2020. A high number of weeds and the dry weight of weeds were also recorded.

5. Conclusions

1. The cultivation of the soybean after it promotes weed infestation. The number and dry weight of weeds was proven to be higher in the monoculture (CM) than in crop rotation (CR), respectively, by 69.4% and 28.6%.
2. Under no-tillage (NT), the number of weeds and dry weight of weeds was, respectively, higher by 42.7% and 36.8% than under conventional tillage (CT).
3. The statistical analysis confirmed that growing soybeans in crop rotation, compared to the monoculture, and, in the no-tillage system, increases the percentage of monocotyledonous species in total dry weight of weeds.
4. In crop rotation, *Echinochloa crus-galli* was the most numerous weeds in the soybean crop, whereas, in a monoculture, it was *Amaranthus retroflexus*. Reducing the numbers of *Amaranthus retroflexus, Chenopodium album, Avena fatua, Solanum nigrum,* and *Elymus repens* was observed in crop rotation.
5. In both tillage systems, *Amaranthus retroflexus* was the weed species that occurred in the greatest numbers in the soybean crop. A significantly lower number of *Avena fatua* and *Solanum nigrum* was found under conventional tillage (CT) when compared to no-tillage (NT).
6. Cropping systems and tillage systems only slightly changed the biological diversity of the weed community in the soybean crop. The Shannon-Wiener index (*H'*) and the Simpson index (*SI*) were shown to have similar values in crop rotation (CR) and the soybean monoculture (CM), whereas the biological diversity of the weed community was found to be only slightly greater under the NT system compared to the CT system.
7. Throughout the study period, soybeans were infected with Ascochyta blight (*Ascochyta* sp.), Septoria leaf spot (*Septoria glycines*), and frogeye leaf spot (*Cercospora sojina*). The severity of disease symptoms in most of the growing seasons was greater in the seasons characterized by a high amount of rainfall.
8. Soybean plants were most infected with the pathogen *Septoria glycines* and least infected with *Ascochyta* sp.
9. No-tillage and monoculture cropping promoted infection of soybeans with fungal diseases in comparison to the conventional system and crop rotation.

Author Contributions: The authors contributed to this article in the following ways: Conceptualization: D.G. Data curation, D.G. and J.Ł. Formal analysis: D.G., M.H., and U.B.-M. Investigation: D.G. and M.H. Supervision: D.G. and M.H. Visualization: D.G. and U.B.-M. Writing—original draft: D.G. and M.H. Writing—review and editing: D.G. M.H. and J.Ł. All authors have read and agreed to the published version of the manuscript.

Funding: The Ministry of Science and Higher Education of Poland as part of statutory activities of Department of Herbology and Plant Cultivation Techniques, University of Life Sciences in Lublin, funded this research.

Conflicts of Interest: The authors declare no conflict of interest.

References

1. Abbasi Surki, A.; Sharifzade, F.; Tavakkol Afshari, R.; Majnoun Hosseini, N.; Gazor, H.R. Optimization of processing parameters of soybean seeds dried in a constant-bed dryer using response surface methodology. *J. Agric. Sci. Tech. Iran.* **2010**, *12*, 409–423.
2. Filoda, G.; Mrówczyński, M.; Matyjaszczyk, E.; Kierzek, R.; Nawracała, J.; Bubniewicz, P.; Fiedler, Ż.; Kornobis, F.; Matysiak, K.; Pruszyński, G.; et al. *Metodyka Integrowanej Ochrony i Produkcji Soi Dla Doradców*; Instytut Ochrony Roślin—Państwowy Instytut Badawczy: Poznań, Poland, 2016. (In Polish)
3. The Act of March 8, 2013 on Crop Protection Products (Dz. U. (Journal of Laws) Item 455, 505. Available online: https://www.gov.pl/web/rolnictwo/akty-prawne3 (accessed on 28 April 2020). (In Polish)
4. Anderson, R.L. Impact of subsurface tillage on weed dynamics in the Central Grein Plains. *Weed Tech.* **2004**, *18*, 186–192. [CrossRef]
5. Chokor, J.U.; Ikuenobe, C.E.; Akaelu, I.A. The Effect of tillage and herbicides (rimsulfuron and codal gold) on weed regeneration. *Inter. J. Soil Sci.* **2008**, *3*, 164–168. [CrossRef]

6. Gruber, S.; Pekrun, C.; Möhring, J.; Claupein, W. Long-term yield and weed response to conservation and stubble tillage in SW Germany. *Soil Till. Res.* **2012**, *121*, 49–56. [CrossRef]
7. Morris, N.L.; Miller, P.C.H.; Orson, J.H.; Froud-Williams, R.J. The adoption of noninversion tillage systems in the United Kingdom and the agronomic impact on soil, crops and the environment—A review. *Soil Till. Res.* **2010**, *108*, 1–15. [CrossRef]
8. Jug, I.; Jug, D.; Sabo, M.; Bojan, S. Winter wheat yield and yield components as affected by soil tillage systems. *Turk. J. Agric. For.* **2011**, *35*, 1–7.
9. Jordan, V.W.L.; Hutcheon, J.A. Influence of Cultivation Practices on Arable Crop Diseases. In *Soil Tillage in Agroecosystems*; El-Titi, A., Ed.; CRC Press: Boca Raton, FL, USA, 2003; pp. 187–207.
10. Poštić's, J.; Cosic, J.; Vrandecic, K.; Jurkovic's, D.; Saleh, A.A.; Leslie, J.F. Diversity of *Fusarium* species isolated from weeds and plant debris in Croatia. *J. Phytopathol.* **2012**, *160*, 76–81. [CrossRef]
11. Perez-Brandan, C.; Arzeno, J.L.; Huidobro, J.; Grümberg, B.; Conforto, C.; Hilton, S.; Bending, G.D.; Meriles, J.M.; Vargas-Gil, S. Long term effect of tillage systems on soil microbiological, chemical and physical parameters and the incidence of charcoal rot by *Macrophomina phaseolina* (Tassi) Goid in soybean. *Crop Prot.* **2012**, *40*, 73–82. [CrossRef]
12. Ohnishi, S.; Miyoshi, T.; Shirai, S. Low temperature stress at different flower developmental stages affects pollen development, pollination, and pod set in soybean. *Environ. Exp. Bot.* **2010**, *69*, 56–62. [CrossRef]
13. Vollmann, J.; Winkler, J.; Fritz, C.N.; Grausgruber, H.; Ruckenbauer, P. Spatial field variations in soybean (*Glycine max* [L.] Merr.) performance trials affect agronomic characters and seed composition. *Eur. J. Agron.* **2000**, *12*, 13–22. [CrossRef]
14. Pittelkow, C.M.; Linquist, B.A.; Lundy, M.E.; Liang, X.; Van Groenigen, K.J.; Lee, J.; Van Gestel, N.; Six, J.; Venterea, R.T.; Van Kessel, C. When does no-till yield more? A global meta-analysis. *Field Crops Res.* **2015**, *183*, 156–168. [CrossRef]
15. Balota, E.L.; Kanashiro, M.; Filho, A.C.; Andrade, D.S.; Dick, R.P. Soil enzyme activities under long-term tillage and crop rotation systems in subtropical agroecosystems. *Braz. J. Microbiol.* **2004**, *35*, 300–306. [CrossRef]
16. Maillard, É.; Angers, D.A.; Chantigny, M.; Lafond, J.; Pageau, D.; Rochette, P.; Lévesque, G.; Leclerc, M.L.; Parent, L.E. Greater accumulation of soil organic carbon after liquid dairy manure application under cereal-forage rotation than cereal monoculture. *Agric. Ecosyst. Environ.* **2016**, *233*, 171–178. [CrossRef]
17. Calzarano, F.; Stagnari, F.; D'Egidio, S.; Pagnani, G.; Galieni, A.; Di Marco, S.; Metruccio, E.G.; Pisante, M. Durum wheat quality, yield and sanitary status under conservation agriculture. *Agriculture* **2018**, *8*, 140. [CrossRef]
18. Sebayang, H.T.; Rifai, A.P. The effect of soil tillage system and weeding time on the growth of weed and yield of soybean (*Glycine max* (L.) Merril). *J. Degrad. Min. Land Manag.* **2018**, *5*, 1237–1243. [CrossRef]
19. Tillmann, M.; Von Tiedemann, A.; Winter, M. Crop rotation effects on incidence and diversity of *Fusarium* species colonizing stem bases and grains of winter wheat. *J. Plant Dis. Protect.* **2017**, *124*, 121–130. [CrossRef]
20. IUSS Working Group WRB. World Reference Base for Soil Resources 2014, update 2015. In *International Soil Classification System for Naming Soils and Creating Legends for Soil Maps*; World Soil Resources Reports; FAO: Rome, Italy, 2015; p. 106.
21. Mirek, Z.; Piękoś-Mirkowa, H.; Zając, A.; Zając, M. Flowering Plants and Pteridophytes of Poland a Checklist. In *Krytyczna Lista Roślin Naczyniowych Polski. W.*; Szafer Institute of Botany, Polish Akademy of Science: Kraków, Poland, 2002; p. 442.
22. Shannon, C.E. A mathematical theory of communications. *Bell Syst. Tech. J.* **1948**, *27*, 379–423. [CrossRef]
23. Simpson, E.H. Measurement of diversity. *Nature* **1949**, *168*, 668. [CrossRef]
24. Łacicowa, B. Metoda laboratoryjna szybkiej oceny odporności jęczmienia na *Helminthosporium sativum*. *Biul. IHAR* **1969**, *3–4*, 61–62. (In Polish)
25. Gomez, K.A.; Gomez, A.A. *Statistical Procedures for Agricultural Research*; John Wiley & Sons: Hoboken, NJ, USA, 1984.
26. Gawęda, D.; Woźniak, A.; Harasim, E. Weed infestations of winter wheat depend on the forecrop and the tillage system. *Acta Agrobot.* **2018**, *71*, 1–9. [CrossRef]
27. Cardina, J.; Herms, C.P.; Doohan, D.J. Crop rotation and tillage system effects on weed seedbanks. *Weed Sci.* **2002**, *50*, 448–460. [CrossRef]

28. Santín-Montanyá, M.I.; Martín-Lammerding, D.; Walter, I.; Zambrana, E.; Tenorio, J.L. Efects of tillage, crop systems and fertilization on weed abundance and diversity in 4-year dry land winter wheat. *Eur. J Agron.* **2013**, *48*, 43–49. [CrossRef]
29. Gawęda, D.; Haliniarz, M.; Cierpiała, R.; Klusek, I. Yield, Weed infestation and seed quality of soybean (*Glycine max* (L.) Merr.) under different tillage systems. *Tarim Bilim. Derg.* **2017**, *23*, 268–275.
30. Sebayang, H.T.; Fatimah, S. The effect of tillage systems and dosages of cow manure on weed and soybeans yield (*Glycine max*, Merrill). *J. Degrad. Min. Land Manag.* **2019**, *7*, 1959–1963. [CrossRef]
31. Woźniak, A.; Rachoń, L. Effect of tillage systems on pea crop infestation with weeds. *Arch. Agron. Soil Sci.* **2019**, *65*, 877–885. [CrossRef]
32. Woźniak, A.; Soroka, M. Effects of a 3-year reduced tillage on the yield and quality of grain and weed infestation of spring triticale (*Triticosecale* Wittmack). *J. Plant Prod.* **2014**, *8*, 231–242.
33. Mahmoodi, S.; Rahimi, A. The critical period of weed control in corn in Birjand region, Iran. *Inter. J. Plant Prod.* **2009**, *3*, 91–96.
34. Mohler, C.L.; Frisch, J.C.; McCulloch, C.E. Vertical movement of weed seed surrogates by tillage implements and natural processes. *Soil Till. Res.* **2006**, *86*, 110–122. [CrossRef]
35. Peigné, J.; Ball, B.C.; Roger-Estrade, J.; David, C. Is conservation tillage suitable for organic farming? A review. *Soil Use Manag.* **2007**, *23*, 129–144. [CrossRef]
36. Ruisi, P.; Frangipane, B.; Amato, G.; Badagliacca, G.; Di Miceli, G.; Plaia, A.; Giambalvo, D. Weed seedbank size and composition in a long-term tillage and crop sequence experiment. *Weed Res.* **2015**, *55*, 320–328. [CrossRef]
37. Małecka, I.; Blecharczyk, A.; Dobrzeniecki, T. Zachwaszczenie zbóż ozimych w zależności od systemu uprawy roli. *Prog. Plant Prot.* **2006**, *46*, 253–255. (In Polish)
38. Chovancova, S.; Illek, F.; Winkler, J. The effect of three tillage treatments on weed infestation in maize monoculture. *Pak. J. Bot.* **2020**, *52*, 697–701. [CrossRef]
39. Małecka-Jankowiak, I.; Blecharczyk, A.; Swędzrzyńska, D.; Sawinska, Z.; Piechota, T. The effect of long-term tillage systems on some soil properties and yield of pea (*Pisum sativum* L.). *Acta Sci. Pol. Agric.* **2016**, *15*, 37–50.
40. Mancinelli, R.; Muleo, R.; Marinari, S.; Radicetti, E. How Soil Ecological Intensification by Means of Cover Crops Affects Nitrogen Use Efficiency in Pepper Cultivation. *Agriculture* **2019**, *9*, 145. [CrossRef]
41. Velykis, A.; Satkus, A. Weed infestation and changes in field pea (*Pisum sativum* L.) yield as affected by reduced tillage of a clay loam soil. *Zemdirbyste* **2010**, *97*, 73–82.
42. Campiglia, E.; Radicetti, E.; Mancinelli, R. Floristic composition and species diversity of weed community after 10 years of different cropping systems and soil tillage in a Mediterranean environment. *Weed Res.* **2018**, *58*, 273–283. [CrossRef]
43. Dainese, M.; Martin, E.A.; Aizen, M.; Albrecht, M.; Bartomeus, I.; Bommarco, R.; Carvalheiro, L.G.; Chaplin-Kramer, R.; Gagic, V.; Garibaldi, L.A.; et al. A global synthesis reveals biodiversity-mediated benefits for crop production. *Sci. Adv.* **2019**, *5*, eaax0121. [CrossRef]
44. García, M.J.; Palma-Bautista, C.; Rojano-Delgado, A.M.; Bracamonte, E.; Portugal, J.; Alcántara-de la Cruz, R.; De Prado, R. The triple amino acid substitution TAP-IVS in the EPSPS gene confers high glyphosate resistance to the superweed *Amaranthus hybridus*. *Int. J. Mol. Sci.* **2019**, *20*, 2396. [CrossRef]
45. Altieri, M.A. The ecological role of biodiversity in agroecosystems. *Agric. Ecosyst. Environ.* **1999**, *74*, 19–31. [CrossRef]
46. Feledyn-Szewczyk, B. Zmiany bioróżnorodności flory segetalnej w systemie ekologicznym w latach 1996–2007. *Res. Appl. Agric. Eng.* **2008**, *53*, 63–68. (In Polish)
47. Scholes, R.J.; Mace, G.M.; Turner, W.; Geller, G.N.; Jürgens, N.; Larigauderie, A.; Muchoney, D.; Walther, B.A.; Mooney, H.A. Toward a global biodiversity observing system. *Science* **2008**, *321*, 1044–1045. [CrossRef] [PubMed]
48. Rands, M.R.; Adams, W.M.; Bennun, L.; Butchart, S.H.; Clements, A.; Coomes, D.; Entwistle, A.; Hodge, I.; Kapos, V.; Scharlemann, J.P.; et al. Biodiversity conservation: Challenges beyond 2010. *Science* **2010**, *329*, 1298–1303. [CrossRef] [PubMed]
49. Clements, D.R.; Weise, R.; Swanton, C.J. Integrated weed management and weed species diversity. *Phytoprotection* **1994**, *75*, 1–18. [CrossRef]

50. Suwara, I.; Stępień, W.; Tymińska, A.; Pruska, K. Wpływ wieloletniego nawożenia mineralnego i zmianowania na zachwaszczenie pszenżyta ozimego. *Fragm. Agron.* **2016**, *33*, 107–116. (In Polish)
51. Sekutowski, T.; Domaradzki, K. Bioróżnorodność gatunkowa chwastów w monokulturze pszenicy ozimej w warunkach stosowania uproszczeń w uprawie roli. *Fragm. Agron.* **2009**, *26*, 160–169. (In Polish)
52. Kim, S.I.; Toon, J.S.; Jung, J.W.; Hong, K.B.; Ahn, Y.J.; Kwon, H.W. Toxicity and repellency of origanum essential oil and its components against *Tribolium castaneum* (*Coleoptera: Tenebrionidae*) adults. *J. Asia Pac. Entomol.* **2010**, *13*, 369–373. [CrossRef]
53. Hepperly, P.R.; Kirkpatrick, B.L.; Sinclair, J.B. *Abutilon theophrasti*: Wild host for three fungal parasites of soybean. *Phytopathology* **1980**, *70*, 307–310. [CrossRef]
54. Srinivas, P.; Ramesh Babu, S.; Ratan, V. Role of sclerotia, plant debris and different hosts on survival of rice sheath blight pathogen, Rhizoctonia solani. *Int. J. Appl Biol. Pharm.* **2014**, *5*, 29–33.
55. Nagaraj, B.T.; Sunkad, G.; Pramesh, D.; Naik, M.K.; Patil, M.B. Host range studies of rice sheath blight fungus *Rhizoctonia solani* (Kuhn). *Int. J. Curr. Microbiol. App. Sci.* **2017**, *6*, 3856–3864. [CrossRef]
56. Chowdhury, A.; Pradhan, S.; Saha, M.; Sanyal, N. Impact of pesticides on soil microbiological parameters and possible bioremediation strategies. *Ind. J. Microbiol.* **2008**, *48*, 114–127. [CrossRef]
57. Feng, P.C.C.; Baley, G.J.; Clinton, W.P.; Bunkers, G.J.; Alibhai, M.F.; Paulitz, T.C.; Kidwell, K.K. Glyphosate inhibits rust diseases in glyphosate-resistant wheat and soybean. *Proc. Natl. Acad. Sci. USA* **2005**, *102*, 17290–17295. [CrossRef] [PubMed]
58. Feng, P.C.C.; Clark, C.; Andrade, G.C.; Balbi, M.C.; Caldwell, P. The control of Asian rust by glyphosate in glyphosate-resistant soybeans. *Pest. Manag. Sci.* **2008**, *64*, 353–359. [CrossRef] [PubMed]
59. Nelson, K.A.; Renner, K.A.; Hammerschmidt, R. Cultivar and herbicide selection affects soybean development and the incidence of Sclerotinia stem rot. *Agron. J.* **2002**, *94*, 1270–1281. [CrossRef]
60. Nelson, K.A.; Renner, K.A.; Hammerschmidt, R. Effects of Protoporphyrinogen oxidase inhibitors on soybean (*Glycine max* L.) response, Sclerotinia sclerotiorum disease development, and phytoalexin production by soybean. *Weed Technol.* **2002**, *16*, 353–359. [CrossRef]
61. Corrêa, M.J.P.; Alves, P.L.; Da, C.A. Effects of herbicides application on photochemical efficiency in conventional and genetically modified soybeans. *Ciênc. Agrotec.* **2010**, *34*, 1136–1145. [CrossRef]
62. Katan, J.; Eshel, Y. Interactions between herbicides and plant pathogens. In *Residue Reviews*; Gunther, F.A., Gunther, J.D., Eds.; Springer: New York, NY, USA, 1973.
63. Araújo, A.S.F.; Monteiro, R.T.R.; Abarkeli, R.B. Effect of glyphosate on the microbial activity of two Brazilian soils. *Chemosphere* **2003**, *52*, 799–804. [CrossRef]
64. Boyette, C.D.; Reddy, K.N.; Hoagland, R.E. Glyphosate and bioherbicide interaction for con-trolling kudzu (*Pueraria lobata*), redvine (*Brunnichia ovata*), and trumpetcreeper (*Campsis radicans*). *Biocontrol. Sci. Tech.* **2006**, *16*, 1067–1077. [CrossRef]
65. Amin, M.; Melkamu, F. Management of Ascochyta Blight (*Ascochyta rabiei*) in chickpea using a new fungicide. *Res. Plant Biol.* **2014**, *2*, 27–32.
66. Bretag, T.W.; MacLeod, W.J.; Kimber, R.B.E.; Moore, K.J.; Knights, E.J.C.; Davidson, J.A. Management of ascochyta blight in chickpeas in Australia. *Australas. Plant Path.* **2008**, *37*, 486–497. [CrossRef]
67. Davidson, J.A.; Kimber, R.B.E. Integrated disease management of ascochyta blight in pulse crops. *Eur. J. Plant Pathol.* **2007**, *119*, 99–110. [CrossRef]
68. Sudheesh, S.; Rodda, M.S.; Davidson, J.; Javid, M.; Stephens, A.; Slater, A.T.; Cogan, N.O.I.; Forster, J.W.; Kaur, S. SNP-based linkage mapping for validation of QTLs for resistance to Ascochyta blight in lentil. *Front. Plant Sci.* **2016**, *7*, 1604. [CrossRef] [PubMed]
69. Małecka, I.; Sawinska, Z.; Blecharczyk, A.; Dytman-Hagedorn, M. Zdrowotność pszenicy ozimej w różnych wariantach uprawy roli. *Prog. Plant Prot. Post. Ochr. Roślin* **2014**, *54*, 246–250. (In Polish)
70. Sułek, A.; Podolska, G.; Jaśkiewicz, B. Plonowanie i zdrowotność dwóch podgatunków pszenicy w zależności od udziału zbóż w strukturze zasiewów w warunkach integrowanej produkcji. *Pol. J. Agron.* **2016**, *27*, 118–125. (In Polish)
71. Strom, N.; Hu, W.; Haarith, D.; Chen, S.; Bushley, K. Interactions between soil properties, fungal communities, the soybean cyst nematode, and crop yield under continuous corn and soybean monoculture. *Appl. Soil Ecol.* **2020**, *147*, 103388. [CrossRef]
72. Kraska, P.; Andruszczak, S.; Kwiecińska-Poppe, E.; Pałys, E. The effect of tillage systems and catch crops on the yield, grain quality and health of spring wheat. *Acta Sci. Pol. Agric.* **2014**, *13*, 21–38.

73. Suproniené, S.; Mankevičiené, A.; Kadžiené, G.; Kačergius, A.; Feiza, V.; Feiziené, D.; Semaškiené, R.; Dabkevičius, Z.; Tamošiūnas, K. The impact of tillage and fertilization on *Fusarium* infection and mycotoxin production in wheat grains. *Zemdirbyste* **2012**, *99*, 265–272.
74. Vrandečić, K.; Jug, D.; Ćosić, J.; Stošić, M.; Poštić, J. The impact of tillage and fertilization on soybean grain infection with fungi. *Rom. Agric. Res.* **2014**, *31*, 139–145.
75. Gawęda, D.; Nowak, A.; Haliniarz, M.; Woźniak, A. Yield and economic effectiveness of soybean grown under different cropping systems. *Int. J. Plant Prod.* **2020**, *53*, 1–11. Available online: https://link.springer.com/article/10.1007/s42106-020-00098-1#article-info (accessed on 29 April 2020). [CrossRef]

Article

Influence of Farming System on Weed Infestation and on Productivity of Narrow-Leaved Lupin (*Lupinus angustifolius* L.)

Agnieszka Faligowska [1,*], Katarzyna Panasiewicz [1], Grażyna Szymańska [1], Karolina Ratajczak [1], Hanna Sulewska [1], Agnieszka Pszczółkowska [2] and Anna Kocira [3]

[1] Department of Agronomy, Faculty of Agronomy and Bioengineering, Poznań University of Life Sciences, Dojazd 11, 60-632 Poznań, Poland; katarzyna.panasiewicz@up.poznan.pl (K.P.); grazyna.szymanska@up.poznan.pl (G.S.); karolina.ratajczak@up.poznan.pl (K.R.); hanna.sulewska@up.poznan.pl (H.S.)

[2] Department of Phytopathology, Entomology and Molecular Diagnostics, University of Warmia and Mazury, 10-719 Olsztyn, Poland; agnieszka.pszczolkowska@uwm.edu.pl

[3] Institute of Agricultural Sciences, State School of Higher Education in Chełm, Pocztowa 54, 22-100 Chełm, Poland; akocira@pwsz.chelm.pl

* Correspondence: agnieszka.faligowska@up.poznan.pl

Received: 3 September 2020; Accepted: 5 October 2020; Published: 8 October 2020

Abstract: Legumes have become important crops, due to an increasing global population and its demand for feed protein. Furthermore, legumes can improve the characteristics of the soil, improve biodiversity levels in crop rotations, and be cultivated in both organic and sustainable farming systems. In this study, a two-factor field experiment was conducted in Gorzyń, Poland in 2011–2015. The first factor was the farming system: low-external inputs (LI; without fertilization and chemical protection), medium-input (MI; medium fertilization level and chemical protection), and high-input (conventional—CONV; high fertilization level and chemical protection). Narrow-leaved lupin cultivar was the second factor; the indeterminate cv. Kalif and the determinate cv. Regent. We evaluated (a) weed infestation levels, (b) seed and protein production, and (c) the economic effects of narrow-leaved lupin cultivation under different farming conditions. A total of 12 weed species were identified, with the lowest weed density level and biomass production observed in CONV, and the greatest weed density level observed in LI. Seed yield was determined by the farming system; the greatest in CONV and significantly lower in LI (by 0.73 t h^{-1}) and MI (by 0.18 t ha^{-1}). Little difference was observed in seed yield between cultivars. The greatest production values for the Kalif and Regent cultivars (996€ and 949€ ha^{-1}, respectively) were recorded in CONV, although LI proved to be the most profitable (with the highest gross agricultural income and lowest total cost of production). LI farming systems, in conjunction with chemical weed control, should be investigated in future studies.

Keywords: lupin; yielding; weed infestation; cultivation intensification; production cost

1. Introduction

With an ever-increasing human population, the agricultural sector is confronted with a number of critical challenges, namely, how to produce sufficient volumes of food for this population, while at the same time preventing the pollution of natural ecosystems [1]. According to Fess et al. [2], high-input farming systems become less sustainable and practical as the global population increases, because of reduced requirement of resources.

Legume crops can be cultivated under both organic and sustainable farming systems [3]. Pulse crops, such as white lupin (*Lupinus albus* L.), yellow lupin (*Lupinus luteus* L.), and narrow-leaved lupin (*Lupinus angustifolius* L.) are native European plants, and could provide an excellent source of plant protein [4,5]. Legumes are also very important in plant production systems, as they fix atmospheric nitrogen (N_2), thereby increasing the soil concentration of a fundamentally important plant nutrient. In European conditions, the most widely cultivated grain legumes (e.g., lupin, faba bean, and field pea) accumulate an average of 130–153 kg N ha^{-1} (from biological fixation) in their aboveground biomass [6]. Indeed, the potential range of N fixation in lupin is even wider, with values of 300 kg N ha^{-1} reported in some studies [7]. Crews and Peoples [8] concluded that the supply of N from legumes (biological) may be a more sustainable source than synthetic (chemical) sources. Moreover, while some countries are very dependent on the latter for food production, many have the capability to substantially reduce synthetic N dependence through the adoption of less meat-intensive diets, and through the reduction of food waste [8]. While many national governments in Europe support the production of legume crops, economic impetus for legume cultivation is lacking, especially in areas where cereal and oilseed crops grow well [9]. However, legume cultivation could provide more plant protein, and also lead to increased biodiversity levels in crop rotations [10]. Modern, high-performance crop varieties are usually bred for high-input farming systems [2], although more environment-friendly cultivation methods have been sought [11,12]. In studies concerning white lupin, the highest income and the lowest cost of production for 1 t of seeds and 1 kg protein were provided by LI, but this type of farming system poses a risk of weed infestation, and reduces seed and protein yields. That is why LI farming system should be supplemented with chemical crop protection against weeds. As such, the low-input technology, along with supplemented chemical weed control, may be considered for higher seed and protein yields of lupin [13].

The objective of this study was to assess the influence of three farming systems (LI, MI, and CONV) on (a) weed infestation, (b) seed and protein yield, and (c) the economic effects of cultivation of two narrow-leaved lupin cultivars.

2. Materials and Methods

2.1. Site Description

The study was part of a long-term field experiment carried out at the Gorzyń Research Station, Poland (52°34′ N, 15°54′ E). The soil type, according to the World Reference Base, is an Albic Luvisol that overlies a gray–brown podzolic. Total N content in the soil was 527 mg kg^{-1} soil, plant available phosphorus (P) was 13.9 mg kg^{-1} soil, and potassium (K) was 10.9 mg kg^{-1} soil. Weather conditions during the study are expressed as the hydrothermal coefficient of water supply according to the Sielianinov index (*K*) (Table 1).

Table 1. Sielianinov index (*K*) during the narrow-leaved lupin growing season (2011–2015), recorded at the Agrometeorological Observatory in Gorzyń, Poland.

Year (Y)	April	May	June	July	August	Sum of Rainfall (mm)
2011	0.25	0.58	0.89	3.18	0.48	287
2012	0.80	0.84	2.26	2.27	1.90	412
2013	0.53	1.63	2.10	0.87	0.49	278
2014	1.99	2.50	0.97	1.18	1.74	390
2015	1.30	0.43	1.02	1.22	0.17	184

K: <0.5 = drought, 0.5–1.0 = semi-drought, 1.0–1.5 = border of optimal moisture, >1.5 = excessive moisture.

The following formula was applied:

$$K = (Mo \times 10)/(D_t \times days)$$

where *K* is the hydrothermal coefficient for an individual month during the growing season, *Mo* is total monthly precipitation, and D_t is the mean daily temperature in a particular month.

During the study, there were considerable differences in the weather conditions during the growing season (Table 1). On average, 2011 and 2015 were least favorable for growth, because of low hydrothermal coefficients (*K*). However, it should be noted that the Sielianinov coefficient was at least 1.0 during the period of greatest demand for water by the plants, i.e., during florescence (June) and pod emergence (July), which means that the narrow-leaved lupin plants had a relatively good supply of water during their most critical development period.

2.2. Experimental Design and Agronomic Management

The field experiment was replicated at the same location every year over a 5-year period (2011–2015). Four replicates of the two-factors were evaluated in a split-plot design. The main plot factor was the farming system: low-external input (LI), medium-input (MI), and high-input (conventional—CONV). The second plot factor was narrow-leaved lupin cultivar (indeterminate cv. Kalif and determinate cv. Regent). A detailed description of the farming systems employed in this study can be found in Table 2. In MI and CONV systems, herbicides were used to target particular plant species: chlorothalonil was used once in 2011, and then three times in each growing season (2012–2015); alfa-cypermethrin was applied once per growing season.

Table 2. Characteristics of the farming systems evaluated in this study.

Agronomic Treatment	Farming System		
	LI	MI	CONV
Seed conditioning	*Bradyrhizobium lupini*	carboxin, thiram (350 mL per 100 kg of seeds) *Bradyrhizobium lupini*	carboxin, thiram (350 mL per 100 kg of seeds) *Bradyrhizobium lupini*
Weed control	mechanical	Mechanical linuron (1.0 l h^{-1}) (direct after sowing)	linuron (1.0 L ha^{-1}) + clomazone (0.2 L ha^{-1}) (direct after sowing) metamitron (1.5 L ha^{-1}) (after emergence)
Soil fertilization (kg ha^{-1})	-	N–15; P–21.8; K–58.1	N–30; P–30.5; K–83
Foliar application of fertilizers	-	-	multiple micro- and macroelements
Disease control	-	chlorothalonil (2.0 l ha^{-1}) (tatrachloroizoftalonitryl) for Anthracnose	chlorothalonil (2.0 L ha^{-1}) (tatrachloroizoftalonitryl) for Anthracnose
Insects control	-	-	alfa-cypermethrin (0.1 L ha^{-1}) (after emergence) for *Sitona* spp.
Desiccation before harvest	-	-	diquat (2.5 L ha^{-1}) (dibromide formula)

Farming system: LI = low external input; MI = medium input; CONV = conventional.

The area of each single experimental plot was 20 m^2. Lupin was cultivated in crop rotation: legumes, winter rape, and cereals. The forecrop was winter wheat (*Triticum aestivum*), which was cultivated in a conventional tillage system, where P and K fertilizers were applied in autumn (80 kg P ha^{-1}, 100 kg K ha^{-1}). Winter wheat N fertilization rates (NH_4NO_3; N 34%) were as follows: 60 kg N ha^{-1} (early spring, before the start of crop growth); 120 kg N ha^{-1} (the second rate during straw shooting phase); and 180 kg N ha^{-1} (the third rate during earing). Each year, the soil was ploughed after harvest of the forecrop (winter wheat) in autumn, and was harrowed in spring before lupin was sown. Recommended sowing rates were as follows: 100 seeds m^{-2} for indeterminate Kalif, and 115 seeds m^{-2} for determinate Regent. Seeds were sown at a depth of 4 cm in rows spaced at 18 cm intervals in early April. In all systems, plots were drilled with a double disk drill (Great Plains, Solid Stand 100 equipped with a fluted coulter for residue cutting, a double disk for seed placement, and a press wheel (3 m wide). Weight of the tractor was 2885 kg). Each year, lupin was harvested in August with a 1.5-m-wide Wintersteiger plot combine harvester.

2.3. Data Collection

During each growing season, weed infestation was assessed on each plot two weeks before harvest. Weeds collected from the site were dried in a laboratory at 80 °C for 48 h. Weed infestation was expressed as the number and dry mass of weeds per unit area (m^2) and as the percentage share of specific species in the total number of weeds in each farming system.

For narrow-leaved lupin, the following traits were assessed: plant density/m^2 before harvest, the biometric traits of 10 randomly selected plants before harvest (the number of pods/plant, number of seeds/plant, and number of seeds/pod), and the mass of 1000 seeds (seeds collected from the harvested seed mass; 2 × 500 seeds were counted and weighed). Seed yield/ha was calculated at the 15% moisture level. Analyses of seed protein content were carried out in the laboratory, according to Kjeldahl (N values were multiplied by 6.25) [14]. Seed protein content was expressed on a dry weight basis (g kg^{-1}), and was recalculated as protein yield (kg ha^{-1}).

Economic analysis of each farming system was evaluated with data from the experimental plots, i.e., machinery operations, inputs, and average yields. Direct costs for all farming systems included seeds, cultivation, seed conditioning, mechanical weed control, and harvest. For the MI and CONV systems, additional direct costs accrued from disease and insect control, and desiccation before harvest (only in CONV). Overall costs included the cost of seeds, fertilizers, chemical crop protection, machinery operations, labor, and services. All calculations were based on 2015 prices. The cost of seeds, fertilizers, and plant protection (herbicides, fungicides, and insecticides) were estimated from agricultural dealers and from national-level market prices [15]. Crop subsidy data were taken from the Agency for Restructuring and Modernization of Agriculture [16]. Subsidies included single area payment (107.59€), additional payment for legumes (241.18€), and direct payments for seeds (30.83€). All prices and costs were recalculated in euro (€) according to the monthly average exchange rate in September 2015 [17].

2.4. Statistical Analysis

The impact of farming system on lupin cultivar traits was examined with two-way analysis of variance (ANOVA) using SAS PROC GLM (SAS Institute Inc. 1996). Least significant difference was verified with Tukey's multiple range test at $p < 0.01$ and $p < 0.05$ significance levels. Relationships between parameters were determined with the Pearson correlation coefficient. Interpretation of Pearson's linear correlation coefficient was conducted according to Stanisz [18]. Correlations between traits in each farming system were determined using SAS PROC CORR.

3. Results

Weed community composition was affected by the farming system (Table 3). A total of 12 weed species were identified: 11 in LI, 8 in MI, and 8 in CONV. POLCO (*Polygonum convolvules* (L.) A. Löve) and CHEAL (*Chenopodium album* L.) were the most frequently observed species across all farming systems, at 43–49% and 27–32%, respectively. Less frequently observed species were LYCAR (*Anchusa arvensis* L.), CAPBP (*Capsella bursa-pastoris* (L.) Medik.), CONAR (*Convolvulus arvensis* L.), GALAP (*Galium aparine* L.), and LAMAM (*Lamium amplexicanle* L.), and accounted for <1% in all farming systems. Lowest weed density and biomass values were recorded in CONV, and the greatest values were recorded in LI (Table 4). Weed dry weight was significantly lower in MI and CONV (approximately 18% and 33%, respectively), compared to LI. The number of weeds/m^2 was 54% greater in LI compared to CONV, but there was no difference between MI and CONV. Cultivar factor did not influence the number of weeds, although weed dry mass was significantly lower (by 18%) under Regent cultivation than Kalif.

Table 3. Effect of farming system on botanical composition of weeds in the study plots (share, %).

Latin Binomial	Bayer Code	Farming System		
		LI	MI	CONV
Agropyron repens (L.) Beauv.	AGRRE	4	2	0
Anchusa arvensis L.	LYCAR	1	0	1
Capsella bursa-pastoris (L.) Medik.	CAPBP	1	0	0
Chenopodium album L.	CHEAL	27	31	32
Convolvulus arvensis L.	CONAR	0	0	1
Echinochola crus-galli (L.) Beauv.	ECHCG	4	8	11
Galinsoga parviflora Cav.	GASPA	2	3	0
Galium aparine L.	GALAP	1	1	0
Lamium amplexicanle L.	LAMAM	1	0	1
Polygonum convolvules (L.) A. Löve	POLCO	49	46	43
Polygonum lapathifolium L.	POLLA	5	1	6
Viola arvensis Murr.	VIOAR	5	8	5
LSD Value		4.72 **		
Number of species	No.	11	8	8

Significance: ** $p < 0.01$. Farming system: LI = low external input; MI = medium input; CONV = conventional.

Table 4. Effect of farming system and cultivar type on level of weed infestation, yield, and seed protein content.

Specification	Farming System			LSD Value
	LI	MI	CONV	
Dry weight of weeds (g)	243.9	199.0	162.7	24.02 **
Number of weeds	97.7	80.8	63.5	17.83 **
Plant density (no. m^2)	72.9	77.7	75.2	NS
Number of pods per plant	6.0	7.3	7.7	1.09 **
Number of seeds per plant	23.2	25.4	27.1	NS
Number of seeds per plant pod	3.9	3.5	3.5	0.24 **
Mass of 1000 seeds (g)	137.1	136.6	137.0	NS
Protein content in seeds (g kg^{-1})	307	304	317	NS
	Cultivar			
	Kalif		**Regent**	
Dry weight of weeds (g)	221.5		182.2	37.82 *
Number of weeds	85.3		76.9	NS
Plant density (no. m^2)	68.9		81.6	3.38 **
Number of pods per plant	7.6		6.4	0.71 **
Number of seeds per plant	27.7		22.8	3.08 **
Number of seeds per plant pod	3.6		3.6	NS
Mass of 1000 seeds (g)	141.5		132.2	3.58 **
Protein content in seeds (g kg^{-1})	313		306	NS

NS: non-significant; * $p < 0.05$ and ** $p < 0.01$. System: LI = low external input; MI = medium input; CONV = conventional.

Farming system intensity substantially affected the numbers of pods and seeds/plant pod (Table 4). However, this factor did not have an impact on plant density, seeds/plant, or the 1000 seed mass. A significantly greater number of pods/plant were observed in the MI and CONV systems, although the number of seeds/plant pod were significantly lower than in LI. Cultivar type was found to modify yield components; a significantly greater plant density was observed in the Regent cultivar, and a significantly lower number of pods and seeds/plant and 1000 seed mass compared to Kalif. There was no significant difference in seed protein content between experimental factors.

No significant interactions were observed between the experimental factors and protein and seed yield (Table 5). The greatest protein yield was observed in CONV, and was significantly lower in LI and MI (by 31.3% and 11%, respectively). Although no significant differences were observed in seed yield between cultivars, the protein yield in Regent was substantially lower (by 54 kg ha^{-1}). Seed yield was strongly affected by farming system, and the greatest yield was recorded in CONV, and was significantly lower in LI (by 0.73 t ha^{-1}) and MI (by 0.18 t ha^{-1}).

Table 5. Narrow-leaved lupin protein yield (kg ha^{-1}) and seed yield (t ha^{-1}).

Cultivar (C)	Farming System (FS)			Mean
	LI	MI	CONV	
		Protein Yield		
Kalif	488	641	714	614
Regent	452	575	654	560
Mean	470	608	684	
LSD value	FS—33.3 **; C—46.4 *; FS × C—NS			
		Seed Yield		
Kalif	1.85	2.45	2.26	2.31
Regent	1.75	2.25	2.44	2.15
Mean	1.80	2.35	2.53	
LSD value	FS—0.13 **; C—NS; FS × C—NS			

NS: non-significant; * $p < 0.05$ and ** $p < 0.01$. System: LI = low external input; MI = medium input; CONV = conventional.

In each farming system, functional relationships were observed between the number of pods/plant and the number of seeds/plant, and seed yield and protein yield (Figure 1). However, correlations between the number of seeds/plant pod and seed yield and the protein yield in LI were also practically functional relationships. Negative relationships were observed between the number of weeds and seed yield and protein yield; strong in LI, poor in MI, and practically functional in CONV.

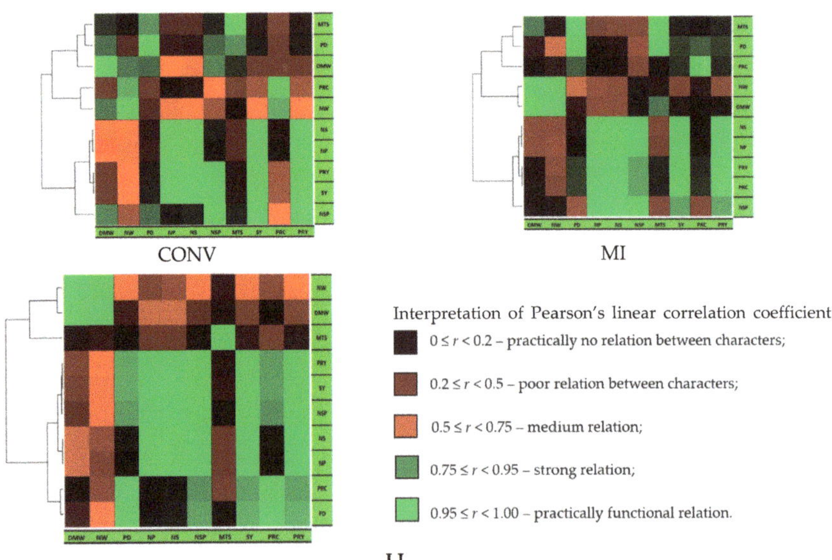

Figure 1. Correlation coefficients between analyzed parameters. System (FS): LI = low input; MI = medium input; CONV = conventional. Parameters compared: DMW = dry mass of weeds; NW = number of weeds; PD = plant density; NP = number of pods/plant; NS = number of seeds/plant; NSP = number of seeds/plant pod; MTS = mass of 1000 seeds; SY = seed yield; PRC = protein content; PRY = protein yield.

In both cultivars, the highest production values were recorded in CONV, which also had the greatest cost of seeds and protein production, and the lowest gross agricultural income (Table 6). The LI system proved to be the most profitable (i.e., gross agricultural income) for both Kalif and Regent cultivation, and the total cost of production was also lowest.

Table 6. Economic analysis of the profitability of narrow-leaved lupin cultivation under different farming systems (all values expressed in euro).

Specification	Kalif			Regent		
	Farming System					
	LI	MI	CONV	LI	MI	CONV
Production value * ha^{-1}	830.17	972.45	996.16	806.45	925.00	948.74
Total cost ha^{-1}	353.87	546.40	886.16	387.98	580.50	920.28
Gross agricultural income ha^{-1}	476.30	426.06	110.00	418.47	344.50	28.46
Cost of 1 t seed production **	186.25	218.56	340.83	215.54	252.39	383.45
Cost of 1 kg protein production ***	0.73	0.85	1.24	0.86	1.01	1.41

System: LI = low external input; MI = medium input; CONV = conventional. * (average yield ha^{-1} × € 237. 14, price of 1 t seeds) + € 379.60 EU subsidies, ** Total cost/ average yield ha^{-1}, *** Total cost/ average protein yield ha^{-1}.

4. Discussion

Our study suggests that weed community assembly was affected by the farming system. A total of 12 weed species were identified, and the lowest weed density/m^2 and biomass/m^2 values were observed with the CONV system, and the greatest were observed in LI. The number of weeds/m^2 was 54% greater in LI compared to CONV, and there was no difference between MI and CONV. The greater weed infestation is due to the absence of herbicide use, and the mechanical treatment employed in the LI system was insufficient to improve lupin competitiveness. Our results are in agreement with Poudel et al. [19], who also found that low-input systems generally have greater aboveground weed biomass and more weed competition in comparison to conventional systems. Others researchers who have analyzed the influence of farming systems on the cultivation of lupin species (yellow, narrow-leaved, and white lupin) have reported significantly greater weed infestation levels in plots that received only mechanical treatment [20]. In the study by Borowska et al. [20], the application of herbicides in medium-input and high-input systems significantly reduced both the level of weed infestation and the weed dry weight of all lupin species. However, additional herbicide application after sowing and post-emergence in the high-input system did not significantly reduce weed dry weight. Chemical weed control should carefully consider the sensitivity of the crop to the herbicide, followed by observation of the application technology, as well as important aspects, such as compound mixture, ambient environmental conditions, and the use of adjuvants [21].

Weed infestation intensity in individual farming systems can be modified not only by the type of treatment applied, but also by the number of lupin plants per area unit. In our study, significantly lower weed dry mass was observed with Regent cultivation, compared to Kalif. We suggest that the determinate cultivar Regent is more competitive than the indeterminate cultivar Kalif, because it exhibited a significantly greater density of plants per area unit, due to higher recommended sowing rates. Farming system did not influence plant density in our study, in contrast to Suliman [22], who showed that conventional tillage and mechanical weed control significantly increased legume plant density.

Our results show that the weed infestation can influence narrow-leaved lupin productivity. Pearson's linear correlation indicated a negative relationship between number of weeds and seed and protein yields, which was strong in LI, poor in MI, and practically functional in CONV. This means that when the number of weeds increase, protein and seed yields decrease, depending on the farming system employed. Weeds in lupin crops have become increasingly difficult to control [23]. In our study, the greatest narrow-leaved lupin protein and seed yields were found in CONV (compared to

LI and MI), although the yield in this system primarily depended on the number of weeds present. Borowska et al. [20] found that seed yield in white and narrow-leaved lupin was significantly greater in the high-input system, while yellow lupin yield was greatest in the high- and medium-input systems. A significant increase in seed yield, along with an increasing intensity of the cultivation in the case of traditional and self-completing cultivars of white and yellow lupin, as well as self-completing cultivars of narrow-leaved lupin, resulted mainly from the development of a higher number of pods. Szymańska et al. [24] also studied the influence of farming system on the development and yield of yellow lupin, and observed that seed yield was 13.1% and 22.0% greater in the MI and CONV systems, respectively, than in LI. Their research also highlighted differences in cultivar yield; the indeterminate cultivar Mister produced more seeds than the determinate cultivar Perkoz (1.95 t ha^{-1} vs. 1.81 t ha^{-1}). In our experiment, there was no significant difference in seed yield between lupin cultivars. We observed significantly lower pod numbers, seeds/plant, and the 1000 seed mass in Regent, compared to Kalif. Determinate lupin cultivars have been reported to exhibit a reduced number of branches to improve the maturation of the plants.

In our study, farming system was also found to modify the yield components. We found a significantly greater number of pods/plant in MI and CONV (compared to LI), but substantially lower numbers of seeds/plant pod. Borowska et al. [20] observed a non-significant difference in the number of pods/plant between medium- and high-input systems, as well as between low- and medium-input systems for all lupin species in their study. Szymańska [24], studying yellow lupin yield components, found that differences in farming system intensity and cultivar type significantly influenced the number of pods and seeds/plant. However, these factors had no influence on the number of seeds/pod or the 1000 seed weight.

Lupin is valued mainly for its high protein content. In our experiment, protein content in seeds ranged from 304 to 317 g kg^{-1}. Protein yield in Regent was lower (by 54 kg ha^{-1}) compared to Kalif, and the greatest protein yield was found in CONV and was significantly lower in LI and MI. A practically functional relationship was observed between protein yield and seed yield in each farming system. This is because of method of calculation of protein yield. In our experiment, the cost of protein and seed production was greatest in CONV, but gross agricultural income was also lowest in this system. According to Czerwińska-Kayzer and Florek [25], gross agricultural income is a basic economic category that is indicative of the profitability of agricultural production. In our study, the LI system (greatest gross agricultural income and lowest total cost of production) proved to be the most profitable for both cultivars. According to Szymańska et al. [24], increased expenditure in CONV farming systems can lead to a decrease in gross agricultural income; in our experiment, the cost of production of 1 tonne of seeds (for on-farm feed production) was lowest in the LI system. Although production was greatest in CONV, the greater cultivation intensity (in yellow lupin) observed in Szymańska et al. [24] was not economically justified (compared to cultivation of narrow-leaved lupin in our experiment). Moreover, work by Panasiewicz et al. [26] showed that the increase in narrow-leaved lupin seed yield produced by a more intensive tillage option (conventional tillage) compensated for the increase in overall costs.

Aside from economic performance, high-input farming systems may lead to increased environmental problems, as a consequence of the fertilizers and pesticides applied during the course of cropping operations. As legumes support biological N_2 fixation, they offer a more environmentally robust and sustainable N source for cropping systems [8]. At zero fertilization levels, weeds (due to extremely strong competition for soil N) might force lupin plants to depend more on N fixation, thereby making the N fixation process more efficient.

5. Conclusions

Our economic analysis indicates that the greatest income and the lowest cost of production for both narrow-leaved lupin cultivars were provided by the LI farming system, although this type of farming system does carry a risk of elevated weed infestation levels and reduced seed and protein yields. Therefore, LI farming systems, used in conjunction with chemical weed control, should be investigated in future studies.

Author Contributions: A.F. and K.P. conceived and designed the experiments; A.F., K.P., G.S., K.R., and H.S. performed the field experiments and analyzed the data; A.F. and K.R. performed statistical analysis; K.P. and G.S. described the materials and methods; H.S., A.P., and A.K. prepared the references; A.F. wrote the manuscript; A.F., K.P., G.S., K.R., H.S., A.P., and A.K. revised the manuscript. All authors have read and agreed to the published version of the manuscript.

Funding: This research was funded by Polish Ministry of Agriculture and Rural Development (Project number HOR 3.3/2011–2015).

Acknowledgments: This study was funded by a grant from the Polish Ministry of Agriculture and Rural Development, Project: *Improving domestic sources of plant protein, their production, trading and use in animal feed*, project No. HOR 3.3/2011–2015.

Conflicts of Interest: The authors declare no conflict of interest.

References

1. Bhardwaj, H.L.; Hamama, A.A.; Van Santen, E. White Lupin Performance and Nutritional Value as Affected by Planting Date and Row Spacing. *Agron. J.* **2004**, *96*, 580–583. [CrossRef]
2. Fess, T.; Kotcon, J.B.; Benedito, V.A. Crop Breeding for Low Input Agriculture: A Sustainable Response to Feed a Growing World Population. *Sustainability* **2011**, *3*, 1742. [CrossRef]
3. Księżak, J.; Staniak, M.; Bojarszczuk, J. The regional differentiation of legumes cropping area in Poland between 2001 and 2007. *Pol. J. Agron.* **2009**, *1*, 25–31.
4. Sujak, A.; Kotlarz, A.; Strobel, W. Compositional and nutritional evaluation of several lupin seeds. *Food Chem.* **2006**, *98*, 711–719. [CrossRef]
5. Annicchiarico, P.; Harzic, N.; Carroni, A.M. Adaptation, diversity, and exploitation of global white lupin (*Lupinus albus* L.) landrace genetic resources. *Field Crop. Res.* **2010**, *119*, 114–124. [CrossRef]
6. Peoples, M.B.; Brockwell, J.; Herridge, D.F.; Rochester, I.J.; Alves, B.J.R.; Urquiaga, S.; Boddey, R.M.; Dakora, F.D.; Bhattarai, S.; Maskey, S.L.; et al. The contributions of nitrogen-fixing crop legumes to the productivity of agricultural systems. *Symbiosis* **2009**, *48*, 1–17. [CrossRef]
7. Sulas, L.; Canu, S.; Ledda, L.; Carroni, A.M.; Salis, M. Yield and nitrogen fixation potential from white lupine grown in rainfed Mediterranean environments. *Sci. Agricola* **2016**, *73*, 338–346. [CrossRef]
8. Crews, T.E.; Peoples, M. Legume versus fertilizer sources of nitrogen: Ecological tradeoffs and human needs. *Agric. Ecosyst. Environ.* **2004**, *102*, 279–297. [CrossRef]
9. Murphy-Bokern, D.; Watson, C. Legume-supported cropping systems for Europe. In Looking Forward 2016. Available online: http://www.legumefutures.de/images/Legume_Futures_Looking_Forward.pdf (accessed on 18 August 2020).
10. Preissel, S.; Reckling, M.; Schläfke, N.; Zander, P. Magnitude and farm-economic value of grain legume pre-crop benefits in Europe: A review. *Field Crop. Res.* **2015**, *175*, 64–79. [CrossRef]
11. Pelzer, E.; Bazot, M.; Makowski, D.; Corre-Hellou, G.; Naudin, C.; Al Rifaï, M.; Baranger, E.; Bedoussac, L.; Biarnès, V.; Boucheny, P.; et al. Pea–wheat intercrops in low-input conditions combine high economic performances and low environmental impacts. *Eur. J. Agron.* **2012**, *40*, 39–53. [CrossRef]
12. Jaśkiewicz, B. The impact of production technology on yields of spring triticale under varied percentages of cereals to total cropped area. *Fragm. Agron.* **2017**, *34*, 7–17. (In Polish)
13. Faligowska, A.; Panasiewicz, K.; Szymańska, G.; Szukala, J.; Koziara, W.; Pszczółkowska, A. Productivity of white lupin (*Lupinus albus* L.) as an effect of diversified farming systems. *Legum. Res. Int. J.* **2017**, *40*, 872–877. [CrossRef]
14. AOAC. *Official Methods of Analysis*; Association of Official Analytical Chemists: Washington, DC, USA, 1975.
15. Ceny Rolnicze. Available online: http://www.cenyrolnicze.pl (accessed on 20 August 2020).
16. Agency for Restructuring and Modernization of Agriculture. Available online: https://www.arimr.gov.pl/ (accessed on 20 August 2020).
17. Central Bank of the Republic of Poland. Available online: http://www.nbp.pl/homen.aspx?f=/kursy/kursyen.htm (accessed on 20 August 2020).
18. Stanisz, A. *Approachable Statistics Course*; StatSoft: Krakow, Poland, 1998; p. 362. (In Polish)

19. Poudel, D.; Horwath, W.; Lanini, W.; Temple, S.; Van Bruggen, A. Comparison of soil N availability and leaching potential, crop yields and weeds in organic, low-input and conventional farming systems in northern California. *Agric. Ecosyst. Environ.* **2002**, *90*, 125–137. [CrossRef]
20. Borowska, M.; Prusinski, J.; Kaszkowiak, E. Production results of intensification of cultivation technologies in three lupin (*Lupinus* L.) species. *Plant Soil Environ.* **2016**, *61*, 426–431. [CrossRef]
21. Dos Reis, A.R.; Vivi, R. Weed Competition in the Soybean Crop Management in Brazil. In *Soybean—Applications and Technology*; IntechOpen: London, UK, 2011; pp. 185–210. Available online: https://www.intechopen.com/books/soybean-applications-and-technology/weed-competition-in-the-soybean-crop-management-in-brazil (accessed on 20 August 2020).
22. Suliman, A.A.M. Contribution of weed control and tillage systems on soil moisture content, growth and forage quality of (Clitoria&Siratro) mixture under-rainfed conditions at Zalingei—Western Darfur state—Sudan. *J. Sci. Technol.* **2008**, *3*, 80–90.
23. Anonymous. *The Biology of Lupinus L. (Lupin or Lupine)*; Australian Government; Version 1: April 2013; Department of Health and Ageing, Office of the Gene Technology Regulator: Canberra, Australia, 2013; p. 64.
24. Szymańska, G.; Faligowska, A.; Panasiewicz, K.; Szukała, J.; Koziara, W. The productivity of two yellow lupine (*Lupinus luteus* L.) cultivars as an effect of different farming systems. *Plant Soil Environ.* **2017**, *63*, 552–557. [CrossRef]
25. Czerwińska-Kayzer, D.; Florek, J. Profitability of selected legumes. Fragm. *Agronomy* **2012**, *29*, 36–44. (In Polish)
26. Panasiewicz, K.; Faligowska, A.; Szymańska, G.; Szukała, J.; Ratajczak, K.; Sulewska, H. The Effect of Various Tillage Systems on Productivity of Narrow-Leaved Lupin-Winter Wheat-Winter Triticale-Winter Barley Rotation. *Agronomy* **2020**, *10*, 304. [CrossRef]

Article

Weed Flora and Soil Seed Bank Composition as Affected by Tillage System in Three-Year Crop Rotation

Beata Feledyn-Szewczyk [1,*], Janusz Smagacz [1], Cezary A. Kwiatkowski [2], Elżbieta Harasim [2] and Andrzej Woźniak [2]

[1] Department of Systems and Economics of Crop Production, Institute of Soil Science and Plant Cultivation—State Research Institute, Czartoryskich 8 Str., 24-100 Puławy, Poland; smagacz@iung.pulawy.pl

[2] Department of Herbology and Plant Cultivation Techniques, University of Life Sciences, Akademicka 13 Str., 20-950 Lublin, Poland; czarkw@poczta.onet.pl (C.A.K.); elzbieta.harasim@up.lublin.pl (E.H.); andrzej.wozniak@up.lublin.pl (A.W.)

* Correspondence: bszewczyk@iung.pulawy.pl; Tel.: +48-081-478-6803

Received: 16 April 2020; Accepted: 20 May 2020; Published: 24 May 2020

Abstract: In recent years, there has been an increasing interest around agricultural science and practice in conservation tillage systems that are compatible with sustainable agriculture. The aim of this study was to assess the qualitative and quantitative changes in weed flora and soil seed bank under reduced tillage and no-till (direct sowing) in comparison with traditional ploughing. In the crop rotation: pea/rape—winter wheat—winter wheat the number and dry weight of weeds increased with the simplification of tillage. The seed bank was the largest under direct sowing and about three times smaller in traditional ploughing. Under direct sowing, most weed seeds were accumulated in the top soil layer 0–5 cm, while in the ploughing system most weed seeds occurred in deeper layers: 5–10 and 10–20 cm. In the reduced and no-till systems, a greater percentage of perennial and invasive species, such as *Conyza canadensis* L., was observed. The results show that it is possible to maintain weed infestation in the no-till system at a level that does not significantly affect winter wheat yield and does not pose a threat of perennial and invasive weeds when effective herbicide protection is applied.

Keywords: reduced tillage; no-till; ploughing; winter wheat; weeds; seed bank; invasive weed species

1. Introduction

Weed infestation in a field consists of the above-ground weed community and the seed stock in and on the soil, called the soil seed bank [1]. The soil seed bank is a reservoir in which the depositing of new seeds takes place systematically and the stock of those seeds changes due to various processes. The germination dynamics of weed seeds and their further development and competitiveness against crops depend on complex habitat, environmental and agrotechnical factors [2–4].

One of the most important agrotechnical factors influencing the weed flora of arable fields is the tillage method [5,6]. Contemporary tillage methods can be divided into three basic groups: (1) traditional tillage with the use of a plough, (2) reduced tillage, intended to eliminate the plough and replacing it with other passive or active tools whose primary task is to loosen the topsoil arable layer without turning it over, and (3) no-till system with direct sowing, including precise placement of the seeds in untilled soil with a specialist seed drill [7,8]. Sustainable agriculture promotes a reduction in the intensity of soil tillage treatments and seeks to replace plough cultivation with conservation tillage that does not turn the soil, or even with direct sowing. These treatments are reflected in scientific research although results on the impact of tillage simplifications on the diversity of the segetal flora and on the soil seed bank are inconclusive [9–13].

The stimulating effect of tillage on weed seed germination has been found in previous studies [14,15]. The exception is tillage at night, as daylight is a factor limiting the germination of the seeds of many weed species [16]. However, soil tillage leads to a decrease in the pool of seeds in the soil bank, which is indicated by their smaller amount in tilled soils as compared to untilled ones [17]. Soil tillage affects the dynamics of weed germination by modifying the seeds depth in the soil [16,18]. The literature shows that different tillage methods influence the accumulation and distribution of seeds in the soil profile. In traditional ploughed tillage systems, the seeds are distributed more or less evenly throughout the entire tillage layer, while in reduced systems a large portion of the seeds are concentrated in the surface layer of soil where they find more favourable conditions for germination [16,19,20]. This results in an increase in weed abundance under reduced tillage compared to ploughing, although the results also depend on the crop and weed species involved [15,21,22]. Moreover, the no-till system increases the number of weeds maturing in the stubble that remains after harvest, the seeds of which fall down, germinate and increase infestation [11]. According to other authors, the increase in weed infestation of the crop and soil occurs in the first few years after the introduction of tillage simplifications [23].

The simplification of soil tillage, including zero tillage, entails not only quantitative, but also qualitative changes in the species composition of weed flora and soil seed bank [4,23–26]. Species diversity is most often declining, although some authors have observed an increase in the biodiversity of weed communities under reduced tillage systems [19]. Some authors draw attention to the threats related to the occurrence of troublesome perennial species, mainly *Elymus repens*, and of invasive species under reduced tillage systems [19,20,27].

Two main methods are used to determine the abundance and species composition of seeds in the soil: the direct seed extraction method and the indirect germination method, called "greenhouse method", or "seedling emergence method" [14,28,29]. The indirect method enables an estimation of the number of live seeds able to germinate under favorable conditions. Gross [30] and Cardina and Sparrow [31] evaluated the effectiveness of both methods for an assessment of the soil seed bank. The research carried out by Gross [30] shows that the number of seeds was much higher in samples determined by the direct method and that this was mainly due to the presence of dead seeds. However, if dead and resting seeds are deducted then the results obtained by the direct method were similar to those obtained by the germination method. According to Cardina and Sparrow [31], the germination method is best suited to predict weed emergence because it is most similar to field conditions.

The aim of the undertaken research was to determine qualitative and quantitative changes in weed flora composition and soil seed bank under a reduced tillage system and no-till cultivation and to investigate whether tillage simplifications lead to an increased threat of invasive species.

The research hypothesis was that simplifications in tillage cause an increase in weed infestation of the crop and the soil as well as leading to qualitative changes in the weed community, such as an increase in the number of invasive species.

2. Materials and Methods

2.1. Characteristics of the Experiment with Tillage Systems

The research was conducted in the years 2010–2013 on 9 fields in total with three of these fields being used for each of the tillage systems: direct sowing, reduced tillage and ploughing, in a farm in Rogów, Lubelskie voivodeship, Poland (N: 51°28′, E: 22°4′) (Figures 1 and 2). The experimental site is located in a moderate continental climatic zone. The long-term average annual total precipitation is 598 mm, with a mean air temperature of 8.5 °C (Table 1).

Growing seasons	Direct sowing			Reduced cultivation			Plough tillage		
	Field 1	Field 2	Field 3	Field 4	Field 5	Field 6	Field 7	Field 8	Field 9
2009/2010	**Pea*** (2010)	**Winter wheat I**	**Winter wheat II**	**Pea*** (2010)	**Winter wheat I**	**Winter wheat II**	**Pea*** (2010)	**Winter wheat I**	**Winter wheat II**
2010/2011	Winter wheat I	Winter wheat II	Pea (2011)	Winter wheat I	Winter wheat II	Pea (2011)	Winter wheat I	Winter wheat II	Pea (2011)
2011/2012	Winter wheat II (frozen, replaced by spring wheat)	Winter rape	Winter wheat I (frozen, replaced by spring wheat)	Winter wheat II (frozen, replaced by spring wheat)	Winter rape	Winter wheat I (frozen, replaced by spring wheat)	Winter wheat II (frozen, replaced by spring wheat)	Winter rape	Winter wheat I (frozen, replaced by spring wheat)
2012/2013	Winter rape	Winter wheat I	Winter wheat II	Winter rape	Winter wheat I	Winter wheat II	Winter rape	Winter wheat I	Winter wheat II

Figure 1. Scheme of the experimental treatment with three tillage systems. * Crops cultivated in the 2009/2010 season are marked in bold.

Figure 2. Winter wheat (I) fields under different tillage systems (fields No. 1, 4 and 7 according to Figure 1, date of sowing: 30 September 2010, date of observation: 16 November 2010).

Table 1. Mean air temperature (°C) and sum of precipitation (mm) in experimental farm in Rogów.

Month	Temperature (°C)				Precipitation (mm)			
	2010/2011	2011/2012	2012/2013	Mean from the Long-Term Period	2010/2011	2011/2012	2012/2013	Mean from the Long-Term Period
IX	12.0	14.9	14.7	13.2	128.7	3.6	29.2	62.4
X	5.5	7.3	8.2	8.6	13.5	28.2	54.8	29.1
XI	6.5	1.7	4.9	3.3	65.6	2.3	22.2	21.8
XII	−4.6	1.8	−4.4	0.6	32.2	50.8	3.3	23.1
I	−1.5	−2.7	−3.9	−1.6	27.1	25.3	66.7	40.2
II	−4.7	−8.3	−0.5	−3.1	20.2	21.3	9.0	20.8
III	1.7	3.9	−2.6	3.1	13.0	29.6	32.6	35.4
IV	9.7	9.3	8.5	9.2	30.2	51.8	54.3	35.3
V	13.4	14.5	15.2	13.9	26.1	66.0	118.8	88.5
VI	17.7	17.5	18.4	17.3	83.7	110.9	106.7	82.4
VII	18.7	20.8	18.6	19.1	214.0	34.3	62.5	93.8
VIII	18.5	18.7	18.7	18.8	46.0	71.4	33.1	65.0
	Mean temperature—8.5 °C				Sum of precipitation—598 mm			

The fields used in the current experiment were set up in autumn 2002 on loess soil, with the granulometric composition of silt loam, characterized by a pH close to neutral (pH KCl = 6.6), a high phosphorus content (158.6 mg P kg^{-1} soil), and a medium potassium level (204.4 mg K kg^{-1} soil). The average humus content in the soil was 1.63%, and the organic carbon content was 0.94%.

In all of the tillage systems the same three-field crop rotation was applied: pea (2010–2011)/winter rape (2011/2012 and 2012/2013)—winter wheat I—winter wheat II (after winter wheat I). All crops in rotation were cultivated in each year. In January 2012, winter wheat froze out and was replaced by spring wheat. Each field was 1 ha in size and the whole experiment had a total of 9 ha (Figure 1).

2.2. Methods of Soil Treatment in Different Tillage Systems

The selection of cultivating machines and tools was varied and depended on the tillage system. In the plough cultivation, immediately after harvesting the forecrop, a shallow cultivation (stubble) was carried out with a disc harrow to a depth of 5–6 cm in order to stop the evaporation of water from the soil and stimulate the germination of weed seeds and volunteers. Then, before sowing winter plants, plowing was carried out using a reversible plough to a depth of 20 cm. Ploughing for peas was completed in late autumn to a depth of 25–30 cm. Sowing of wheat and rapeseed was carried out in autumn on optimal dates for this region of the country, while peas were sown in spring using a seedbed cultivator (disc harrow + cultivating roller + seeder + seed harrow).

In the reduced tillage system, after plant harvest, post-cultivation was carried out with a rotary cultivator to a depth of 5 cm. This treatment was then repeated after about 3–4 weeks to limit the emergence of weeds and volunteers. Winter crops were sown at optimal dates, while peas were sown in spring using a seedbed cultivator.

In the no-till system, the preparation of the field for sowing was reduced to the use of a herbicide based on glyphosate in order to limit weed infestation, followed by direct sowing of wheat and rapeseed in autumn, and peas in spring with a special seeder for direct sowing.

2.3. Fertilization

The fertilization of crops was not differentiated depending on the tillage system. Immediately after harvesting the forecrop plant and before the post-harvest cultivation, phosphorus, potassium, nitrogen and sulfur fertilization was applied. Top dressing with nitrogen was completed in spring, taking into account the crop requirements, as shown in Table 2.

Table 2. Mineral fertilization (kg ha^{-1}) for wheat, peas and rapeseed.

Growing Seasons	Wheat I and II	Growing Seasons	Pea/Rape
2009/2010 2010/2011	pre-sowing: phosphorus (P_2O_5)—72, potassium (K_2O)-72, nitrogen (NH_4)-24, S (SO_3)-27; top dressing with nitrogen (NH_4NO_3)-160, including: start of vegetation in spring-70, shooting at the blade-50, heading-40.	2010	Pea: pre-sowing: phosphorus (P_2O_5)—72; potassium (K_2O)—72; nitrogen (NH_4)—24; S (SO_3)—27; in spring immediately before sowing pea—N (NH_4NO_3)—20.
		2011	Pea: pre-sowing: phosphorus (P_2O_5)—72; potassium (K_2O)—72; nitrogen (NH_4)—24; S (SO_3)—27; in spring immediately before sowing pea—N (NH_4NO_3)—20.
2011/2012	winter wheat froze out and was replaced by spring wheat. pre-sowing: phosphorus (P_2O_5)—60, potassium (K_2O)—60, nitrogen (NH_4)—20, S (SO_3)—23.5; top dressing with nitrogen (NH_4NO_3)—160, including: starting vegetation in spring—80, shooting at the blade—50, heading—40.	2011/2012	Winter rape: pre-sowing: phosphorus (P_2O_5)—60; potassium (K_2O)—60; nitrogen (NH_4)—20; S (SO_3)—23.5; top dressing with nitrogen (NH_4NO_3)—160, including: starting vegetation in spring—100, budding phase—60.
2012/2013	pre-sowing: phosphorus (P_2O_5)—60, potassium (K_2O)—60, nitrogen (NH_4)—20, S (SO_3)—23.5; top dressing with nitrogen (NH_4NO_3)—140, including: starting vegetation in spring—80, shooting at the blade—60.	2012/2013	Winter rape: pre-sowing: phosphorus (P_2O_5)—60; potassium (K_2O)—60; nitrogen (NH_4)—20; S (SO_3)—23.5; top dressing with nitrogen (NH_4NO_3)—160, including: starting vegetation in spring—100, budding phase—60.

2.4. Weed Control

In each growing season, intensive chemical control was used to reduce weed infestation of the crops. From 2 to 4 herbicides, including glyphosate, were used on each crop in growing season, the same herbicides in all tillage systems (Table 3).

Table 3. Herbicides used to reduce weed infestation in wheat, pea and rape.

Growing Seasons	Wheat I and II	Growing Seasons	Pea/Rape
2009/2010	Roundup 360 SL (glyphosate 360 g l^{-1}; 28.77%)—1 × 3.0 l ha^{-1} Lintur 70WG (dicamba 65.9%, triasulfuron 4.1%)—1 × 0, 15 l ha^{-1} + Axial 100EC (pinoxaden 100 g l^{-1})—1 × 0.3 l ha^{-1}	2010	Pea: Barox 460 SL (bentazone 400 g l^{-1}; MCPA 60 g l^{-1})—1 × 2.5 l ha^{-1} Roundup 360 SL (glyphosate 360 g l^{-1}; 28.77%)—1 × 4.0 l ha^{-1} (desiccation of peas)
2010/2011	Roundup Energy 450 SL (glyphosate 450 g l^{-1}; 34.5%)—1 × 3.0 l ha^{-1} Maraton 375 SC (pendimethalin 250 g l^{-1}; isoproturon 125 g l^{-1})—1 × 2.5 l ha^{-1} Granstar 75 WG (methyl tribenuron 75%)—1 × 15 g ha^{-1} + Starane 250 EC (fluroxypyr 250 g l^{-1}; 24.77%)—1 × 0.3 l ha^{-1}	2011	Pea: Basagran 480 SL (bentazone 480 g l^{-1})—1 × 2.5 l ha^{-1} Roundup 360 SL (glyphosate 360 g l^{-1}; 28.7%)—1 × 4.0 l ha^{-1}) (desiccation of peas)
2011/2012	Puma Universal 069 EW (fenoxaprop-P-ethyl 69 g l^{-1})—1.0 l ha^{-1} + Mustang 306 SE (2,4-D 300 g l^{-1}, florasulam 6.25 g l^{-1})—1 × 0.8 l ha^{-1}	2011/2012	Winter rape: Butisan Star 416 SC (metazachlor 333 g l^{-1}; quinmerac 83 g l^{-1})— 1 × 2.0 l ha^{-1} + Command 480 EC (clomazone 480 g l^{-1})—1 × 0.15 l ha^{-1} Targa Super 05 EC (quizalofop 5%)—1 × 0.7 l ha^{-1}
2012/2013	Roundup Energy 450 SL (glyphosate 450 g l^{-1}; 51%)—1 × 3.0 l ha^{-1} Axial 100EC ((pinoxaden 100 g l^{-1})—1 × 0.4 l ha^{-1} Granstar 75 WG (methyl tribenuron 75%)—1 × 10 g ha^{-1} + Starane 250 EC ((2,4-D 300 g l^{-1}, florasulam 6.25 g l^{-1})—1 × 0.2 l ha^{-1}	2012/2013	Winter rape: Roundup 360 SL (glyphosate 360 g l^{-1}; 28.77%)—1 × 4.0 l ha^{-1} Targa Super 05 EC (quizalofop 5%)—1 × 0.7 l ha^{-1})—1 × 0.6 l ha^{-1} Butisan Star 416 SC (metazachlor 333 g l^{-1}; quinmerac 83 g l^{-1}—1 × 1.2 l ha^{-1} + Navigator 360 SL (clopyralid 240 g l^{-1} 20.31%; picloram 80 g l^{-1}, 6.77%, aminopyralid 40 g l^{-1} 3.38%)—1 × 0.2 l ha^{-1}

2.5. Weed Flora Analysis

Analyses of species composition, number of weeds and air-dried matter of weeds were determined after collection using the weed-picking frame method. The weeds were collected from a frame with dimensions of 0.5 × 1 m, with five replications in each crop field. To avoid the edge effect, the frames were placed a few meters from the edges of plots.

Tillage effects on weed population composition were assessed in each field from July 2010 to April 2013 (3 growing seasons) on 5 terms:

1. before the crop harvest,
2. after the crop harvest and soil operations in the plough and reduced tillage systems,
3. before sowing of winter wheat, after pre-sowing soil treatment in the plough and reduced tillage systems,
4. late autumn, during the tillering stage of winter wheat,
5. spring, for winter wheat after the start of the growing season, for spring wheat in the tillering stage.

Plant species were identified according to the method of Rutkowski [32]. Dry matter of weeds and wheat was determined after drying at 40 °C for 7 days.

2.6. Soil Seed Bank Analysis

The seed bank was determined using the "greenhouse method" (seedling emergence method). Soil samples for the evaluation of the soil seed bank were taken in 3 years (2010–2012) from all fields after harvesting the crop, but before the cultivation operations. The soil was collected using a soil cylinder with a diameter of 8 cm, from levels: 0–5 cm, 5–10 cm and 10–20 cm, in 5 repetitions in each field (Figure 3A,B). Soil samples were placed into pots of about 20 cm in diameter that were partly filled with sand, which served as a drain, and covered with agro-textile (Figure 3C). The pots were placed in a vegetation hall and watered regularly for 11 months. Sprouting weeds were identified by species and counted every 1–2 months. The number of germinating weed seeds (viable seed bank) from the surface of each cylinder (50 cm^2) were scaled up to that for 1 m^2. Weed seed atlases were used for species determination [33].

Figure 3. Method of sampling for the determination of the soil seedbank. (**A**)—soil collection method using a cylinder, (**B**)—soil sample; (**C**)—seedling emergence method to assess the soil seed bank

2.7. Statistical Analyses

2.7.1. Diversity Indicators

The structures of weed communities and soil seed bank were analyzed using diversity indices: Shannon's diversity index: $H' = -\sum Pi \ln Pi$ [34] and Simpson's dominance index: $SI = \sum Pi2$ [35], where Pi is the probability of species occurrence in the sample. The values of the indicators were calculated using the Multi-Variate Statistical Package (MVSP) 3.1 program, Kovach Computing Services, Anglesey, United Kingdom [36].

2.7.2. Similarity Indices

The qualitative and quantitative Sorensen's similarity indices were used to compare weed flora communities and soil seed banks in the different tillage systems [37]:

$$\text{Sorensen's qualitative index} = 2C/A + B \times 100\%$$

where:

A—the number of species in one of the two communities compared,
B—the number of species in the second community compared,
C—the number of common species in the compared communities.

$$\text{Sorensen's quantitive index} = 2\ Nt/Na + Nb \times 100\%$$

where:

Nt—sum of the smallest numbers of common species in the compared variants,
Na—the number of all weeds in one of the compared variants,
Nb—the number of all weeds in the second variant compared.

The values of the indicators were calculated using the Multi-Variate Statistical Package (MVSP) 3.1 program, Kovach Computing Services, Anglesey, United Kingdom [36].

2.7.3. Assessment of the Significance of Differences

In order to check the normality of the distributions, the Shapiro–Wilk test was used. The data on weed species richness and abundance did not meet the requirements for parametric tests. Therefore, the nonparametric Kruskal–Wallis test was used for the identification of significant differences between samples at $p \leq 0.05$ using Statistica 10 software (StatSoft, Kraków, Poland).

3. Results and Discussion

3.1. Above-Ground Weed Flora in the Different Tillage Systems

The average number of weeds and their dry matter did not differ significantly in the three tillage systems studied, but a tendency of increasing weed infestation in line with tillage simplification was found (Figure 4). The number of weeds (on average from all of the assessments of crops in rotation) in the plough tillage system was 22.2 plants m^{-2}, under the reduced tillage conditions it was 20% higher, while under direct sowing it was 33% higher. The dry weight of weeds in the plough system was 9.0 g m^{-2}, while in the reduced tillage system it was higher by 23% and in the no-till system by 28% in comparison to the plough system (Figure 4). The number and weight of weeds observed in all the cultivation systems was low due to intensive herbicide protection, including glyphosate (Table 3), which was very effective in weed control in all compared tillage systems. A higher weed abundance and density under reduced tillage as compared to plough has been confirmed by earlier studies [38–40]. Demjanová et al. [41] indicated that reduced soil tillage in maize was accompanied with significantly higher weed biomass compared to moldboard ploughing. Similarly, in the studies by Woźniak et al. [42] and Vakali et al. [43], a greater weed infestation of spring wheat and barley occurred in no-till and reduced tillage than in conventionally ploughed fields.

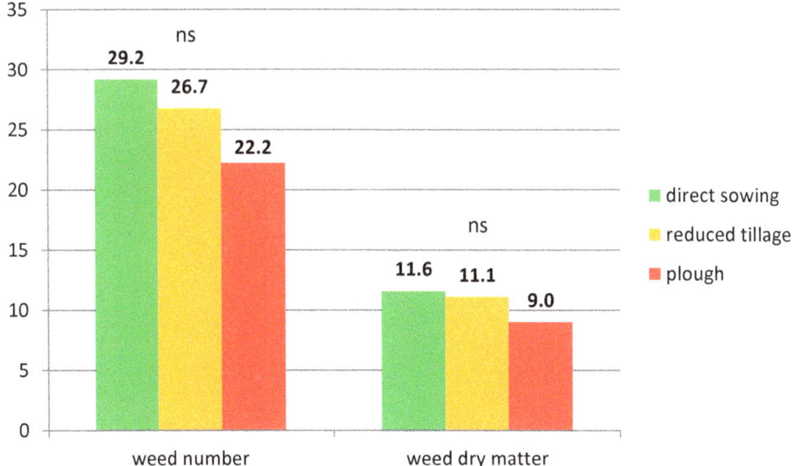

Figure 4. Average number (plants m^{-2}) and dry weight of weeds (g·m^{-2}) in different tillage systems (average of all assessment terms in 3 growing seasons and crops in rotation); ns—no significant differences according to the Kruskal–Wallis test at $p \leq 0.05$. Years × tillage system interaction not significant.

When analysing the weed infestation of individual crops, a significantly higher number and weight of weeds were found in winter wheat (I) under the direct sowing system as compared to the plough system (Figure 5A,B). In the wheat grown after wheat I (wheat II), the differences between tillage systems in the number and weight of weeds were insignificant. Armengot et al. [21] also found that total weed coverage was higher under reduced tillage, although this result was not consistent for different crops. In the study by Woźniak [44], no-plough tillage significantly increased the number and air-dried weight of weeds in winter wheat canopy as compared to ploughed tillage. In the studies by Starczewski and Czarnocki [45], triticale tillage simplifications resulted in a tendency to increase weed infestation, but a significant increase in their fresh and dry weight was only observed on no-till plots. In research of these authors, in the fields where soil tillage simplifications were applied, an increase in weed infestation was observed despite of the use of herbicides, as is the case in this present research.

A

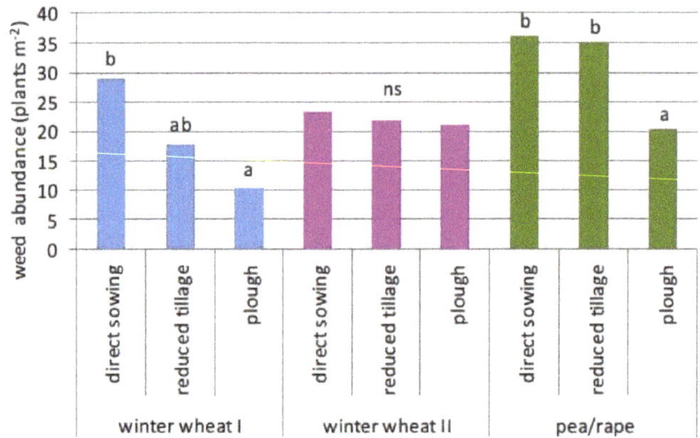

B

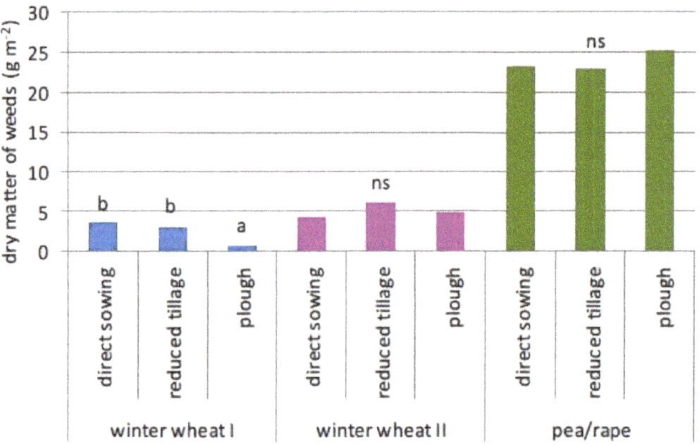

Figure 5. Weed abundance (**A**) and dry matter of weeds (**B**) in crops cultivated in different tillage systems (mean from the years 2010–2013). Within each crop, different letters indicate significant differences between tillage systems according to the Kruskal–Wallis test at $p \leq 0.05$; ns—no significant differences. Years × tillage system interaction not significant.

Examples of the dynamics of weed infestation of winter wheat I (spring wheat in 2012) at different research dates and tillage systems are presented in Figure 6A,C. These tendencies were not constant, and various dependencies on weed abundance in the different tillage systems were found in particular study dates and years. In the first year of the study (2010/2011) the plough tillage system was characterized by the lowest number of weeds (Figure 6A), while in the last term of analysis (2012/2013) it was the highest

(Figure 6C). The large number of weeds at the last term in the 2011/2012 season was associated with the poor wintering of winter wheat and low crop competitiveness (Figure 6B).

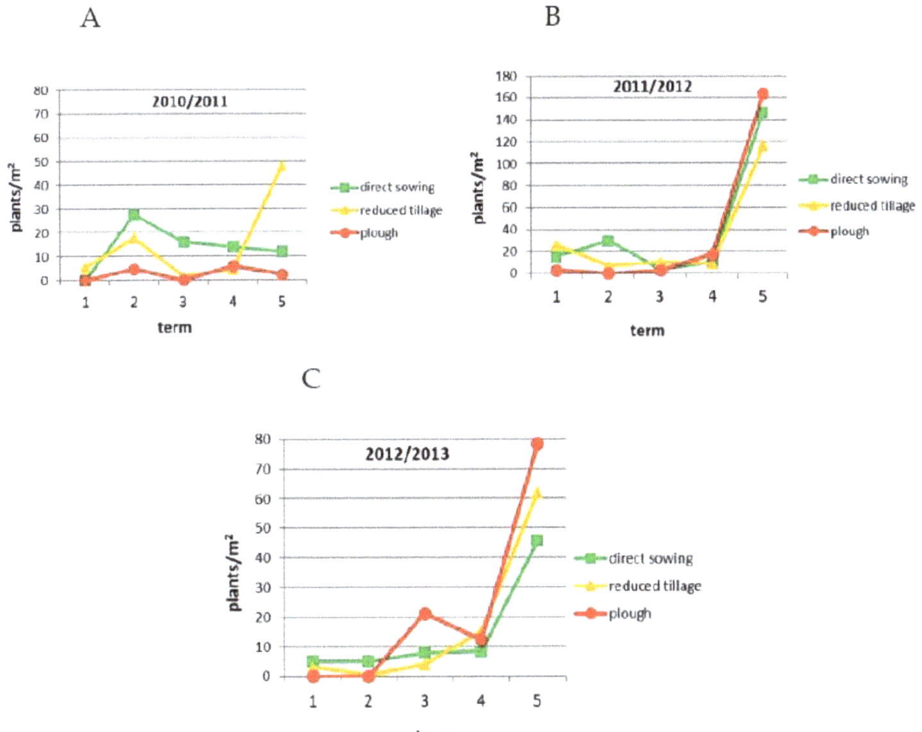

Figure 6. The dynamics of weed abundance in winter wheat under different tillage systems. (**A**)—2010/2011 season; (**B**)—2011/2012 season; (**C**)—2012/2013 season. Terms of weed flora assessment: 1—before the crop harvest, 2—after the crop harvest and soil operations in the plough and reduced tillage systems, 3—before sowing of winter wheat, 4—late autumn, 5—spring.

Analyses of weed flora biodiversity occurring in crops grown under different tillage systems showed a similar number of species (38–40), but differences in species composition occurred (Table 5). In the plough system the dominant species were *Fallopia convolvulus* and *Galeopsis tetrahit*, while under reduced tillage they were: *Fallopia convolvulus* and *Viola arvensis* (Tables 4 and 5). Plough cultivation was accompanied by a larger number of *Stellaria media*, *Thlaspi arvense*, *Sinapis arvensis*, *Anthemis arvensis* and *Brassica napus* volunteers than in the other systems.

In crops sown directly, there was a larger number of *Galium aparine*, *Veronica persica*, *Capsella bursa-pastoris*, *Lamium purpureum*, *Echinochloa crus-galli* and *Setaria glauca* than in the other cultivation systems (Tables 4 and 5). There was also a higher occurrence of perennial species: *Equisetum arvense*, *Plantago major* and *Sonchus arvensis* in comparison to the traditional ploughing. Only in the direct sowing system were *Sonchus oleraceus*, *Melandium album* and *Anagallis arvensis* found, while only in the plough system were *Polygonum persicaria*, *Rumex obtusifolius*, *Lapsana communis* and *Arctium lappa* found. Cardina et al.'s [11] study confirmed that, in no-tillage plots, the density of *Veronica* sp. was higher compared to moldboard plough plots.

Reduced cultivation was characterised by a higher occurrence of Fallopia convolvulus, Viola arvensis, Chenopodium album, Fumaria officinalis, Tripleurospermum inodorum, Descurainia sophia,

Apera spica-venti, Galinsoga parviflora, Consolida regalis and the exclusive occurrence of Veronica hederifolia, Raphanus raphanistrum and Senecio vulgaris (Tables 4 and 5). Similarly, Clements et al. [46] found that Chenopodium album dominated the aboveground weed population in chisel ploughing in comparison with moldboard ploughing and no-till.

Invasive alien weed species (IAS), which pose a threat to the biodiversity and the economy or human health, were more numerous in the reduced tillage systems compared to ploughing (Table 5). Most species belonged to the lowest invasiveness category (I) according to Tokarska-Guzik et al. [47]. It is worth emphasizing that, despite the greater number of invasive weeds in reduced and no-till systems, the number of weeds was not large (5.5 plants m^{-2} in total in no-till system and 2.1 plants m^{-2} in reduced tillage vs. 0.4 plants m^{-2} in plough system) (Table 5), due to effective chemical control using herbicides (Table 3). Other authors also draw attention to the risks associated with the presence of invasive species in reduced tillage systems [13]. In the research of Woźniak [44], the replacement of ploughing with harrowing caused an increase of two dominating species in the weed population: *Echinochloa crus-galli* and *Viola arvensis*.

Weed biodiversity as measured by Shannon's index was the highest under reduced tillage—2.56, due to the most even share of weed species in the community (Table 5). The direct sowing system was characterized by the lowest value of the Shannon's diversity index—2.28 and a high value of the Simpson's dominance index—0.15, due to the domination of two species: *Galium aparine* and *Fallopia convolvulus*. Travlos et al.'s [13] review states that conservation tillage systems seem to be associated with higher weed richness and diversity, as the elimination of ploughing creates more enhancing conditions for some weed species. However, they found some cases where reduced tillage systems led to less diverse weed communities compared to more intensive tillage systems due to the domination of some weed species.

According to Legere et al. [27], the tillage method has little impact on weed diversity as expressed by diversity indices, but plays a key role in shaping the species composition of the weed community. Similarly, Sans et al. [48], in their research on tillage systems under the conditions of organic farming, did not find significant differences in Shannon's diversity index values for weed communities in cereal crops, but observed a significantly higher degree of weed coverage in the reduced tillage system compared to the plough system. Plaza et al. [12], based on a 23-year long experiment, found that no large differences between tillage systems had arisen that was related to weed diversity. They only found that no-till appeared to support more species than the two other tillage systems. In addition, their species richness and Shannon diversity index varied greatly through the years in all the tillage systems. This indicates that research on weed diversity and changes as a result of different tillage methods should be carried out over many years, as too short a period of observation may lead to the divergent results that can be found in the literature. This was confirmed by Woźniak [44], who found that Shannon's diversity index values for weed species composition in winter wheat were similar for different tillage systems and more diverse between study years.

Table 4. Dominant and characteristic species in weed communities under different tillage systems.

Species	Direct Sowing	Reduced Tillage	Plough
Dominant	*Galium aparine* L *Fallopia convolvulus* (L.) Á. Löve *Veronica persica* Poir. *Capsella bursa-pastoris* (L.) Medik	*Fallopia convolvulus* (L.) Á. Löve *Viola arvensis* Murray	*Fallopia convolvulus* (L.) Á. Löve *Galeopsis tetrahit* L.
Occurring more abundantly than in other tillage systems	*Lamium purpureum* L. *Geranium pussillum* Burm. F. ex L. *Echinochloa crus-galli* (L.) P. Beauv. *Setaria glauca* (L.) P. Beauv. *Equisetum arvense* L. *Plantago major* L. *Sonchus arvensis* L.	*Chenopodium album* L. *Fumaria officinalis* L. *Tripleurospermum inodorum* (L.) Schultz-Bip. *Descurainia sophia* (L.) Webb ex Prantl *Apera spica-venti* L. *Galinsoga parviflora* Cav *Consolida regalis* S.F. Gray	*Stellaria media* (L.) Vill. *Thlaspi arvense* L. *Sinapis arvensis* L. *Anthemis arvensis* L. *Brassica napus* L. volunteers
Occurring only in a given tillage system (characteristic or incidental species)	*Sonchus oleraceus* L. *Melandrium album* (Mill.) Garcke*Anagallis arvensis* L.	*Veronica hederifolia* L. *Raphanus raphanistrum* L.	*Polygonum persicaria* L. *Rumex obtusifolius* L. *Lapsana communis* L. *Arctium lappa* L.

Table 5. Species composition and number of weeds (plants m^{-2}) in different tillage systems (on average for rotation and study years 2010–2013).

	Weed Species	A/P *	Tillage System			Mean
			Direct Sowing	Reduced Tillage	Plough	
1.	*Fallopia convolvulus* (L.) Á. Löve	A	5.70	8.90	7.73	7.443
2.	*Viola arvensis* Murray	A	3.80	4.25	3.83	3.960
3.	*Galium aparine* L.	A	6.14	2.60	1.48	3.406
4.	*Galeopsis tetrahit* L.	A	1.50	2.26	4.03	2.597
5.	*Veronica persica* Poir. I	A	4.85 I	1.10 I	0.10 I	2.011 I
6.	*Chenopodium album* L.	A	0.84	3.05	0.78	1.555
7.	*Capsella bursa-pastoris* (L.) Medik	A	2.74	0.42	0.50	1.219
8.	*Stellaria media* (L.) Vill.	A	0.35	0.41	0.73	0.496
9.	*Fumaria officinalis* L.	A	0.36	0.53	0.45	0.444
10.	*Tripleurospermum inodorum* (L.) Schultz-Bip.	A	0.16	0.63	0.48	0.422
11.	*Brassica napus* L.	A	0.04	0.06	0.70	0.267
12.	*Descurainia sophia* (L.) Webb ex Prantl	A	0.31	0.37	0.08	0.250
13.	*Apera spica-venti* L.	A	0.18	0.43	0.13	0.244
14.	*Galinsoga parviflora* Cav. I	A	0.01 I	0.63 I	0.00 I	0.211 I
15.	*Thlaspi arvense* L.	A	0.18	0.06	0.31	0.183
16.	*Erigeron annuus* L. II	A	0.14 II	0.19 II	0.19 II	0.175 II
17.	*Lamium purpureum* L.	A	0.38	0.06	0.02	0.154
18.	*Echinochloa crus-galli* (L.) P. Beauv. I	A	0.28 I	0.05 I	0.11 I	0.144 I
19.	*Geranium pussillum* Burm. F. ex L.	A	0.35	0.04	0.00	0.128
20.	*Polygonum aviculare* L.	A	0.04	0.13	0.05	0.072
21.	*Equisetum arvense* L.	P	0.19	0.02	0.01	0.071
22.	*Setaria glauca* (L.) P. Beauv. I	A	0.16 I	0.02 I	0.02 I	0.064 I
23.	*Consolida regalis* S.F. Gray	A	0.03	0.16	0.01	0.064
24.	*Sinapis arvensis* L.	A	0.00	0.07	0.12	0.061
25.	*Euphorbia helioscopia* L.	A	0.16	0.00	0.01	0.056
26.	*Avena fatua* L. I	A	0.01 I	0.08 I	0.01 I	0.033 I
27.	*Convolvulus arvensis* L.	P	0.03	0.05	0.02	0.033
28.	*Sonchus arvensis* L.	P	0.05	0.01	0.03	0.028
29.	*Paphaver rhoeas* L.	A	0.06	0.01	0.02	0.028
30.	*Polygonum persicaria* L.	A	0.00	0.00	0.08	0.025
31.	*Cirsium arvense* (L.) Scop.	P	0.00	0.04	0.03	0.022
32.	*Plantago major* L.	P	0.04	0.02	0.00	0.019
33.	*Anthemis arvensis* L.	A	0.01	0.00	0.05	0.019
34.	*Erodium cicutarium* (L.) L'Her	A	0.01	0.05	0.00	0.019
35.	*Mercurialis annua* L.	A	0.01	0.00	0.05	0.019
36.	*Centaurea cyanus* L.	A	0.00	0.01	0.03	0.014
37.	*Lamium amplexicaule* L.	A	0.00	0.02	0.02	0.011
38.	*Taraxacum officinale* Weber	P	0.01	0.01	0.01	0.008
39.	*Gnaphalium uliginosum* L.	A	0.01	0.00	0.02	0.008
40.	*Poa annua* L.	A	0.02	0.00	0.01	0.008
41.	*Plantago lanceolata* L.	P	0.00	0.01	0.01	0.006
42.	*Anagallis arvensis* L.	A	0.02	0.00	0.00	0.006
43.	*Veronica hederifolia* L.	A	0.00	0.02	0.00	0.006
44.	*Raphanus raphanistrum* L.	A	0.00	0.02	0.00	0.006
45.	*Senecio vulgaris* L.	A	0.01	0.01	0.00	0.006
46.	*Rumex obtusifolius* L.	P	0.00	0.00	0.02	0.006
47.	*Melandrium album* (Mill.) Garcke	A	0.01	0.00	0.00	0.003
48.	*Lapsana communis* L.	A	0.00	0.00	0.01	0.003
49.	*Arctium lappa* L.	A	0.00	0.00	0.01	0.003
50.	*Sonchus oleraceus* L.	A	0.01	0.00	0.00	0.003
Sum			29.17	26.73	22.22	26.04
Total number of species			39	38	40	50
Shannon's diversity index (H')			2.28	2.56	2.35	2.58
Simpson's dominance index (SI)			0.15	0.11	0.15	0.11

* A—annual or biennial; P—perennial. I–IV invasive species, category of invasiveness from I (the lowest invasiveness) to IV (the highest invasiveness) according to Tokarska-Guzik et al. [47].

The share of perennial species in the weed community in the no-tillage system was about two times higher than in the plough and reduced tillage systems (Figure 7). However, the high level of agricultural culture and the use of herbicides, especially glyphosate, prevented the excessive occurrence of perennial weeds, also in the no-till system. This has also been confirmed by other authors [49]. According to some authors, tillage simplification promotes an increase in perennial weeds in the field, mainly *Elymus repens* [19,50]. Starczewski and Czarnocki [45] found that, under a cultivation reduction system for winter triticale, perennial weeds were a major problem, while traditional ploughing effectively eliminated them. Bilalis et al. [25] found that three annual species prevailed in the conventional and minimum tillage systems, while one perennial species prevailed in the no-tillage system. Not using glyphosate in the no-tillage system leads to an increase in the abundance of perennial weeds, especially *Cirsium arvense*, *Elymus repens* and *Convolvulus arvensis* [51].

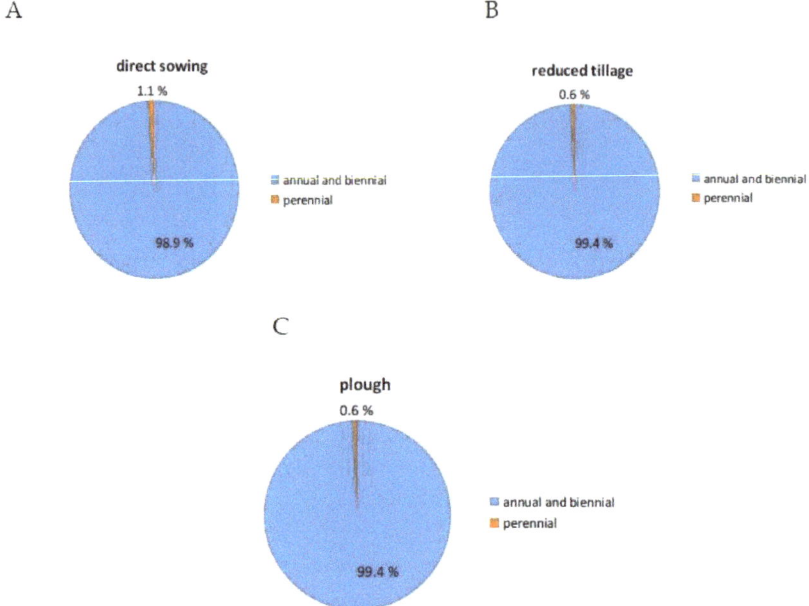

Figure 7. Percentage of annual/biennial and perennial species in the total density of weed communities under different tillage systems.

The comparison of weed communities occurring in plants grown under different tillage systems showed that there was a bigger qualitative than quantitative similarity, i.e., a bigger similarity in species composition than in the number of common species (Table 6), which is also confirmed by the results of Zanin et al. [24] and Duer and Feledyn-Szewczyk [52]. Sorensen's qualitative and quantitative similarity indices showed that weed communities in the direct sowing and ploughing systems were the least similar.

Table 6. Sorensen's qualitative and quantitative similarity indices for weed communities in different tillage systems.

Sorensen's Quantitative Similarity Index (%)		Sorensen's Qualitative Similarity Indices (%)		
		Direct Sowing	Reduced Tillage	Plough
	Direct sowing	-	81	79
	Reduced tillage	55	-	80
	Plough	42	63	-

3.2. Soil Seed Bank in Different Soil Tillage Systems

The number of weed seeds was dependent on the tillage system and soil layer. The average number of weed seeds in the 0–20 cm soil layer was significantly the highest under the no-till system (6518 seeds m^{-2}) and over 3 times lower in the fields where traditional ploughing was performed

(2080 seeds m^{-2}) (Figure 8). In Vanasse and Leroux's [53] study, ridge-tilled fields had a larger soil seedbank (2992 seeds m^{-2}) than moldboard-plowed fields (1481 seeds m^{-2}) in the top 15 cm layer. They explained that the results were due to the larger perennial seedbank in reduced tillage fields at both the 0–5 cm and 5–15 cm depths.

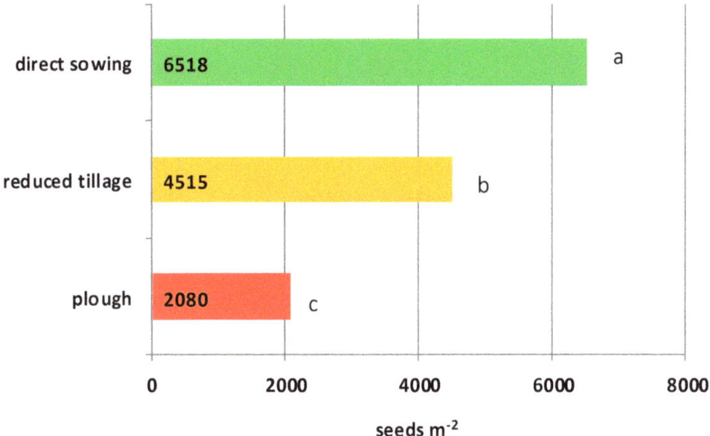

Figure 8. Seed number in 0–20 cm soil layer in different tillage systems (mean from crop rotation and years 2010–2012). Different letters indicate significant differences between tillage systems according to the Kruskal–Wallis test at $p \leq 0.05$.

The vertical distribution of weed seeds in the soil depended on the tillage system. In fields where the soil profile was not affected by tillage (direct sowing), the vast majority of seeds were at a depth of up to 5 cm and their number decreased rapidly in line with the depth (Figures 9 and 10). The distribution of seeds in the soil profile of the ploughing system was reversed, where as a result of soil inversion, a large part of the seeds from the soil surface was moved to the layers 5–10 and 10–20 cm. In the fields where simplified tillage was applied, the seed distribution was variable. Reduced tillage caused an increase in the number of diasporas in the 0–10 cm layer and a decrease in the number of weed seeds in the deeper soil layer (10–20 cm). Piskier and Sekutowski [54] found that in corn cultivation, the highest weed seeds number were recorded in the soil samples collected from the 0–5 cm soil layer in both the no-tillage and reduced tillage systems. In conventional tillage (plough) they noted that weed seeds were more evenly distributed throughout all of the 0–20 cm soil layer. This is in line with the results of Clements et al. [46], who found that more than 60% of the weed seedbank was concentrated in the upper 5 cm of soil in chisel plow and no-till systems. They additionally found that the seedbank of the moldboard plough system was more uniformly distributed over depth, but, contrary to our results, was larger than in the other systems.

In all tillage systems, a decrease in the weed seed stock was found in subsequent years of the study, with the most significant decrease being in the direct sowing system and the smallest decrease being in the plough system (Figure 11). This is probably the result of intensive chemical weed control for years, including glyphosate, as presented in Table 3, which caused decrease in seeds reservoir. In Fracchiolla et al.'s study [26], the seed bank was clearly impoverished after the long-term applications of preemergence herbicides, both in terms of richness and of diversity.

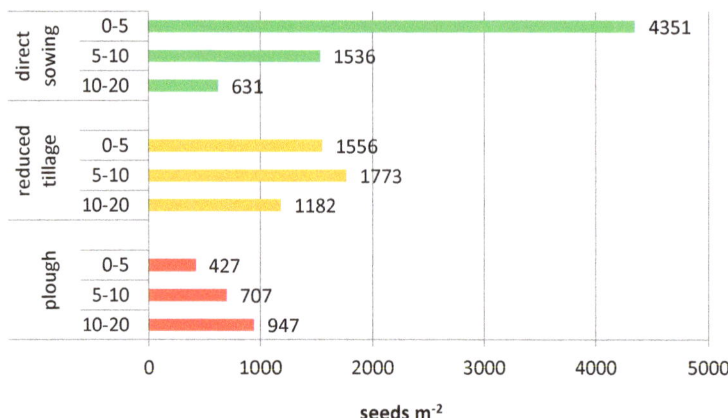

Figure 9. Seed number in three soil layers: 0–5, 5–10, 10–20 cm in different tillage systems (mean from crop rotations and years 2010–2012).

Figure 10. Weed seedling emergence from the soil layer 0–5 cm in different tillage systems. (**A**)—direct sowing, (**B**)—reduced tillage, (**C**)—plough.

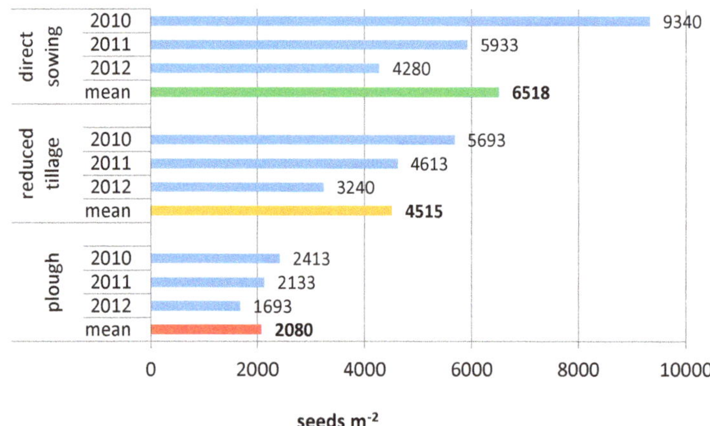

Figure 11. Seed number in the soil layer 0–20 cm in different tillage systems and years.

The greatest diversity of species during the three years of research was noted in the seed bank under direct sowing (30 species in total), which could also be associated with a much higher number of seeds occurring in the 0–20 cm layer compared to the other systems (Table 7). A similar number of weed seed species (29) was found in the seedbank of fields where simplified tillage was applied, while the smallest number was recorded in the plough system (22 species).

Table 7. Species composition and number of weed seeds (seeds m^{-2}) in the soil layer 0–20 cm in different tillage systems (average for the years of research and crops in rotation).

	Weed Species	A/P *	Tillage System			Mean
			Direct Sowing	Reduced Tillage	Plough	
1.	*Viola arvensis* Murray	A	1160.0	1631.1	342.2	1044.4
2.	*Apera spica-venti* L.	A	1340.0	720.0	217.8	759.3
3.	*Capsella bursa-pastoris* (L.) Medik	A	1822.2	97.8	208.9	709.6
4.	*Veronica persica* Poir. I	A	942.2 I	293.3 I	31.1 I	422.2 I
5.	*Chenopodium album* L.	A	173.3	444.4	208.9	275.6
6.	*Fallopia convolvulus* (L.) Á. Löve	A	102.2	368.9	226.7	232.6
7.	*Tripleurospermum inodorum* (L.) Schultz-Bip.	A	204.4	164.4	328.9	232.6
8.	*Fumaria officinalis* L.	A	351.1	115.6	31.1	165.9
9.	*Galeopsis tetrahit* L.	A	40.0	195.6	142.2	125.9
10.	*Stellaria media* (L.) Vill.	A	115.6	142.2	115.6	124.4
11.	*Thlaspi arvense* L.	A	13.3	8.9	128.9	50.4
12.	*Galium aparine* L.	A	44.4	75.6	26.7	48.9
13.	*Consolida regalis* S.F. Gray	A	26.7	97.8	0.0	41.5
14.	*Conyza canadensis* (L.) Cronquist IV	A	31.1 IV	35.6 IV	22.2 IV	29.6 IV
15.	*Echinochloa crus-galli* (L.) P. Beaur. I	A	8.9 I	26.7 I	4.4 I	13.3 I
16.	*Gnaphalium uliginosum* L.	A	4.4	17.8	13.3	11.9
17.	*Sonchus arvensis* L.	P	22.2	8.9	0.0	10.4
18.	*Plantago major* L.	P	17.8	8.9	4.4	10.4
19.	*Paphaver rhoeas* L.	A	13.3	4.4	4.4	7.4
20.	*Descurainia sophia* (L.) Webb ex Prantl	A	13.3	4.4	0.0	5.9
21.	*Cirsium arvense* (L.) Scop.	P	8.9	8.9	0.0	5.9
22.	*Erodium cicutarium* (L.) L'Her	A	8.9	4.4	4.4	5.9
23.	*Lamium amplexicaule* L.	A	8.9	0.0	4.4	4.4
24.	*Brassica napus* L.	A	0.0	8.9	4.4	4.4
25.	*Spergula arvensis* L.	A	8.9	4.4	0.0	4.4
26.	*Taraxacum officinale* Weber	P	0.0	8.9	0.0	3.0
27.	*Myosotis arvensis* (L.) Hill	A	0.0	4.4	4.4	3.0
28.	*Amaranthus retroflexus* L. I	A	8.9 I	0.0 I	0.0 I	3.0 I
29.	*Lapsana communis* L.	A	4.4	0.0	4.4	3.0
30.	*Sinapis arvensis* L.	A	8.9	0.0	0.0	3.0
31.	*Veronica arvensis* L.	A	0.0	4.4	0.0	1.5
32.	*Geranium pussillum* Burm. F. ex L.	A	4.4	0.0	0.0	1.5
33.	*Melandrium album* (Mill.) Garcke	A	4.4	0.0	0.0	1.5
34.	*Galinsoga parviflora* Cav. I	A	4.4 I	0.0 I	0.0 I	1.5 I
35.	*Plantago lanceolata* L.	P	0.0	4.4	0.0	1.5
36.	*Euphorbia helioscopia* L.	A	0.0	4.4	0.0	1.5
Sum			6518	4515	2080	4371
Total number of species			30	29	22	36
Shannon's diversity index (H')			2.04	2.18	2.41	2.32
Simpson's dominance index (SI)			0.18	0.18	0.11	0.14

* A—annual or biennial; P—perennial. I–IV invasive species, category of invasiveness from I (the lowest invasiveness) to IV (the highest invasiveness) according to Tokarska-Guzik et al. [47].

Despite the highest number of weed seed species being in the no-till system, the weed seed pool was characterized by a low value of Shannon's diversity index (2.04) and a high value of Simpson's domination index (0.18), indicating the dominance of particular weed species: *Capsella bursa-pastoris, Apera spica venti, Viola arvensis* and *Veronica persica* (Table 7). The results of Conn's [38] research confirm the high abundance of common, annual weed species with small seeds, such as *Apera spica-venti, Viola arvensis, Capsella bursa pastoris,* and *Veronica persica* under reduced tillage and direct sowing systems. A larger number of *Apera spica-venti* and *Viola arvensis* in the reduced tillage system compared to the traditional one was also observed by Sekutowski and Smagacz [24], whereas Krawczyk et al. [55] observed a decreasing number of these two species along with tillage simplifications.

In our study, the share of weed seeds in the soil in the plough system was the most evenly distributed and it is difficult to indicate dominant species, as evidenced by the high value of Shannon's diversity index (2.4) and the low value of Simpson's dominance index (0.11). Diversity indices used by Cardina et al. [15] and Borin et al. [56], comparing the structure of the soil seed bank under traditional and simplified tillage, showed that an increase in the frequency of tillage operations resulted in a decrease in species diversity of weed seeds. Sekutowski and Smagacz [23] pointed out the lower values of Shannon's diversity index and the higher values of Simpson's dominance index for the aboveground weed flora and soil seed bank in reduced tillage systems compared to the plough system. Feldman et al. [9] found a bigger species diversity in the soil seed bank of systems that cause less soil disturbance, i.e., in the reduced and zero tillage systems. Cardina et al. [11], on the basis of 35 years of research, concluded that the amount of seeds in the soil is more affected by crop rotations than by tillage methods (plough, simplified, or no-till). The authors also demonstrated a significant interaction of crop rotations and tillage system.

Some studies have revealed a direct correlation between the presence of specific weed species and the tillage system [3]. In our research, under the reduced and no-till systems, there was a higher percentage of seeds of invasive species in the soil, such as: *Veronica persica, Conyza canadensis, Echinochloa crus-galli, Amaranthus retroflexus* and *Galinsoga parviflora* (Table 7), although the latter two species occurred only in the no-till system. Similar to above-ground weed flora, most invasive species in soil seed bank belonged to the lowest (I) category of invasiveness [45]. Only *Conyza canadensis* represented the species of the highest invasiveness (IV category). Fracchoiolla et al. [26] also observed that seed bank of *Conyza canadensis* was higher in reduced tillage plots in comparison with conventional tillage fields. The study carried out by Sheley et al. [57] showed that invasive weeds are inclined to recover rapidly when tillage is interrupted. Shifts in plant communities are usually described or quantified by means of the various existing abundance and diversity indices.

Under the reduced and no-till systems there was a higher number of seeds of perennial species: *Sonchus arvensis, Plantago major* and *Descurainia sophia*, which is considered a ruderal species [58]. Similarly, in the study of Thomas et al. [59], perennial species such as *Sonchus arvensis* and *Cirsium arvense* were associated with reduced and no-till systems, while annual species were associated with a wider range of tillage systems.

The depth at which the seeds are placed in the soil is a factor that indirectly influences their germination capacity, since this involves differences in temperature, oxygen and light, i.e., the intensity of the factors directly affecting the germination process. Particularly sensitive to depth are fine-size seeds, for example, *Stellaria media*, which germinate best from a depth of 0.6 cm to 1.2 cm [33]. The seeds of *Descurainia sophia* germinate both in light and in the dark, although the access of light stimulates this process. The germination capacity of *Descurainia sophia* and *Apera spica-venti* seeds depends significantly on the sowing depth and they germinate the best from the soil surface [58]. The most numerous emergences of *Echinochloa crus-galli* were recorded when the seeds were placed at a depth of 1.5 to 3.0 cm [58]. Grundy et al. [60] observed that chamomile weeds, *Stellaria media, Veronica persica, Veronica arvensis* and *Polygonum aviculare* responded with a deterioration of germination capacity according to increasing soil depth, while *Chenopodium album* germinated less from the soil surface. *Galeopsis tetrahit* and *Thlaspi arvense* germinate 50% more when exposed to light, hence the higher amount of these weed seeds germinating in the ploughed and reduced tillage systems compared to the direct sowing system were observed.

The soil seed banks in the tested tillage systems were similar in terms of species composition, as indicated by similar values of the qualitative Sorensen's similarity index (77–78) (Table 8). Similar to the aboveground flora, the qualitative similarity of the seed stock in soil was higher than the quantitative similarity (Tables 6 and 8). Bigger differences between the tillage systems occurred in the values of the quantitative similarity index (36–57). The least similar were the seed banks in the plough system and direct sowing, similarly as for the aboveground weed flora (Tables 6 and 8).

The results show that the consequence of the application of tillage simplifications is an increase in the number of seeds in the top layer of soil, which leads to an increase in the weed population in the crop canopy. Moreover, tillage reduction leads to an increase in the abundance of perennial and invasive weed species. Maintaining weed infestation at a level that does not significantly affect winter wheat yields (below 50 plants m^{-2} and 50 g m^{-2} of weed dry matter) is possible with adequately effective herbicide protection. Under effective weed control, an increase in the number of seeds in the topsoil does not adversely affect the competition between weed and crop. Dorado et al.'s [61] results also confirm the necessity of spring weed control in winter cereals when a no-till system is used.

A comparison between moldboard and chisel ploughing presented by Ball [62] indicated that weed seeds of predominant species were more prevalent near the soil surface after chisel ploughing. They note that the number of predominant annual weed seed over the three-year period increased more rapidly in the seedbank after chisel ploughing compared to moldboard ploughing and that required effective weed control to produce a decline in the seedbank number. The complete soil cleansing of all weed seeds is not possible due to the complex nature of dormancy and its interaction with environmental factors, although it is possible to reduce the number significantly under the influence of agrotechnical treatments and by a reduction in the supply of new seeds [16]. Attempts are being made to manipulate the dynamics of seed dormancy [1,24,63]. There is ongoing research for methods to interrupt seed dormancy, for example by applying chemical germination stimulants to the soil in order to later effectively destroy all the weeds at the same time [64]. According to Travlos et al. [13], further research is essential in order to understand the complex relationships of weed species and how they are affected by different tillage systems.

Table 8. Sorensen's qualitative and quantitative similarity indices for soil seed bank communities in different tillage systems.

Sorensen's Quantitative Similarity Index (%)		Sorensen's Qualitative Similarity Index (%)		
		Direct sowing	Reduced tillage	Plough
	Direct sowing	-	78	77
	Reduced tillage	57	-	78
	Plough	36	51	-

4. Conclusions

Weed infestation of wheat and pea/rape fields increased with the simplification of tillage. The seed bank was the largest under direct sowing, and was about three times smaller in traditional ploughing. Under direct sowing, most weed seeds were accumulated in the top soil layer 0–5 cm, while in the ploughing system most weed seeds occurred in deeper layers: 5–10 and 10–20 cm. In the systems of reduced and no-till, a greater share of invasive species was observed, e.g., *Veronica persica*, *Galinsoga parviflora*, *Echinochloa crus-galli*, *Setaria glauca* (above-ground weed flora), *Conyza canadensis*, *Galinsoga parviflora*, *Amarantus retroflexus* (weed seed bank) and perennial species, mainly *Sonchus arvensis* L. and *Plantago major* L. The lowest similarity was found between weed communities and soil seed bank in the direct sowing and ploughing system. We conclude that maintaining weed infestation at a level that does not significantly affect winter wheat yields yet eliminates the risk of perennial and invasive weeds in the no-till system is possible with adequate effective herbicide protection. Weed management programs must take this information into account.

Author Contributions: Conceptualization, J.S. and B.F.-S.; methodology, B.F.-S.; validation, C.A.K., E.H. and A.W.; investigation, B.F.-S. and J.S.; data curation, B.F.-S.; writing—original draft preparation, B.F.-S. and J.S.; writing—review and editing, B.F.-S., C.A.K., E.H. and A.W.; visualization, B.F.-S., E.H.; supervision, J.S. and B.F.-S.; project administration, J.S.; funding acquisition, J.S. All authors have read and agreed to the published version of the manuscript.

Funding: The study was carried out under the project "Development of a soil cultivation system for sustainable agriculture" WND-POIG.01.03.01-00-042/09. The publication was prepared under the multi-annual program of IUNG-PIB, Task No 2.1. (2016-2020).

Acknowledgments: The authors thank Adam Berbeć, Ewa Markowska-Strzemska, and Marek Woźniak for their technical support while conducting the experiments and assessments.

Conflicts of Interest: The authors declare no conflict of interest. The funders had no role in the design of the study; in the collection, analyses, or interpretation of data; in the writing of the manuscript; or in the decision to publish the results.

References

1. Radosevich, S.; Holt, J.; Ghersa, C. *Weed Ecology: Implications for Management*; John Wiley&Sons, Inc.: Hoboken, NJ, USA, 1997.
2. Patterson, D.T. Effects of environmental stress on weed/crop interactions. *Weed Sci.* **1995**, *43*, 483–490. [CrossRef]
3. Buhler, D.D.; Kohler, K.A.; Thompson, R.L. Weed seed bank dynamics during a five-year crop rotation. *Weed Technol.* **2001**, *15*, 170–176. [CrossRef]
4. Fracchiolla, M.; Terzi, M.; Frabboni, L.; Caramia, D.; Lasorella, C.; De Giorgio, D.; Montemurro, P.; Cazzato, E. Influence of different soil management practices on ground-flora vegetation in an almond orchard. *Renew. Agric. Food Syst.* 2015. [CrossRef]
5. Wilson, R.G. *Biology of Weed Seeds in the Soil. W: Weed Management in Agroecosystems: Ecological Approaches*; Altieri, M.A., Liebman, M., Eds.; CRC Press, Inc.: Boca Raton, FL, USA, 1988; Volume 3, pp. 25–39.
6. Sekutowski, T.; Smagacz, J. Share of anthropophytes in the crop sequence: Winter wheat—Maize—Spring wheat depending on tillage system. *Acta Agrobot.* **2014**, *67*, 117–122. [CrossRef]
7. Koller, K. *"Techniques of Soil Tillage," in Soil Tillage in Agroecosystems*; Titi, A.E., Ed.; CRC Press: Boca Raton, FL, USA, 2003; pp. 1–25.
8. Smagacz, J. *Conservation Soil Tillage—Tendencies and Importance in Contemporary Agriculture. Monografie i Rozprawy Naukowe*; IUNG-PIB Puławy: Puławy, Poland, 2018; Volume 59, p. 126. (In Polish)
9. Feldman, S.R.; Torres, C.A.; Lewis, P. The effect of different tillage systems on the composition of the seedbank. *Weed Res.* **1997**, *37*, 71–76. [CrossRef]
10. Stevenson, F.C.; Legere, A.; Simard, R.R.; Angers, D.A.; Pageau, D.; Lafond, J. Weed species diversity in spring barley varies with crop rotation and tillage, but not with nutrient source. *Weed Sci.* **1997**, *45*, 798–806. [CrossRef]
11. Cardina, J.; Herms, C.P.; Doohan, D.J. Crop rotation and tillage system effects on weed seedbanks. *Weed Sci.* **2002**, *50*, 448–460. [CrossRef]
12. Plaza, E.H.; Kozak, M.; Navarrete, L.; Gonzalez-Andujar, J.L. Tillage system did not affect weed diversity in a 23-year experiment in Mediterranean dryland. *Agric. Ecosyst. Environ.* **2010**, *140*, 102–105. [CrossRef]
13. Travlos, I.S.; Cheimona, N.; Roussis, I.; Bilalis, D.J. Weed-Species abundance and diversity indices in relation to tillage systems and fertilization front. *Environ. Sci.* **2018**. [CrossRef]
14. Froud-Williams, R.J.; Chancellor, R.J.; Drennan, D.S.H. Influence of cultivation regime upon buried weed seeds in arable cropping systems. *J. Appl. Ecol.* **1983**, *20*, 199–208. [CrossRef]
15. Cardina, J.; Regnier, E.; Harrison, K. Long-term tillage effects on seed banks in three Ohio soils. *Weed Sci.* **1991**, *39*, 186–194. [CrossRef]
16. Benech-Arnold, R.L.; Sanchez, R.A.; Forcella, F.; Kruk, B.C.; Ghersa, C.M. Environmental control of dormancy in weed seed banks in soil. *Field Crops Res.* **2000**, *67*, 105–122. [CrossRef]
17. Dyer, W.E. Exploiting weed seed dormancy and germination requirements through agronomic practices. *Weed Sci.* **1995**, *43*, 498–503. [CrossRef]
18. Ghersa, C.M.; Martinez-Ghersa, M.A. Ecological correlates of weed seed size and persistence in the soil under different tilling systems: Implications for weed management. *Field Crops Res.* **2000**, *67*, 141–148. [CrossRef]
19. Mayor, J.-P.; Maillard, A. Resultats d'un essai de culture sans labour depuis plus de 20 ans a changins. IV. Stock semencier et maitrise de la flore adventice. *Revue Suisse d'agric.* **1995**, *27*, 229–236.
20. Tørresen, K.S.; Skuterud, R.; Tandsaether, H.J.; Hagemo, M.B. Long-term experiments with reduced tillage in spring cereals. I. Effects on weed flora, weed seedbank and grain yield. *Crop Protect.* **2003**, *22*, 185–200. [CrossRef]
21. Armengot, L.; Berner, A.; Blanco-Moreno, J.M.; Mäder, P.; Sans, F.X. Long-term feasibility of reduced tillage in organic farming. *Agron. Sustain. Develop.* **2015**, *35*, 339–346. [CrossRef]
22. Armengot, L.; Blanco-Moreno, J.M.; Bàrberi, P.; Bocci, G.; Carlesi, S.; Aendekerk, R.; Berner, A.; Celette, F.; Grosse, M.; Huiting, H.; et al. Tillage as a driver of change in weed communities: A functional perspective. *Agric. Ecosyst. Environ.* **2016**, *222*, 276–285. [CrossRef]
23. Sekutowski, T.; Smagacz, J. Similarity between soil seed bank and current weed infestation in winter wheat cultivation. *J. Res. Appl. Agric. Eng.* **2011**, *56*, 93–98.
24. Zanin, G.; Otto, S.; Riello, L.; Borin, M. Ecological interpretation of weed flora dynamics under different tillage systems. *Agriculture, Ecosyst. Environ.* **1997**, *66*, 177–188. [CrossRef]

25. Bilalis, D.; Efthimiadis, P.; Sidiras, N. Effect of three tillage systems on weed flora in a 3-year rotation with four crops. *J. Agron. Crop Sci.* **2001**, *186*, 135–141. [CrossRef]
26. Fracchiolla, M.; Stellacci, A.M.; Cazzato, E.; Tedone, L.; Alhajj Ali, S.; De Mastro, G. Effects of Conservative Tillage and Nitrogen Management on Weed Seed Bank after a Seven-Year Durum Wheat—Faba Bean Rotation. *Plants* **2018**, *7*, 82. [CrossRef] [PubMed]
27. Legere, A.C.S.F.; Stevenson, F.C.; Benoit, D.L. Diversity and assembly of weed communities: Contrasting responses across cropping systems. *Weed Res.* **2005**, *45*, 303–315. [CrossRef]
28. Simpson, R.L.; Allessio Leck, M.; Parker, V.T. Seed Banks: General Concepts and Methodological Issues. In *Ecology of Soil Seed Banks*; Allessio Leck, M., Parker, V.T., Simpson, R.L., Eds.; Academic Press, Inc.: Cambridge, MA, USA, 1989; Volume 1, pp. 3–8.
29. Forcella, F.; Eradat-Oskoui, K.; Wagner, S.W. Application of weed seedbank ecology to low-input crop management. *Ecol. Appl.* **1993**, *3*, 74–83. [CrossRef] [PubMed]
30. Gross, K.L. A comparison of methods for estimating seed numbers in the soil. *J. Ecol.* **1990**, *78*, 1079–1093. [CrossRef]
31. Cardina, J.; Sparrow, D.H. A comparison of methods to predict weed seedling populations from the soil seedbank. *Weed Sci.* **1996**, *44*, 46–51. [CrossRef]
32. Rutkowski, L. *Key to Identification of Vascular Plants in Lowland Poland*; PWN: Warsaw, Poland, 2007.
33. Grzesiuk, S.; Kulka, K. *Seed Physiology and Biochemistry*; PWRiL: Warsaw, Poland, 1981. (In Polish)
34. Shannon, C.E. A mathematical theory of communications. *Bell Syst. Tech. J.* **1948**, *27*, 379–423. [CrossRef]
35. Simpson, E.H. Measurement of diversity. *Nature* **1949**, *168*, 668. [CrossRef]
36. Kovach, W.L. *MVSP Version 3*; Kovach Computing Services: Anglesey, UK, 2011; p. 112.
37. Magurran, A.E. *Ecological Diversity and Its Measurement*; Princeton University Press: Princeton, NJ, USA, 1988. [CrossRef]
38. Conn, J.S. Weed seed bank affected by tillage intensity for barley in Alaska. *Soil Tillage Res.* **2006**, *90*, 156–161. [CrossRef]
39. Peigné, J.; Ball, B.C.; Roger-Estrade, J.; David, C. Is conservation tillage suitable for organic farming? A review. *Soil Use Manag.* **2007**, *23*, 129–144. [CrossRef]
40. Gruber, S.; Claupein, W. Effect of tillage intensity on weed infestation in organic farming. *Soil Tillage Res.* **2009**, *105*, 104–111. [CrossRef]
41. Demjanová, E.; Macák, M.; Dalovic, I.; Majernik, F.; Tyr, S.; Smatana, S. Effects of tillage systems and crop rotation on weed density, weed species composition and weed biomass in maize. *Agron. Res.* **2009**, *7*, 785–792.
42. Woźniak, A.; Wesołowski, M.; Soroka, M. Effect of long-term reduced tillage on grain yield, grain quality and weed infestation of spring wheat. *J. Agr. Sci. Tech.* **2015**, *17*, 899–908.
43. Vakali, C.; Zaller, J.G.; Köpke, U. Reduced tillage effects on soil properties and growth of cereals and associated weeds under organic farming. *Soil Tillage Res.* **2011**, *111*, 133–141. [CrossRef]
44. Woźniak, A. Effect of tillage system on the structure of weed infestation of winter wheat. *Span. J. Agric. Res.* **2019**, *16*, e1009. [CrossRef]
45. Starczewski, J.; Czarnocki, S. Tillage method and triticale weed infestation and yielding. *Acta Sci. Pol. Agric.* **2004**, *3*, 69–76. (In Polish)
46. Clements, D.R.; Benoit, D.L.; Murphy, S.D.; Swanton, C.J. *Tillage Effects on Weed Seed Return and Seedbank Composition*; Cambridge University Press: Cambridge, UK, 1996; pp. 314–322. [CrossRef]
47. Tokarska-Guzik, B.; Dajdok, Z.; Zając, M.; Zając, A.; Urbisz, A.; Danielewicz, W.; Hołdyński, C. *The Plants of Foreign Origin in Poland, With Particular Regard to Invasive Species*; The General Directorate for Environmental Protection: Warsaw, Poland, 2012; p. 197. Available online: http://www.gdos.gov.pl/files/artykuly/5050/Rosliny_obcego_pochodzenia_w_PL_poprawione.pdf (accessed on 4 May 2020). (In Polish)
48. Sans, F.X.; Berner, A.; Armengot, L.; Mäder, P. Tillage effects on weed communities in an organic winter wheat–sunflower–spelt cropping sequence. *Weed Res.* **2011**, *51*, 413–421. [CrossRef]
49. Puricelli, E.; Tuesca, D. Weed density and diversity under glyphosate-resistant crop sequences. *Crop Prot.* **2005**, *24*, 533–542. [CrossRef]
50. Froud-Williams, R.J.; Drennan, D.S.H.; Chancellor, R.J. Influence of cultivation regime on weed floras of arable cropping systems. *J. Appl. Ecol.* **1983**, *20*, 187–197. [CrossRef]
51. Woźniak, A.; Soroka, M. Biodiversity of weeds in pea cultivated in various tillage systems. *Rom. Agric. Res.* **2015**, *32*, 231–237.

52. Duer, I.; Feledyn-Szewczyk, B. Interaction of crop production system with weed species composition and soil seed bank in winter wheat. *Roczn. Nauk Rol.* **2000**, *115*, 109–130. (In Polish)
53. Vanasse, A.; Leroux, G.D. Floristic diversity, size and vertical distribution of the weed seedbank in ridge and conventional tillage systems. *Weed Sci.* **2000**, *48*, 454–460. [CrossRef]
54. Piskier, T.; Sekutowski, T.R. Effect of simplified tillage on the number and distribution of weed seeds in soil. *J. Res. Appl. Agric. Eng.* **2013**, *58*, 109–117.
55. Krawczyk, R.; Kubsik, K.; Mrówczyński, M.; Kaczmarek, S. Effect of soil variation and soil tillage systems on soil weed seedbank. *Prog. Plant Prot./Post. Ochr. Roślin* **2008**, *48*, 276–280. (In Polish)
56. Borin, M.; Zanin, G.; Zuin, M.C. The Comparison of Seed Banks in Conventional and Ridge-Tilled Soils In North-Eastern Italy. In *Soil Management in Sustainable Agriculture*; Cook, H.F., Lee, H.C., Eds.; Wye College Press: Ashford, UK, 1995; pp. 544–555.
57. Sheley, R.; James, J.; Rinella, M.; Blumenthal, D.; DiTomaso, J.; Briske, D. *Invasive Plant Management and Anticipated Conservation Benefits: A Scientific Assessment. Conservation Benefits of Rangeland Practices—Assessment, Recommendations, And Knowledge Gaps*; Briske, D.D., Ed.; Allen Press: Lawrence, Kansas, 2011; pp. 291–335. Available online: https://www.nrcs.usda.gov/Internet/FSE_DOCUMENTS/stelprdb1045802.pdf (accessed on 20 May 2020).
58. Kapeluszny, J.; Haliniarz, M. Selected elements of germination biology of flaxweed [*Descurainia sophia* Webb. ex Prantl.] and rye brome [*Bromus secalinus* L.]). *Sect. E Agric.* **2007**, *62*, 226–233. (In Polish)
59. Thomas, A.G.; Derksen, D.A.; Blackshaw, R.E.; Van Acker, R.C.; Légère, A.; Watson, P.R.; Turnbull, G.C. A multistudy approach to understanding weed population shifts in medium- to long-term tillage systems. *Weed Sci.* **2017**, *52*, 874–880. [CrossRef]
60. Grundy, A.C.; Mead, A.; Bond, W. Modelling the effect of weed—Seed distribution in the soil profile on seedling emergence. *Weed Res.* **1996**, *36*, 375–384. [CrossRef]
61. Dorado, J.; Del Monte, J.P.; López-Fando, C. Weed seedbank response to crop rotation and tillage in semiarid agroecosystems. *Weed Sci.* **1999**, *47*, 67–73. [CrossRef]
62. Ball, D.A. Weed Seedbank Response to Tillage, Herbicides, and Crop Rotation Sequence. *Weed Sci.* **1992**, *40*, 654–659. [CrossRef]
63. Kremer, R.J. Management of weed seed banks with microorganisms. *Ecol. Appl.* **1993**, *3*, 42–52. [CrossRef]
64. Hurtt, W.; Taylorson, R.B. 1986: Chemical manipulation of weed emergence. *Weed Res.* **1986**, *26*, 259–267. [CrossRef]

agriculture

MDPI

Article

Diversity of Segetal Flora in *Salix viminalis* L. Crops Established on Former Arable and Fallow Lands in Central Poland

Maria Janicka [1], Aneta Kutkowska [1,*] and Jakub Paderewski [2]

[1] Agronomy Department, Faculty of Agriculture and Biology, Institute of Agriculture, Warsaw University of Life Sciences—SGGW, Nowoursynowska 159, 02-776 Warsaw, Poland; maria_janicka@sggw.edu.pl

[2] Biometry Department, Faculty of Agriculture and Biology, Institute of Agriculture, Warsaw University of Life Sciences—SGGW, Nowoursynowska 159, 02-776 Warsaw, Poland; jakub_paderewski@sggw.edu.pl

* Correspondence: anetakutkowska@onet.pl

Abstract: The flora of willow (*Salix viminalis* L.) plantations consists of various plant groups, including plants related to arable land, called segetal plants. Knowledge of this flora is important for maintaining biodiversity in agroecosystems. The aim of the study was to assess the segetal flora of the willow plantations in central Poland, depending on the land use before the establishment of the plantations (arable land or fallow) and the age of the plantations. Moreover, the aim was also to check for the presence of invasive, medicinal, poisonous and melliferous species. The vegetation accompanying willow was identified based on an analysis of 60 phytosociological relevés performed using the Braun-Blanquet method. For each species, the following parameters were determined: the phytosociological class; family; geographical and historical group; apophyte origin; biological stability; life-form; and status as an invasive, medicinal (herbs), poisonous or melliferous species. The results were statistically processed. Segetal species accounted for 38% of the flora accompanying willow. The plantations on former arable land were richer in segetal species than those on fallow. Mostly, short-lived and native species dominated. In line with the age of the plantations, the number of segetal species decreased. The share of apophytes increased, and anthropophytes decreased. Furthermore, many valuable plants were found among the flora accompanying willow.

Keywords: *Salix viminalis* L. crops; energy crops; segetal flora; dynamic of flora; age of plantation; biodiversity; willow plantation; invasive species; medicinal species; melliferous species

check for updates

Citation: Janicka, M.; Kutkowska, A.; Paderewski, J. Diversity of Segetal Flora in *Salix viminalis* L. Crops Established on Former Arable and Fallow Lands in Central Poland. *Agriculture* **2021**, *11*, 25. https://doi.org/10.3390/agriculture11010025

Received: 7 October 2020
Accepted: 28 December 2020
Published: 2 January 2021

Publisher's Note: MDPI stays neutral with regard to jurisdictional claims in published maps and institutional affiliations.

1. Introduction

Climate change caused by increasing emissions of greenhouse gases into the atmosphere, the depletion of fossil fuel resources and the rising demand for energy [1,2] has led the world to take an interest in renewable energy sources [3,4]. Legal regulations coming into being oblige states to increase the proportion of energy derived from renewable sources [5–8]. According to Directive 2009/28/EC of the European Parliament and of the Council (of 23 April 2009 on the promotion of the use of energy from renewable sources, repealing Directives 2001/77/EC and 2003/30/EC), by 2020, the share of energy from renewable sources (RES) in Poland should be 15% of total energy consumption. In 2018, this share was 11.3% of gross final energy consumption and was 73% of the expected share of RES [9].

One of the sources of renewable energy is biomass obtained from energy plants such as shrubs and trees that grow quickly after cutting, e.g., poplar (*Populus* sp.), willow (*Salix viminalis* L.) and black locust (*Robinia pseudoacacia* L.); perennial forbs, e.g., Virginia mallow *Sida hermaphrodita* (L.) Rusby; and perennial grasses, e.g., *Miscanthus sp.* From these plants, in the northern and central parts of Europe, including Poland, the main energy crop is willow [10]. Poland's willow plantations were established mainly after

2000, and at present, willow is cultivated in Poland on about 8700 ha [11], including about 230 ha in Łódź Province ([12]; authors' estimations). Plantations of *Salix viminalis* L. are established on various soils, especially poor, light (class IV–VI) arable land, including fallow lands, which can be managed in this way [13,14]. These plantations are located in field depressions, among arable land; they are bordered by small rivers, forests and grasslands, thus enriching the agricultural landscape and becoming an important element of the environment for numerous fauna and flora [15–17]. Moreover, the cultivation of *Salix viminalis* L. for energy purposes is one of the activities contributing to stopping climate change (carbon sequestration) [18,19]

The flora of the energy willow plantation includes different groups of plants: meadow, woodland, shrub, ruderal and segetal species, as well as shrubs and tree seedlings. Segetal plants, which are the subject of this work, are plants mainly associated with arable land, especially with cereal crops and root crops. Their occurrence is influenced by a number of factors: climatic, geo-environmental, soil and anthropogenic [20–22]. The introduction of new cultivars of crops; the use of seed material well cleared of diaspora weeds; and improvements in tillage technology, herbicides and mineral fertilizers, as well as changes in the structure of sown crops and the simplification of crop rotation, all define the structure of segetal flora [23–27]. Its biodiversity decreases with the intensification of agricultural production and the emergence of monocultural, large-area crops, and as research findings show, many species have already disappeared irretrievably [28]. On *Salix viminalis* L. plantations, segetal plants have different conditions for their development than in typical fields due to the different technology for cultivating this species. The long period of time (2–3 years) between successive willow harvests often has an effect on the limited light conditions, and also on the thermal and humidity conditions, which prevents the plant community from stabilizing [29]. Moreover, the flora of willow plantations grown for energy purposes depends on factors such as crop age and soil conditions [30,31], as well as the land use before the plantation was established [30,32].

Although *Salix* sp. has been cultivated in Poland for many years, to date, segetal plants occurring on willow plantations have only been dealt with sporadically. The vegetation accompanying willow was usually assessed in the first years of cultivation and presented as a whole [33–35]. The flora of older *Salix viminalis* L. plantations is particularly poorly researched. Knowledge of segetal plants present on willow plantations is important for maintaining the biodiversity of native flora in agroecosystems [27]. Moreover, they are an important element of agrobiocenoses, as many are valuable medicinal and melliferous species [36,37]. The aims of this study were as follows: (1) to determine and analyse, in a broad sense, the species composition of segetal flora on plantations of *Salix viminalis* L. in central Poland, depending on the land use before establishing plantations (arable land and fallow land); (2) to assess the formation of segetal flora depending on the age of *Salix viminalis* L. plantations; (3) to analyse the flora accompanying energy willow crops and the choice of factors that are important for biodiversity plant species: invasive, medicinal, poisonous and melliferous.

2. Materials and Methods

2.1. Study Area and Characteristics of the Plantations

The study area is located in Łódź Province in the central part of Poland. It is, like most of the Łódź region, situated on a plain area. The study was carried out on eight commercial *Salix viminalis* L. plantations, in five localities: Dmosin (1 plantation: N 51.542775; E 19.451714; altitude: 157 m a.s.l. (above sea level)), Okołowice (1 plantation: N 51.431806; E 19.193768; altitude: 180 m a.s.l.), Piwaki (2 plantations: N 51.124068; E 19.453671; altitude: 224 m a.s.l. and N 51.131197; E 19.461013; altitude: 222 m a.s.l.), Podole (2 plantations: N 51.532193; E 19.30404; altitude: 166 m a.s.l. and N 51.533847; E 19.303854; altitude: 162 m. a.s.l.), Świątniki (2 plantations: N 51.425982; E 19.191872; altitude: 188 m a.s.l. and N 51.424868; E 19.194597; altitude: 183 m a.s.l.) (Figure 1). In total, the study was carried out in approx. 60% of all the localities of Łódź Province, where energy willow plantations, established on former arable (land under temporary crops) and fallow (arable land temporarily, for at least two growing seasons, excluded from agricultural production) lands, were located.

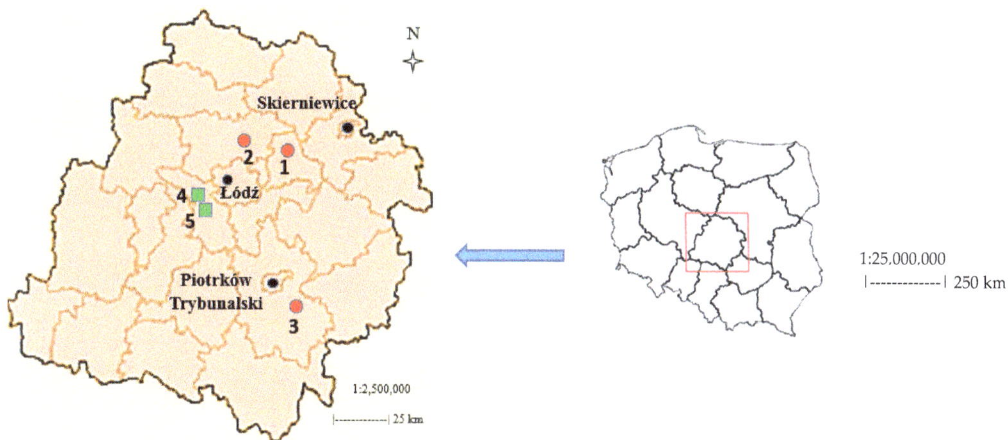

Figure 1. Location of the study area in Poland and Łódź Province. Explanations: ● Plantations established on former arable lands (AR): 1—Dmosin; 2—Podole; 3—Piwaki. ■ Plantations established on former fallow lands (FA): 4—Okołowice; 5—Świątniki.

The energy willow plantations that were studied differed in terms of age and the type of land use before their establishment. Five plantations were established on former arable lands (in 2004–2005) and three, on former fallow lands (in 2006–2008) (Figure 1). Plantations established on former arable land are hereinafter abbreviated as AR, and plantations established on former fallow land, as FA. Before the establishment of the *Salix viminalis* L. plantations (AR and FA), cereals (rye and oats) were mainly cultivated and an extensive farming system was used. One-year sprouts, at a density of 26,000 sprouts per ha, were planted on each plantation. Before planting the sprouts, the lands were ploughed and harrowed. The AR plantations were located among arable land, whereas the FA plantations were located among arable land and mixed forests. In subsequent years, all the plantations were managed in a similar manner. During the study period, they were not fertilized and no pesticides were used. There were also no treatments such as ploughing or mowing. Willow was harvested every 2–3 years in the same years on all the plantations.

The study was carried out between 2011 and 2015 and in 2018 on the same plantations in the subsequent years of research (Table 1). The species composition of the segetal flora depending on the land use before establishing the plantation was analysed on the basis of an assessment of AR plantations in 2011–2012 and 2013–2014 (6–7 years old and 9–10 years old, respectively) and FA plantations in 2014–2015 (6–7 years old) and 2018 (10 years old). The different study periods in the FA plantations (6–7 and 9–10 years old) compared to the AR, which was due to the later year of their establishment, did not affect the results (the number of species or coverage), because the vegetation developed in similar weather conditions (Table 2). Only 2013 was characterized by a higher sum of precipitation, although only single relevés (three) were included from this year.

Table 1. Number of phytosociological relevés carried out in individual localities of energy willow plantations.

Year of the Study	Age of Plantations	Type of Plantations/Locality				
		Plantations established on former arable lands (AR)				
		Dmosin	Podole 1	Podole 2	Piwaki 1	Piwaki 2
2011–2012	6–7 years old	6	1	1	1	1
2013–2014	9–10 years old	5	2	1	1	1
		Plantations established on former fallow land (FA)				
		Okołowice	Świątniki 1	Świątniki 2		
2011–2012	3–4 years old	5	2	3		
2013	5 years old	5	3	2		
2014–2015	6–7 years old	6	3	1		
2018	9–10 years old	6	3	1		

Table 2. Weather conditions in the growing seasons in the years 2011–2015 and 2018 (Meteorological Station in Bratoszewice).

Year Month	Average Monthly Air Temperature (°C)						
	2011	**2012**	**2013**	**2014**	**2015**	**2018**	**1971–2000**
IV	10.5	9.2	7.3	10.2	8.1	13.2	7.7
V	13.7	14.8	14.2	13.3	12.9	16.5	13.4
VI	17.9	16.5	17.6	15.7	16.4	18.3	16.1
VII	17.7	19.9	19.4	20.6	19.5	20.2	17.7
VIII	18.6	18.7	18.7	17.6	22.1	20.5	17.6
IX	15.0	14.3	11.7	14.6	14.6	15.3	13.0
X	8.9	8.2	9.8	9.9	7.3	9.9	8.2
IV-X	14.6	14.5	14.1	14.5	14.4	16.2	13.3
	Sums of Monthly Precipitation (mm)						
Month	**2011**	**2012**	**2013**	**2014**	**2015**	**2018**	**1971–2000**
IV	22.3	54.2	48.8	43.3	32.8	34.7	36.0
V	46.1	21.4	112.0	106.3	37.2	44.2	51.0
VI	58.8	70.9	131.4	61.2	29.7	24.7	68.0
VII	165.2	117.6	68.4	50.4	70.9	146.6	88.0
VIII	92.4	41.1	78.2	84.9	17.9	83.9	61.0
IX	6.5	58.1	89.8	31.9	61.1	29.4	51.0
X	23.1	36.7	25.1	24.4	52.6	58.2	40.0
IV-X	414.4	400.0	553.7	402.4	302.2	421.7	395.0

The dynamics of segetal flora were analysed on FA plantations starting in 2011 (3–4 years old) for 7 consecutive years (2011–2018). The paper presents the results of flora from these plantations when they were 3–4 years old, 5 years old, 6–7 years old and 9–10 years old.

2.2. Soil Conditions

All the *Salix viminalis* L. plantations were located on soils that, according to Polish Soil Classification 2019 [38] and WRB Classification 2015 [39], were categorised as Cambisols. According to "Particle size distribution and textural classes of soils and mineral material-Classification of Polish Society of Soil Science 2008" [40], these soils were sand (FA) and loamy sand (AR). The soils had a very acid reaction (pH $_{KCl}$ 4.1–4.4), with a medium to high phosphorous content (53.1–81.9 mg·kg^{-1} soil), but were very poor and poor in their potassium (30.7–63.0 mg·kg^{-1} soil) and magnesium (11.0–17.0 mg·kg^{-1} soil) concentrations. The content of organic matter ranged from 1.97 to 2.28%.

The soils on which the AR plantations were located belonged to the soil agricultural complexes 5 and 6, whereas the soils, on which the FA plantations were located were classified into complexes 6 and 7. Soil agriculture complex 5 (soils suitable for rye) includes soils slightly sensitive to droughts. This is underlying loam material that reduces the permeability and retains the water supply in the soil, thus providing favourable conditions for plants. Soil agriculture complex 6 (soils less suitable for rye) includes soils often too dry. On complex 6 soils, plants only provide satisfactory yields in wet years. Soil agriculture complex 7 (soils very bad for rye) includes very dry sandy soils.

The information about the soil type and soil agriculture complex is based on agricultural maps on a scale of 1:5000 obtained from the Voivodeship Geodesy Office in Łódź (Solna St. 14, 91–423 Łódź, Poland), the Łódź Voivodeship Geoportal [41] and Polish Soil Classification 2019 [38]. Soil samples were taken with an Egner–Riehm stick (up to a depth of 20 cm) in accordance with the methods for soil study material. Topsoil analysis was carried out at the Regional Chemistry–Agriculture Station in Łódź and Warsaw. The contents of available phosphorus and available potassium were determined according to the Egner–Riehm (DL) method in calcium lactate extract, and the magnesium content was determined according to the Schatschabel method in calcium chloride extract. In addition, the soil pH in KCl and the content of organic matter according to the Tiurin method were identified. The soil texture according to research procedure (PB 40 ed. 3, 14.02.2011) was also identified.

2.3. Weather Conditions

The weather conditions during the study period were determined based on the dates recorded at the Meteorological Station of the Institute of Soil Science and Plant Cultivation, located in Bratoszewice, near Łódź. The long-term data were recorded for Łódź. The sum of precipitation in the vegetation period in the years 2011, 2012, 2014 and 2018 was similar and was 400.0–421.7 mm (Table 2). However, in 2013, it was about 40% higher than the long-term mean, and in 2015, it was about 24% lower than the long-term mean (Table 1). The greatest sum of precipitation was noted in May and June of that year (2015). The mean air temperatures between April and October in 2011–2015 were similar to each other (14.1–14.6 °C) and about 0.8–1.3 °C higher than the long-term mean. In 2018, the temperature was about 2.9 °C higher than the long-term mean. Detailed data regarding the average monthly air temperatures and the sums of monthly precipitation are shown in Table 2.

2.4. Methods

The vegetation accompanying willow (*Salix viminalis* L.) energy crops was identified based on an analysis of 60 phytosociological relevés (descriptions of vegetation patches—phytocenoses), i.e., ten relevés carried out in each group of plantations mentioned above. The number of phytosociological relevés carried out on an individual plantation depended on its area and soil diversity; i.e., if the plantation was located on two different soil complexes, separate relevés were taken on each of them. Details of the number of relevés taken on individual plantation is presented in Table 1. Each relevé had an area of 100 m^2, was roughly square-shaped, and was made using the Braun-Blanquet [42] method. This method simultaneously captures the number and degree of coverage of a given species in what is called a phytosociological relevé. When producing such a relevé, the first stage involved selecting a uniform, typical patch of vegetation, i.e., one without any unusual places, e.g., hills and hollows. In turn, a list was made of all the species that were present in the area, and the coverage of a given plant species was estimated (plant image on the surface) using the Braun-Blanquet scale (1–5, +, r); the meaning of individual symbols is presented in Table 3 [43,44]. The research was mainly carried out in June and July each year.

Table 3. Quantitative scale of Braun-Blanquet coverings.

Symbol	Cover (%)
5	75–100
4	50–75
3	25–50
2	5–25
1	Less than 5 % of the test area
+	Rarely, with slight coverage
r	Very rarely, one or more specimens

Plant communities were distinguished by characteristic and differential species according to Matuszkiewicz [45], affiliated to phytosociological classes, and analysed in each group where they occurred. Subsequently, the number and proportions of species were determined. The proportion of each plant species was determined based on constancy class (S): V—80–100% of all phytosociological relevés; IV—60–80%; III—40–60%; II—20–40%; I—0.01–20%. The cover coefficient (D) was calculated according to the following formula: the sum of the average percentage of species cover that occurred in all the phytosociological relevés divided by the total number of phytosociological relevés and multiplied by 100 [43].

For each species, the following parameters were determined: family; geographical and historical groups; apophyte (i.e., synanthropic species of local origin) origin; biological stability; life-form; and status as an invasive, medicinal (herb), poisonous or melliferous species. The geographical and historical groups, apophyte origins, biological stability and life-form were mainly identified based on the following sources: Anioł-Kwiatkowska [46], Korniak [47], Mirek et al. [48], Rutkowski [49], Sowa and Warcholińska [50], Szafer et al. [51], Zając and Zając [52,53], and Zając [54]. Invasive species status was determined based on Tokarska-Guzik et al. [55]. Medicinal and poisonous plants were identified based on Rutkowski [49]. Melliferous plants were determined based on Sulborska [56]. Tree seedlings were not classified as melliferous plants. The Latin names of the vascular plants are in accordance with those presented by Mirek et al. [48], and the phytosociological classifications follow Matuszkiewicz [45].

2.5. Statistical Analysis

Canonical correspondence analysis (CCA, constrained ordination) was used to visualize the main differences in species composition occurring in the six groups of plantations (the age of the plantations and the type of land use before establishing the plantations) that were studied [57,58]. The relative species abundance was analysed by this method. The explanatory variable (categorical) was the phytosociological relevé's affiliation to one of the six groups. The significance of constraints (with plantation types treated as the dummy variables) was tested by means of the permutation test [59].

Cluster analysis based on the mean species coverage was conducted for each of the six plantation groups. The Euclidean measure and Ward method were used. The coverage of each species was averaged for each of the plantation types. This 2-dimensional table (plantation type by species) was analysed by cluster analysis. The similarities between the age influence and previous land use influence on the plant composition are shown in the dendrogram chart.

The species richness was calculated according to the first-order jackknife species richness estimator [60,61]. This statistic introduces a correction for species not recorded in the abundance table during the study.

These statistical analyses were conducted with the use of the R software [62]. The CCA analysis was performed by the "cca" function contained in the R "vegan" package. The cluster analysis was performed by the "dist" and "hclust" functions. The "specpool" function from the "vegan" package was used for the estimated species richness.

3. Results

The flora of all eight *Salix viminalis* L. plantations that were studied totalled 114 vascular plant species. A large proportion of them were segetal species, i.e., 43 species (38%). The characteristics of this group of plants on 6–7-year-old and on 9–10-year-old energy willow plantations established on former arable (AR) and fallow (FA) lands are presented below. Then, the dynamics of the segetal flora on 3–10-year-old willow plantations established on former fallow lands were analysed in terms of four age groups (3–4, 5, 6–7 and 9–10 years old). In turn, the invasive, medicinal, poisonous and melliferous species accompanying the *Salix viminalis* L. plantations were discussed.

3.1. Characteristics of the Segetal Flora Depending on the Land Use before Establishing Plantations

The flora of the *Salix viminalis* L. plantations (6–7-year-old and 9–10-year-old) established on former arable (AR) and fallow land (FA) had a total of 103 vascular plant species. The species richness estimated by the jackknife method for the 6–7-year-old and 9–10-year-old plantations established on the arable land (AR) was 87.1 and 82.1, respectively. That is, the estimated species number included missed species was a little higher for 6–7-year-old plantations. Among the observed species, 61 and 56 (for 6–7-year-old and 9–10-year-old plantations, respectively) segetal species, which occurred mainly in cereal and root crops, accounted for 37% (38 species) (Table 4). On the plantations, the species characteristics for cereal crops such as *Anthoxanthum aristatum* Boiss., *Matricaria maritima subsp. inodora* (L.) Dostál, *Vicia hirsuta* (L.) Gray, *Vicia tetrasperma* (L.) Schreb. and *Vicia villosa* Roth. were recorded, while the species characteristics were also recorded for root crops including *Chenopodium album* L. and *Sonchus arvensis* L. The segetal species belonged to 15 botanical families. The most numerous family was *Asteraceae* (eight species), *Fabaceae* (seven species) and *Poaceae* (five species) (Table 4).

The study shows that more segetal species were on the AR plantations than on the FA (Table 4). On 6–7-year-old AR plantations, their number was 30% higher than on the FA, but on 9–10-year-old plantations, the difference was greater (41%) (Table 4). In both groups of 6–7-year-old plantations (AR and FA), the same 12 species were recorded, i.e., 32% of the segetal flora, and on older plantations (9–10 years old), eight species (21% of the segetal flora) were the same. These were mostly perennial plants, mainly reproducing vegetatively through rhizomes or stolons, such as *Achillea millefolium* L. s.str., *Artemisia vulgaris* L., *Cirsium arvense* (L.) *Scop*., *Convolvulus arvensis* L., *Elymus repens* (L.), *Rumex acetosella* L., and two short-lived (1–2 years) species, i.e., *Trifolium arvense* (L.) and invasive species *Conyza canadensis* (L.) Cronquist, characterised by high seed production. The remaining species occurred differently on both types of plantations (Table 4).

In line with the age of the plantations, the number of segetal plants decreased in both groups of plantations—on the AR, from 27 to 22, and on the FA, from 19 to 13 species. Only in younger plantations (6–7 years old) were there species such as *Anthoxanthum aristatum* Boiss., *Galeopsis bifida* Boenn., *Matricaria maritima subsp. inodora* (L.) Dostál, *Medicago lupulina* L., *Setaria pumila* (Poir.), *Vicia tetrasperma* (L.) and *Vicia villosa* Roth. They were mainly short-lived, photophilous species. On plantations established on former arable land, 37% of the species disappeared, while nearly half (47%) disappeared on formerly fallow land. They were replaced with new species (found only on older 9–10-year-old plantations), which constituted 23% of each group of plantations (Table 4). They were also mainly short-lived species reproducing from seeds, e.g., *Polygonum hydropiper* L., *Sonchus asper* L. and *Veronica arvensis* L. on AR and *Thlaspi arvense* L. and *Chenopodium album* L. on FA.

In the segetal flora, regardless of the type of land use before establishing the energy willow plantations, species of native origin (apophytes) dominated. With older plantations, the proportion of native species increased and that of anthropophytes decreased; this was more visible on the FA plantations (Figure 2).

Table 4. Constancy classes, botanical families, phytosociological classes, life-forms, and geographical and historical groups of segetal species occurring on 6–7-year-old and 9–10-year-old *Salix viminalis* L. plantations established on former arable land and fallow.

Species	Arable Land		Fallow		Botanical Family	Phytosociological Class	Life-Form	Geographical and Historical Group
	Plantation Age in Years							
	6–7	9–10	6–7	9–10				
Elymus repens (L.) Gould	V	IV	I	II	*Poaceae*	Agr int-rep	G	Anw
Artemisia vulgaris L.	IV	II	IV	II	*Asteraceae*	Artemi	C	Al
Cirsium arvense (L.) Scop	IV	II	I	I	*Asteraceae*	Artemi	G	Al
Achillea millefolium L.s. str.	III	III	IV	IV	*Asteraceae*	Moll-Arr	H	Ał
Conyza canadensis (L.) Cronquist	III	I	II	II	*Asteraceae*	Stel med	T	An
Poa annua L.	III	I			*Poaceae*	Artemi	T	Ał
Vicia hirsuta (L.) Gray	II	III	III		*Fabaceae*	Stel med	T	An
Digitaria ischaemum (Schreb.) H.L. Mühl.	II	I			*Poaceae*	Stel med	T	An
Equisetum arvense L.	II	I			*Equisetaceae*	Agr int-rep	G	Ał
Convolvulus arvensis L.	II	I	I	I	*Convolvulaceae*	Agr int-rep	H	Amk
Matricaria maritima subsp.inodora (L.) Dostál	II				*Asteraceae*	Stel med	T	An
Galeopsis tetrahit L.	I	II			*Lamiaceae*	Stel med	T	Al
Rumex acetosella L.	I	I	III	II	*Polygonaceae*	Stel med	G	Ap
Chenopodium album L.	I			I	*Chenopodiaceae*	Stel med	T	Anw
Galeopsis bifida Boenn.	I		I		*Lamiaceae*	t	T	Al
Galium aparine L.	I	I			*Rubiaceae*	Artemi	T	Al
Medicago lupulina L.	I				*Fabaceae*	t	T	Amk
Melandrium album (Mill.) Garcke	I	I			*Caryophyllaceae*	Artemi	T	Ał
Myosotis arvensis (L.) Hill	I	I			*Boraginaceae*	Stel med	T	An
Plantago major L.	I				*Plataginaceae*	Moll-Arr	H	Al
Setaria pumila (Poir.) Roem. and Schult.	I		I		*Poaceae*	Stel med	T	An
Sonchus oleraceus L.	I				*Asteraceae*	Stel med	T	An
Stachys palustris L.	I				*Lamiaceae*	Moll-Arr	G	Ał
Stellaria media (L.) Vill.	I	I			*Caryophyllaceae*	Stel med	T	Ał
Trifolium arvense L.	I	I	II	I	*Fabaceae*	Koele-Coryne	T	Ap
Vicia angustifolia L.	I			I	*Fabaceae*	Stel med	T	An
Vicia cracca L.	I		I	I	*Fabaceae*	Moll-Arr	H	Ał
Anthoxanthum aristatum Boiss.			I		*Poaceae*	Stel med	T	An
Cardaminopsis arenosa (L.) Hayek			II	I	*Brassicaceae*	t	H	Ap
Cerastium holosteoides Fr. emend. Hyl.		I			*Caryophyllaceae*	Moll-Arr	T	Ał
Polygonum hydropiper L.		I	II		*Polygonaceae*	Bident	T	Anw

Table 4. *Cont.*

Species	Arable Land		Fallow		Botanical Family	Phytosociological Class	Life-Form	Geographical and Historical Group
	Plantation Age in Years							
	6–7	9–10	6–7	9–10				
Sonchus arvensis L.		I	I		*Asteraceae*	Stel med	H	Anw
Sonchus asper (L.) Hill		I			*Asteraceae*	Stel med	T	An
Thlaspi arvense L.				I	*Brassicaceae*	Stel med	T	An
Veronica arvensis L.		I	I		*Scrophulariaceae*	t	T	An
Vicia tetrasperma (L.) Schreb.			I		*Fabaceae*	Stel med	T	An
Vicia villosa Roth			I		*Fabaceae*	Stel med	T	An
Viola arvensis Murray			II		*Violaceae*	Stel med	T	An
Number of species	27	22	19	13	-	-		

Explanations: I–V constancy classes, empty space—species did not occur. Phytosociological classes: Agr int-rep—Agropyretea intermedio-repentis; Artemi—Artemisietea vulgaris; Bident—Bidentetea tripartiti; Koele-Coryne—Koelerio glaucae-Corynephoretea canescentis; Moll-Arr—Molinio-Arrhenatheretea; Stel med- Stellarietea mediae; t—Sporadic species. Life-forms according to Raunkiaer: C—herbaceous chamaephyte; G—geophyte; H—hemicryptophyte; T—terophyte. Geographical and historical groups: Al—woody-shrub apophyte; Ał—meadow apophyte; Amk— xerothermic apophyte; Anw—waterside and wetside apophyte; Ap—sandyside apophyte; An—antropophyte.

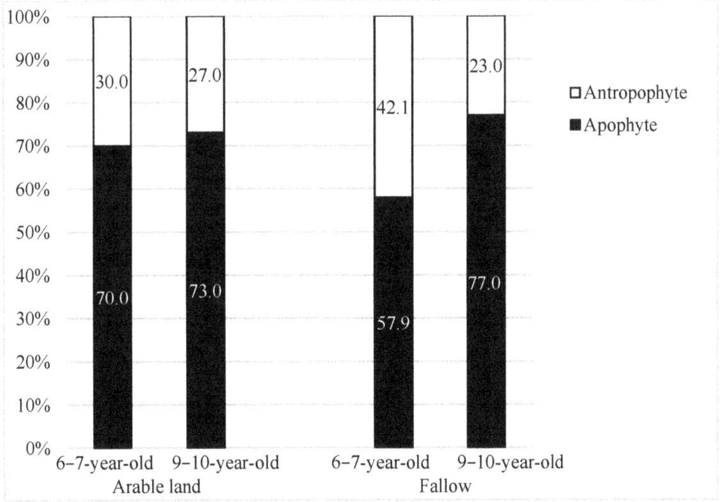

Figure 2. The origin of segetal species (%) occurring on 6–7-year-old and 9–10-year-old *Salix viminalis* L. plantations established on former arable land and fallow.

Woody-shrub apophytes (native species of natural forest and shrub community origin, well-established in anthropogenic habitats) dominated in the structure of the agrophytocenoses in both types of plantation. On AR, their share was slightly higher than on FA (Figure 3). Besides woodland-shrub apophytes, meadow apophytes make up a high proportion on AR plantations, and sandyside apophytes (native species of sandy habitat, where soils are light, dry, low in nutrients and often acidic) constitute a high proportion on FA plantations. Among these, the following have the highest coverage: *Achillea millefolium* L. s.str and *Artemisia vulgaris* L. (Table 5).

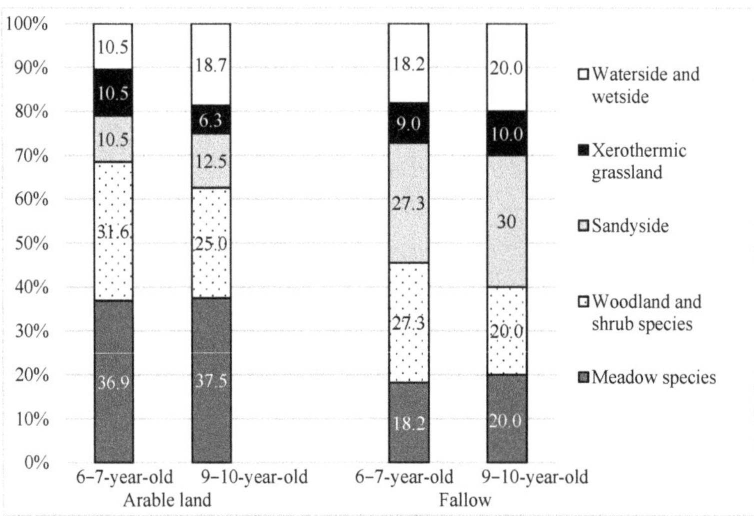

Figure 3. The origin of segetal apophytes (%) occurring on 6–7-year-old and 9–10-year-old *Salix viminalis* L. plantations established on former arable land and fallow.

Table 5. Constancy classes (S) and cover coefficient (D) of segetal species, which occurred with the highest constancy classes in all groups of *Salix viminalis* L. plantations established on fallow and arable land.

Species	Fallow								Arable Land			
	Plantation Age in Years											
	3–4		5		6–7		9–10		6–7		9–10	
	S	D	S	D	S	D	S	D	S	D	S	D
Rumex acetosella L.	IV	204	III	153	III	202	II	102	I	50	I	50
Artemisia vulgaris L.	III	152	III	54	IV	253	II	4	IV	203	II	151
Elymus repens (L.) Gould	IV	650	IV	725	I	425	II	325	V	2451	IV	1750
Achillea millefolium L. s. str.	IV	551	V	780	IV	650	IV	725	III	201	III	277
Vicia hirsuta L. (Gray)	II	3	IV	204	III	250	0	0	II	276	III	54
Viola arvensis Murray	II	4	III	5	II	151	0	0	0	0	0	0
Convolvulus arvensis L.	II	53	IV	55	I	51	I	1	II	101	I	50
Trifolium arvense L.	II	101	III	327	II	151	I	2	I	175	I	50
Conyza canadensis (L.) Cronquist	I	2	I	3	II	52	II	150	III	152	I	100
Cirsium arvense (L.) Scop	I	50	0	0	I	1	I	50	IV	203	II	150
Poa annua L.	0	0	0	0	0	0	0	0	III	152	I	175

The study shows considerable diversity in the flora in terms of the share of life forms and biological stability. Regardless of the type of former land use, before the establishment of willow plantations, therophytes dominated in the communities (Figure 4). However, on AR, their proportion was higher than on FA, especially on older plantations (9–10 years old) by approx. 35% (i.e., by nine species). Taking into account the biological stability of plants, on AR plantations, over 63% of the species were short-lived species (Figure 5), whereas on the FA ones, the proportion of short-lived species was lower, especially on older plantations. The shading of the soil caused by the willow plants and the lack of arable land use (ploughing) meant that perennial plants dominated in line with the age, and on 9–10-year-old FA, these plants constituted approx. 54% of plants. On AR plantations, the proportion of perennial plants was at a similar level.

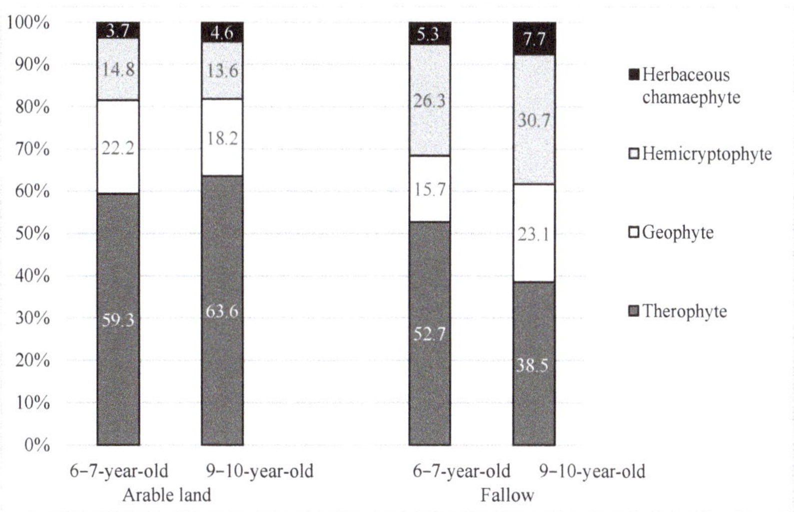

Figure 4. Raunkier life-forms (%) of segetal species occurring on 6–7-year-old and 9–10-year-old *Salix viminalis* L. plantations established on former arable land and fallow.

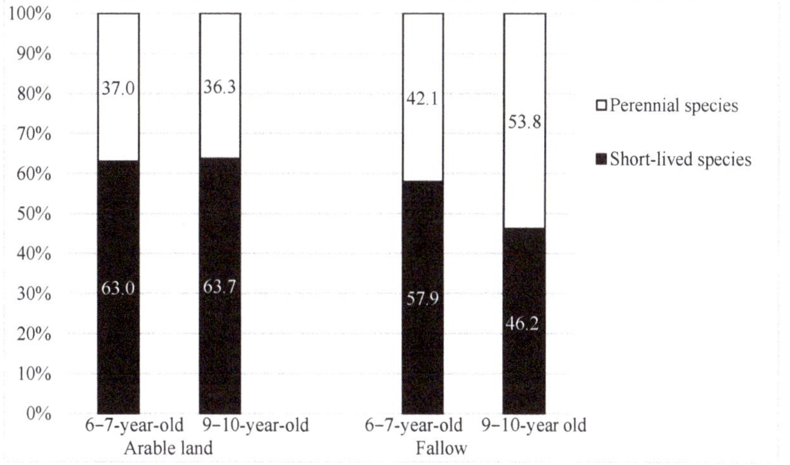

Figure 5. Biological stability (%) of segetal species occurring on 6–7-year-old and 9–10-year-old *Salix viminalis* L. plantations established on former arable land and fallow.

The vast majority of the segetal plants on both types of plantations belonged to the *Stellarietea mediae* class, which is characteristic of arable land (with an average of about 43%) (Table 4). Species from the *Artemisietea vulgaris* class characteristic of nitrophilous communities, perennials and climbers found in ruderal habitats were noted more often on AR plantations than on FA (Figure 6). The next groups, in order, were species from the *Molinio-Arrhenatheretea* class, characteristic of semi-natural and anthropogenic meadow communities, and species from the *Agropyretea intermedio-repentis* class, characteristic of rhizome and stolon plants. Moreover, the willow plantations were accompanied by several species from other classes: *Bidentetea tripartiti; Koelerio glauce—Corynephoretea canescentis;* and species not classified as being in any of the phytosociological classes but enriching the floristic composition of communities (Figure 6).

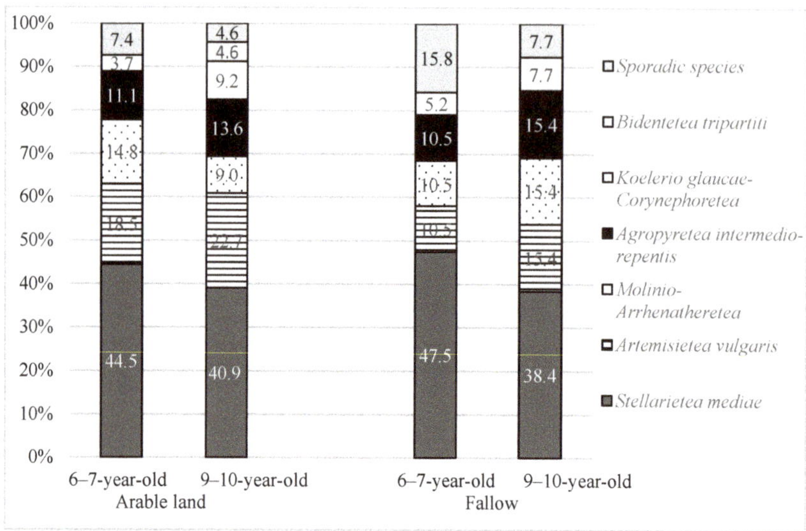

Figure 6. Phytosociological classes (%) of segetal species occurring on 6–7-year-old and 9–10-year-old *Salix viminalis* L. plantations established on former arable land and fallow.

The species from the *Agropyretea intermedio-repentis* class had the highest share of soil coverage on AR plantations, among which *Elymus repens* (L.) Gould played an important role. On the other hand, on FA, the highest share was of species from the *Molinio-Arrhenatheretea* and *Stellarietea mediae* classes, the coverage of which was mainly determined by *Achillea millefolium* L. s. str. and *Rumex acetosella* L. (Tables 5 and 6). Species from the *Stellarietea mediae* class, despite making up a considerable proportion, were characterised by a low cover coefficient. Among the species from this class, *Rumex acetosella* L. occurred more frequently and with greater coverage on FA plantations, and *Vicia hirsuta* (L.) Gray and *Conyza canadensis* (L.) Cronquist. occurred more frequently on AR plantations (Table 5).

Table 6. Cover coefficients (D) of phytosociological classes and their shares (%) in phytosociological classes on *Salix viminalis* L. plantations established on fallow and arable land.

Class	Fallow						Arable Land					
	Plantation Age in Years											
	6–7		9–10		Total (6–10)		6–7		9–10		Total (6–10)	
	D	% D	D	% D	D	% D	D	% D	D	% D	D	% D
Stellarietea mediae	857	32.4	255	18.7	1112	27.8	692	16.0	457	13.3	1149	14.8
Artemisietea vulgaris	254	9.6	54	4	308	7.7	712	16.5	626	18.2	1338	17.2
Molinio-Arrhenatheretea	751	28.4	727	53.3	1478	36.9	254	5.8	278	8	532	6.8
Agropyretea intermedio-repentis	476	18	326	23.9	802	20	2503	57.7	1801	52.4	4304	55.4
Koelerio glaucae-Corynephoretea canescentis	151	5.8	2	0.1	153	3.8	175	4	50	1.5	225	2.9
Bidentetea tripartiti	0	0	0	0	0	0	0	0	50	1.5	50	0.6
Sporadic species	153	5.8	1	0	154	3.8	2	0	176	5.1	178	2.3
Total	2642	100	1365	100	4007	100	4338	100	3438	100	7776	100

An analysis of the constancy classes and the cover coefficients showed that most of the segetal species in both types of plantations were in the I and II constancy classes. Only 25% of species on FA and 28% on AR were in higher constancy classes, i.e., III, IV and V (Table 4). It was found that segetal species achieved much higher cover coefficients on AR plantations (D = 7776) compared to FA (D = 4007) (Table 5).

The species that were in higher constancy classes and had cover coefficients above 600 were perennial plants that produced numerous rhizomes and stolons: *Achillea millefolium* L. s.str. on FA (sum of D = 1375) and *Elymus repens* (L.) Gould on AR (sum of D = 4201). In accordance with the ages of the plantations, all the species except *Achillea millefolium* L. s.str. were lowered in their constancy classes and cover coefficients. Due to the many years of willow cultivation, some species of the *Stellarietea mediae* class, usually accompanying cultivation on arable land, did not occur on 9–10-year-old plantations, e.g., *Matricaria maritima subsp. inodora* (L.) Dostál, *Setaria pumilla* (Poir.) Roem and Schult., and *Vicia tetrasperma* (L.) Schreb. (Table 3). The coverage with species from this class decreased in both types of plantation (Table 4).

3.2. Changes in the Segetal Flora on Salix viminalis L. Plantations Established on Fallow, Depending on the Age of the Plantation

With the aim of determining changes in the species composition of the segetal flora accompanying the willow with time, the flora of *Salix viminalis* L. plantations of different ages, i.e., 3–10 years old, and established on fallow land were analysed. The species richnesses estimated by the jackknife method were 66.3, 72.9, 73.6 and 64.0 for 3–4-, 5-, 6–7- and 9–10-year-old plantations, respectively. A total of 91 vascular plants species were recorded on these plantations, including 34 segetal species, which accounted for 37% of the total flora. The species belonged to six phytosociological classes and 14 botanical families. The most numerous were *Asteraceae* (six species), *Fabaceae* (six species) and *Poaceae* (five species). The remaining families consisted of 1–3 species (Table 7). On the youngest (3–5 years old) willow plantations, the proportion of segetal species was the highest (approx. 40.0%) and decreased with the age of the plantation; on 6–7-year-old plantations, it constituted 34.5%, and on 9–10-year-old ones, 27.6%. The species that were not confirmed in the flora of the 9–10-year-old plantations were short-lived anthropophytes: *Vicia hirsuta* (L.) Gray, *Vicia villosa* Roth, and *Viola arvensis* Murray.

Table 7. Segetal species that occurred on 3–10-year-old *Salix viminalis* L. plantations established on fallow, and their affiliations to botanical families and phytosociological classes.

Name of Species	Plantation Age in Years				Family	Phytosociological Classes
	3–4	5	6–7	9–10		
Achillea millefolium L. s.str.	+	+	+	+	*Asteraceae*	*Molinio-Arrhenatheretea*
Anchusa arvensis (L.) M. Bieb.		+			*Boraginaceae*	*Stellarietea mediae*
Anthoxanthum aristatum Boiss.			+		*Poaceae*	*Stellarietea mediae*
Apera spica-venti (L.) P. Beauv.		+			*Poaceae*	*Stellarietea mediae*
Arenaria serpyllifolia L.		+			*Caryophyllaceae*	*Sporadic species*
Artemisia vulgaris L.	+	+	+	+	*Asteraceae*	*Artemisieta vulgaris*
Cardaminopsis arenosa (L.) Hayek			+	+	*Brassicaceae*	*Sporadic species*
Chenopodium album L.		+		+	*Chenopodiaceae*	*Stellarietea mediae*
Chenopodium polyspermum L.	+				*Chenopodiaceae*	*Stellarietea mediae*
Cirsium arvense (L.) Scope	+		+	+	*Asteraceae*	*Artemisieta vulgaris*
Convolvulus arvensis L.	+	+	+	+	*Convolvulaceae*	*Agropyretea intermedio-repentis*
Conyza canadensis (L.) Cronquist	+	+	+	+	*Asteraceae*	*Stellarietea mediae*
Elymus repens (L.) Gould	+	+	+	+	*Poaceae*	*Agropyretea intermedio-repentis*
Equisetum arvense L.	+	+			*Equisetaceae*	*Agropyretea intermedio-repentis*
Galeopsis bifida Boenn.			+		*Lamiaceae*	*Sporadic species*
Galeopsis tetrahit L.		+			*Lamiaceae*	*Stellarietea mediae*
Galium aparine L.	+	+			*Rubiaceae*	*Artemisieta vulgaris*
Matricaria maritima subsp. inodora (L.) Dostál	+				*Asteraceae*	*Stellarietea mediae*
Melandrium album (Mill.) Gracke		+			*Caryophyllaceae*	*Artemisieta vulgaris*
Myosotis arvensis (L.) Hill	+	+			*Boraginaceae*	*Stellarietea mediae*
Rumex acetosella L.	+	+	+	+	*Polygonaceae*	*Stellarietea mediae*
Setaria pumila (Poir.) Roem & Schult.	+		+		*Poaceae*	*Stellarietea mediae*
Setaria viridis (L.) P. Beauv.	+	+			*Poaceae*	*Stellarietea mediae*
Sonchus arvensis L.			+		*Asteraceae*	*Stellarietea mediae*

Table 7. *Cont.*

Name of Species	Plantation Age in Years				Family	Phytosociological Classes
	3–4	5	6–7	9–10		
Stachys palustris L.	+				*Lamiaceae*	*Molinio-Arrhenatheretea*
Thlaspi arvense L.				+	*Brassicaceae*	*Stellarietea mediae*
Trifolium arvense L.	+	+	+	+	*Fabaceae*	*Koelerio glauce-Corynephoretea*
Veronica arvensis L.			+		*Scrophulariaceae*	*Sporadic species*
Vicia angustifolia L.	+	+		+	*Fabaceae*	*Stellarietea mediae*
Vicia cracca L.	+	+	+	+	*Fabaceae*	*Molinio-Arrhenatheretea*
Vicia hirsuta (L.) Gray	+	+	+		*Fabaceae*	*Stellarietea mediae*
Vicia tetrasperma (L.) Schreb.			+		*Fabaceae*	*Stellarietea mediae*
Vicia villosa Roth	+	+	+		*Fabaceae*	*Stellarietea mediae*
Viola arvensis Murray	+	+	+		*Violaceae*	*Stellarietea mediae*
Number of species	21	22	19	13		

+ means the species was present; empty space—species did not occur.

The species that occurred in all the age groups of the plantations studied were *Achillea millefolium* L. s.str., *Artemisia vulgaris* L., *Convolvulus arvensis* L., *Elymus repens* (L.), *Rumex acetosella* L., *Trifolium arvense* (L.), *Vicia cracca* and the invasive species *Conyza canadensis* (L.) Cronquist. Among these, four occurred in higher constancy classes (IV and III) (Table 5): *Achillea millefolium* L. s.str, *Artemisia vulgaris* L., *Elymus repens* (L.) and *Rumex acetosella* L.

On all the willow plantations, native species predominated over foreign species. Their share increased with the age of the plantations and ranged from 57% on 3–4-year-old plantations to 77% on 9–10-year-old ones (Table 8, Figure 2). Among the apophytes on 3-4-year-old plantations, species of meadow origin (33.3%) and woodland-shrub origin (25%) predominated. On 9–10-year-old plantations, their share decreased and amounted to 20% of each plantation (Table 8, Figure 3).

Table 8. Characteristics of segetal flora occurring on different age (3–4-, 5-, 6–7- and 9–10-year-old) *Salix viminalis* L. plantations established on fallow.

Category	Plantation Age in Years							
	3–4		5		6–7		9–10	
	No.	%	No.	%	No.	%	No.	%
Geographical and Historical Groups								
Antropophytes	9	43	9	41	8	42.1	3	23
Apophytes	12	57	13	59	11	57.9	10	77
Apophyte origin								
Meadow species	4	33.3	4	30.8	2	18.2	2	20
Woodland and shrub species	3	25.0	3	23.0	3	27.3	2	20
Sandyside	2	16.7	2	15.4	3	27.3	3	30
Xerothermic grasslands	1	8.3	2	15.4	1	9.0	1	10
Waterside and wetside	2	16.7	2	15.4	2	18.2	2	20
Biological stability								
Perennial species	9	43	7	32	8	42.1	7	53.8
Short-lived species	12	57	15	68	11	57.9	6	46.2
Life-form								
Hemicryptophyte	3	14.3	3	13.6	5	26.3	4	30.7
Therophyte	12	57.1	15	68.2	10	52.7	5	38.5
Geophyte	5	23.8	3	13.6	3	15.7	3	23.1
Herbaceous chamaephyte	1	4.8	1	4.6	1	5.3	1	7.7

An analysis of the biological stability showed that on 3–7-year-old plantations, short-lived species dominated (60% on average), while on older 9–10-year-old plantations, perennial species did (53.8%). In the biological spectrum, therophytes dominated on the 3–5-year-old plantations (constituting, on average, about 62%). Their share decreased in line with the age of the plantations, and on 9–10-year-old plantations, it was 38.5%. On the other hand, the share of hemicryptophytes increased, on average, from 17 to 30.7% (Table 8, Figure 2).

In all the groups of plantations (from 3 to 10 years old), species from the *Stellarietea mediae* class made up the highest proportions, with the following classes next, in order: *Artemisieta vulgaris*, *Agropyretea intermedio-repentis* and *Molinio-Arrhenatheretea*. On 9–10-year-old plantations, the proportion of species from the *Stellarietea mediae* class decreased, and species from the classes *Artemisieta vulgaris*, *Molinio-Arrhenatheretea* and *Agropyretea intermedio-repentis* increased (Table 9).

Table 9. The share of phytosociological classes (%) and cover coefficients (D) and their shares (D%) on 3–4-year-old, 5-year-old, 6–7-year-old and 9–10-year-old *Salix viminalis* L. plantations established on fallow.

Class	Plantation Age in Years											
	3–4			5			6–7			9–10		
	%	D%	D	%	D%	D	%	D%	D	%	D%	D
Stellarietea mediae	52.4	14.2	274	54.6	19.6	475	47.5	32.4	857	38.4	18.7	255
Artemisietea vulgaris	14.3	13.2	256	13.6	2.3	56	10.5	9.6	254	15.4	4	54
Molinio-Arrhenatheretea	14.3	33.7	653	9	32.3	782	10.5	28.4	751	15.4	53.3	727
Agropyretea intermedio-repentis	14.3	33.7	651	13.6	32.3	781	10.5	18	476	15.4	23.9	326
Koelerio glaucae-Corynephoretea canescentis	4.7	5.2	101	4.6	13.5	327	5.2	5.8	151	7.7	0.1	2
Sporadic species	0	0	0	4.6	0	1	15.8	5.8	153	7.7	0	1
Total	100	100	1935	100	100	2422	100	100	2642	100	100	1365

Most of the segetal species were characterised by low constancy classes (I–II). Only eight species were in higher constancy classes (III–V): *Achillea millefolium* L. s.str., *Artemisia vulgaris* L., *Convolvulus arvensis* L., *Elymus repens* (L.), *Rumex acetosella* L., *Trifolium arvense* (L.), *Vicia hirsuta* and *Viola arvensis* (Table 5). In total, in all the study periods, the species from the *Molinio-Arrhenatheretea* class had the highest cover coefficient (D-2913), and the lowest was seen in the species from the *Koelerio glauce-Corynephoretea canescentis* class (D-581). On older plantations (over 7 years old), a great reduction in the D index value was found in the *Stellarietea mediae* and *Artemisieta vulgaris* classes (Table 9). On 9–10-year-old plantations, the number of species in higher constancy classes visibly decreased, or the species disappeared (Table 5).

3.3. Similarities and Differences between the Species Compositions of Six Groups of Salix viminalis L. Plantations Proved by Statistical Methods

The flora of all six groups of *Salix viminalis* L. plantations totalled 114 vascular plant species. There were 47 (9–10-year-old FA plantations) to 61 (6–7-year-old AR plantations) species. Among these species, there were 13 to 27 segetal plant species (Figure 7).

Cluster analysis allowed the groups of plantations in question to be connected. Two main groups were distinguished. The first group contains the plantations established on former arable land, and the second one consists of plantations established on formerly fallow land. The abundance of segetal species was similar in each of these groups (Figure 8).

The constrained components retained 14.9 percent of the relative species abundance (Figure 9). Of that 14.9 percent, 7.3 percent were related to the first component (CCA 1) and 3.1 percent, to the second one (CCA 2). The permutation test of the significance of the plantation types confirmed the significance of the CCA model (the P p-value statistic was equal to 0.001), and the first two axes (the P p-value statistic was equal to 0.001 and 0.026 for the first axis and for the second axis, respectively, whereas the third axis was not significant at the 0.05 level).

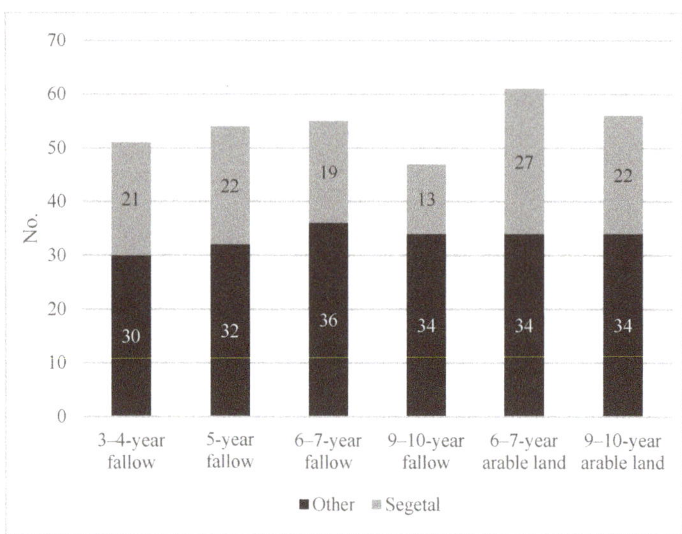

Figure 7. The numbers of species accompanying *Salix viminalis* L., including segetal species, recorded on 3–10-year-old plantations established on former arable land and fallow.

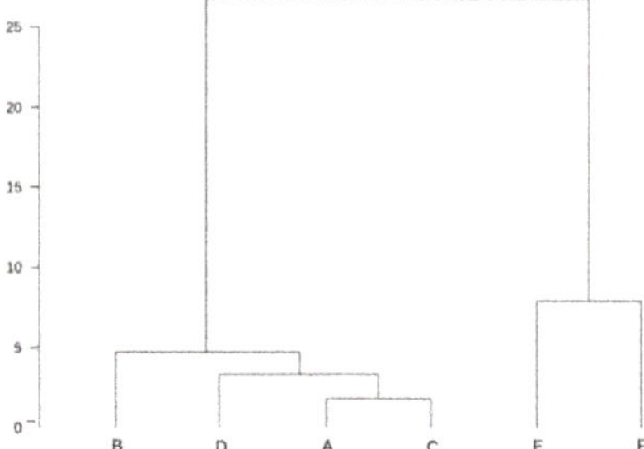

Figure 8. The dendrogram constructed by the Ward method of cluster analysis for mean segetal species coverage of six plantation groups (A–F). A—3–4-year-old FA plantations; B—5-year-old FA plantations; C—6–7-year-old FA plantations; D—9–10-year-old FA plantations; E—6–7-year-old AR plantations; F—9–10-year-old AR plantations.

The similarities between the groups of relevés according to the CCA analysis suggests that the plant communities on the plantations established on the arable lands had similar species abundances. The plantations established on fallow lands also had similar relative species abundance. This is in accordance with the cluster analysis (Figure 8). The CCA analysis also showed that the oldest plantations (9–10 years old) established on fallow areas were different from the rest of the plantations studied but were more similar to younger plantations on FA than those on AR (Figure 9).

3.4. Invasive Species Accompanying *Salix viminalis* L. Plantations

Among the 114 species of vascular flora of *Salix viminalis* L. crops, there were nine invasive species (Table 10). These were four segetal species: *Anthoxanthum aristatum* Boiss., *Conyza canadensis* (L.) Cronquist., *Setaria pumila* (Poir.) Roem and Schult., *Setaria viridis* (L.) P. Beauv. and five other species, i.e., *Echinocystis lobata* (F. Michx.) Torr. and A. Gray, *Erigeron annuus* (L.) Pers, *Padus serotina* (Ehrh.) Borkh, *Solidago canadensis* L. and *Solidago gigantea* Aiton.

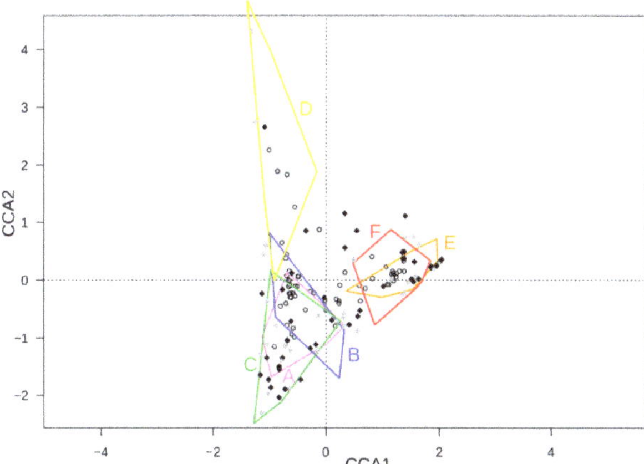

Figure 9. The biplot for the canonical correspondence analysis for relative species abundance (black rhombus—segetal species; grey rhombus—other species; circle—phytosociological relevés) in the relevés with plantation group (A–F) as the explanatory variable; 7.3 percent of the variability was related to the first CCA axis, and 3.1 percent, to the second CCA axis. A—3–4-year-old FA plantations; B—5-year-old FA plantations; C—6–7-year-old FA plantations; D—9–10-year-old FA plantations; E—6–7-year-old AR plantations; F—9–10-year-old AR plantations.

Table 10. Constancy classes (S) and cover coefficients (D) of invasive (I), medicinal (L.), poisonous (P) and melliferous (M) species occurring in flora accompanying *Salix viminalis* L. plantations established on fallow and arable land.

Species	Fallow								Arable Land				Properties
	Plantation Age in Years												
	3–4		5		6–7		9–10		6–7		9–10		
	S	D	S	D	S	D	S	D	S	D	S	D	
Achillea millefolium L. s.str.	IV	551	V	780	IV	650	IV	725	III	201	III	277	L
Anchusa arvensis (L.) M. Bieb.	-	-	I	50	-	-	-	-	-	-	-	-	L
Anthoxanthum aristatum Boiss.	-	-	-	-	I	50	-	-	-	-	-	-	I
Arctium minus (Hill) Bernh.	-	-	-	-	-	-	-	-	-	-	I	51	L
Artemisia absinthium L.	-	-	-	-	-	-	-	-	I	50	I	1	L,P
Artemisia campestris L.	-	-	-	-	II	52	I	1	-	-	-	-	L,P
Artemisia vulgaris L.	III	152	+		IV	253	II	4	IV	203	II	151	L,P
Centaurea jacea L.	-	-	II	100	I	1	-	-	-	-	-	-	M
Chenopodium album L.	-	-	+	-	-	-	I	1	I	1			M
Convolvulus arvensis L.	II	53	+		I	51	I	1	II	101	I	50	P
Conyza canadensis (L.) Cronquist	I	2	II	3	II	52	II	150	III	152	I	100	I,L
Crataegus monogyna Jacq.	-	-	-	-	-	-	-	-	-	-	I	1	L
Daucus carota L.	III	103	II	101	III	152	III	251	-	-	-	-	L
Echinocystis lobata (F.Michx)Torr. and A.Gray	-	-	-	-	-	-	-	-	-	-	I	50	I
Erigeron annuus (L.) Pers.	II	151	III	5	I	2	II	3	II	101	I	175	I
Frangula alnus Mill.	-	-	-	-	-	-	I	1	-	-	-	-	L,P
Fraxinus excelsior L.	-	-	-	-	I	1	-	-	-	-	-	-	L
Galeopsis bifida Boenn.	-	-	-	-	I	50	-	-	I	1	-	-	L
Galeopsis tetrahit L.	-	-	I	2	-	-	-	-	I	1	II	101	L,M
Geum urbanum L.	-	-	-	-	-	-	-	-	III	201	II	150	L
Helichrysum arenarium (L.) Moench	-	-	I	1	-	-	-	-	-	-	-	-	L
Hieracium pilosella L.	IV	600	III	451	IV	775	III	375	I	50	I	50	L
Hypericum perforatum L.	II	2	II	3	II	53	I	1	I	50	I	100	L,M
Jasione montana L.	III	5	I	2	I	51	I	2	I	50	-	-	M
Lotus corniculatus L.	-	-	-	-	I	1	-	-	-	-	-	-	M

Table 10. *Cont.*

Species	Fallow								Arable Land				Properties
	Plantation Age in Years												
	3–4		5		6–7		9–10		6–7		9–10		
	S	D	S	D	S	D	S	D	S	D	S	D	
Melilotus alba Medik.	-	-	-	-	-	-	-	-	I	1	I	1	L,M
Oenothera biennis L.	II	151	III	103	IV	253	III	153	-	-	-	-	M
Padus serotina (Ehrh.) Borkh.	II	2	IV	6	IV	351	V	901	-	-	-	-	I
Plantago lanceolata L.	I	50	-	-	I	50	I	1	-	-	-	-	L
Plantago major L.	-	-	-	-	-	-	-	-	I	1	-	-	L
Plantago media L.	-	-	-	-	-	-	I	1	I	1	-	-	L,M
Potentilla anserina L.	-	-	-	-	-	-	-	-	-	-	I	50	L
Potentilla argentea L.	-	-	-	-	I	1	-	-	-	-	I	50	L
Quercus petraea (Mat.) Liebl.	I	1	-	-	I	1	I	2			1	1	L
Quercus robur L.	I	1	III	r	II	53	II	52	I	r	-	-	L
Rubus caesius L.	-	-	-	-	-	-	I	50	-	-	-	-	M
Rumex acetosa L.	-	-	-	-	-	-	II	4	I	100	I	225	M
Rumex acetosella L.	IV	204	III	153	III	202	II	102	I	50	I	50	M
Senecio jacobaea L.	I	51	I	r	II	101	III	54	II	52	1	175	L,P
Setaria pumila (Poir.) Roem. and Schult.	I	1			I	51	-	-	I	1	-	-	I
Setaria viridis (L.) P. Beauv.	I	2	I	1	-	-	-	-	-	-	-	-	I
Solidago canadensis L.	I	51	II	151	III	326	IV	427	II	52	II	228	I,L,M
Solidago gigantea Aiton	-	-	-	-	-	-	I	100	-	-	-	-	I,L,M
Sonchus arvensis L.	-	-	-	-	I	50	-	-	-	-	I	1	P
Sonchus asper (L.) Hill.	-	-	-	-	-	-	-	-	-	-	I	1	P
Sonchus oleraceus L.	-	-	-	-	-	-	-	-	I	1	-	-	P
Sorbus aucuparia L. emend. Hedl.	-	-	-	-	-	-	-	-	I	1	-	-	L
Stachys palustris L.	I	50			-	-	-	-	I	2	-	-	L,M
Taraxacum officinale F.H. Wigg.	III	5	II	3	II	52	III	103	IV	279	+		L,M
Thlaspi arvense L.	-	-	-	-	-	-	I	1	-	-	-	-	L
Trifolium pratense L.	-	-	-	-	I	1	-	-	-	-	-	-	M
Trifolium repens L.	-	-	-	-	-	-	-	-	I	50	-	-	M
Urtica dioica L.	-	-	-	-	-	-	I	50	II	53	II	150	L
Verbascum densiflorum Bertol.	I	50	I	50	I	2	-	-	-	-	-	-	M
Vicia cracca L.	II	52	I	2	I	100	I	2	I	50	-	-	M
Viola arvensis Murray	II	4	III	5	II	151	-	-	-	-	-	-	L

Regardless of the type of land use, before the willow plantations were established (AR and FA), the following species occurred: *Conyza canadensis* (L.) Cronquist., *Erigeron annuus* (L.) Pers, *Setaria pumila* (Poir.) Roem and Schult., and *Solidago canadensis* L. Two of these species were always present, regardless of the age of the plantation: *Conyza canadensis* (L.) Cronquist. and *Solidago canadensis* L.

In line with the age of the plantations, *Padus serotina* (Ehrh.) Borkh., a species that grows quickly after being cut down and whose fruits are willingly eaten by birds and spread over long distances, increased its coverage on FA plantations. However, on both types of plantations (AR and FA), with time, the coverage of *Solidago canadensis* L. increased. It is an expansive, perennial species with a high productivity of seeds dispersed by the wind, thus forming dense canopies, which prevented the development of other plants. On the FA plantations, this species had higher coverage (Table 10).

3.5. *Medicinal, Poisonous and Melliferous Species of Flora Accompanying the Salix viminalis L. Plantations*

Among the flora accompanying the willow energy crops, 34 medicinal, 9 poisonous and 20 melliferous plant species were found (Table 10). These groups of plants accounted for 50% of all the plant species occurring on the willow plantations. In the groups of medicinal and mellifluous plants, segetal species made up considerable proportions, i.e., 29% and 25%, respectively.

The vast majority of the plants from the groups discussed here were perennial, native species that had low degrees of stability and low coverage (Table 10). Only *Achillea millefolium* L. s.str. had higher degrees of constancy and higher coverage. Of the remaining species, *Artemisia vulgaris* L. occurred with higher degrees of constancy but with low coverage.

Among the plants in this group, it is worth noting those species providing valuable raw materials for medicines: *Achillea millefolium* L. s.str., *Hypericum perforatum* L., *Taraxacum officinale* F.H.Wigg. and *Urtica dioica* L. [63]. Poisonous species included *Artemisia absinthium* L., *Convolvulus arvensis* L., *Senecio jacobaea* L., *Sonchus arvensis* L. and *Sonchus asper* (L.) Hill. It should be emphasized that with the development of various fields of science, it becomes increasingly difficult to assess whether a species has healing properties or is poisonous.

Among the melliferous plants on the willow plantations, the following species were reported: *Melilotus alba* Medik., a species providing large amounts of nectar and pollen with a high nutritional value, and *Verbascum densiflorum* Bertol., a species valuable for nectar because of its extended flowering. Moreover, *Chenopodium album* L., *Rumex acetosella* L., *Stachys palustris* L. and *Vicia cracca* L. also have melliferous properties.

4. Discussion

The vascular flora of the *Salix viminalis* L. plantations established in central Poland, amounting to 114 species, is rich and diverse compared to the willow flora found in other areas [29,64,65]. This large species diversity results from the different geographical location, the soil conditions in which the plantations were established and their locations with regard to diverse ecosystems and different types of landscape. Discussion on segetal flora is difficult because in the literature, the vegetation accompanying the energy willow is usually presented as a whole, without separating out segetal plants, and detailed data are often not included.

The 38% share of segetal species (43 taxa) on the *Salix viminalis* L. plantations found in this author's own research is high compared to the results obtained by Krechowski et al. [66], who recorded only 8.2% of typical cereal and root crop species on plantations of *Salix viminalis* L. of various ages, established on former arable and meadow land. This share is similar to the results obtained by Feledyn-Szewczyk et al. [27] in their research on the flora accompanying 10 species of energy crops, including *Salix viminalis* L. These authors found the occurrence of segetal species to be 38–39%.

The results of the author's own research showed that the species composition of segetal plants and their structure depend on the type of land use before the plantations were established. This is consistent with earlier studies by Piórek et al. [67] and Trąba et al. [29]. The lower number of segetal species that were found on FA plantations than on AR may be the result of a longer period of the non-use of cultivation treatments that stimulate plant diasporas. Trzcińska-Tacik [68] and Luzuriaga et al. [69] emphasize the importance of agrotechnical factors in shaping the species composition of arable crops. The cessation of mechanical soil cultivation causes the accumulation of diasporas in the soil. It was found that the soil seed bank is much larger in fallow soil compared to in the soil of arable land [70,71]. However, in the plant canopy, these relationships may be opposite [71]. It was found that over the years, the species similarity (in the plant canopy) of fallow land and arable land decreases [72]. In our research, in line with the age of the plantation, the influence of the land use before the establishment of a *Salix viminalis* L. plantation on the composition of flora became progressively smaller. This confirms the findings obtained by Baum et al. [73], according to which the history of a field may only have an impact on the phytodiversity of short rotation coppice (SRC) plantations in the initial period after establishing the plantation. Similarly, for energy poplar (*Populus* sp.) crops, an effect of earlier land use was only found in young plantations [32].

The transformations in the flora on energy willow plantations in the following years indicate that the structure of vegetation is mainly influenced by the nutrient content and light conditions [73]. The growth and development of willow plants causes greater soil shading and intensifies the competition between plants. As a result, with time, there is a decrease in the number of species accompanying the willow plantation, including segetal plants, as shown by our study results. It should be emphasized that among the segetal plants, apart from the common species, species rarely found on willow plantations in other regions of Poland were also noted, such as *Anthoxanthum aristatum* Boiss., *Digitaria ischaemum* (Schreb.) H.L. Mühl., and *Sonchus oleraceus* L. [29,74,75].

The predominance of native species (apophytes) among the segetal plants; their proportion, which increases with the age of the plantation; and, at the same time, the decreasing proportion of anthropophytes, regardless of the type of land use before the plantations were established, all confirmed the directions of flora development in *Salix viminalis* L. crops shown by other authors [33,74,76,77]. Moreover, these transformations confirmed that with a lack of arable land use (ploughing and soil cultivation treatments), perennial energy willow monoculture does not favour the development of alien species [78].

Many years of willow cultivation resulted in a large proportion of woodland and shrub apophytes in the agrophytocenosis structure. Besides, meadow apophytes (on AR plantations) and sandysite apophytes (on FA plantations) constituted a great proportion of the agrophytocenoses. Similarly, Korniak et al. [74] and Janicka et al. [77], studying willow, and Birmele et al. [79], analysing poplar, found a predominance of meadow, woodland and shrub species.

The results of our studies confirm changes in the biological stability of segetal flora on plantations established on fallow lands. An increase in the proportion of perennial species in the whole flora accompanying energy willow plantations and, at the same time, a decrease in the proportion of short-lived species in line with the age of the plantations were shown previously (in plantations more than 3 years old) by Korniak et al. [74] and Trąba et al. [29]. On the older plantations, mainly perennial, rhizomatous and stolon species developed, which was also found by Korniak et al. [74] and Baum et al. [30]. The frequent occurrence of *Elymus repens* (L.) Gould., which grows intensively on willow energy plantations, is also consistent with the findings of Rowe et al. [80] and Baum et al. [30].

The phytosociological affiliation of the segetal flora presented in this paper is fundamentally different to the phytosociological structure of the whole flora of *Salix viminalis* L. presented in the literature. The segetal species of the plantations assessed here mainly belonged to the *Stellarietea mediae* class, while in the willow flora considered as a whole, species from the *Molinio-Arrhenatheretea* class predominated, and the proportion of plants from the *Stellarietea mediae* class was much smaller [66,81]. This fact may be a result of the age of the plantations in question. According to Trąba et al. [29], species from the *Stellarietea mediae* class on perennial plantations do not have suitable conditions for development, due to the fact that the soil surface is largely covered with plants. The dominance of the *Stellarietea mediae* class on willow and poplar plantations was found by Ziaja and Wnuk et al. [76], but these were young plantations (2–5 years old).

It was found that most of the segetal species on the *Salix viminalis* L. plantations occurred in low constancy classes (I and II), which is consistent with the results of research on willow flora conducted by, for example, Korniak [74], Trąba [29] and Krechowski [66]. These authors claim that this indicates a poor balance of *Salix viminalis* L. flora. This is due to the fact that willow is cut systematically (every 2–3 years), causing a change in light, thermal and humidity conditions, which makes it impossible to balance the flora [29].

An important part of the flora accompanying the *Salix viminalis* L. plantations is medicinal, poisonous and melliferous species. They are important for nature and enrich the biodiversity of agricultural areas, especially in the era of plant protection products being used intensively in agriculture [37,82]. What is worrying, though, is the presence of invasive species, especially those with high constancy classes and coverage, such as *Conyza canadensis* (L.) Cronquist., *Erigeron annuus* (L.) Pers., *Solidago canadensis* L. and

Padus serotina (Ehrh.) Borkh. What needs to be particularly emphasized is that *Conyza canadensis* (L.) Cronquist. is an invasive species in many different habitats [83], and in Poland, it now has the status of an epoecophyte (a species settled in segetal and ruderal habitats). The presence of this species on energy willow plantations was previously confirmed by Wróbel et al. [84], Duer and Feledyn-Szewczyk [85], Feledyn-Szewczyk et al. [27] and Janicka et al. [31], and in *Sida hermaphrodita* (L.) Rusby by Bacieczko and Borcz [86]. Feledyn-Szewczyk [87] also recorded a considerable proportion of this species in the willow harvested each year. A particularly undesirable element of the flora is *Solidago canadensis* L., which was found on FA plantations. *Solidago spp.* is known to be a species typical for fallow lands [88]. The presence of these and other non-native and invasive plant species on willow plantations should be monitored. In conclusion, it should be emphasized that research on the biodiversity of energy willow and other plants used for bioenergy should be continued in the future, as these crops are important in the context of climate change and the increase in greenhouse gas emissions to the atmosphere.

5. Conclusions

Segetal species, occurring mainly in cereal and root crops, constitute a considerable proportion of the flora accompanying *Salix viminalis* L. (38%). It was found that their number depends on the type of land use before the establishment of the plantations. The plantations established on arable land were richer in segetal species than plantations established on fallow. However, both types of plantations were dominated by the same plant species (with the highest constancy classes and coverage), such as *Achillea millefolium* L. s.str, *Artemisia vulgaris* L. and *Elymus repens* (L.).

Regardless of the type of former land use, before the willow plantations were established, species of local origin were predominant among the segetal plants. In line with the age of the plantations, there was a decrease in the number of segetal species, especially from the *Stellarietea mediae* class. The proportion of anthropophytes also decreased, and apophytes increased. The dominant phytosociological classes were *Molinio-Arrhenatheretea* and *Agropyretea intermedio-repentis*.

Among the flora accompanying the willow, medicinal and melliferous plants important for humanity and biodiversity were noted, as well as invasive alien species, whose presence threatens native flora and should be monitored. A considerable part of the medicinal species was made up of segetal species.

To sum up, *Salix viminalis* L. plantations promote the preservation of many arable land plant species and contribute to maintaining the mosaic of the agricultural landscape and also the biodiversity of agroecosystems.

Author Contributions: Conceptualization, M.J. and A.K.; methodology, M.J. and A.K.; validation, M.J., A.K. and J.P.; formal analysis, J.P. and A.K.; investigation, A.K. and M.J.; data curation, A.K.; writing—original draft preparation, A.K. and M.J.; writing—review and editing, M.J., A.K. and J.P.; visualization, A.K. and M.J.; supervision, M.J.; project administration, M.J. All authors have read and agreed to the published version of the manuscript.

Funding: This research received no external funding.

Institutional Review Board Statement: Not applicable.

Informed Consent Statement: Not applicable.

Data Availability Statement: Data are available by contacting the authors.

Conflicts of Interest: The authors declare no conflict of interest.

References

1. Eurostat. Available online: https://ec.europa.eu/eurostat (accessed on 20 December 2019).
2. Renewables Global Status Report. Available online: https://www.energia.org/renewables-2018-global-status-report-ren21/ (accessed on 20 December 2019).
3. Faber, A. Natural effects of cultivation of energy crops. *Studia i Raporty IUNG-PIB* **2008**, *11*, 43–53.

4. Scioe-Murg, O.M.; Ciolac, R.; Tonea, E.; Maria, C.C.; Martin, S. The impact of willow growings in energy forestry systems upon the environment. *Lucrări Științifice Manag. Agric.* **2015**, *17*, 268–272.
5. Kyoto Protocol. Available online: http://prawo.sejm.gov.pl/isap.nsf/download.xsp/WDU20052031684/O/D20051684.pdf (accessed on 5 January 2020).
6. Polish Energy Policy Until 2030. Available online: http://isap.sejm.gov.pl/isap.nsf/download.xsp/WMP20100020011/O/M20100011.pdf (accessed on 2 October 2020). (In Polish)
7. Directive 2009/28/EC of the European Parliament and of the Council of 23 April 2009. Available online: http://data.europa.eu/eli/dir/2001/77/oj (accessed on 1 October 2020).
8. Regulation (EC) No 1099/2008 of the European Parliament and of the Council of 22 October 2008. Available online: https://eur-lex.europa.eu/eli/reg/2008/1099/oj (accessed on 2 October 2020).
9. GUS. Energy from Renewable Energy Sources in 2018. 2019. Available online: https://stat.gov.pl/obszary-tematyczne/srodowisko-energia/energia/energia-ze-zrodel-odnawialnych-w-2018-roku,10,2.html (accessed on 28 September 2020).
10. Stolarski, M.J. Plantations of Fast-Growing Trees and Shrubs as an Alternative to Wood from the Forest—Private Resource Bases. 2015. Available online: http://www.npl.ibles.pl/sites/default/files/referat/mariusz_stolarski_uwm_plantacje_drzew_i_krzewow_szybko_rosnacych_jako.pdf (accessed on 28 January 2019). (In Polish)
11. Stolarski, M.J.; Niksa, D.; Krzyżaniak, M.; Tworkowski, J.; Szczukowski, S. Willow productivity from small- and large scale experimental plantations in Poland form 2000 to 2017. *Renew. Sustain. Energy Rev.* **2019**, *101*, 461–475. [CrossRef]
12. Grzybek, A. Bioenergy in Poland. Energy Crops in Poland—The Current State. 2015. Available online: http://www.pimot.eu/attachments/article/753/Anna%20Grzybek%20-%20ITP%20POLBIOM.%20Bioenergia%20w%20Polsce.%20Uprawy%20energetyczne%20w%20Polsce%20-%20stan%20obecny.pdf (accessed on 6 April 2019). (In Polish)
13. Jadczyszyn, J.; Faber, A.; Zaliwski, A. Designation of Areas Potentially Suitable for the Cultivation of Willow and Virginia Mallow for Energy Purposes in Poland. *Studia Rap.* **2008**, *11*, 55–66. (In Polish)
14. Błażej, J. Status and prospects of willow energy plantations in the Podkarpackie Voivodeship. *Pamiet. Pulawski* **2009**, *150*, 56–63. (Summary In English)
15. Fry, D.A.; Slater, F.M. Early rotation short rotation willow coppice as a winter food resource for birds. *Biomass Bioener.* **2011**, *35*, 2545–2553. [CrossRef]
16. Chauvat, M.; Perez, G.; Hedde, M.; Lamy, I. Establishment of bioenergy crops on metal contaminated soils stimulates belowground fauna. *Biomass Bioener.* **2014**, *62*, 207–211. [CrossRef]
17. Brunbjerg, A.K.; Høye, T.T.; Eskildsen, A.; Nygaard, B.; Damgaard, C.F.; Ejrnæs, R. The collapse of marsh fritillary (*Euphydryas aurinia*) populations associated with declining host plant abundance. *Biol. Conserv.* **2017**, *211*, 117–124. [CrossRef]
18. Ledin, S. Environmental consequences when growing short rotation forests in Sweden. *Biomass Bioenergy* **1998**, *15*, 49–55. [CrossRef]
19. Grelle, A.; Aronsson, P.; Weilien, P.; Klemedtsson, L.; Lindroth, A. Large carbon-sink potential by Kyoto forests in Sweden-a case study on willow plantations. *Tellus* **2007**, *59*, 910–918. [CrossRef]
20. Holzner, W. Weed species and weed communities. *Vegetatio* **1978**, *38*, 13–20. [CrossRef]
21. Fanarillo, E.; Petit, S.; Dessaint, F.; Rosati, L.; Abbate, G. Species composition, richness, and diversity of weed communities of winter arable land in relation to geo-environmental factors: A gradient analysis in mainland Italy. *Botany* **2020**, *98*, 381–392. [CrossRef]
22. Kutkowska, A.; Janicka, M.; Paderewski, J. The characteristics of *Salix viminalis* L. crop flora established on soils with different phosphorous contents. *Soil Sci. Ann.* **2020**, *71*, 252–264. [CrossRef]
23. Lossová, Z.; Chytrý, M.; Cimalová, S.; Kropáč, Z.; Otýpková, Z.; Pyšek, P.; Tichý, L. Weed vegetation of arable land in Central Europe: Gradient of diversity and species composition. *J. Veg. Sci.* **2004**, *15*, 415–422. [CrossRef]
24. Fried, G.; Norton, R.L.; Reboud, X. Environmental and management factors determining weed species composition and diversity in France. *Agric. Ecosyst. Environ.* **2008**, *128*, 68–76. [CrossRef]
25. Dobrzański, A.; Adamczewski, K. The influence of weed control on agrophytocenosis biodiversity. *Prog. Plant Prot.* **2009**, *49*, 982–995. (Summary In English)
26. Wanic, M.; Parzonka, M.; Załuski, D. Biodiversity of weed communities in common wheat and spelt following various forecrops. *Acta Agrobot.* **2018**, *71*, 1751. [CrossRef]
27. Feledyn-Szewczyk, B.; Matyka, M.; Staniak, M. Comparision ot the effect of perennial energy crops and agricultural crops on weed flora diversity. *Agronomy* **2019**, *9*, 695. [CrossRef]
28. Kędziora, A.; Karg, J. Risks to biological diversity. *Nauka* **2010**, *4*, 107–114. (Summary In English)
29. Trąba, Cz.; Majda, J.; Wolański, P. Plant communities accompanying plantations of *Salix viminalis* L. in podkarpackie voievodeship. *Pamiet. Pulawski* **2009**, *150*, 323–336. (Summary In English)
30. Baum, S.; Weih, M.; Bolte, A. Stand age characteristics and soil properties affect species composition of vascular plants in short rotation coppice plantations. *BioRisk* **2012**, *7*, 51–71. [CrossRef]
31. Janicka, M.; Kutkowska, A.; Paderewski, J. Diversity of vascular flora accompanying *Salix viminalis* L. crops depending on soil conditions. *Glob. Ecol. Conserv.* **2020**, *23*, e01068. [CrossRef]
32. Archaux, F.; Chevalier, R.; Berthelot, A. Towards practices favourable to plant diversity in hybrid poplar plantations. *For. Ecol. Manag.* **2010**, *259*, 2410–2417. [CrossRef]

33. Wojciechowski, W.; Sowiński, J.; Zawieja, J. The effect of age of willow plantation on weed infestation in the Sudety Mountains. *Pamiet. Pulawski* **2009**, *150*, 351–358. (Summary In English)
34. Fehér, A.; Halmová, D.; Končeková, L. Gradient analysis of importance of spontaneously occurring vascular plant species in energy tree and grass stands. *Acta Reg. Environ.* **2013**, *2*, 31–33. [CrossRef]
35. Pučka, I.; Lazdiņa, D.; Bebre, I. Ground flora in plantations of three years old short rotation willow coppice. *Agron. Res.* **2016**, *14*, 1450–1466.
36. Hochół, T. Weeds or plants accompanying crops? *Pamiet. Pulawski* **2003**, *134*, 89–96. (Summary In English)
37. Grimau, L.; Gómez, M.; Figueroa, R.; Pizarro, R.; Gabriel Núñez, G.; Montenegro, G. The importance of weeds as melliferous flora in central Chile. *Environ. Ecol.* **2014**, *41*, 387–394. [CrossRef]
38. Kabała, C.; Bednarek, R.; Białousz, S.; Bielska, A.; Charzyński, P.; Chodorowski, J.; Chojnicki, J.; Czępińska-Kamińska, D.; Drewnik, M.; Glina, B.; et al. *Systematic of Polish Soil*, 6th ed.; Wydawnictwo Uniwersytetu Przyrodniczego we Wrocławiu, Instytut Nauk o Glebie i Ochrony Środowiska Uniwersytetu Przyrodniczego we Wrocławiu, Polskie Towarzystwo Gleboznawcze. Komisja Genezy, Klasyfikacji i Kartografii Gleb: Wrocław-Warszawa, Poland, 2019; pp. 1–292.
39. IUSS Working Group WRB. *World Reference Base for Soil Resources 2014*; Update 2015 International Soil Classification System for Naming Soils and Creating Legends for Soil Maps. World Soil Re-sources Reports No. 106; FAO: Rome, Italy, 2015.
40. Polish Society of Soil Science. Particle Size Distribution and Textural Classes of Soils and Mineral Materials-Classification of Polish Society of Soil Science 2008. *Soil Sci. Ann.* **2009**, *60*, 5–16. (In Polish with English abstract)
41. Łódź Voivodeship Geoportal. Available online: http://www.geoportal.lodzkie.pl (accessed on 24 January 2020).
42. Braun-Blanquet, J. *Pflanzensoziologie: Grundzüge der Vegetationskunde*; Springer: Vienna, Austria; New York, NY, USA, 1964.
43. Pawłowski, B.; Pawłowski, B. Composition and structure of plant communities and methods of their research. In *Polish Vegetation*, 2nd ed.; Szafer, W., Zarzycki, K., Eds.; PWN: Warszawa, Poland, 1972; pp. 237–269. (In Polish)
44. Dzwonko, Z. *Guidebook to Phytosociological Studies*, 2nd ed.; Sorus, S.C., Ed.; Instytut Botaniki Uniwersytetu Jagiellońskiego: Poznań- Kraków, Poland, 2008; pp. 1–304. (In Polish)
45. Matuszkiewicz, W. *Guide to the Determination of Polish Plant Communities*; Wydawnictwo Naukowe PWN: Warszawa, Polska, 2012; pp. 1–537. (In Polish)
46. Anioł-Kwiatkowska, J. Flora and synanthropic communities of Legnica, Lubin, Polkowice. *Acta Univ. Wratislav. Prace Bot.* **1974**, *229*, 1–151. (In Polish)
47. Korniak, T. Segetal flora of north-eastern Poland, its spatial differentation and current changes. *Acta Acad. Agric. Tech. Olst. Agric.* **1992**, *53* (Suppl. A), 1–77. (Summary In English)
48. Mirek, Z.; Piękoś-Mirkowa, H.; Zając, A.; Zając, M. Flowering plants and pteridophytes of Poland—A checklist. In *Biodiversity of Poland*; Mirek, Z., Szafer, W., Eds.; Institute of Botany, Polish Adacemy of Sciences: Kraków, Poland, 2002; Volume 1, pp. 1–442.
49. Rutkowski, L. *Key for the Determination of Lowland Poland Vascular Plants*, 2nd ed.; PWN: Warszawa, Polska, 2008; pp. 1–814. (In Polish)
50. Sowa, R.; Warcholińska, U. Synanthropic flora of Sulejów and Podklasztorze. *Acta Univ. Lodz. Folia Bot.* **1981**, *1*, 77–131. (Summary In English).
51. Szafer, W.; Kulczyński, S.; Pawłowski, B. *Polish Vegetation*, 3rd ed.; PWN: Warszawa, Poland, 1969; pp. 1–1020. (In Polish)
52. Zając, E.U.; Zając, A. The list of archeophytes occurring in Poland. *Zesz. Nauk. Uniw. Jagiellońskiego Prace Bot.* **1975**, *395*, 7–16. (Summary In English)
53. Zając, M.; Zajac, A. A tentative list of segetal and ruderal apophytes in Poland. *Zesz. Nauk. Uniw. Jagiellońskiego Prace Bot.* **1992**, *24*, 7–23.
54. Zając, A. The origin of the archaeophytes occurring in Poland. *Rozpr. Habilitacyjne Uniw. Jagiellońskiego* **1979**, *29*, 1–218. (In Polish)
55. Tokarska-Guzik, B.; Dajdok, Z.; Zajac, M.; Zajac, A.; Urbisz, A.; Danielewicz, W.; Hołdyński, C. *Plants of Alien Origin in Poland with Particular Emphasis on Invasive Species*; Generalna Dyrekcja Ochrony Środowiska: Warszawa, Poland, 2012; pp. 1–197.
56. Sulborska, A. *Melliferous Plants*; Bee & Honey Sp. z o.o.: Klecza Dolna, Poland, 2019; pp. 1–754.
57. Ter Braak, C.J.F. Cannonical correspondence analysis: A new eigenvector technique for multivariate direct gradient analysis. *Ecology* **1986**, *67*, 1167–1179. [CrossRef]
58. Angolini, C.; Nucci, A.; Frignani, F.; Landi, M. Using multivariate analyses to assess effects of fluvial type one plant, species distribution in Mediterranean river. *Wetlands* **2011**, *31*, 167–177. [CrossRef]
59. Legendre, P.; Oksanen, J.; ter Braak, C.J. Testing the significance of canonical axes in redundancy analysis. *Methods Ecol. Evol.* **2011**, *2*, 269–277. [CrossRef]
60. Walther, B.A.; Morand, S. Comparative performance of species richness estimation methods. *Parasitology* **1998**, *116*, 395–405. [CrossRef]
61. Petersen, F.T.; Meier, R.; Larsen, M.N. Testing species richness estimation methods using museum label data on the Danish Asilidae. *Biodivers. Conserv.* **2003**, *12*, 687–701. [CrossRef]
62. R Core Team. *R: A Language and Environment for Statistical Computing*; R Foundation for Statistical Computing: Vienna, Austria, 2020.
63. Podlech, D. *Pocket Encyclopedia*, 3rd ed.; Wydawnictwo MUZA SA: Warszawa, Poland, 1997; pp. 1–253.
64. Cunningham, M.D.; Bishop, J.D.; McKay, H.V.; Sage, R.B. *ABRE Monitoring-Ecology of Short Rotation Coppice*; URN04/961; Department of Trade and Industry: London, UK, 2004; pp. 1–157.

65. Welc, M.; Lundkvist, A.; Nordh, N.-E.; Verwijst, T. Weed community trajectories in cereal and willow cultivations after termination of a willow short rotation coppice. *Agron. Res.* **2017**, *15*, 1795–1814.
66. Krechowski, J.; Piórek, K.; Ciosek, M.T. Vegetation accompanying energetic willow (*Salix viminalis* L.) plantations in Radzyń Podlaski. *Pamiet. Pulawski* **2009**, *150*, 195–205. (Summary In English)
67. Piórek, K.; Krechowski, M.T.; Ciosek, M.T.; Sikorski, R. Effect of the selected factors on floristic composition of communities developing in energetic willow. *Pamiet. Pulawski* **2009**, *150*, 219–232. (Summary In English)
68. Trzcińska-Tacik, H. Importance of field weeds species diversity. *Pamiet. Pulawski* **2003**, *134*, 253–262. (Summary In English)
69. Luzuriaga, A.L.; Escudero, A.; Olano, J.M.; Loidi, J. Regenerative role of seed bank following an intense soil disturbance. *Acta Oecologica* **2005**, *27*, 57–66. [CrossRef]
70. Zawieja, J. Increashe of potential weed infestation threat on fields temporarily eliminated from cultivation. *Fragm. Agron.* **2006**, *23*, 126–138.
71. Sekutowski, T.R.; Włodek, S.; Biskupski, A.; Sienkiewicz-Cholewa, U. Comparision of the content of seeds and plants of the goldenrod (*Solidago* Sp.) in the fallow and adjacent field. *Zesz. Nauk. Uniw. Przyr. we Wrocławiu-Rol.* **2012**, *584*, 99–112.
72. Kurus, J. Weed infestation of the fields neighbouring with multiyear fallow lands on two types of soil. *Fragm. Agron.* **2010**, *27*, 84–93.
73. Baum, S.; Weih, M.; Bolte, A. Floristic diversity in Short Rotation Coppice (SRC) plantations: Comparison between soil seed bank and recent vegetation. *Appl. Agric. For. Res.* **2013**, *63*, 221–228. [CrossRef]
74. Korniak, T.; Hołdyński, Cz.; Wąsowicz, K. Changes in the weed flora of willow plantations in north-eastern Poland. *Pamiet. Pulawski* **2009**, *150*, 159–170. (Summary In English)
75. Skrajna, T.; Skrzyczyńska, J.; Rzymowska, Z.; Affek-Starczewska, A. Composition and structure of communities infestin of Salix sp. in northern part of Południowopodlaska Lowland. *Pamiet. Pulawski* **2009**, *150*, 255–264.
76. Ziaja, M.; Wnuk, Z. Weed infestation of energetic crop in Leszczawa Dolna (podkarpackie voivodeship). *Pamiet. Pulawski* **2009**, *150*, 367–375. (Summary In English)
77. Janicka, M.; Kutkowska, A.; Paderewski, J. Diversity of vascular flora in *Salix viminalis* L. crops depending on the harvest cycle. *Annu. Set Environ. Prot.* **2019**, *21*, 1175–1201.
78. Anioł-Kwiatkowska, J.; Kącki, Z.; Śliwiński, M. A comparision of species composition of three energy willow crops. *Pamiet. Pulawski* **2009**, *150*, 19–33. (Summary In English)
79. Birmele, J.; Kopp, G.; Brodbeck, F.; Konold, W.; Sauter, U.H. Successional changes of phytodiversity on a short rotation coppice plantation in Oberschwaben, Germany. *Front. Plant Sci.* **2015**, *6*, 1–8. [CrossRef]
80. Rowe, R.L.; Hanley, M.E.; Goulson, D.; Clarke. D.J.; Doncaster, C.P.; Taylor, G. Potential benefits of commercial willow Short Rotation Coppice (SRC) for farm-scale plant and invertebrate communities in the agri-environment. *Biomass Bioener.* **2011**, *25*, 325–336. [CrossRef]
81. Jezierska-Domaradzka, A.; Domaradzki, K. Accompanying vegetation of *Salix viminalis* L. on meadow habitat in Muchów (Kaczawskie Highland). *Pamiet. Pulawski* **2009**, *150*, 129–135, (Summary In English).
82. Kwiatkowski, C.A.; Haliniarz, M.; Harasim, E. Weed infestation and health of organically grown Chamomile (*Chamomilla recutita* (L.) Rausch.) depending on selected foliar sprays and row spacing. *Agriculture* **2020**, *10*, 168. [CrossRef]
83. Djudjević, L.; Mitrović, M.; Gajić, G.; Jarić, S.; Kostić, O.; Oberan, L.; Pavlović, P. An allelopathic investigation of the domination of the introduced invasive *Conyza canadensis* L. *Flora* **2011**, *206*, 921–927. [CrossRef]
84. Wróbel, M.; Wróbel, J.; Gregorczyk, A. The effect of chemical soil properties on weed infestation structure in willow (*Salix* L.) Short-Rotation Coppice. *Pol. J. Environ. Stud.* **2012**, *21*, 1893–1899.
85. Duer, I.; Feledyn-Szewczyk, B. Botanical monitoring in perennial bioenergy crops. *Pamiet. Pulawski* **2009**, *150*, 105–119. (Summary In English)
86. Bacieczko, W.; Borcz, A. Segetal flora of the plantations of Virginia fanpetals *Sida hermaphrodita* (L.) Rusby in Łobez commune (West Pomerania). *Folia Pomeranae Univ. Technol. Stetin. Agric. Aliment. Piscaria Zootech.* **2016**, *38*, 17–36. [CrossRef]
87. Feledyn-Szewczyk, B. The influence of agricultural land use on weed flora diversity. *Monogr. Rozpr. Nauk.* **2013**, *36*, 1–184. (Summary In English)
88. Rola, J.; Rola, H. *Solidago* spp. as bioindicator of fallow occurence on arable area. *Fragm. Agron.* **2010**, *27*, 122–131. (Summary In English)

Article

Evaluation of the Weed Infestation, Grain Health, and Productivity Parameters of Two Spelt Wheat Cultivars Depending on Crop Protection Intensification and Seeding Densities

Małgorzata Haliniarz [1], Dorota Gawęda [1,*], Bożena Nowakowicz-Dębek [2], Agnieszka Najda [3], Sylwia Chojnacka [1], Justyna Łukasz [1], Łukasz Wlazło [2] and Monika Różańska-Boczula [4]

1 Department of Herbology and Plant Cultivation Techniques, Faculty of Agrobioengineering, University of Life Sciences in Lublin, Akademicka 13, 20-950 Lublin, Poland; malgorzata.haliniarz@up.lublin.pl (M.H.); schojnacka2@gmail.com (S.C.); justynalukasz1992@gmail.com (J.Ł.)
2 Department of Animal Hygiene and Environmental Hazards, Faculty of Animal Sciences and Bioeconomy, University of Life Sciences in Lublin, Akademicka 13, 20-950 Lublin, Poland; bozena.nowakowicz@up.lublin.pl (B.N.-D.); lukasz.wlazlo@up.lublin.pl (Ł.W.)
3 Department of Vegetable Crops and Medicinal Plants, Faculty of Horticulture and Landscape Architecture, University of Life Sciences in Lublin, Akademicka 15, 20-950 Lublin, Poland; agnieszka.najda@up.lublin.pl
4 Department of Applied Mathematics and Computer Science, Faculty of Production Engineering, University of Life Sciences in Lublin, Głęboka 28, 20-612 Lublin, Poland; monika.boczula@up.lublin.pl
* Correspondence: dorota.gaweda@up.lublin.pl; Tel.: +48-81-445-67-69

Received: 15 May 2020; Accepted: 11 June 2020; Published: 15 June 2020

Abstract: Spelt wheat is one of the oldest wheat with very high nutritional value. It does not have particular climatic requirements and tolerates adverse environmental conditions well. The versatile advantages of spelt wheat make it attractive to farmers, plant breeders, food technologists, and consumers. The aim of this study was to determine the effect of different crop protection systems and seeding densities on yield, weed infestation, and grain health of the spelt wheat cultivars "Rokosz" and "Schwabenspelz". The research showed that the spelt wheat cultivars studied responded differently to production intensification. The use of crop protection chemicals in the crop of the cultivar "Rokosz" resulted in lower weed infestation and in obtaining higher yields. In the case of the cultivar "Schwabenspelz", production intensification did not have a significant effect on its productivity and quantitative weed infestation parameters. Therefore, this cultivar can be recommended for cultivation in farms with extensive farming methods, for example, in organic farms. In both cultivars studied, an increase in seeding density and chemical plant protection with fungicide caused lower grain contamination with mycotoxins, and the content of individual mycotoxins did not exceed the maximum levels set for grain intended for food and animal feed purposes.

Keywords: spelt wheat; cultivars; crop protection methods; seeding rate; weed infestation; yield; fungi; mycotoxins

1. Introduction

Spelt wheat (*Triticum aestivum* ssp. *spelta*) probably originates from the region of southeastern Asia and is one of the oldest wheat subspecies [1,2]. The beginnings of its cultivation date back to the Stone Age. Archaeological investigations reveal that, starting with the Bronze Age, spelt was the most important consumption crop in Europe for many years [3,4]. However, its small production potential and difficulties with grain dehulling became the reason why the cultivation of this crop was abandoned in favor of more productive common wheat varieties [4,5].

Changing trends both in cultivation of agricultural crops and in human nutrition have been observed in recent years. People have started to attach a great importance to increased biological diversity of agricultural plant communities, reduced use of industrial production means in agriculture, and consumption of products with health promoting properties [3,4,6]. During the last 20 years, due to its nutritional qualities, spelt wheat was again introduced into cultivation, primarily through the organic farming system [7–11]. In Poland, similar to other European countries, the cultivation of spelt wheat is concentrated in organic farms and the interest in this cereal is continually growing [12]. The estimated cropped area of spelt in 2015 was about 225 ha [13]. The productivity of this species in Poland varies greatly. Grain yield can exceed 6 t ha^{-1} [14], while under conditions unfavorable for growing common wheat, yields can be even higher than those obtained for common wheat [12]. In comparison with common wheat, spelt grain contains more protein, gluten, and fat, but less dietary fiber. It is a rich source of silicon, minerals, and group B vitamins. Moreover, it contains substances with antioxidant properties [5,15–18]. Spelt wheat does not have high climatic and soil requirements, and tolerates adverse environmental conditions well [19–22]. It is characterized by a rather high level of resistance to ear and grain diseases caused by fungal pathogens [8,23] and, due to its high tillering ability, competes with weeds better than common wheat [24]. These characteristics allow this crop to be grown without excessive use of crop protection products and mineral fertilizers, and hence it is considered to be environmentally friendly [25]. Studies conducted by different researchers have shown great variations in productivity and quality parameters between different spelt wheat varieties [18,21,26,27], as well as their varying response to soil and climatic conditions [22] and agronomic factors [18,27–29]. Most research is focused on organic cultivation of spelt [17,21,22,30], but there is little information how spelt wheat responds to production intensification. According to Rachoń et al. [31] and Andruszczak et al. [32], higher grain yields were obtained under full chemical protection conditions compared to extensive cultivation conditions. Therefore, a study was undertaken to explain how crop protection intensification and different seeding densities affect yield, yield components, weed infestation, and grain health of two winter wheat spelt cultivars. Taking into account the genetic characteristics of spelt wheat, the experiment assumed that (i) the absence of any crop protection or mechanical protection and application of herbicide alone would not contribute to a significant decrease in grain yield and health, compared to full chemical protection; (ii) pro-ecological technology involving the complete abandonment of crop protection or the use of only harrowing would not deteriorate the productivity of spelt wheat and its health, in comparison with chemical protection; (iii) in treatments with a higher seeding rate, the denser spelt crop would be more competitive against weeds, which would contribute to reduced weed infestation and would not increase grain infection with fungal pathogens, in consequence allowing a satisfactory yield to be obtained; and (iv) the response of spelt wheat to production technologies is cultivar-dependent.

2. Materials and Methods

2.1. Location of the Experiment—Soil and Climatic Conditions

The field study was carried out for three growing seasons, 2012/2013, 2013/2014, and 2014/2015, at the Czesławice Experimental Farm (51°18′23″ N, 22°16′2″ E), belonging to the Lublin University of Life Sciences, Poland. The experiment on growing spelt wheat was established on a loess-derived Luvisol. The arable layer of the soil was characterized by high availability of phosphorus (P, 76.3–77.7 mg kg^{-1} soil) and potassium (K, 117.8–132.3 mg kg^{-1} soil), as well as medium availability of magnesium (Mg, 79–85 mg kg^{-1} soil), slightly acidic pH (in 1 M KCl, 6.1–6.4), and a low humus content of 1.59–1.63%.

The 2013/2014 season could be described as very wet since the total rainfall was higher by 281 mm than in the 2014/2015 season, 147.1 mm higher than in 2012/2013, and 206.3 mm higher than the

1963–2010 average (Table 1). In the years 2013 and 2014, the amount of rainfall in May and June much exceeded the long-term average (LTA). In August 2014, there was 32.5 mm more rain than the LTA. In 2015 the amount of rainfall in May was high, but June and August were very dry with only 13.5 and 5.9 mm of rain, respectively. The temperature ranges in the 2013/2014 and 2014/2015 growing seasons were similar. The mean temperature in these years was higher by 0.9 °C in the 2013/2014 season and by 1 °C in 2014/2015. In the 2012/2013 season, the mean temperature was the same as the long-term average. In all the growing seasons, July was warmer than over the period 1963–2010.

Table 1. The rainfall and mean monthly air temperature in the growing season of spelt wheat, recorded by the Meteorological Station in Czesławice (Poland). LTA is the long term average.

Years	Months												September–August
	September	**October**	**November**	**December**	**January**	**February**	**March**	**April**	**May**	**June**	**July**	**August**	
	Rainfalls (mm)												Sum
2012/2013	40.4	110.7	29	17.4	60.3	30	37.6	53.2	103.3	108.3	44.3	26.6	661.1
2013/2014	49.5	7.3	60.6	13.7	54.5	5.8	49.1	63.9	230.2	110.2	61.4	102	808.2
2014/2015	21.8	27.5	24.1	57.8	50.9	15.8	48.6	39.1	169.6	13.5	52.6	5.9	527.2
LTA 1963–2010	59.5	45.6	41	36.9	30.3	29.2	31.3	42.4	63.5	72.7	80	69.5	601.9
	Temperature (°C)												Mean
2012/2013	14.6	7.7	5	−3.4	−4.4	−1.3	−2.6	7.4	14.9	18.1	18.7	18.7	7.8
2013/2014	11.3	9.5	4.9	1.7	−2.9	0.3	4.9	8.9	13	15.2	19.6	18.3	8.7
2014/2015	14.0	9.7	4.6	−0.1	1.0	−1.1	2.8	6.5	11.5	16.1	19	21.9	8.8
LTA 1963–2010	13.1	7.9	2.9	−1.3	−3.0	−1.7	1.8	7.7	13.6	16.5	18.3	17.7	7.8

2.2. Experimental Design and Agronomic Practices

A field experiment was set up using a split-block design in three replicates. Two winter spelt wheat cultivars, the German cultivar "Schwabenspelz" and the Polish cultivar "Rokosz", were evaluated in this study. The experiment included the following factors:

I—crop protection

1. pro-ecological:

A—untreated control plots (no herbicide, fungicide, or insecticide treatment);
B—harrowing in early spring at the beginning of the growing season;

2. chemical:

C—application of the herbicides Sekator 125 OD (Bayer AG, Leverkusen, Germany, active ingredients (a.i.) amidosulfuron, iodosulfuron, mefenpyr-diethyl) at a rate of 150 mL ha^{-1} and Attribut 70 WG (Bayer AG, Leverkusen, Germany; a.i. propoxycarbazone-sodium) at a rate of 60 g ha^{-1} at BBCH (**B**iologische **B**undesanstalt, **B**undessortenamt und **CH**emische Industrie) stages 22–24 of spelt wheat;

D—application of complete chemical protection—the herbicides Sekator 125 OD at a rate of 150 mL ha^{-1} and Attribut 70 WG at a rate of 60 g ha^{-1} at BBCH 22–24 of spelt wheat; the growth retardant Cerone 480 SL (Bayer AG, Leverkusen, Germany, a.i. ethephon) at BBCH 30–31 at a rate of 0.75 L ha^{-1}, the fungicide Wirtuoz 520 EC (Corteva, Inc., Wilmington, DE, USA, a.i. prochloraz, tebuconazole, proquinazid) at BBCH 24–25 and 32–33 at a rate of 1 L ha^{-1}, as well as the insecticide Decis 2,5 EC (Bayer AG, Leverkusen, Germany, a.i. deltamethrin) at BBCH 37–39 at an amount of 0.25 L ha^{-1}.

II—seeding density:

1. optimum—130 kg ha^{-1} of cv. "Rokosz" seeds and 200 kg ha^{-1} of cv. "Schwabenspelz" spikelets;
2. increased—200 kg ha^{-1} of cv. "Rokosz" seeds and 350 kg ha^{-1} of cv. "Schwabenspelz" spikelets.

Mineral fertilization was applied at the following rates: N, 50 kg ha^{-1}; P, 26.2 kg ha^{-1}; and K, 58.1 kg ha^{-1}. All phosphorus and potassium fertilizers and part of the nitrogen fertilizers (20 kg N) were applied before sowing the spelt wheat. The remaining portion of nitrogen fertilizers was applied right after the beginning of the growing season. The sown and harvested plot area was 13.5 m^2, per repetition. Winter wheat was the previous crop for spelt wheat. Tillage was typical for common wheat cultivation. The spelt wheat was sown in the third decade of September 2012–2014 and harvested in the second decade of August 2013–2015.

2.3. Evaluation of Yield and Weed Infestation

Approximately 3–4 weeks after herbicide treatment and at the dough stage of spelt wheat (BBCH 83–85), evaluation of weed infestation of the spelt crop was made using the dry-weight-rank system. The evaluation involved determination of the botanical composition of weeds, as well as their number and dry weight per 1 m^2. The sampling area was delineated with a 0.5 × 1 m quadrat frame in two randomly selected places in each plot. Weed nomenclature followed Mirek et al. [33].

Plant height and ear length were determined based on 30 randomly selected plants from each plot. Weight of grains per ear was determined based on the average results for 30 randomly selected ears per plot. Ear density was calculated in two randomly chosen places with an area of 0.5 m^2 per plot. Yield was estimated based on grain harvested from the individual plots and expressed on a per hectare basis.

2.4. Determination of Fungal Contamination of Grain

Samples of about 20 g of spelt grain (from each plot) were homogenized by hand for 5 min and then sedimented for 15 minutes in sterile dilution liquid with approximately 100 µL of Tween 80 (Sigma-Aldrich, St. Louis, MO, USA). The dilutions were prepared and inoculated on sterile Sabouraud Chloramphenicol Agar (BTL Poland Ltd., Warsaw, Poland). To determine the total number of fungi, the plates were incubated at a temperature of 25 ± 0.2 °C for 7 days according to the relevant standard [34,35]. The colonies were transferred to microcultures and species were determined using Watanabe's key [36]. Each sample was analyzed in triplicate and the results presented as the number of fungi per 1 g of the tested material (cfu g^{-1}).

2.5. Determination of Mycotoxin Contamination of Grain

The content of deoxynivalenol (DON) and its derivatives: 3-O-acetylDON (3AcDON), 15-O-acetylDON (15AcDON), and zearalenone (UAE) were determined according to the authors' method developed in the Laboratory of Vegetable and Herb Raw Material Quality Research: "Determination of deoxynivalenol and its derivatives, nivalenol and zearalenone by HPLC method with purification by immunological affinity column in plant raw materials". The presence of trichothecene mycotoxins such as T-2 toxin (T-2) and its derivatives HT-2 toxin (HT-2), nivalenol (NIV), DAS (diacetoxyscirpenol), and fuzarenone X were analyzed by gas chromatography following the procedure given by Valle-Algarra et al. [37].

2.5.1. Sample Preparation and Extraction

A 5 g sample of spelt grain previously finely ground with a laboratory mill (FW135, ChemLand, Stargard Szczeciński, Poland) was placed in a test tube and treated with 20 mL of acetonitrile (ACN): H2O so lution (85:15). The diluted samples were shaken with a centrifuge (30 min, 21.000× g, Universal 32, Hettich, Tuttlingen, Germany) and left for 6 h, after which time it was again shaken for 30 min. The extract was filtered through a 0.2 Pm membrane syringe filter (Costar X, Corning Inc., Salt Lake City, UT, USA). Then, 5 mL of filtrate was collected for thorough purification and applied to a MycoSep 225 Trich column (Romer Laboratories, Union, MO, USA). Two mL of purified eluents were evaporated to dryness at 50 °C, and the residue dissolved in 0.5 mL of mobile phase (10% aqueous acetonitrile) after which the mycotoxin content was determined.

2.5.2. HPLC Analysis Conditions

A LaChrom–Merck HPLC liquid chromatograph (Merck, Darmstadt, Germany) was equipped with a DAD diode detector (L-7450), binary pump (L-7100), degasser (L-7612), thermostat (L-7360), Rheodyne injector, and steel column LiChrospher 100 RP C18 (250 × 4 mm) filled with stationary phase (particle diameter—dp = 5 μM). UV detection was performed at 220 nm (deoxynivalenol, 3-O-acetylDON, 15-O-acetylDON) and 220–440 (zearalenone) at 28 °C and a flow rate of 0.800 mL min^{-1}. The mobile phase was aqueous acetonitrile. Separation was achieved in an isocratic water: acetonitrile (8:92) analysis by deoxynivalenol, 3-O-acetylDON, 15-O-acetylDON, and (63:37) by analysis of zearalenone. To increase the separation, 1% acetic acid was added to the mobile phase. Identification of analyzed compounds was made on the basis of retention time. Quantitative assessment was carried out using the external standard method, after creating a calibration curve using pure reference substances.

2.5.3. GC/MS Analysis Conditions

Separation of trichothecenes, nivalenol, and fuzarenone X was carried out using a Varian CP-3800 gas chromatograph (Varian Medical Systems, Palo Alto, CA, USA) equipped with an detector ECD (instant connect Electron Capture Detector) and columns VF-1ms 30 m × 0.32 mm ID (Column Internal Diameter), DF (film thickness) = 0.25 μm from Varian. The temperature of the dispenser chamber was 250 °C and the detector 300 °C. The following oven temperature gradient was used: 80 °C for 1 min, increment 40 °C for 1 min to 160 °C, 3 °C for 1 min to 178 °C, 2 °C for 1 min to 240 °C, 40 °C for 1 min to 270 °C, 270 °C for 1 min. Nitrogen was used as carrier gas at a flow rate of 1 mL for min.

2.5.4. Chemical Reagents

Organic solvents with spectroscopic purity were obtained from Merck (Darmstadt, Germany). All deoxynivalenol (DON) purity ≥ 95% (reference standard), 3-O-acetylDON (3AcDON) purity ≥ 95% (reference standard), 15-O-acetylDON (15AcDON) purity ≥ 95% (reference standard), zearalenone (UAE) purity ≥ 95% (reference standard), T-2 toxin (T-2) purity ≥ 95% (reference standard) and its derivatives HT-2 toxin (HT-2) purity ≥ 95% (reference standard), nivalenol (NIV) purity ≥ 95% (reference standard), DAS—diacetoxyscirpenol purity ≥ 95% (reference standard) and fuzarenone X purity ≥ 95% (reference standard) were from Sigma-Aldrich (St. Louis, MO, USA).

2.6. Statistical Analysis

The results were statistically analyzed using analysis of variance (Statistica PL 13.3, StatSoft Polska, Krakow, Poland). The distribution conformity with normal distribution was verified with the Shapiro–Wilk test, while homogeneity of the variance was tested with Levene's test. In the case of any feature where a significant interaction was found between the factor and year, interaction effects were analyzed separately for each year of the study. If no significant interactions were found between the factors, then Tukey's multiple comparison test was used to compare differences between the means for main factors (crop protection, seeding density, cultivar). Confidence half-intervals were calculated by Tukey's test at a significance level of 0.05. Significant differences between the two groups of treatments, pro-ecological protection (treatments A and B) and chemical protection (treatments C and D), were evaluated based on the Scheffe test ($\alpha = 0.05$). Total number of fungi in grain is presented in log CFU g^{-1}.

3. Results

3.1. Weed Infestation of Spelt Wheat

The statistical analysis of the weed infestation parameters for the crop of the "Rokosz" spelt wheat cultivar, as evaluated 3–4 weeks after herbicide application, revealed significant differences in the traits between years (Table 2). The highest number of weeds was found in the first year of the study in the treatments where no herbicide was used (treatments A and B) and in the second year of the study in the control treatment. Weeds produced the highest biomass in 2013 in treatments A and B. Harrowing done as the only weed control operation (treatment B) proved to be ineffective in reducing weed infestation because the number and dry weight of weeds were similar to those found in the control treatment (A) in each year of the study. However, it was only in 2013 that weed infestation in this treatment was significantly higher than in the treatments where the herbicide was applied. In the years 2014 and 2015, as well as in treatments C and D in 2013, weed infestation of the crop was at a similar level. At the dough stage of wheat, crop protection method and year were not found to significantly affect number of weeds. The highest weight of weeds was determined in the first year of the study in the treatments without herbicide application (A and B). In the other treatments, the weight of weeds was similar. The results for the three-year study showed that in cv. "Rokosz", herbicide application had a significant impact on weed infestation of the crop. At the first evaluation of weed infestation, the average number of weeds in the chemical protection treatments (C and D) was more than one-half lower, while their dry weight was four times lower than in the pro-ecological treatments A and B. At the dough stage of cv. "Rokosz", similar relationships were observed as those made 3–4 weeks after herbicide treatment.

Table 2. The weed infestation indicators in the spelt wheat cv. "Rokosz" showing the interaction between crop protection and years of research, and comparison of the pro-ecological and chemical crop protection. NW I—number of weeds 3–4 weeks after herbicide treatment, WW I—air-dry weight of weeds 3–4 weeks after herbicide treatment, NW II—number of weeds at the dough stage of spelt wheat (BBCH 83–85), WW II—air-dry weight of weeds at the dough stage of spelt wheat (BBCH 83–85). Different letters denote significant differences ($p \leq 0.05$). The same letter means not significantly different values.

Years	Crop Protection	NW I (no. m^{-2})	WW I (g m^{-2})	NW II (no. m^{-2})	WW II (g m^{-2})
2013	A	90.8 a	68.66 a	86.3 a	158.06 a
	B	78.7 ab	62.12 a	58.3 a	119.1 ab
	C	27.2 c	5.68 b	18 a	21.7 c
	D	28.5 c	8.18 b	21.2 a	25.33 c
2014	A	53 abc	25.4 b	56 a	39.39 bc
	B	37.7 bc	13.71 b	42 a	19.37 c
	C	28 c	7.79 b	32.3 a	13.17 c
	D	31.3 bc	6.51 b	42 a	19.87 c
2015	A	21.3c	10.99b	12 a	13.13 c
	B	22.7c	13.6b	13.3 a	17.25 c
	C	13.3c	7.02b	7.8 a	5.09 c
	D	13.3c	8.31b	8.7 a	9.11 c
Comparison of the pro-ecological and chemical crop protection (mean for 2013–2015)					
pro-ecological (A, B)		50,7 a	32.41 a	44.7 a	61.05 a
chemical (C, D)		23,6 b	7.25 b	21.7 b	15.71 b

In the cv. "Schwabenspelz" crop, the evaluation of weed infestation performed 3–4 weeks after herbicide treatment demonstrated that significantly the lowest number of weeds was found under chemical protection conditions (treatments C and D) (Table 3). Nonetheless, the number of weeds in the harrowing treatment and in the treatment where the herbicide was applied alone (treatment C) was similar. After application of chemical protection (treatments C and D), the dry weight of weeds was lower by 76.4% and 69.4% than in the control treatment (A). The statistical calculations showed that at the first evaluation of weed infestation, the average number and dry weight of weeds found in the chemical treatments (C and D) were significantly lower than those determined for the pro-ecological treatments (A and B) by 42.7% and 70.5%, respectively. At the dough stage of wheat, crop protection method was not shown to have an effect on the number and dry weight of weeds in the cv. "Schwabenspelz" crop.

Table 3. The impact of crop protection system on weed infestation indicators in the spelt wheat cv. "Schwabenspelz" (mean for 2013–2015). NW I—number of weeds 3–4 weeks after herbicide treatment, WW I—air-dry weight of weeds 3–4 weeks after herbicide treatment, NW II—number of weeds at the dough stage of spelt wheat (BBCH 83–85), WW II—air-dry weight of weeds at the dough stage of spelt wheat (BBCH 83–85). Different letters denote significant differences ($p \leq 0.05$). The same letter means not significantly different values.

Crop Protection	NW I (no. m^{-2})	WW I (g m^{-2})	NW II (no. m^{-2})	WW II (g m^{-2})
A	44.3 a	26.35 a	33.7 a	32.58 a
B	43.8 ab	21.95 a	27.8 a	25.22 a
C	26.2 bc	6.21 b	16.8 a	8.42 a
D	24.2 c	8.05 b	22.9 a	12.89 a
Comparison of the pro-ecological and chemical crop protection				
pro-ecological (A, B)	44 a	24.15 a	30.7 a	28.9 a
chemical (C, D)	25.2 b	7.13 b	19.9 a	10.66 a

On average over the period 2013–2015, seeding density was not found to have a significant influence on the number and dry weight of weeds in the cv. "Rokosz" the cv. "Schwabenspelz" crop at both times that weed infestation was evaluated (Figure 1). Nonetheless, it could be observed that a higher seeding density had a more beneficial effect on weed infestation, as demonstrated by the lower number and weight of weeds in the crop.

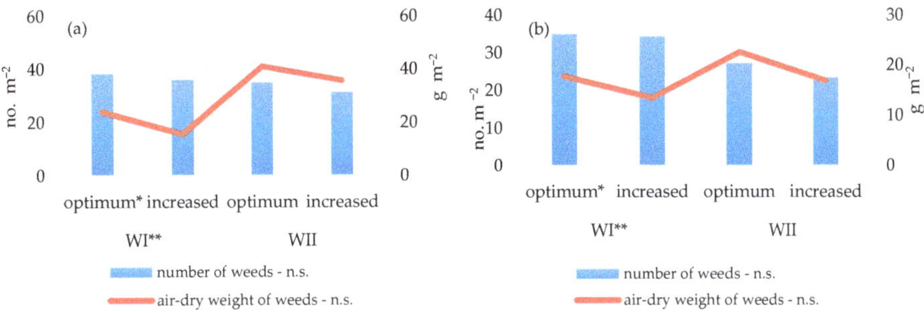

Figure 1. The impact of seeding density on weed infestation indicators in the spelt wheat cultivars "Rokosz" (**a**) and "Schwabenspelz" (**b**) (mean for 2013–2015). Note: * seeding density, **WI—number and air-dry weight of weeds 3–4 weeks after herbicide treatment, WII—number and air-dry weight of weeds at the dough stage of spelt wheat (BBCH 83–85), n.s.—not significant differences.

The analysis of the quantitative weed infestation parameters did not reveal significant differences in the number of weeds between the studied cultivars, but it proved a significantly lower biomass of weeds in the tall cultivar "Schwabenspelz". At the first evaluation of weed infestation, 3–4 weeks after herbicide treatment, the weed biomass in the crop of this cultivar was determined to be 15.64 g m^{-2} and it was 21% lower than for cv. "Rokosz". During the evaluation of weed infestation at the dough stage of spelt wheat, the difference in the biomass of weeds occurring in the crops of the cultivars studied was even greater, since for cv. "Rokosz" it was almost twice higher than for cv. "Schwabenspelz". At the first evaluation of weed infestation, *Viola arvensis*, *Geranium pusillum*, and *Apera spica-venti* were predominant in the spelt wheat crop (Figure 2a,b). In both cultivars, chemical crop protection significantly reduced the number of *Apera spica-venti* and *Viola arvensis* (Figure 2a). Seeding density did not cause differences in numbers of the dominant weed species in the crop (Figure 2b). At the dough stage (BBCH 83–85) of spelt wheat, the number of *Viola arvensis* plants was similar among all the treatments, while the number of *Apera spica-venti* was significantly lower on chemical crop protection treatments, compared to treatments A and B. (Figure 3a). In the cv. "Schwabenspelz", an increased seeding density reduced significantly the numbers of *Apera spica-venti* (Figure 3b).

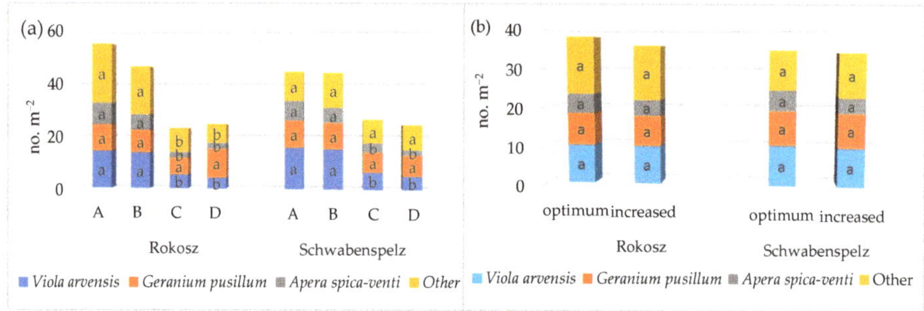

Figure 2. The impact of crop protection system on number of dominant weeds species 3–4 weeks after herbicide treatment in the spelt wheat cultivars "Rokosz" and "Schwabenspelz" (mean for 2013–2015). Different letters denote significant differences ($p \leq 0.05$) of number of weed species among different crop protection methods (A, B, C, D) (**a**) and seeding density (**b**). The same letter means not significantly different values.

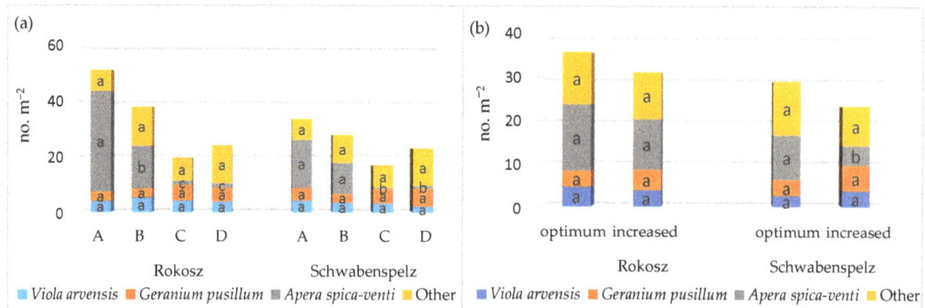

Figure 3. The impact of crop protection system on number of dominant weeds species at the dough stage (BBCH 83–85) in the spelt wheat cultivars "Rokosz" and "Schwabenspelz" (mean for 2013–2015). Different letters denote significant differences ($p \leq 0.05$) of number of weed species among different crop protection methods (A, B, C, D) (**a**) and seeding density (**b**). The same letter means not significantly different values.

3.2. Yield and Yield Components of Spelt Wheat

On average over the three-year study period, the crop protection system did not cause significant differences in the yield components of the cultivar "Rokosz" (Table 4). The comparison of pro-ecological crop protection systems (A, B) and chemical crop protection did not reveal any significant differences in the yield components.

Table 4. The impact of crop protection system on selected yield components of the spelt wheat cv. "Rokosz" (mean for 2013–2015). HP—height of plant, LE—length of ear, NE—number of ears per m^2, WG—weight of grains per ear. The same letter means not significantly different values.

Crop Protection	HP (cm)	LE (cm)	NE (no. m^{-2})	WG (g)
A	106a	8.3a	417a	1.1a
B	104.2a	8.4a	421a	1.15a
C	105.5a	8.6a	413a	1.15a
D	99.9a	8.5a	459a	1.33a
Comparison of the pro-ecological and chemical crop protection				
pro-ecological (A, B)	105.1a	8.4a	419a	1.13a
chemical (C, D)	102.7a	8.5a	436a	1.24a

The statistical analysis showed cv. "Rokosz" grain yield to be significantly dependent on crop protection system and year (Table 5). In the first year of the experiment, the highest grain yield was obtained from the treatment where full chemical crop protection was used (treatment D). In the years 2014 and 2015, crop protection intensification did not contribute to increased yields of cv. "Rokosz". In 2015, higher yields were obtained in all crop protection treatments than in 2014, while treatments A, B, and C were higher than in 2013. The statistical analysis, performed based on the means for the three-year study period, revealed that the average grain yield from the chemical protection treatments (C, D) was 19.1% higher than in the pro-ecological treatments (A, B).

Table 5. The grain yield of the spelt wheat cv. "Rokosz". The interaction between crop protection and years of research, and comparison of the pro-ecological and chemical crop protection. Different letters denote significant differences ($p \leq 0.05$).

Crop Protection	2013	2014	2015
		t ha^{-1}	
A	2.43 d	3.23 cd	5.62 ab
B	2.73 d	3.31 cd	5.66 ab
C	3.57 cd	3.19 cd	5.73 a
D	4.7 abc	4.05 bcd	6.1 a
Comparison of the pro-ecological and chemical crop protection (mean for 2013–2015)			
pro-ecological (A, B)		3.83 a	
chemical (C, D)		4.56 b	

As far as the analyzed yield components for cv. "Schwabenspelz" are concerned, the crop protection methods significantly modified plant height, but they had no effect on ear length, density, or grain weight per ear (Table 6). Wheat in the treatment that included only harrowing produced the highest plants (treatment B); the plant height was significantly lower in the treatment with full chemical protection (D). The different crop protection methods used did not affect yield of the spelt wheat cultivar "Schwabenspelz". The Scheffe test did not show significant differences in grain yield between pro-ecological technologies and chemical protection of the cv. "Schwabenspelz" crop.

Table 6. The impact of crop protection system on selected yield components and grain yield of the spelt wheat cv. "Schwabenspelz" (mean for 2013–2015). HP—height of plant, LE—length of ear, NE—number of ears per m^2, WG—weight of grains per ear, Y—yield for the crop protection methods A–D.

Crop Protection	HP (cm)	LE (cm)	NE (no. m^{-2})	WG (g)	Y (t ha^{-1})
A	127.9 ab	11.5 a	357 a	1.03 a	3.12 a
B	131.4 a	11.5 a	353 a	1.14 a	3.02 a
C	128.9 ab	11.2 a	372 a	1.09 a	3.16 a
D	119.3 b	11.4 a	389 a	1.13 a	3.56 a
Comparison of the pro-ecological and chemical crop protection					
pro-ecological (A, B)	129.6 a	11.5 a	355 a	1.09 a	3.07 a
chemical (C, D)	124.1 a	11.3 a	381 a	1.11 a	3.36 a

Different letters denote significant differences ($p \leq 0.05$). The same letter means not significantly different values.

On average over the period 2013–2015, seeding density significantly modified only the ear length of the wheat cultivar "Rokosz", but significantly the highest value of this parameter was achieved at an optimal seeding density (Table 7). The seeding density factor did not cause significant differences in the other biometric characteristics and yield of cv. "Rokosz". Also, seeding density was not found to have a significant effect on the yield components and grain yield of cv. "Schwabenspelz".

Table 7. The impact of seeding density on selected yield components and grain yield of the spelt wheat cultivars "Rokosz" and "Schwabenspelz" (mean for 2013–2015). HP–height of plant, LE–length of ear, NE–number of ears per m^2, WG–weight of grains per ear, Y-yield.

Seeding Density	HP (cm)	LE (cm)	NE (no. m^{-2})	WG (g)	Y (t ha^{-1})
cv. "Rokosz"					
optimum	105.7 a	8.7 a	412 a	1.22 a	4.25 a
increased	102.2 a	8.3 b	443 a	1.15 a	4.14 a
cv. "Schwabenspelz"					
optimum	125.9 a	11.3 a	350 a	1.11 a	3.24 a
increased	127.9 a	11.5 a	386 a	1.09 a	3.25 a

Different letters denote significant differences ($p \leq 0.05$). The same letter means not significantly different values.

Comparing the cultivars studied, it was found that the spelt wheat cultivar "Schwabenspelz" was characterized by significantly higher plants (by 22.7%) and longer ears (by 29.5%) relative to cv. "Rokosz". The comparison of the spelt wheat cultivars concerning the other biometric characteristics did not show significant differences. The grain yield of the spelt wheat cultivar "Rokosz" was significantly higher than that of cv. "Schwabenspelz", on average by 31.8%.

3.3. Grain Health

The results showed that the total number of fungi in cv. "Rokosz" grain ranged from 3.11 log CFU g^{-1} in treatment D with a increased seeding density to 3.87 log CFU g^{-1} in treatment A with an optimal seeding density, while in cv. "Schwabenspelz" grain ranged from 3.82 log CFU g^{-1} in treatment D with a higher seeding density to 4.46 log CFU g^{-1} in treatment B with an optimal seeding density (Figure 4a,b). In the grain with chemical crop protection (treatments C and D) in cv. "Rokosz" the total number of fungi was statistically significantly lower ($p \leq 0.05$). The fungi *Penicillium* sp., *Penicillium expansum*, *Candida* sp., *Aureobasidium* sp., *Aspergillus* sp., *Aspergillus flavus*, *Aspergillus terreus*, *Cladosporium* sp., *Cladosporium cladosporides*, *Ulocladium* sp., *Fusarium* sp., *Rhodotorula* sp., *Rhizopus* sp., *Trichophyton* sp., and *Phoma* sp. were found in different ratio in spelt wheat (Figures 5 and 6). In cv. "Rokosz" grain, *Fusarium* sp. (19.4–76.7%) and *Aspergillus* sp. (11.9–38.9%) were the most abundant species, while in cv. "Schwabenspelz" grain the most abundant were *Cladosporium* sp. (7.23–53.3%) and *Fusarium* sp. (4.76–38.5%).

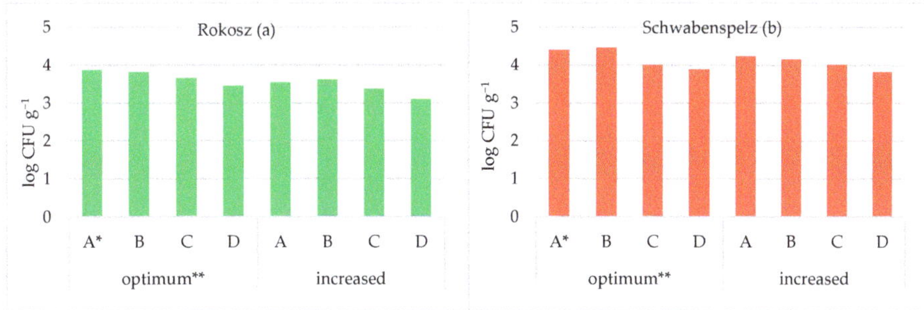

Figure 4. Total number of fungi in the grain of spelt wheat cv. "Rokosz" (**a**) and cv. "Schwabenspelz" (**b**) (mean for 2014–2015). Note: * crop protection methods, ** seeding density.

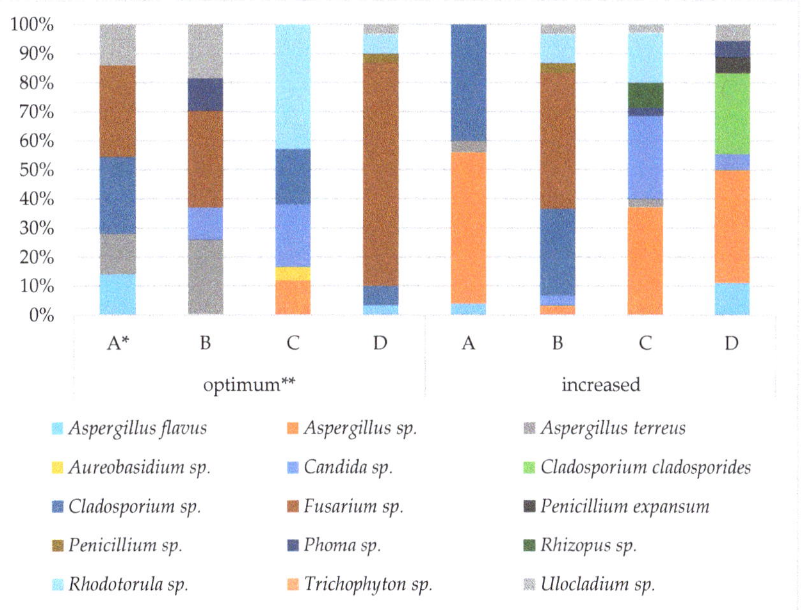

Figure 5. Identified mold fungi in the grain of spelt wheat cv. "Rokosz" (%) (mean for 2014–2015). Note: * crop protection methods, ** seeding density.

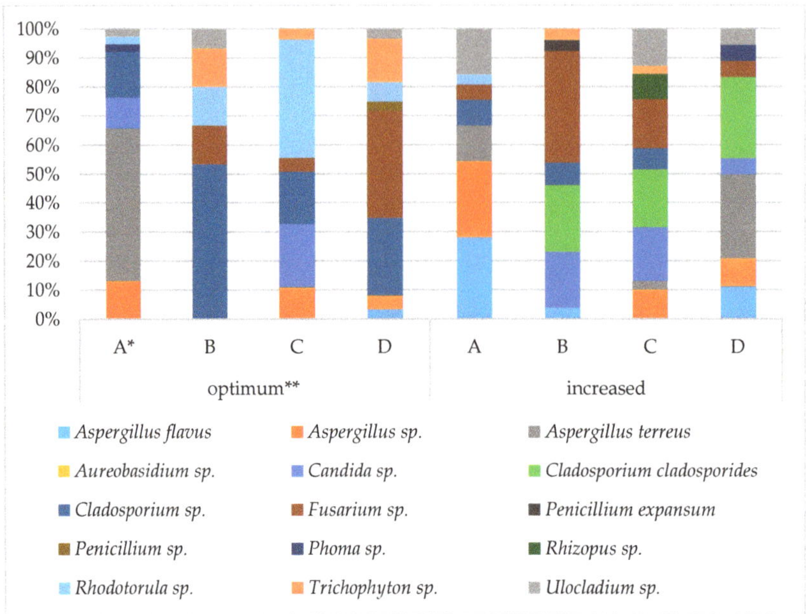

Figure 6. Identified mold fungi in grain of spelt wheat cv. "Schwabenspelz" (%) (mean for 2014–2015). Note: * crop protection methods, ** seeding density.

Both total mycotoxin content and the content of individual toxins in spelt grain were modified by the crop protection system and seeding density (Tables 8 and 9). In cv. "Rokosz" grain, significantly the highest number of mycotoxins was found in treatment C after application of herbicide crop protection alone, while the lowest number of mycotoxins was recorded in treatment D (after application of the fungicide) (Table 8). In grain harvested from the pro-ecological treatments (A and B), the number mycotoxins was at a similar level, but the amount of DAS, DON, ZEA, HT-2, and 15-acDON mycotoxins differed significantly. A denser seeding resulted in reduced grain contamination with mycotoxins. In cv. "Schwabenspelz", the lowest number of mycotoxins was found in the treatment with full chemical crop protection (treatment D) (Table 9). In the other treatments, on the other hand, grain contamination with mycotoxins was at a similar level and significantly higher than in treatment D. Similarly as in the case of cv. "Rokosz", a denser seeding contributed to a decrease in the number of mycotoxins in grain.

Regardless of the experimental factors, the amount of mycotoxins in grain of the spelt wheat cultivars studied did not differ significantly. One can only observe slightly greater susceptibility of cv. "Rokosz" to contamination with mycotoxins. In grain of this cultivar, the amounts of DAS and DON mycotoxins and of HT-2 toxin were found to be significantly greater than in "Schwabenspelz". Cv. "Rokosz" was found to have the highest content of NIV, 3-AcDON, and DAS mycotoxins, while cv. "Schwabenspelz" had the largest amounts of 3-AcDON, NIV, and 15-AcDON.

Table 8. Mycotoxin content in wheat grain of the spelt wheat cv. "Rokosz" (concentration of mycotoxins μg kg^{-1} of grain) (mean for 2014–2015). NIV—nivalenol , DAS—diacetoxyscirpenol, 15- AcDON—15-acetyldeoxynivalenol, DON—deoxynivalenol, ZEA—zearalenone, 3-AcDON—3-acetyldeoxynivalenol.

Experimental Factors	NIV	DAS	15-AcDON	T-2	DON	ZEA	HT-2 Toxin	3-AcDON	Fuzarenon X	Sum
Crop protection										
A	243b	167b	78b	13a	203a	6b	5c	100b	51c	866b
B	277b	216a	145a	11a	89b	51a	49b	47c	24c	909b
C	317a	69c	126a	20a	75b	7b	154a	299a	205a	1272a
D	16c	0.5d	75b	-	11c	-	-	295a	150b	548c
Seeding density										
optimum	345a	87b	132a	6a	141a	9a	100a	193a	99a	1111a
increased	82b	139a	79b	17a	48b	23a	4b	177a	117a	685b

Different letters denote significant differences ($p \leq 0.05$). The same letter means not significantly different values.

Table 9. Mycotoxin content in wheat grain of the spelt wheat cv. "Schwabenspelz" (concentration of mycotoxins μg kg^{-1} of grain) (mean for 2014–2015). NIV—nivalenol, DAS—diacetoxyscirpenol, 15- AcDON—15-acetyldeoxynivalenol, DON—deoxynivalenol, ZEA—zearalenone, 3-AcDON—3-acetyldeoxynivalenol.

Experimental Factors	NIV	DAS	15-AcDON	T-2	DON	ZEA	HT-2 Toxin	3-AcDON	Fuzarenon X	Sum
Crop protection										
A	227a	47a	135a	31	6b	-	-	259a	132a	836a
B	245a	61a	61b	-	54a	-	27a	176b	90c	712a
C	221a	60a	110a	-	14a	-	11a	206ab	107bc	727a
D	39b	5b	110a	-	11a	-	-	165b	128ab	457b
Seeding density										
optimum	258a	73a	110a	16	16a	-	5a	264a	163a	904a
increased	108b	13b	98a	-	26a	-	14b	139b	65b	462b

Different letters denote significant differences ($p \leq 0.05$). The same letter means not significantly different values.

4. Discussion

Many authors indicate that the appropriate selection of a weed competitive variety is a major no-cost method of reducing weed infestation in crops [24,38–40]. In the opinion of Korres and Froud-Williams [41], as well as Mason and Spaner [42], high plants with intense tillering and forming a dense crop contribute to a reduced occurrence of weeds. This theory was confirmed by Kraska et al. [14], according to whom the selection of an appropriate cultivar had a significant impact on weed infestation of crops. Weeds produced significantly the lowest biomass in the tall spelt cultivar "Oberkulmer Rotkorn", while the highest biomass developed in the shortest spelt cultivar "Badengold". Although the research presented in this paper did not confirm statistically significant differences in the number of weeds found in the crops of the evaluated cultivars, at each time weed infestation was evaluated a significantly lower dry weight of weeds was determined in the tall cultivar "Schwabenspelz". Due to the higher competitive pressure of this cultivar, weeds were more shaded and had less space for growth, and hence they produced a lower biomass. Other research results confirm that spelt wheat is characterized by high competitiveness against weeds, but they do not indicate unequivocally that this trait depends directly on plant height [24,43]. According to the study by Andruszczak et al. [40], the spelt cultivar "Frankenkorn", which was not significantly lower than many evaluated cultivars [43], was the most competitive against weeds, whereas according to Feledyn-Szewczyk [24] and Andruszczak et al. [39] the lowest weed infestation was found in the tall cultivar "Schwabenkorn" [43]. In the present study, the averages for the three-year study indicate that chemical crop protection (treatments C and D), based on its effect evaluated 3–4 weeks after application, significantly reduced the number and dry weight of weeds in the crops of both spelt wheat cultivars. The effect of the applied herbicide could be seen particularly in the first year of the study, in which the high rainfall in October caused intensive emergence of weeds. In 2014, the rainfall at the end of April and in May could have reduced the effectiveness of the herbicide. In the last year of the study, on the other hand, weed infestation of the crop was so low that the effect of herbicide action was not very visible. Herbicide application had no significant impact on weed infestation of cv. "Schwabenspelz" before harvest, but in the cv. "Rokosz" crop significantly lower weed infestation was shown in the treatments where herbicides were applied in comparison with the pro-ecological protection methods. In the studies by Andruszczak et al. [39,40], herbicide treatment significantly decreased the dry weight of weeds, but had no effect on their number. The results of the present study confirm the opinion of many authors who stress the low effectiveness of harrowing and its high dependence on weather conditions [44,45]. In the cv. "Rokosz" crop, weed infestation of the crop was not found to be significantly lower in the harrowing treatment compared to the control treatment in any year of the study. Nevertheless, in the first two growing seasons the number and biomass of weeds were slightly lower than in treatment A, whereas in 2015 weed infestation was slightly higher than in the harrowing treatment. In the cv. "Schwabenspelz" crop, harrowing also reduced weed infestation relative to the control treatment. According to Hansen et al. [46] and Lundkvist [47], the effectiveness of harrowing as a weed control method can be increased by performing this operation at an early time and repeating it two or three times, as well as by selecting weed-competitive crop plant cultivars. Despite the opinion of Korres and Froud-Williams [41] that an increase in seeding density of wheat contributes to reduced weed infestation of crops, this study did not demonstrate such a correlation for spelt wheat. It was attributable to the fact that a denser seeding did not cause a significant increase in ear density and therefore did not increase significantly the crop density.

The yield potential of crops is genetically driven and therefore selection of a variety with expected parameters is very important in crop production [14,22,32]. Moreover, in field cultivation a crop is affected by variable atmospheric conditions, which determine crop growth and development and, in consequence, yields obtained [16,22,28]. Spelt wheat is considered to be a species resistant to stress caused by adverse environmental conditions [25]. In the present study, the average grain yield for the period 2013–2015 was lower than those obtained by some researchers conducting studies in Poland [14,28]. This is due to adverse weather conditions in the first two years of the experiment, especially unevenly distributed rainfall in May and June, which contributed to the lower crop productivity. In a study by Wojtkowiak and Stępień [28], the average yield of cv. "Schwabenkorn" in 2013 was 7.47 t ha^{-1}, while a year earlier it was 46% lower. Jablonskytė-Raščė et al. [16] also found large differences in spelt yield in individual years of the study. This confirms the thesis that relatively small and unstable yields are a factor that limits the cultivation of spelt wheat [14]. Andruszczak [43] and Jablonskytė-Raščė et al. [16] found spelt wheat yields at a similar level compared to that obtained in the present study. In the study by Andruszczak [43], the yield of cv. "Schwabenspelz" was higher than in the present experiment, standing at 4.18 t ha^{-1}. According to a study conducted by Glamočlija et al. [21] in Serbia, on the other hand, yields of organically grown spelt wheat were lower, ranging from 2.33 to 3.35 t ha^{-1}. The study by Pospišil et al. [48,49], on the other hand, demonstrates that agronomic factors have a great impact on spelt wheat yield.

In many papers, it is stressed that spelt wheat is tolerant to different growing conditions and does not require intensive protection [9,21]. The results of conducted research indicate that the effect of crop protection intensification on this species' yield is dependent on the variety. In the case of the Polish cultivar "Rokosz", a 19% higher yield was obtained after application of chemical crop protection (C, D) compared to pro-ecological cropping (A, B). Cv. "Schwabenspelz", on the other hand, did not respond with a substantial increase in yield after application of crop protection products, but the yields obtained were slightly higher than for the pro-ecologically grown crop. Andruszczak [43] demonstrated that crop protection intensification had a beneficial effect on the productivity of spelt wheat, whereas in the research carried out by Pospišil et al. [48] the impact of fungicide application on spelt wheat varied between years.

Similar to the studies conducted by Andruszczak [4] and Pospišil et al. [48,49], in the present experiment seeding density did not have a significant effect on most of the yield components and yield of spelt wheat. This finding is related to the phenomenon of the so-called crop self-regulation, due to which the number of plants per unit area is not directly proportional to the seeding rate. With increasing seeding rate, the phenomenon of self-regulation of the number of plants in a crop intensifies in order not to allow excessive density. At lower seeding densities, the tillering ability of the crop increases [12].

Wheat grains are colonized by various fungi, including plant pathogens and mycotoxin-producing species. The species composition of seed-colonizing fungi can be modified by various agronomic practices and weather conditions [23,50,51]. In this study, spelt grain was colonized by many fungi, but mainly by *Fusarium* sp., *Cladosporium* sp., and *Aspergillus* sp. The main mycotoxin producers are fungi of the genera *Aspergillus*, *Penicillium*, and *Fusarium* [52,53]. Species of the genus *Fusarium* differ in their epidemiology, pathogenicity, and ability to produce mycotoxins. Most *Fusarium* species can produce one or more mycotoxins exhibiting different degrees of toxicity [54,55]. Deoxynivalenol (DON), T-2, HT-2, and nivalenol (NIV) are the most common in cereals [56,57]. In the spelt grain evaluated, similar to products investigated by Serrano et al. [58], the mycotoxin NIV was found to have the highest content. In the opinion of Serrano et al. [58], special attention should be paid to this mycotoxin due to the toxicological effects of its occurrence and its large amount in cereal products. Spelt wheat is characterized by a higher level of resistance to fungal ear and grain infections than common wheat [8,59]. In grain of the spelt cultivars studied, all nine mycotoxins were found to occur, whereas Juan et al. [60] identified in spelt only four mycotoxins: DON, NIV, fuzarenon X (FUS-X), and enniatin A (EN A), whose amounts were comparable to the content of these compounds in the cultivar "Schwabenspelz". According to Mankevičienė et al. [23], lower DON, ZEA, and T-2/HT-2 concentrations were found in spelt grain without glumes than in spelt grain with glumes, in glumes, and in spring wheat grain. Likewise, Vučkovič et al. [8] detected higher levels of mycotoxins in hulls compared to grains. Conditions that are most favorable for the development of fungi and contamination of grain with mycotoxins include high humidity, frequent rainfall, particularly before harvest, and relatively warm night temperatures [61,62]. In the opinion of Blandino et al. [63,64], high plant density is associated with higher grain infection, while in the present study, grain of both cultivars contained fewer mycotoxins under conditions of denser seeding compared to less dense seeding. Such results could have been attributable to several factors. Firstly, the denser seeding did not cause an increase in ear density in the crop. Furthermore, despite that it was not confirmed statistically, under conditions of the higher seeding rate, the weed infestation of the crop before harvest of spelt wheat was slightly lower. Weeds occurring in the crop significantly affect crop health because they can change moisture conditions in the crop and host disease-causing fungi [65]. Lower infestation of spelt wheat resulted in lower fungal infection of grain and thus contributed to lower mycotoxin content. The effect of plant density on the amount of mycotoxins was investigated by many authors, but due to the diversity of factors affecting grain infection, their results are not unambiguous [66–68]. Mycotoxin content also depends on the cultivar [67,69]. In grain of both cultivars, in line with the expectations, the lowest amount of mycotoxins was found in the treatment with fungicide application. In cv. "Rokosz", most mycotoxins were found after application of the herbicide alone (treatment C). In cv. "Schwabenspelz", the amount of mycotoxins in grain obtained from treatments A, B, and C was at a similar level. Grain of the evaluated spelt wheat cultivars, both in the pro-ecological treatments (A and B) and in the treatments with application of crop protection products, can be considered as safe and good because the content of individual mycotoxins did not exceed the maximum levels set for grain intended for food and animal feed purposes [70].

5. Conclusions

The present study showed how production intensification affected the productivity, weed infestation, and grain health of the spelt wheat cultivars "Rokosz" and "Schwabenspelz". The cultivar "Rokosz" was significantly less infested with weeds and produced significantly higher yields under chemical crop protection conditions than in the case of pro-ecological crop protection. The effect of crop protection methods on the quantitative weed infestation parameters and yield of this cultivar varied between years. Grain harvested from the treatment with herbicide application alone was characterized by the highest contamination with mycotoxins, while it was found to be lowest in the case of grain of grain harvested from treatment D. In the case of the cultivar "Schwabenspelz", no significant differences in yield were revealed between the pro-ecological and chemical crop protection treatments, and the effect of crop protection treatments on yield was not dependent on weather conditions. Cv. "Schwabenspelz" proved to be more competitive to weeds. At the dough stage of this crop plant, no significant differences in weed infestation were found between the pro-ecological treatments and those in which an herbicide was used. The lowest content of mycotoxins was determined in grain harvested from treatment D. In cv. "Schwabenspelz" grain, in comparison with the cultivar "Rokosz", the amount of mycotoxins DAS, DON, and HT-2 was determined be significantly lower, while the mycotoxin ZEA was not found at all.

Seeding density did not have a significant effect on most of the yield components, yield, and weed infestation of the spelt wheat crop. In both cultivars investigated, a higher seeding density contributed to reduced grain contamination with mycotoxins.

Due to the favorable response of the cultivar "Rokosz" to application of chemical crop protection, it is advisable to grow this species in conventional farms using integrated crop protection. The cultivar "Schwabenspelz", on the other hand, can be recommended for cultivation in organic farms on account of its competitive ability and resistance.

Author Contributions: The authors contributed to this article in the following ways—conceptualization: M.H., D.G., and B.N.-D.; data curation: D.G., S.C., and J.Ł.; formal analysis: M.H., D.G., B.N.-D., A.N., Ł.W., and M.R.-B.; funding acquisition: M.H.; investigation: M.H., D.G., B.N.-D., and A.N.; methodology: M.H., D.G., B.N.-D., A.N., and Ł.W.; project administration: M.H.; supervision: M.H. and D.G.; writing—original draft: M.H., D.G., B.N.-D., S.C., and J. Ł.; writing—review and editing: M.H., D.G., S.C., and M.R.-B. All authors have read and agreed to the published version of the manuscript.

Funding: Research supported by the Ministry of Science and Higher Education of Poland as the part of statutory activities of Department of Herbology and Plant Cultivation Techniques, University of Life Sciences in Lublin.

Conflicts of Interest: The authors declare no conflict of interest.

References

1. Szumiło, G.; Rachoń, L. Plonowanie i jakość ziarna jarej pszenicy orkisz w zależności od zastosowanego materiału siewnego. *Pol. J. Agron.* **2015**, *20*, 9–14. (in Polish).
2. Stankowski, S.; Pużyński, S.; Sobolewska, M.; Biel, W. Effect of weed control and swing rate on the baking quality of spelt in comparison with common wheat. *Bulg. J. Agric. Sci.* **2016**, *22*, 604–610.
3. Moudrý, J.; Konvalina, P.; Stehno, Z.; Capouchová, I.; Moudrý, J., Jr. Ancient wheat species can extendbiodiversity of cultivated crops. *Sci. Res. Essays* **2011**, *6*, 4273–4280.
4. Andruszczak, S. Spelt wheat grain yield and nutritional value response to sowing rate and nitrogen fertilization. *J. Anim. Plant Sci.* **2018**, *28*, 1476–1484.
5. Kohajdová, Z.; Karovicová, J. Nutritional value and banking applications of pelt wheat. *Acta Sci. Pol. Technol. Aliment.* **2008**, *7*, 5–14.

6. Biel, W.; Jaroszewska, A.; Stankowski, S.; Sadkiewicz, J.; Bośko, P. Effects of genotype and weed control on the nutrient composition of Winter spelt (*Triticum aestivum* ssp. *spelta*) and common wheat (*Triticum aestivum* ssp. *vulgare*). *Acta Agr. Scand. B—S. P.* **2016**, *66*, 27–35.
7. Konvalina, P.; Capouchova, I.; Stehno, Z.; Moudry, J. Agronomic characteristics of the spring forms of the wheat landraces (einkorn, emmer, spelt, intermediate bread wheat) grown in organic farming. *J. Agrobiol.* **2010**, *27*, 9–17. [CrossRef]
8. Vučković, J.; Bodroža-Solarov, M.; Vujić, D.; Bočarov-Stančić, A.; Bagi, F. The protective effect of hulls on the occurrence of *Alternaria* mycotoxins in spelt wheat. *J. Sci. Food Agric.* **2013**, *93*, 1996–2001. [CrossRef]
9. Kwiatkowski, C.A.; Wesołowski, M.; Pałys, E.; Kraska, P.; Haliniarz, M.; Nowak, A.; Andruszczak, S.; Kwiecińska-Poppe, E. Aspekty proekologicznego gospodarowania w agroekosystemach. *Wyd. Perfekta* **2014**, 169. (in Polish).
10. Biel, W.; Stankowski, S.; Sobolewska, M.; Radkiewicz, J.; Jaroszewska, A.; Pużyński, S. Effect of selected agronomic factors on the banking quality of winter spelt strains and cultivars (*Triticum aestivum* ssp. *spelta*) in comparison with common wheat (*Triticum aestivum* ssp. *vulgare*). *Rom. Agric. Res.* **2016**, *33*, 1–8.
11. Petrenko, V.; Spychaj, R.; Prysiazhniuk, O.; Sheiko, T.; Khudolii, L. Evaluation of three wheat species (*Triticum aestivum* L., *T. spelta* L., *T. dicoccum* (Schrank) Schuebl) commonly used in organic cropping systems, considering selected parameters of technological quality. *Rom. Agric. Res.* **2018**, *35*, 255–264.
12. Sulewska, H.; Koziara, W.; Panasiewicz, K.; Ptaszyńska, G. Plonowanie dwóch odmian ozimych orkiszu pszennego w zależności od terminu i ilości wysiewu w warunkach Środkowej Wielkopolski. *J. Agr. Eng. Res.* **2008**, *53*, 85–91.
13. Litwińczuk, M. Przyszłość orkiszu na Zamojszczyźnie. Available online: https://www.agropolska.pl/uprawa/zboza/przyszlosc-orkiszu-na-zamojszczyznie,209.html (accessed on 6 June 2020).
14. Kraska, P.; Andruszczak, S.; Kwiecińska-Poppe, E. Reaction of spelt wheat cultivars (*Triticum aestivum* ssp. *spelta*) to foliar applications of fertilizers. *Agron. Sci.* **2019**, *74*, 37–47. [CrossRef]
15. Gawlik-Dziki, U.; Świeca, M.; Dziki, D. Comparison of phenolic acids profile and antioxidant potential of six varieties of spelt (*Triticum spelta* L.). *J. Agric. Food Chem.* **2012**, *60*, 4603–4612. [CrossRef] [PubMed]
16. Jablonskytė-Raščė, D.; Maikštėnienė, S.; Mankevičienė, A. Evaluation of productivity and quality of common wheat (*Triticum aestivum* L.) and spelt (*Triticum spelta* L.) in relation to nutrition conditions. *Zemdirbyste* **2013**, *100*, 45–56. [CrossRef]
17. Kwiatkowski, C.; Haliniarz, M.; Tomczyńska-Mleko, M.; Mleko, S.; Kawecka-Radomska, M. The content of dietary fiber, amino acids, dihydroxyphenols and some macro- and micronutrients in grain of conventionally and organically grown common wheat, spelt wheat and proso millet. *Agric. Food Sci.* **2015**, *24*, 195–205. [CrossRef]
18. Kraska, P.; Andruszczak, S.; Dziki, D.; Stocki, M.; Stocka, N.; Kwicińska-Poppe, E.; Różyło, K.; Gierasimiuk, P. Green grain of spelt (*Triticum aestivum* ssp. *spelta*) harvested at the stage of milk-dough as a rich source of valuable nutrients. *Emir. J. Food Agric.* **2019**, *31*, 263–270.
19. Sliesaravičius, A.; Pekarskas, J.; Rutkovienė, V.; Baranauskis, K. Grain yield and disease resistance of winter cereal varieties and application of biological agent in organic agriculture. *Agron. Res.* **2006**, *4*, 371–378.
20. Schober, T.J.; Beana, S.R.; Kuhn, M. Gluten proteins from spelt (*Triticum aestivum* ssp. *spelta*) cultivars: Arheological and size-exclusion high-performance liquid chromatography study. *J. Cereal Sci.* **2006**, *44*, 161–173. [CrossRef]
21. Glamočlija, D.; Žarković, B.; Dražić, S.; Radovanović, V.; Popović, V.; Urgenović, V. Morphological and productive characteristics spelt wheat on the chernozem and degraded soil. In *Proceedings of XXVII Conference of Agronomists, Veterinarians, Technologists and Agricultural Economists, Belgrade, Serbia, 20–21 February 2013*; Simić, D., Beskorovajni, R., Eds.; Institut PKB Agroekonomik: Belgrade, Serbia, 2013; pp. 23–31.
22. Ugrenović, V.; Bodroža Solarov, M.; Pezo, L.; Disalov, J.; Popović, V.; Marić, B.; Filipović, V. Analysis of spelt variability (*Triticum spelta* L.) grown in different conditions of Serbia by organic conditions. *Genetika* **2018**, *50*, 635–646. [CrossRef]
23. Mankevičienė, A.; Jablonskytė-Raščė, D.; Maikštėnienė, S. Occurrence of mycotoxins in spelt and common wheat grain and their products. *Food Addit. Contam. Part A Chem. Anal. Control. Expo. Risk Assess.* **2014**, *31*, 132–138. [CrossRef] [PubMed]

24. Feledyn-Szewczyk, B. Porównanie zdolności konkurencyjnych w stosunku do chwastów oraz plonów ziarna pszenicy orkisz (*Triticum aestivum* ssp. *spelta*) z odmianami pszenicy zwyczajnej (*Triticum aestivum* ssp. *vulgare*) w ekologicznym systemie produkcji. *Folia Pomer. Univ. Technol. Stetin. Agric. Aliment. Pisc. Zootech.* **2012**, *293*, 13–26. (in Polish).
25. Babenko, L.M.; Hospodarenko, H.M.; Rozhkov, R.V.; Pariy, Y.F.; Pariy, M.F.; Babenko, A.V.; Kosakivska, I.V. *Triticum spelta*: Origin, biological characteristics and perspectives for use in breeding and agriculture. *Regul. Mech. Biosyst.* **2018**, *9*, 250–257. [CrossRef]
26. Longin, C.F.H.; Würschum, T. Genetic variability, heritability and correlation among agronomic and disease resistance traits in a diversity panel and elite breeding material of spelt wheat. *Plant. Breed.* **2014**, *133*, 459–464. [CrossRef]
27. Korkhova, M.; Panfilova, A.; Chernova, A.; Rozhok, O. The effect of pre-sowing seed treatment with biopreparations on productivity of cultivars of *Triticum spelta* L. *Agrolife Sci. J.* **2019**, *8*, 120–127.
28. Wojtkowiak, K.; Stępień, A. Nutritive value of spelt (*Triticum aestivum* spp. *spelta* L.) as influenced by the foliar application of copper, zinc and manganese. *Zemdirbyste* **2015**, *102*, 389–396. [CrossRef]
29. Wiwart, M.; Szafranska, A.; Wachowska, U.; Suchowilska, E. Quality parameters and rheological dough properties of 15 Spelt (*Triticum spelta* L.) varieties cultivated today. *Cereal Chem.* **2017**, *94*, 1037–1044.
30. Zorovski, P.; Popov, V.; Georgieva, T. Growth and development of *Triticum monococcum* L., *Triticum dicoccum* Sch. and *Triticum spelta* L. in organic farming conditions. *Contemp. Agric. Serb. J. Agric. Sci.* **2018**, *67*, 45–50. [CrossRef]
31. Rachoń, L.; Szumiło, G.; Nita, Z. Plonowanie ozimych rodów *Triticum durum* i *Triticum aestivum* ssp. spelta w warunkach okolic Lublina. *Agron. Sci.* **2020**, *64*, 101–109. (in Polish).
32. Andruszczak, S.; Kwiecińska-Poppe, E.; Kraska, P.; Pałys, E. Yield of winter cultivars of spelt wheat (*Triticum aestivum* ssp. spelta L.) cultivated under diversified conditions of mineral fertilization and chemical protection. *Acta Sci. Pol. Agric.* **2011**, *10*, 5–14.
33. Mirek, Z.; Piękoś-Mirkowa, H.; Zając, A.; Zając, M. *Flowering plants and pteridophytes of Poland a checklist. Krytyczna lista roślin naczyniowych Polski*; W. Szafer Institute of Botany, Polish Academy of Sciences: Kraków, Poland, 2002; Volume 1, p. 442.
34. *PN-ISO 21527–2: " Mikrobiologia żywności i Pasz. Horyzontalna Metoda Oznaczania Liczby Drożdży i pleśni"*; Polish Committee for Standardization: Warsaw, Poland, 2009.
35. Woźniak, A.; Nowakowicz-Dębek, B.; Stępniowska, A.; Wlazło, Ł. Effect of ozonation on microbiological and chemical traits of wheat grain Plant. *Soil Environ.* **2016**, *62*, 552–557. [CrossRef]
36. Watanabe, T. *Pictorial Atlas of Soil and Seed Fungi*; CRS Press LLC.: Boca Raton, FL, USA, 2002; p. 486.
37. Valle-Algarra, F.M.; Medina, A.; Gimeno-Adelantado, J.V.; Llorens, A.; Jimenez, M.; Mateo, R. Comparative assessment of solid-phase extraction clean-up procedures, GC columns and perfluoroacylation reagents for determination of type B trichothecenes in wheat by GCECD. *Talanta* **2005**, *66*, 194–201. [CrossRef] [PubMed]
38. Beres, B.L.; Harker, K.N.; Clayton, G.W.; Bremer, E.; Blackshaw, R.E.; Graf, R.J. Weed-competitive ability of spring and winter cereals in the Northern Great Plains. *Weed Tech.* **2010**, *24*, 108–116. [CrossRef]
39. Andruszczak, S.; Kraska, P.; Kwiecińska-Poppe, E.; Pałys, E. Weed infestation of crops of winter spelt wheat (*Triticum aestivum* ssp. *spelta*) cultivars grown under different conditions of mineral fertilization and chemical plant protection. *Acta Agrobot.* **2012**, *65*, 109–118. [CrossRef]
40. Andruszczak, S.; Kraska, P.; Kwiecińska-Poppe, E.; Pałys, E. The effect of tillage system and herbicide application on weed infestation of crops of winter spelt wheat (*Triticum aestivum* ssp. *spelta* L.) cultivars. *Acta Agrobot.* **2013**, *66*, 173–184. [CrossRef]
41. Korres, N.E.; Froud-Williams, R.J. Effect of winter wheat cultivars and seed rate on the biological characteristics of naturally occurring weed flora. *Weed Res.* **2002**, *42*, 417–428. [CrossRef]
42. Mason, H.E.; Spaner, D. Competitive ability of wheat in conventional and organic management systems: A review of the literature. *Can. J. Plant. Sci.* **2006**, *86*, 333–343. [CrossRef]
43. Andruszczak, S. Reaction of winter spelt cultivars to reduced tillage system and chemical plant protection. *Zemdirbyste* **2017**, *104*, 15–22. [CrossRef]
44. Kurstjens, D.A.G.; Kropff, A.M. The impact of uprooting and soil-covering on the effectiveness of weed harrowing. *Weed Res.* **2001**, *41*, 211–228. [CrossRef]
45. Kapeluszny, J.; Dyńska, M.; Haliniarz, M. Wpływ sposobów pielęgnacji na zachwaszczenie dwóch linii pszenicy twardej (*Triticum durum* Desf.). *Prog. Plant Prot.* **2012**, *52*, 287–293.

46. Hansen, P.K.; Rasmussen, I.A.; Holst, N.; Candreasen, C. Tolerance of four spring barley (*Hordeum vulgare*) varieties to weed harrowing. *Weed Res.* **2007**, *47*, 241–251. [CrossRef]
47. Lundkvist, A. Effects of pre- and post-emergence weed harrowing on annual weeds in peas and spring cereals. *Weed Res.* **2009**, *49*, 409–416. [CrossRef]
48. Pospišil, A.; Pospišil, M.; Svečnjak, Z.; Matotan, S. Influence of crop management upon the agronomic traits of spelt (*Triticum spelta* L.). *Plant Soil Environ.* **2011**, *57*, 435–440. [CrossRef]
49. Pospišil, A.; Pospišil, M.; Brčić, M. Influence of seeding rate and nitrogen topdressing upon the agronomic traits of spelt (*Triticum spelta* L.). *Rom. Agric. Res.* **2016**, *33*, 235–240.
50. Bankina, B.; Bimšteine, G.; Neusa-Luca, I.; Roga, A.; Fridmanis, D. What influences the composition of fungi in wheat grains? *Acta Agrobot.* **2017**, *70*, 17–26. [CrossRef]
51. Karron, E.; Runno-Paurson, E.; Lõiveke, H.; Islamov, B.; Kütt, M.L.; Talve, T.; Lauringson, E.; Hõrak, H.; Edesi, L.; Niinemets, Ü. Application of widely used fungicides does not necessarily affect grain yield, and incidence of *Fusarium* spp. and mycotoxins DON, HT-2 and T-2 in spring barley in northern climates. *Kvas. Prumys* **2020**, *66*, 215–223. [CrossRef]
52. Binder, E.M.; Tan, L.M.; Chin, L.J.; Handl, J.; Richard, J. Worldwide occurence of mycotoxins in commodieties, feeds and feeds ingredients. *Anim. Feed Sci. Tech.* **2007**, *137*, 265–282. [CrossRef]
53. Sultan, Y.; Magan, N. Mycotoxigenic fungi in peanuts from different geographic regions of Egypt. *Mycotox. Res.* **2010**, *26*, 133–140. [CrossRef]
54. Bottalico, A.; Perrone, G. Toxigenic Fusarium species and mycotoxins associated with head blight in small-grain cereals in Europe. *Eur J. Plant. Pathol.* **2002**, *108*, 611–624. [CrossRef]
55. Orlando, B.; Grignon, G.; Vitry, C.; Kashefifard, K.; Valade, R. Fusarium species and enniatin mycotoxins in wheat, durum wheat, triticale and barley harvested in France. *Mycotoxin Res.* **2019**, *35*, 369–380. [CrossRef]
56. Yli-Mattila, T.; Rämo, S.; Hussien, T.; Carlobos-Lopez, A.L.; Cumagun, C.J.R. Molecular quantification and genetic diversity of toxigenic *Fusarium* species in northern Europe as compared with those in southern Europe. *Microorganisms* **2013**, *1*, 162–174. [CrossRef] [PubMed]
57. Hietaniemi, V.; Rämö, S.; Yli-Mattila, T.; Jestoi, M.; Peltonen, S.; Kartio, M.; Sieviläinen, E.; Koivisto, T.; Parikka, P. Updated survey of Fusarium species and toxins in Finnish cereal grains. *Food Addit. Contam. Part A Chem. Anal. Control. Expo. Risk Assess.* **2016**, *33*, 831–848. [CrossRef] [PubMed]
58. Serrano, A.B.; Font, G.; Ruiz, M.J.; Ferrer, E. Co-occurrence and risk assessment of mycotoxins in food and diet from Mediterranean area. *Food Chem.* **2012**, *135*, 423–429. [CrossRef]
59. Wiwart, M.; Perkowski, J.; Jackowiak, H.; Packa, D.; Borusiewicz, A.; Buoeko, M. Response of some cultivars of spring spelt (*Triticum spelta*) to *Fusarium culmorum* infection. *Die Bodenkultur* **2004**, *55*, 29–36.
60. Juan, C.; Ritieni, A.; Mañes, J. Occurrence of *Fusarium* mycotoxins in Italian cereal and cereal products from organic farming. *Food Chem.* **2013**, *141*, 1747–1755. [CrossRef]
61. Beres, B.L.; Brûlé-Babel, A.L.; Ye, Z.; Graf, R.J.; Turkington, T.K.; Harding, M.W.; Kutcher, H.R.; Hooker, D.C. Exploring Genotype × Environment × Management synergies to manage *Fusarium* head blight in wheat. *Can. J. Plant. Pathol.* **2018**, *40*, 179–188. [CrossRef]
62. Bernhoft, A.; Torp, M.; Clasen, P.E.; Løes, A.K.; Kristoffersen, A.B. Influence of agronomic and climatic factors on Fusarium infestation and mycotoxin contamination of cereals in Norway. *Food Addit. Contam.* **2012**, *29*, 1129–1140. [CrossRef]
63. Blandino, M.; Reyneri, A.; Vanara, F. Effect of plant density on toxigenic fungal infection and mycotoxin contamination of maize kernels. *Field Crop. Res.* **2008**, *106*, 234–241. [CrossRef]
64. Blandino, M.; Reyneri, A.; Colombari, G.; Pietri, A. Comparison of integrated field programmes for the reduction of fumonisin contamination in maize kernels. *Field Crop. Res.* **2009**, *111*, 284–289. [CrossRef]
65. Edwards, S.G. Influence of agricultural practices on *Fusarium* infection of cereals and subsequent contamination of grain by trichothecene mycotoxins. *Toxicol. Lett.* **2004**, *153*, 29–35. [CrossRef]
66. Bertuzzi, T.; Romani, M.; Rastelli, S.; Giorni, P. Mycotoxins and Related Fungi in Italian Paddy Rice During the Growing Season and Storage. *Toxins* **2019**, *11*, 151. [CrossRef] [PubMed]
67. Gorczyca, A.; Oleksy, A.; Gala-Czekaj, D.; Urbaniak, M.; Laskowska, M.; Waśkiewicz, A.; Stępień, Ł. *Fusarium* head blight incidence and mycotoxin accumulation in three durum wheat cultivars in relation to sowing date and density. *Sci. Nat.* **2018**, *105*, 2. [CrossRef] [PubMed]

68. van der Burgt, G.J.H.M.; Timmermans, B.G.H.; Scholberg, J.M.S.; Osman, A.M. *Fusarium* head blight and deoxynivalenol contamination in wheat as affected by nitrogen fertilization. *NJAS—Wagen. J. Life Sci.* **2011**, *58*, 123–129. [CrossRef]
69. Bérubé, M.E.; Vanasse, A.; Rioux, S.; Bourget, N.; Dion, Y.; Tremblay, G. Effect of glyphosate on *Fusarium* head blight in wheat and barley under different soil tillages. *Plant. Dis.* **2012**, *96*, 338–344. [CrossRef] [PubMed]
70. Commission Regulation (EC) No 1126/2007. Polish Nutritional Standards 2014. Available online: https://eur-lex.europa.eu/legal-content/PL/TXT/PDF/?uri=CELEX:32007R1126&from=EN (accessed on 17 April 2020).

Article

Segetal Diversity in Selected Legume Crops Depending on Soil Tillage

Jolanta Bojarszczuk * and **Janusz Podleśny**

Department of Forage Crop Production, Institute of Soil Science and Plant Cultivation—State Research Institute, Czartoryskich 8 Str., 24-100 Puławy, Poland; jp@iung.pulawy.pl

* Correspondence: jbojarszczuk@iung.pulawy.pl; Tel.: +48-81-478-6796

Received: 29 October 2020; Accepted: 10 December 2020; Published: 14 December 2020

Abstract: The aim of the paper was to determine weed infestation expressed by weeds number and weed weight and other index under a three different tillage system: no-tillage (NT), reduced tillage (RT), and ploughing tillage (CT) in two legume species crops: pea and narrowed-leaved lupine. The research proved that growing legume under no-tillage conditions caused the increasing weed infestation. Weather conditions in each of the study years were shown to influence the weed infestation. The dry weight of weeds was higher in narrow-leaved lupine by 7% in flowering stage assessment and by 6% before harvest than in pea crop. The weeds number in the conventional tillage system in the flowering stage in pea and lupine crops was 24 and 26 plants·m^{-2}, respectively, under the reduced tillage conditions it was 33 and 29% higher, while under no-tillage it was 58 and 67% higher. In all tillage systems the dominant species were *Chenopodium album* L., *Viola arvensis* L., *Anthemis arvensis* L., and *Cirsium arvense* L. The results prove that soil tillage system affect weed infestation of legume crops.

Keywords: tillage system; weed infestation; biological diversity; reduced tillage; no-till; ploughing; legume

---

1. Introduction

The tillage method is one of the most important agrotechnical factors affecting the weed flora of cultivated crops [1,2]. Soil tillage systems can be divided into three basic groups: (1) conventional (traditional) tillage with the use of a plough, (2) reduced tillage, and (3) no-till system with direct sowing [3].

Conservation tillage, compared with ploughing tillage, causes decreasing labour and energy costs in the process of production. This system favorably affects the condition of soil environment [4,5]. It also reduces water erosion, temperature variations, and increases content of organic substances in the soil [6,7]. However, it may cause an increase in weed infestation [8,9].

Reduced tillage systems, which may cause changes at species composition and the number of occurring weeds, favor the development of monocotyledonous and perennial weeds. It also leads to the faster spread of perennial than annual weeds. In reduced tillage systems, there is lower movement of seeds than in the conventional (plough) system; therefore, most of the weed seeds are generally placed in the upper soil layer [10,11]. The method of tillage effects the depth of the weed seeds placement in the soil, and also on their accesses to the light. This has importance in relation to species which require short access to light with the aim of stimulating their germination [12]. Loosening of the topsoil in the cultivation without plowing favors the germination of weed seeds, whereas an effect of the traditional plough is that a considerable part of seeds is transferred to the deeper layers of soil, which greatly complicates their germination and emergence. The number, and species composition, of weed infestation depends also on crop species [13] and even on its variety [14].

Most of studies conducted so far concerned with the effect of soil tillage systems on weed infestation of crops, refer to cereals [15–19]. However, in the last years, the study directed to the evaluation of the possibility to use reduced tillage in the other crops, especially in the cultivation of legumes has been carried out [20–22].

The presented problem has great meaning in regard to the growing interest of farmers with reduced methods of various crop species cultivation, including legumes. Crops with legumes are particularly prone to weed infestation, and their weed control is a great problem because of the low number of herbicides approved to use at their cultivation.

The aim of the study was to determine weed species composition under a different tillage system in legume crops.

2. Materials and Methods

2.1. Field Experiment and Cultivation Management

The field experiment was carried out for three growing seasons: 2017–2019, at the Agricultural Experimental Unit in Grabów [51°21′18″ N 21°40′09″ E] (Masovian Voivodeship, Poland) belonging to the Institute of Soil Science and Plant Cultivation—State Research Institute in Puławy (Lublin voivodeship) (Figure 1). The experimental site is located in a moderate continental climatic zone. The experimental factors were as follows: (A) the legume species: pea (Batuta cultivar and narrow-leaved lupine (Regent cultivar); (B) tillage system: direct sowing (NT), reduced tillage (RT), and ploughing (CT). The experiment was set up in split-plot design with four replications, on a soil belonging to a good wheat complex, class IIIb. The soil was characterized by the following nutrient content: (mg·100 kg^{-1} soil): p 15.7–22.0; K 9.2–20.4 and Mg 5.9–7.6. The organic carbon content was 0.74–0.83%. Soil pH, as determined in 1-N KCl, was 5.4–6.3. The previous crop were cereals.

Figure 1. Localization of the study site. Source: Public Domain, https://commons.wikimedia.org/w/index.php?curid=89531 [23].

The experimental area was 720 m^2. A single plot area of 30 m^2 and for harvest—15 m^2. The density (units·m^{-2}) of legume was 200, in row intercropping–100. In all the years of the study the seeds were sown in the first 10 days of April. Mineral fertilization was applied at following rates: N—15 kg· ha^{-1}, P_2O_5—50 kg·ha^{-1}, 75 kg·ha^{-1}. All fertilizers were applied before sowing the legumes. In each growing season, chemical preparations was used to reduce weed infestation of the crops. Two herbicides were used on each crop in growing season, the same herbicides in all tillage systems (2017—Boxer 800 SC (prosulfocarb 800 g l^{-1}; 78.43%)—1 × 2.5 l·ha^{-1}, Stomp Aqua 416 SC (pendimethalin 455 g l^{-1})—1 × 2.0 lha^{-1}, in 2018 and 2019—Boxer 800 SC—1 × 3.0 lha^{-1}, Stomp Aqua 416 SC—1 × 3.5 lha^{-1}, Fusilade Super 150 EC (fluazifop-P-buty 150 g l^{-1}; 15.8%—1 × 1.0 lha^{-1}). Plants were harvested at full maturity stage of pea (BBCH 89) in the third 10 days of July and narrow-leaved lupine (BBCH 89) in the first 10 days of August.

2.2. Analysis of Weed Flora

Analyses of species composition, number of weeds, and dry matter of weeds were determined after collection using the weed-picking frame method. The weeds were collected from a frame with dimensions of 0.5 × 1 m, with four replications in each crop field.

Tillage effects on weed population composition were assessed in each field on two terms:

1. The tillering stage of legume (BBCH 61–65).
2. The seed ripening stage (before the legume harvest) (BBCH 89–92).

Weed species nomenclature followed Mirek et al. [24]. Dry matter of weeds was determined after drying at 55 °C for 4 days.

For comparison of weed infestation of legume species in dependent on cropping method, the biomass index was determined and calculated for three years (2017–2019) according to the formula by Patriquin [25]:

$$biomass\ index = \frac{crop\ biomass \times 100}{weed\ biomass + crop\ biomass} \quad (1)$$

2.3. Diversity Indicators

The weed structure composition in the studied crops was also described using two index: the Shannon–Wiener index index (*H'*) and the Simpson dominance index (*SI*). The Shannon's index is an indicator of species diversity. Its value determines the probability that two sampled individuals will belong to different species. The Simpson index (*SI*) describes the probability of occurring two individuals of the same species.

The Shannon–Wiener diversity index (*H'*) and the Simpson dominance index (*SI*) were calculated according to the following formula [26,27]:

$H' = -\Sigma\ pi\ ln\ pi$

$SI = \Sigma\ pi^2$

where:

$pi = n/N$,

n—number of individuals in species,

N—total number of individuals in the sampling area, and

ln—the natural log.

2.4. Weather Conditions

In the experiment period (2017–2019), weather conditions varied between all the years (Table 1). In 2017 the highest amount of rainfall was noted in spring in April, exceeded by 77% the long-term average. In June and the first 10 days of July was recorded a small amount of precipitation (32.6 and 9.7 mm, respectively) and was lower than the long-term average by 54.1 and 65.0%, respectively. In the first 10 days of August there were very small amount of precipitation (0.9 mm). In 2018, the amount of precipitation in May (97.4 mm) and July (118.5 mm) exceeded the average from multi-years by 70.9 and 41.1% respectively, which favored the yields of legume. During April and June the total precipitation was only 65% and 63% of the long-term average, respectively. In 2019 total precipitation during growing season of legumes was the smallest (252.4) compared to earlier two years. In July was recorded a small amount of precipitation and was lower than the long-term average by 30.0%.

Table 1. Course of weather conditions during the vegetation periods.

Specification	Month						Sum/ Average III-VIII
	III	IV	V	VI	VII	VIII	
				2017			
Precipitation (mm)	35.8	69.1	34.4	32.6	86.3	55.3	313.5
Temperature °C	5.7	7.5	13.9	18.1	18.6	19.6	13.9
				2018			
Precipitation (mm)	14.1	25.3	97.4	44.6	118.5	70.6	370.5
Temperature °C	−0.1	13.3	17.0	18.4	20.4	20.2	14.9
				2019			
Precipitation (mm)	22.2	37.5	51.5	51.2	20.2	69.8	252.4
Temperature °C	5.4	9.8	13.1	21.7	18.7	20.2	14.8
* Average precipitation from multi-year (mm)	30.0	39.0	57.0	71.0	84.0	75.0	356.0
Mean temperature from multi-year °C	1.6	7.7	13.4	16.7	18.3	17.3	12.5

* Mean for 1871–2000.

2.5. Statistical Analysis

The date presented are the mean values from the years 2017–2019, as a result of a similar reaction of the plant examined to different soil tillage system during three study years. The results were statistically analyzed with the use of the variance analysis using Statistica v.10.0 program. Tukey's multiple comparison test was used to compare differences between the means for soil tillage system while confidence intervals for the means of LSD ($\alpha = 0.05$) were used.

3. Results and Discussion

Tillage system caused significant differences in weed infestation of the legume crops (Tables 2 and 3; Figures 2 and 3). On average over the period of the study, in both legume species (pea and lupine) cropped in no-tillage system (NT), dry weight of weeds assessed in the flowering stage was higher by 55 and 14%, respectively, than in plough tillage (CT). While in the second term of assessment the relation was: 50 and 15%, respectively. The weeds air dry weight did not differ between reduced tillage (RT) and no-tillage system in pea crop in term before harvest only in the third year of the study. The higher weed air-dry weight was found in term before harvest legume. Analysis of weed infestation in both species crop showed that the dry weight of weeds was higher in narrow-leaved lupine by 7% in flowering stage assessment and by 6% before harvest than in pea crop.

Table 2. Dry weight of weeds (g·m^{-2}) in different tillage system in the flowering stage in 2017–2019.

Tillage System	Crop Species					
	Pea			Narrow-Leaved Lupine		
	2017	2018	2019	2017	2018	2019
NT	17.2 ^{c}	28.2 ^{c}	22.9 ^{c}	18.2 ^{c}	30.5 ^{c}	27.1 ^{c}
RT	15.1 ^{b}	25.6 ^{b}	19.2 ^{b}	16.0 ^{b}	22.0 ^{b}	20.1 ^{b}
CT	12.0 ^{a}	16.9 ^{a}	15.2 ^{a}	13.5 ^{a}	19.2 ^{a}	18.0 ^{a}
Mean	14.8	23.6	19.1	15.9	23.9	21.7

NT—no-tillage, RT—reduced tillage, CT—plough tillage. Different letters denote significant differences ($p < 0.05$) of the air-dry weight of weed among different tillage system (CT and NT and RT). The same letter means it is not significantly different.

Table 3. Dry weight of weeds (g·m^{-2}) in different tillage system in term before harvest in 2017–2019.

Tillage System	Crop Species					
	Pea			Narrow-Leaved Lupine		
	2017	2018	2019	2017	2018	2019
NT	27.9 ^{c}	24.6 ^{c}	25.9 ^{c}	29.0 ^{c}	25.9 ^{c}	28.7 ^{c}
RT	22.2 ^{b}	20.0 ^{b}	24.3 ^{b}	20.2 ^{b}	21.6 ^{b}	25.9 ^{b}
CT	16.9 ^{a}	14.2 ^{a}	13.2 ^{a}	17.8 ^{a}	15.8 ^{a}	14.9 ^{a}
Mean	22.3	19.6	21.1	22.3	21.1	23.2

NT—no-tillage, RT—reduced tillage, CT—plough tillage. Different letters denote significant differences ($p < 0.05$) of the air-dry weight of weed among different tillage system (CT and NT and RT). The same letter means it is not significantly different.

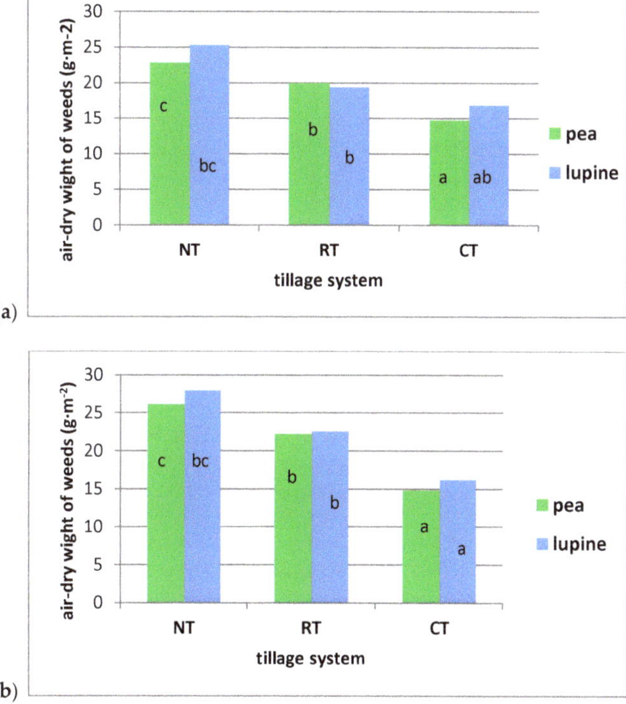

Figure 2. Average (for the years of research) air-dry weight of weeds (g·m^{-2}) in different tillage system in term: (**a**) flowering stage and (**b**) before harvest. NT—no-tillage, RT—reduced tillage, CT—plough tillage.

Weed infestation of the legume crop expressed as dry weight of weeds in estimated in the flowering stage was highest in the second year of the experiment (2018), which was characterized by a higher amount of precipitation than the other years of the experiment, while in term of estimate before harvest for no-tillage (NT) and plough tillage (CT) system in the first year of the study (2017) and for reduced tillage (RT) in the third year of the study (2019).

The average number of weeds differ significantly in the three tillage systems studied (Tables 4 and 5).

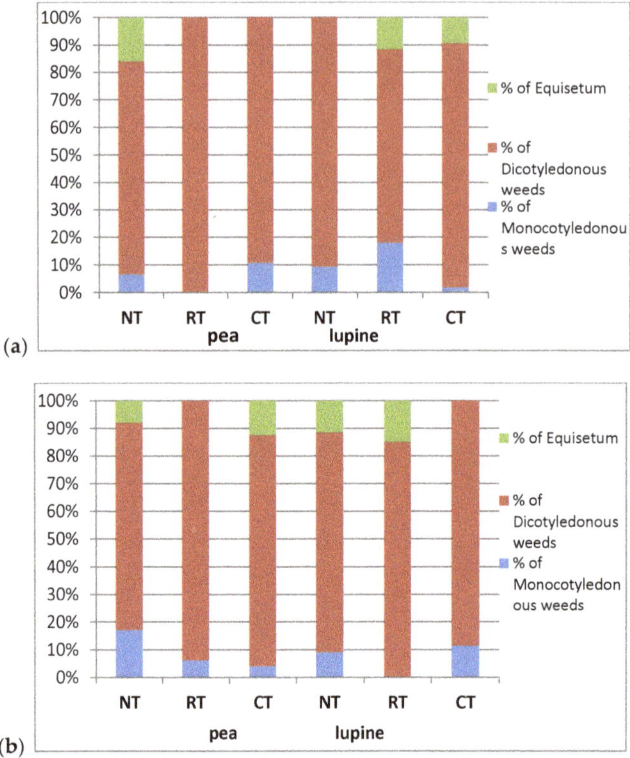

Figure 3. Percentage of monocotyledonous, dicotyledonous, and equisetum in total number of weeds in different tillage system in term: (**a**) in flowering stage (%) and (**b**) before harvest (mean for 2017–2019). NT—no-tillage, RT—reduced tillage, CT—plough tillage.

The number of weeds (for all years of the study) in the CT system in the flowering stage in pea and lupine crops was 24 and 26 plants·m^{-2}, respectively, under the reduced tillage conditions it was 33 and 29% higher, while under NT system it was 58 and 67% higher. While in term of assessment before harvest the number of weeds in the CT system was 23 and 26.5 plants per m^2 for pea and legume, respectively, under reduced tillage system it was 22 and 15% higher, while under NT system 40 and 49% higher.

A higher weed infestation under reduced soil tillage treatment as compared to plough (conventional tillage) has been confirmed by studies of Peigné et al. and Gruber and Claupein [28,29]. Demjanová et al. [30] found that soil simplification of soil was caused significantly higher weed matter compared ploughing. Similarly, Woźniak et al. [31] and Vakali et al. [32] found that a higher number of weeds in cereal crops occurred in no-tillage and reduced tillage systems than in conventionally ploughed crops. Armengot et al. [33] also recorded that total weed was higher under reduced tillage. In the study by Woźniak [34], no-tillage system significantly increased the number and weed air-dried matter in wheat crop as compared to conventional tillage.

Table 4. Weed species composition in pea and lupine crop depending on tillage systems in flowering stage (plants·m^{-2}) (average for the years of research).

Weed Species	A/P *	Pea			Narrow-Leaved Lupine		
		Tillage System					
		NT	RT	CT	NT	RT	CT
Echinochloa crus-galli (L.) *P. Beauv.*	A	2.5 ^{b}	2.0 ^{b}	1.0 ^{a}	2.0 ^{a}	-	3.0 ^{b}
Elymus repens (L.) Gould	P	4.0	-	-	2.0	-	-
Sum of Monocotyledonous weeds		6.5 ^{c}	2.0 ^{b}	1.0 ^{a}	4.0 ^{a}	-	3.0 ^{b}
Anthemis arvensis L.	A						
Aphanes arvensis L.	A		3.0				2.0
Centaurea cyanus L.	A	-	-	4.5		5.0 ^{b}	3.0 ^{a}
Chenopodium album L.	A	5.0 ^{b}	3.0 ^{a}	5.5 ^{b}	5.0 ^{a}		5.0 ^{a}
Cirsium arvense (L.) Scop	P	4.5 ^{a}	5.5 ^{b}	4.0 ^{a}	0.5 ^{a}	5.0 ^{b}	7.0 ^{c}
Convolvulus arvensis	P						
Geranium dissectum L.	A	4.5	-	-	4.0		
Plantago major L.	A	2.5 ^{a}	2.0 ^{a}		4.0		
Polygonum aviculare L.	A		3.0			4.5 ^{b}	1.5 ^{a}
Polygonum persicaria L.	A	3.5 ^{a}		3.5 ^{a}	4.5 ^{b}	4.0 ^{b}	2.0 ^{a}
Senecio vulgaris L.	A	1.5				4.5	
Sonchus arvensis L.	A	4.0 ^{b}	6.0 ^{c}	2.5 ^{a}	1.5 ^{a}	6.0 ^{b}	2.5 ^{a}
Sonchus asper L.	A				5.5	4.5	
Stellaria media (L.) Vill.	A	3.0 ^{a}	4.0 ^{a}	-	5.5		
Viola arvensis Murr.	A	-	3.5	-	4.0		
Sum of Dicotyledonous weeds		28.5 ^{b}	30.0 ^{b}	20.0 ^{a}	34.5 ^{b}	28.5	23.0 ^{a}
Eguisetum arvense L.	P	3.0 ^{a}	-	3.0 ^{a}	5.0 ^{a}	5.0 ^{a}	-
Total		38.0 ^{c}	32.0 ^{b}	24.0 ^{a}	43.5 ^{c}	33.5 ^{b}	26.0 ^{a}
Number of species		11	9	7	12	8	8

* A—annual, P—perennial, NT—no-tillage, RT—reduced tillage, CT—plough tillage. Different letters denote significant differences ($p < 0.05$) of the number of weed species among different tillage system (CT and NT and RT). The same letter means it is not significantly different.

When analyzing the weed infestation of two legume species, a significantly higher number were found in lupine crop under the NT system as compared to the CT system (Tables 4 and 5) The differences between those two species were insignificant.

Weed flora biodiversity occurring in legume grown under different tillage systems showed a similar weed species composition but same differences in number of species occurred (Tables 4 and 5). In all tillage systems the dominant species were *Chenopodium album* L., *Viola arvensis* L., *Anthemis arvensis* L., and *Cirsium arvense* L. Reduced tillage was accompanied by a larger number of *Centaurea cyanus* L. volunteer than in the two other systems.

In crops grown in no-tillage system, there was a larger number of weed species than in the other cultivation systems (Tables 4 and 5). There was also a higher number of perennial species: *Equisetum arvense* L., *Plantago major* L., and *Sonchus arvensis* L. in comparison to the traditional ploughing system. Reduced cultivation was characterized by a higher occurrence of *Fallopia convolvulus*, *Viola arvensis* L., *Chenopodium album* L., *and Cirsium arvense* L. (Tables 4 and 5). While in crops in ploughing system, there was a larger number of *Chenopodium album* L., *Cirsium arvense* L., and *Polygonum persicaria* L. Further, Clements et al. [35] found that *Chenopodium album* L. dominated the weed population in ploughing in comparison with ploughing and no-tillage system.

Table 5. Weed species composition in pea and lupine crop depending on tillage systems before harvest (plants·m^{-2}) (average for the years of research).

Weed Species	A/P *	Pea			Narrow-Leaved Lupine		
		Tillage System					
		NT	RT	CT	NT	RT	CT
Echinochloa crus-galli (L.) P. Beauv	A	2.0 ^{a}	-	2.5^{a}	3.5 ^{b}	5.5 ^{c}	0.5 ^{a}
Elymus repens (L.) Gould	P	0.3	-	-	-	-	-
Sum of Monocotyledonous weeds		2.3 ^{a}	-	2.5 ^{a}	3.5 ^{b}	5.5 ^{c}	0.5 ^{a}
Anthemis arvensis L.	A	-	6.0	-	4.0	-	5.0
Artemisia vulgaris L.	P	-	1.5 ^{a}	5.0 ^{b}	1.0 ^{a}	-	4.0 ^{b}
Centaurea cyanus L.	A	7.5	-	-	3.0 ^{a}	5.0 ^{b}	-
Chenopodium album L.	A	1.5 ^{a}	-	6.0 ^{b}	4.0 ^{b}	3.5 ^{b}	2.5 ^{a}
Cirsium arvense (L.) Scop	P	7.0 ^{b}	-	4.5 ^{a}	-	-	5.0
Fallopia convolvulus (L.) Á. Löve	A	-	5.5	-	1.5 ^{a}	5.5 ^{b}	-
Galium aparine L.	A	-	-	-	2.5	-	-
Geranium dissectum L.	A	1.5 ^{a}	1.5 ^{a}	-	5.0 ^{b}	3.0 ^{a}	-
Plantago major L.)	A	4.0 ^{a}	6.5 ^{b}	-	4.5 ^{a}	6.5 ^{b}	-
Solanum nigrum L	A	2.0	-	-	-	4.5	-
Viola arvensis Murr.	A	3.0 ^{a}	7.0 ^{c}	5.0 ^{b}	8.0 ^{b}	5.0 ^{a}	7.0 ^{b}
Sum of Dicotyledonous weeds		26.5	28.0	20.5	33.5	21.5	23.5
Eguisetum arvense L.	P	5.5	-	-	-	3.5 ^{a}	2.5 ^{a}
Total		34.3 ^{c}	28.0 ^{b}	23.0 ^{a}	37.0 ^{c}	30.5 ^{b}	26.5 ^{a}
Number of species		10	6	5	10	9	7

* A—annual; P—perennial. NT—no-tillage, RT—reduced tillage, CT—plough tillage. Different letters denote significant differences ($p < 0.05$) of the number of weed species among different tillage system (CT and NT and RT). The same letter means it is not significantly different.

The statistical analysis showed that the examined factors modified the share of monocotyledonous and dicotyledonous species in total number of weed. Average for the all the years of the study the highest percentage in weeds composition was dicotyledonous species regardless on soil tillage systems and term of assessment. In no-tillage system in the first term of assessment, the share of dicotyledonous species was significantly lowest, whereas the share of the monocotyledonous species was lower than in simplificated systems. The no-tillage system increased the percentage of monocotyledonous species and decreased the percentage of dicotyledonous species in the structure of weeds. While in the term before harvest, the highest percentage of dicotyledonous was in no-tillage system in lupine crop and the lowest in the pea crop in the same of tillage system (Figure 3).

The agrophytocenosis of legume crop found in flowering stage included 18 weed species while 14 belonged to annual weeds (77.8%). Twelve weed species (three perennial and nine annual) where noted in legume grown in no-tillage system. While in the second term of assessment (before harvest) the agrophytocenosis included 14 weed species (4 perennial weeds).

The share of perennial weed species in the no-tillage system was higher than in the ploughing and reduced tillage systems. It is mean that soil simplification promotes an increase in perennial weeds in the field (Tables 4 and 5). Study of Bilalis et al. [36] stated that three annual weed species occurred in large number in the ploughing and reduced tillage systems, while one perennial weed species often occurred in the no-tillage system.

A total of 10 weed species (mean for two terms of assessment) occurred in legume crop grown in no-tillage (NT) system and 8 in reduced tillage (RT) and 7 in conventional tillage (CT).

The evaluation of the biological diversity of the segetal flora in flowering stage showed that the Shannon–Wiener index (H') and the Simpson index (SI) measured for pea crop differ significantly in no-tillage system and reduced tillage, but significantly differences was found between conventional tillage and other tillage systems. Analysis of segetal flora diversity in term before harvest showed no-significantly differences between all tillage system in both legume species. Weed biodiversity of legume crop as measured by Shannon's index was the highest under no-tillage system—2.20 and 1.71

(average for two terms) for pea and lupine crop respectively, due to the most even percentage of weed species in the community. The no-tillage system was characterized by the lowest value of Simpson dominance index—average for two terms and for both legume species—0.12 (Figures 4 and 5).

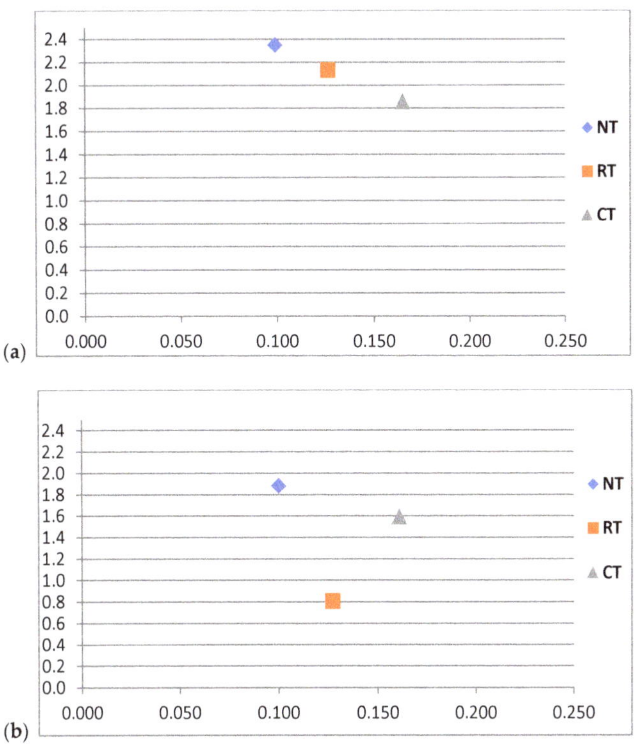

Figure 4. Shannon–Wiener's diversity index (*H'*) and Simpson's dominance index (*SI*) of the weed community in (**a**) pea, (**b**) lupine crop in flowering stage depending on the tillage system (mean for 2017–2019). NT—no-tillage, RT—reduced tillage, CT—conventional tillage.

The highest value of Simpson's index was found under conventional tillage—especially in flowering stage—0.17 and 0.21 for pea and lupine respectively, due to the domination of three weed species: *Chenopodium album* L., *Cirsium arvense* L., and *Centaurea cyanus* L.

The study results of Santín-Montanyá et al. [37] and Gawęda et al. [38] paper, show that weed community diversity, as expressed by the Shannon index, increased under the NT system. Sebayang and Rifai [39], point out that the weed matter in the soybean crop did not differ significantly between conventional tillage, reduced, and no-tillage. Nonetheless, many authors recorded that the no-tillage system increases weed infestation of crops [40–42]. Results of our study have been confirmed by Feledyn-Szewczyk et al. [43], who found that weed biodiversity as expressed by diversity indices was the highest under reduced soil tillage. Study of Legere et al. [44], the tillage system has little impact on weed diversity but according to those authors soil tillage system plays an important role in shaping the weed species composition. Furthermore, Sans et al. [45], did not find significant differences in Shannon's diversity index values for weed communities in cereal crops. However those authors point out a significantly higher degree of weed infestation in the reduced tillage system compared to the conventional system.

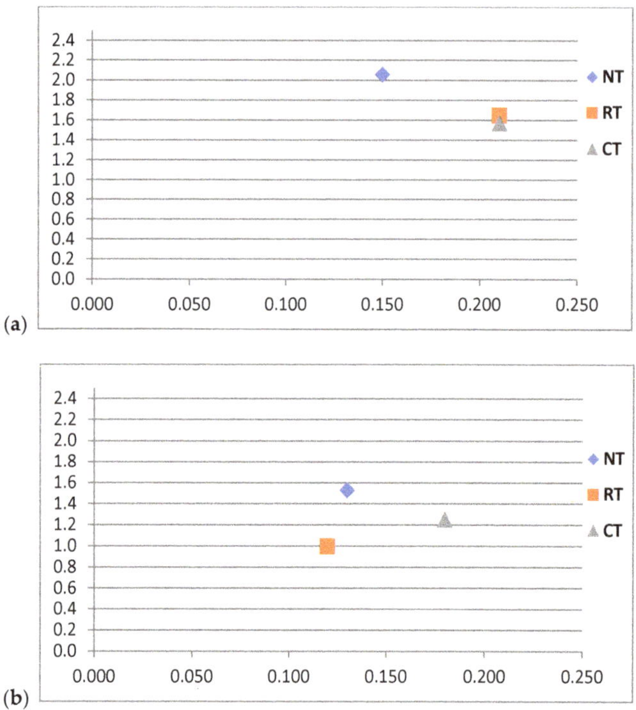

Figure 5. Shannon–Wiener's diversity (*H'*) and Simpson's dominance (*SI*) indices of the weed community in (**a**) pea and (**b**) lupine crop before harvest depending on the tillage system (mean for 2017–2019). NT—no-tillage, RT—reduced tillage, CT—conventional tillage.

Another authors [20,46] stated that tillage system modifies soil properties, and thus influences plant growth and development. Campiglia et al. [47] and Gawęda et al. [40] found increasing number of perennial weeds in no-tillage system. Velykis and Satkus [48] point out that no-tillage system caused a change in the composition of weed species compared to the conventional system. It caused the increase of the occurrence of *Galium aparine* and *Chenopodium album* in the legume crop. Gawęda et al. [40] found that the NT positively influences the segetal diversity in the soybean crop compared to the CT system.

Biomass index was additional factor indicating the degree of weed infestation. The lowest values of this index were found for no-tillage system of both legume species crop (average 91.6%), which shows a high percentage of weed matter in the total biomass of legume crop per unit area, while the highest values of the biomass index were recorded for pea crop grown in conventional tillage (94.6%), yet they did not significantly differ from the other tillage system, where the value of this index was also very high (Figure 6). On average, from all of the years of the study, the highest index of relation between legume seeds yield and weeds dry matter was in no-tillage soil system, where the lowest yields in such as cropping method was noted (Figure 6). In this system, the highest air-dried weed matter was noted. While the lowest index in legume (pea and lupine) grown in ploughing system.

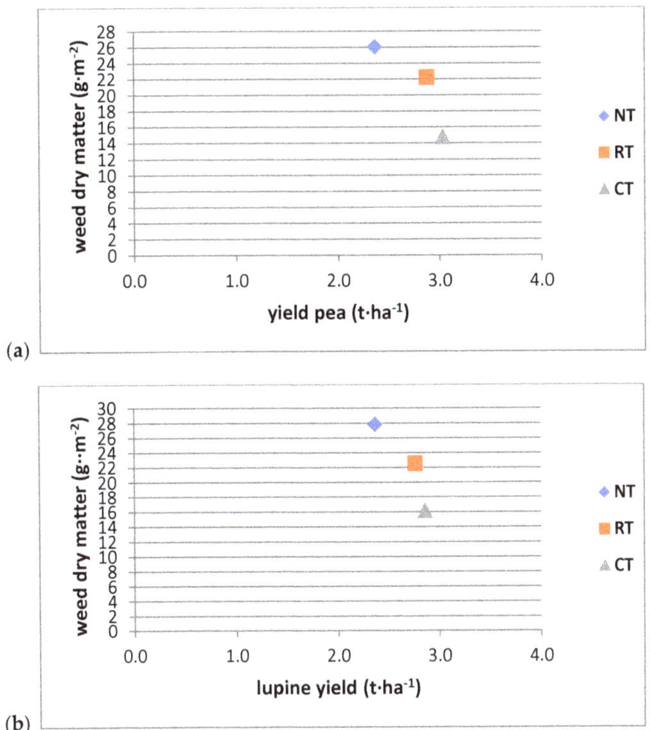

Figure 6. Relation between weeds dry matter seeds and yield of (**a**) pea, (**b**) legume.

4. Conclusions

Simplifications in soil tillage increased the contribution of weeds in legume crops. A higher number of weeds per 1 m^2 was recorded in ploughless tillage (no-tillage system), whereas 35% less on the plot with plough tillage. Weed infestation expressed by weeds number was higher by 9.1% in the first term (in the flowering stage) than in the second term (ripening of pods and seeds). The number of weeds in both the evaluation terms was also determined by the applied tillage method. As in the flowering stage as before harvest of legume crops, the highest infestation occurred in no-tillage soil system.

The dry weight of weeds was higher in narrow-leaved lupine by 7% in flowering stage assessment and by 6% before harvest than in pea crop. The number of weeds in the plough tillage system in the flowering stage in pea and lupine crops was 24 and 26 plants·m^{-2}, respectively, under the reduced tillage conditions it was 33 and 29% higher, while under no-tillage it was 58 and 67% higher.

Diversified soil tillage influenced the biodiversity of weed in legume crops field. In the first term of evaluation, 12 weed species (average for both legume species) were recorded on the plot with no-tillage soil, while 40% less weed species were occurred in plough (conventional) tillage.

Shannon's diversity index had highest values and Simpson's dominance index had lowest values in no-tillage systems, which showed their high biodiversity importance. The evaluation of the biological diversity of the segetal flora in flowering stage showed that the Shannon–Wiener index (*H'*) and the Simpson index (*SI*) measured for pea crop differ significantly in no-tillage system and reduced tillage, but significantly differences was found between conventional tillage and other tillage systems.

In all tillage systems the most numerous species were *Chenopodium album* L., *Viola arvensis* L., *Anthemis arvensis* L., and *Cirsium arvense* L. Reduced soil tillage was accompanied by a larger number of *Centaurea cyanus* L. volunteer than in the other systems.

Author Contributions: Conceptualization, J.B. and J.P.; methodology, J.B.; validation, J.B.; investigation, J.B. and J.P.; data curation, J.P.; writing—original draft preparation, J.B.; writing—review and editing, J.B.; visualization, J.B. and J.P.; supervision, J.B. and J.P. Both authors have read and agreed to the published version of the manuscript.

Funding: This research was funded by Ministry of Agriculture and Rural Development in Poland the task 2.3 of the Multiannual Program of the Institute of Soil Science and Plant Cultivation–State Research Institute in Puławy assigned for the years 2016–2020.

Acknowledgments: This research was supported by the Ministry of Agriculture and Rural Development in Poland.

Conflicts of Interest: The authors declare no conflict of interest. The funders had no role in the design of the study; in the collection, analyses, or interpretation of data; in the writing of the manuscript; or in the decision to publish the results.

References

1. Wilson, R.G. *Biology of Weed Seeds in the Soil. W: Weed Management in Agroecosystems: Ecological Approaches*; Altieri, M.A., Liebman, M., Eds.; CRC Press, Inc.: Boca Raton, FL, USA, 1988; Volume 3, pp. 25–39.
2. Sekutowski, T.; Smagacz, J. Share of anthropophytes in the crop sequence: Winter wheat—Maize—Spring wheat depending on tillage system. *Acta Agrobot.* **2014**, *67*, 117–122. [CrossRef]
3. Koller, K. Techniques of Soil Tillage. In *Soil Tillage in Agroecosystems*; Titi, A.E., Ed.; CRC Press: Boca Raton, FL, USA, 2003; pp. 1–25.
4. Dzienia, S.; Zimny, L.; Weber, R. Najnowsze kierunki w uprawie roli i technice siewu [The newest trends in soil tillage and techniques of sowing]. *Fragm. Agron.* **2006**, *2*, 227–241. (In Polish)
5. Weber, R. Wpływ okresu stosowania systemów bezpłużnych na właściwości gleby [Soil properties as affected by duration of using no-tillage systems]. *Postepy Nauk Rol.* **2010**, *1*, 63–75. (In Polish)
6. Holland, J.M. The environmental consequences of adopting conservation tillage in Europe: Reviewing the evidence. *Agric. Ecosyst. Environ.* **2004**, *103*, 1–25. [CrossRef]
7. Kraska, P. Effect of conservation tillage and catch crops on some chemical properties of rendzina soil. *Acta Sci. Pol. Agric.* **2011**, *10*, 77–92.
8. Jedruszczak, M.; Palys, E.; Kraska, P. Conservation tillage and weeds—Sustainability implication. ISTRO. In Proceedings of the 17th Triennial Conference, Sustainability—Its Impact on Soil Management and Environment, Kiel, Germany, 28 August–3 September 2006; pp. 558–565.
9. Pullaro, T.C.; Marino, P.C.; Jackson, D.M.; Harrison, H.F.; Keinath, A.P. Effects of killed cover crop mulch on weeds, weed seeds, and herbivores. *Agric. Ecosyst. Environ.* **2006**, *115*, 97–104. [CrossRef]
10. Piskier, T.; Sekutowski, T.R. Wpływ uproszczeń w uprawie roli na liczebność oraz rozmieszczenie nasion chwastów w glebie. *J. Res. Appl. Agric. Eng.* **2013**, *58*, 109–117.
11. Wrzesińska, E.; Pużyński, S.; Komorowska, A. The effect of tillage systems on soil seedbank. *Acta Agrobot.* **2013**, *66*, 113–118. [CrossRef]
12. Podleśny, J.; Podleśna, A. The effect of soil tillage in darkness on weed infestation. *Fragm. Agron.* **1997**, *2b*, 539–542.
13. Bojarszczuk, J.; Podleśny, J.; Nowak, J. The assessment of the diversity of weed flora communities in crops cultivated in selected farms in Lubelskie voivodeship. *Prog. Plant Prot.* **2018**, *58*, 216–223.
14. Lemerle, D.; Smith, A.; Verbeek, B.; Rudd, S.; Martin, P. Breeding for competitive cultivars of wheat. In Proceedings of the III International Weed Science Congress, Foz do Iguassu, Brazil, 6–11 June 2000; pp. 11–75.
15. Jug, I.; Jug, D.; Sabo, M.; Stipesevc, B.; Stosic, M. Winter wheat yield and yield components as affected by soil tillage systems. *Turk. J. Agric. For.* **2011**, *35*, 1–7.
16. De Vita, P.; Di Paolo, E.; Fecondo, G.; Di Fonzo, N.; Pisante, M. No-tillage and conventional tillage effects on durum wheat yield, grain quality and soil moisture content in southern Italy. *Soil Tillage Res.* **2007**, *92*, 69–78. [CrossRef]
17. Rieger, S.; Richner, W.; Streit, B.; Frossard, E.; Liedgens, M. Growth, yield, and yield components of winter wheat and the effects of tillage intensity, proceeding crops, and N fertilization. *Eur. J. Agron.* **2008**, *28*, 405–411. [CrossRef]
18. Streit, B.; Rieger, S.; Stamp, P.; Richner, W. Weed populations in winter wheat as affected by crop sequence, intensity of tillage and time of herbicide application in a cool and humid climate. *Weed Res.* **2002**, *43*, 20–32. [CrossRef]

19. Ozpinar, S. Effects of tillage systems on weed population and economics for winter wheat production under the Mediterranean dryland conditions. *Soil Tillage Res.* **2006**, *87*, 1–8. [CrossRef]
20. Woźniak, A.; Soroka, M.; Stępniowska, A.; Makarski, B. Chemical composition of pea (*Pisum sativum* L.) seeds depending on tillage systems. *J. Elem.* **2006**, *19*, 1143–1152. [CrossRef]
21. Małecka-Jankowiak, I.; Blecharczyk, A.; Swędrzyńska, D.; Sawinska, J.; Piechota, T. The effect of long-term tillage systems on some soil properties and yield of pea. *Acta Sci. Pol. Agric.* **2016**, *15*, 37–50.
22. Yin, X.; Al-Kaisi, M.M. Periodic response of soybean yields and economic returns to long term no tillage. *Agron. J.* **2004**, *96*, 723–733. [CrossRef]
23. Available online: https://commons.wikimedia.org/w/index.php?curid=89531 (accessed on 1 August 2018).
24. Mirek, Z.; Piękoś-Mirkowa, H.; Zając, A.; Zając, M. Flowering Plants and Pteridophytes of Poland a Checklist. In *Krytyczna Lista Roślin Naczyniowych Polski. W.*; Szafer Institute of Botany, Polish Academy of Science: Kraków, Poland, 2002; p. 442.
25. Patriquin, D.G.; Bains, D.; Lewis, J.; Macdougall, A. Weed control in organic farming systems. In *Weed Control in Agroecosystems: Ecological Approaches*; Altieri, M.A., Liebman, M., Eds.; CRS Press: Boca Raton, FL, USA, 1988; pp. 303–317.
26. Shannon, C.E. A mathematical theory of communications. *Bell Syst. Tech. J.* **1948**, *27*, 379–423. [CrossRef]
27. Simpson, E.H. Measurement of diversity. *Nature* **1949**, *168*, 668. [CrossRef]
28. Peigné, J.; Ball, B.C.; Roger-Estrade, J.; David, C. Is conservation tillage suitable for organic farming? A review. *Soil Use Manag.* **2007**, *23*, 129–144. [CrossRef]
29. Gruber, S.; Claupein, W. Effect of tillage intensity on weed infestation in organic farming. *Soil Tillage Res.* **2009**, *105*, 104–111. [CrossRef]
30. Demjanová, E.; Macák, M.; Dalovic, I.; Majernik, F.; Tyr, S.; Smatana, S. Effects of tillage systems and crop rotation on weed density, weed species composition and weed biomass in maize. *Agron. Res.* **2009**, *7*, 785–792.
31. Woźniak, A.; Wesołowski, M.; Soroka, M. Effect of long-term reduced tillage on grain yield, grain quality and weed infestation of spring wheat. *J. Agric. Sci. Technol.* **2015**, *17*, 899–908.
32. Vakali, C.; Zaller, J.G.; Köpke, U. Reduced tillage effects on soil properties and growth of cereals and associated weeds under organic farming. *Soil Tillage Res.* **2011**, *111*, 133–141. [CrossRef]
33. Armengot, L.; Berner, A.; Blanco-Moreno, J.M.; Mäder, P.; Sans, F.X. Long-term feasibility of reduced tillage in organic farming. *Agron. Sustain. Dev.* **2015**, *35*, 339–346. [CrossRef]
34. Woźniak, A. Effect of tillage system on the structure of weed infestation of winter wheat. *Span. J. Agric. Res.* **2019**, *16*, 1009. [CrossRef]
35. Clements, D.R.; Benoit, D.L.; Murphy, S.D.; Swanton, C.J. *Tillage Effects on Weed Seed Return and Seedbank Composition*; Cambridge University Press: Cambridge, UK, 1996; pp. 314–322.
36. Bilalis, D.; Efthimiadis, P.; Sidiras, N. Effect of three tillage systems on weed flora in a 3-year rotation with four crops. *J. Agron. Crop Sci.* **2001**, *186*, 135–141. [CrossRef]
37. Santín-Montanyá, M.I.; Martín-Lammerding, D.; Walter, I.; Zambrana, E.; Tenorio, J.L. Effects of tillage, crop systems and fertilization on weed abundance and diversity in 4-year dry land winter wheat. *Eur. J. Agron.* **2013**, *48*, 43–49. [CrossRef]
38. Gawęda, D.; Haliniarz, M.; Bronowicka-Mielniczuk, U.; Łukasz, J. Weed infestation and health of the soybean crop depending on cropping system and tillage system. *Agriculture* **2020**, *10*, 208. [CrossRef]
39. Sebayang, H.T.; Rifai, A.P. The effect of soil tillage system and weeding time on the growth of weed and yield of soybean (*Glycine max* (L.) Merril). *J. Degrad. Min. Lands Manag.* **2018**, *5*, 1237–1243. [CrossRef]
40. Gawęda, D.; Haliniarz, M.; Cierpiała, R.; Klusek, I. Yield, Weed infestation and seed quality of soybean (*Glycine max* (L.) Merr.) under different tillage systems. *J. Agric. Sci.* **2017**, *23*, 268–275.
41. Sebayang, H.T.; Fatimah, S. The effect of tillage systems and dosages of cow manure on weed and soybeans yield (*Glycine max*, Merrill). *J. Degrad. Min. Lands Manag.* **2019**, *7*, 1959–1963. [CrossRef]
42. Woźniak, A.; Rachoń, L. Effect of tillage systems on pea crop infestation with weeds. *Arch. Agron. Soil Sci.* **2019**, *65*, 877–885. [CrossRef]
43. Feledyn-Szewczyk, B.; Smagacz, J.; Kwiatkowski, C.A.; Harasim, E.; Woźniak, A. Weed flora and soil seed bank composition as affected by tillage system in three-year crop rotation. *Agriculture* **2020**, *10*, 186. [CrossRef]
44. Legere, A.C.S.F.; Stevenson, F.C.; Benoit, D.L. Diversity and assembly of weed communities: Contrasting responses across cropping systems. *Weed Res.* **2005**, *45*, 303–315. [CrossRef]

45. Sans, F.X.; Berner, A.; Armengot, L.; Mäder, P. Tillage effects on weed communities in an organic winter wheat–sunflower–spelt cropping sequence. *Weed Res.* **2011**, *51*, 413–421. [CrossRef]
46. Mancinelli, R.; Muleo, R.; Marinari, S.; Radicetti, E. How Soil Ecological Intensification by Means of Cover Crops Affects Nitrogen Use Effciency in Pepper Cultivation. *Agriculture* **2019**, *9*, 145. [CrossRef]
47. Campiglia, E.; Radicetti, E.; Mancinelli, R. Floristic composition and species diversity of weed community after 10 years of different cropping systems and soil tillage in a Mediterranean environment. *Weed Res.* **2018**, *58*, 273–283. [CrossRef]
48. Velykis, A.; Satkus, A. Weed infestation and changes in field pea (*Pisum sativum* L.) yield as affected by reduced tillage of a clay loam soil. *Zemdirbyste* **2010**, *97*, 73–82.

Publisher's Note: MDPI stays neutral with regard to jurisdictional claims in published maps and institutional affiliations.

Article

Biodiversity of Weeds in Fields of Grain in South-Eastern Poland

Barbara Sawicka [1], Barbara Krochmal-Marczak [2,*], Piotr Barbaś [3], Piotr Pszczółkowski [4] and Marek Ćwintal [1]

1 Department of Plant Production Technology and Commodity Sciences, University of Life Sciences in Lublin, Akademicka 15, 20-950 Lublin, Poland; barbara.sawicka@up.lublin.pl (B.S.); helena.cwintal@up.lublin.pl (M.Ć.)
2 Department of Food Production and Safety, Krosno State College, Dmochowskiego 12, 38-400 Krosno, Poland
3 Department of Potato Agronomy, IHAR–PIB, Branch in Jadwisin, Szaniawskiego 15, 05-140 Serock, Poland; p.barbas@ihar.edu.pl
4 Experimental Station for Cultivar Assessment of Central Crop Research Centre, Uhnin, 21-211 Dębowa Kłoda, Poland; p.pszczolkowski.inspektor@coboru.pl
* Correspondence: barbara.marczak@kpu.krosno.pl; Tel.: +48-508-086-387

Received: 30 October 2020; Accepted: 25 November 2020; Published: 27 November 2020

Abstract: Analysis of weed infestation of selected fields of grain (winter wheat, spring wheat, spring triticale) was conducted between 2013 and 2016 in five commercial farms in south-eastern Poland (49°52′ N, 21°46′ E) based on a quantitative and qualitative (quadrat) method and an agro-phytosociological method. The quadrat analysis was conducted prior to weeding procedures, and the agro-phytosociological analysis by grain harvest. The biodiversity of weed communities was measured with the Shannon and Simpson indices. The degree of weed infestation of grain species was significantly differentiated by weeding procedures carried out by farmers. The highest share of weeds in grain crops included dicot weeds (80.6–86.4% of all species, depending on location), and the remaining weed groups were a much smaller issue. The greatest weed infestation was found in spring triticale, and the smallest in winter wheat. The highest Shannon biodiversity index was recorded in the field of triticale, and the lowest in the field of winter wheat. The Simpson index points to the greatest biodiversity in fields of triticale and the smallest in fields of spring wheat. The conducted research will help categorize segetal flora characteristics for a given crop, determine its quantity and species composition, and evaluate biodiversity of weeds in fields of grain.

Keywords: segetal flora; weed quantity; weed mass; grain species

1. Introduction

Weeds are part of agroecosystem biodiversity. In fact, weed infestation of crops is shaped by habitat conditions. The properties of weeds that make them more abundant than crops are mainly rapid initial growth rate, high efficiency of CO_2, water and nutrient assimilation; adaptability to changing environment, high multiplication rate, adaptation of fruit and seeds to long-distance dispersal, nonconcomitant seed germination (polymorphism of dormancy), seed longevity [1,2]. The presence of various weed species in a crop field increases the diversity of soil microflora and microfauna, is conducive to the residence of a number of antagonistic insects [3,4]. Country-level biodiversity includes genetic diversity, which determines inherent variability within the population and between populations of organisms; a genetic signature that relates to energy for survival and reproduction in different climates; genetic diversity which determines species diversity of the studied ecosystem or planet. In evaluating biodiversity, the variability of taxons in the ecosystem must be taken into account.

Biodiversity increases in parallel to the number of species and balances their share, but also depends on their uniform distribution, interspecies diversity, group diversity, or ecosystem biodiversity [2,5,6]. The growth stage of individual biocoenoses influences species variety in the given ecosystem. The more mature the phytocenosis, the more plant species it contains [7]. Genetic, species, and ecological diversities interact. The connections between the three measures of natural abundance make it hard to separate them [2,4,5]. The key role in limiting or maintaining biodiversity is played by agricultural activity, which determines the intensity of cultivation and agrotechnical procedures. Other important factors include geography and topography, and the related climate conditions. In fact, weed infestation of crops is shaped by habitat conditions. The number and species structure of segetal weeds depend e.g., on soil quality and properties, i.e., granulometric composition, fertility, pH, and water air relationships in soil [1,3,8]. The factors that have the greatest influence on weed infestation include agrotechnical weed control procedures, such as crop rotation, cultivation, selection of species and cultivars, sowing time, sowing quantity, row spacing, soil mulching [3,8,9]. Weed biodiversity also has a number of biological functions in and around fields. Moreover, it plays a significant role in the nutrient cycle and use, as well as in maintaining the balance of crops attacked by diseases and pests [10]. However, weed infestation in fields of grain is a serious problem in plant production. The quantitative relations between weed species can change at different grain growth stages and over the years. This demonstrates the adaptability of weeds to agrotechnical cycles [7,11]. Weeds compete with other plants for nutrients, water, and light, and in consequence cause high crop losses [1,3,8,12]. They assimilate much more water and nutrients than crops. Therefore, they can be particularly competitive in case of elements deficiency in soil. Another factor affecting crop loss is the collection of weed seeds together with crops. It has a significant influence on lowering the quality of agricultural products. The growing share of grains in the crop structure in Poland—over 75% [13]—facilitates the growth of segetal plants in fields of crops. Weeds, which are undesirable plants in crop fields, affect both the quantity and quality of yield. Regular mechanic weeding procedures do not eliminate weeds; they remain on arable soils more or less long-lastingly by means of appropriate multiplication and adaptability to changing environmental conditions. The conditions vary depending on the type of crops [3,7]. In fields of grain, weeds occur in relatively long-lasting multispecies communities. EU regulations require the monitoring of biodiversity in Member States, including biodiversity in rural areas and the efficiency of various action plans [6,14]. Poland is obliged to protect and monitor the environment and biodiversity in ways specified in legal regulations and based on scientific knowledge. Evaluation and monitoring of biodiversity can take place in various spatial scales, which influences the selection of methods and indicators. Hence, the aim of this paper was to evaluate weed infestation in fields of selected winter and spring grains on farms in south-eastern Poland. The conducted research will enable the categorization of segetal flora characteristic for various species of grain crops, the determination of its quantity and species composition, and the evaluation of biodiversity in this region of Europe.

2. Materials and Methods

Analysis of weed infestation of selected fields of grain (winter wheat, spring wheat, spring triticale) was conducted between 2013 and 2016 in five commercial farms in south-eastern Poland in the Strzyżowski District (49°52′ N, 21°46′ E). In terms of geomorphology, the area of the Strzyżów District is located in the Alpine Zone, the Carpathian Province, the Western Carpathians Sub-Province, the Outer Carpathians Macroregion, the Mesoregion of the Strzyżowski Foothills and Dynowski Foothills. This area is almost entirely located in the Wisłok catchment—the second-order watercourse feeding the Vistula [15]. In the area of the district, there are podzolic soils and marshes located on the plains of the Wisłok valley as well as brown and podzolic soils produced as a result of weathering of the Carpathian flysch rocks, which are predominant in the areas located higher. According to the general classification, these soils belong to the soils of mountainous areas with medium valuation classes. The parent rocks are the subsoil, flysch rocks or crumb rock and loess slope clays, while the soils are mostly of the dust type, most often used as arable soils [15,16].

2.1. Agrotechnical Treatments

The forecrop of grains was diversified, but included mainly grains and rape (Table 1).

Table 1. Forecrop and fertilization used on the farms under study.

Cultivated Plant	Forecrop	Fertilization
		*(A)
Winter wheat	Rape	N—160 kg·ha^{-1}, P_2O_5—66 kg·ha^{-1}, K_2O—66 kg·ha^{-1}, S—40 kg·ha^{-1}, MgO—9 kg·ha^{-1}
Spring wheat	Corn	N—120 kg·ha^{-1}, P_2O_5—56 kg·ha^{-1}, K_2O—50 kg·ha^{-1}, S—45 kg·ha^{-1}, MgO—9 kg·ha^{-1}
Spring triticale	Winter rape	N—50 kg·ha^{-1}, P_2O_5—40 kg·ha^{-1}, K_2O—40 kg·ha^{-1}
		(B)
Winter wheat	Corn	N—120 kg·ha^{-1}, P_2O_5—60 kg·ha^{-1}, K_2O—60 kg·ha^{-1}
Spring wheat	Potato	N—90 kg·ha^{-1}, P_2O_5—40 kg·ha^{-1}, K_2O—40 kg·ha^{-1}, S—40 kg·ha^{-1}, MgO—15 kg·ha^{-1}
Spring triticale	Winter rape	N—50 kg·ha^{-1}, P_2O_5—30 kg·ha^{-1}, K_2O—30 kg·ha^{-1}
		(C)
Winter wheat	Rape	N—60 kg·ha^{-1}, P_2O_5—40 kg·ha^{-1}, K_2O—40 kg·ha^{-1}
Spring wheat	Corn	N—120 kg·ha^{-1}, P_2O_5—45 kg; K_2O—45 kg·ha^{-1}, S—45 kg·ha^{-1}, MgO—20 kg·ha^{-1}
Spring triticale	Rape	N—60 kg·ha^{-1}, P_2O_5—55 kg·ha^{-1}, K_2O—60 kg·ha^{-1}
		(D)
Winter wheat	Potato	N—90 kg·ha^{-1}, P_2O_5—50 kg·ha^{-1}, K_2O—50 kg·ha^{-1}
Spring wheat	Fodder beet	N—70 kg·ha^{-1}, P_2O_5—42 kg·ha^{-1}, K_2O—42 kg·ha^{-1}
Spring triticale	Rape	N—50 kg·ha^{-1}, P_2O_5—34 kg·ha^{-1}, K_2O—34 kg·ha^{-1}
		(E)
Winter wheat	Oat	N—130 kg·ha^{-1}, P_2O_5—50 kg·ha^{-1}, K_2O—50 kg·ha^{-1}
Spring wheat	Winter rape	N—90 kg·ha^{-1}, P_2O_5—40 kg·ha^{-1}, K_2O—40 kg·ha^{-1}
Spring triticale	Winter rape	N—80 kg·ha^{-1}, P_2O_5—45 kg·ha^{-1}, K_2O—50 kg·ha^{-1}

* farm indicators: A—Lubla$_1$, B—Dobrzechów, C—Lubla$_2$, D—Wiśniowa, E—Markuszowa.

Forecrop harvest was followed by post-harvest tillage. Moreover, in autumn each year prior to spring grain cultivation, mineral phosphorus, and potassium fertilization was applied, followed by prewinter ploughing. Nitrogen fertilizers were sown in spring by mixing with soil by means of a cultivation aggregate (cultivator + cage roller). Fertilizer doses used on the farms (Table 1) were determined by the content of those ingredients in the soil. Certified C1 seeds were sown according to normal agricultural practice [17] in the third quarter of September, whereas spring wheat and spring triticale were sown in the third decade of March. The norm for winter and spring wheat sowing was 400 pcs m^{-2}, and for spring triticale 450 pcs m^{-2}. On all farms, the same grain cultivars were used: winter wheat—cv. "Bamberka", spring wheat—cv. "Bogatka", spring triticale—cv. "Borowik". Grain seeds were treated with Baytan Universal 094 FS in the amount of 400 mL of the preparation and 200 mL of water per 100 kg of seeds, and Baytan Trio 180 FS in the amount of 200 mL of the preparation and 400 mL of water per 100 kg of seeds. In the all farms mechanical and chemical weed control was used. Herbicide spraying took place with the use of a tractor-mounted field sprayer. Plant protection against diseases and pests was used according to the recommendations of the Institute of Plant Protection—National Research Institute [18] (Table 2).

Table 2. Mechanical and chemical cultivation was used on the farms under study (A–E).

Mechanical Treatments	Chemical Treatments
Shallow ploughing, presowing, ploughing	Apyros 75 WG (sulfosulfuron) + Starane 333 EC (fluroksypyr) 26.5 g·ha^{-1} + 0.8 dm^3 ha^{-1}
Cultivator, prewinter ploughing, spring harrowing	Chwastox MP 600 SL (mekoprop-P) + Agritox Turbo 750 SL(MCPA + dicamba) 1 dm^3 ha^{-1} + 1.25 dm^3 ha^{-1}
Shallow ploughing, harrowing, presowing, ploughing	Chwastox MP 600 SL (mekoprop-P) + Starane 330 EC (fluroksypyr) 1 dm^3 ha^{-1} + 0.8 dm^3 ha^{-1}

2.2. Cultivation

All farms used mechanical and chemical weed control by means of agrotechnical weeding procedures and chemical plant protection agents (Table 2).

The active substances of the herbicides used on the farms under study are shown in Table 3.

Table 3. Active substances of herbicides used in cultivation.

Preparation Trade Names	Active Substances	Content of Active Substances	Dosage	Utility Forms	Grace *
			Herbicides		
Apyros 75 WG	Sulfosulfuron	75%	26.5g·ha^{-1}	Granules for water suspension	Not applicable
Starane 333 EC	Fluroxypyr	31.56%	0.8 dm^3 ha^{-1}	Concentrate for water suspension	Not applicable
Chwastox MP 600 SL	Mecoprop-P	52.9%	1 dm^3 ha^{-1}	Concentrate for water suspension	Not applicable
Agritox Turbo 750 SL	MCPA + Dicamba	55.71% 7.59%	1.25 dm^3 ha^{-1}	Concentrate for water suspension	Not applicable

* Sources: [19].

2.3. Soil Assessment

Before the start of the experiment, soil samples were collected from each farm: 20 samples each from the topsoil (0–20 cm), making up a collective 0.5 kg sample [20]. Samples were tested for the granulometric composition of the soil, the content of assimilable phosphorus, potassium and magnesium, and soil pH [21–23]. The granulometric composition and physico-chemical parameters of the soil were analyzed in the certified laboratory of the Regional Chemical and Agricultural Research Laboratory in Rzeszów, with the following methods:

- the granulometric composition of soil was determined by means of laser diffraction [16];
- pH: in a suspension of 1 mol KCl dm^{-3} and in a suspension of H_2O with a potentiometric method [16];
- organic carbon content Corg.—Tiurin's method [22];
- assimilable magnesium content—Schachtschabel method [24];
- assimilable phosphorus and potassium content—Egner–Riehm method [21,22].

The results of soil analysis were evaluated based on limit values established by the Institute of Soil Science and Plant Cultivation—National Research Institute in Puławy [24].

2.4. Analysis of Weed Infestation in Grains

The evaluation of weed infestation in grain crops on the farms under study was based mainly on the quadrat method and the agro-phytosociological method. The evaluation of crop and weed growth stage was based on the BBCH-scale [25]. The mass of the weed species was measured with the weight method. The quadrat method was applied prior to using weed control methods, whereas agro-phytosociological weed assessment was carried out prior to grain harvest. The quadrat analysis of weed was carried out on a surface of 1 m^2 in four replicates [26].

2.4.1. Quadrat Method

The research involved the assessment of species composition, species quantity, and measurement of fresh and air-dried weed mass [26].

The number of weeds per 1 m^2 was measured using the following formula:

$$L_{ch} = \frac{(L1 + L2 + L3 + L4)}{(lp * pr)},$$

where:

L_{ch} = number of weeds of a given species per 1 m^2;
L_1; L_2; L_3; L_4 = number of plants in a quadrat in subsequent measurements [No.];
l_p = number of measurements;
p_r = quadrat surface [m^2] [26].

The weight analysis of weeds allowed to determine the mass of all weed species in a given field. The weed floristic list was compiled. To that end, a thorough search in the entire field was conducted, and a list of all weed species from the field was established and their names entered in an analysis sheet.

Sampling. In the crop plot, all weeds were removed from a surface covered by a randomly placed quadrat. Then, the plants were placed in plastic containers and labeled for reference. This was carried out four times in each field. The plants collected during sampling were sorted and divided into groups of individual weed species; fresh mass of each species was measured, and the results were recorded in the analysis sheet. Masses of all plants of a given weed species were totaled and the average from four replicates was calculated for each crop plot. All results were recorded in the sheet. The mean weight of weeds was calculated as the arithmetic mean of four replicates.

The measurement of air-dried mass in fresh plant material involved drying the sample until air-dried in a natural or artificial way at a temperature of up to 60 °C with forced air flow [26]. The paper refers to air-dried weed mass.

2.4.2. Agro-Phytosociological Method

This method was used to evaluate weed infestation in crops prior to harvest. The agro-phytosociological analysis was based on determining soil coverage by crops and weeds. The analysis included soil coverage in %—separate for crops and weeds. Crop and weed growth stages were determined based on the BBCH scale. The highest possible crop coverage was 100%. In order to determine soil coverage by weeds, a floristic list was compiled including all weed species from the given field, divided by class (monocot, dicot and fern) [25]. At the top of the list were the species that were the most numerous in each class. The analysis was carried out in each field, including their entire surface, except for the borders to eliminate the border effect on the analysis results. In determining soil coverage, weeds were considered separately from crops. For each weed class, soil coverage was calculated as a percentage [26].

2.4.3. Weed Community Structure

The structure of communities in the grain crops under study was described with two ecological indices: Shannon biodiversity index (H') and Simpson dominance index (SI). Shannon index was calculated based on the Shannon–Weaver formula (1) [27],

$$H = -\sum_{j=1}^{n} pilnpi \tag{1}$$

where: Pi = $\frac{ni}{N}$—the ratio of weeds of a given species ni to the total number of weeds *N* on the sample area.

Simpson index *SI* accounts for the number of species and the relative quantity of each species, and was calculated with a modified formula: D = $\frac{\sum n(n-1)}{N(N-1)}$, where n—number of specimens of a given species; N—number of all specimens of all species [28]. This index assumes values from 0 to 1, in which values close to 1 point to a significant dominance of one or several species and small diversity of the community [26]. Furthermore, the end result was transformed into the so-called Simpson's Index of Diversity: 1—D, and then into the Simpson's Reciprocal Index: = $\frac{1}{D}$ [26,28].

2.4.4. Soil Conditions

The research was conducted in soil consisting of flysch sediments, referred to as Carpathian loess or carbonate-free loess-like soil. Classification of these soils: autogenic—luvisols—lessive—pseudogley [29]. The granulometric composition of the soils points to sandy loam (Table 4).

Table 4. Granulometric composition of soil (%) (average from 2013–2016).

Farms *	Percentage of Fraction with Diameter (mm Diameter)			Grain Size Subgroup
	Sand	Silt	Clay	
	2.0–0.05 mm	0.05–0.002 mm	<0.002 mm	
A	72.95	18.47	8.58	Sandy loam
B	65.08	25.21	9.71	Sandy loam
C	75.05	18.56	6.39	Sandy loam
D	70.40	19.94	9.66	Sandy loam
E	72.60	16.97	10.43	Sandy loam
Average	71.22	19.83	8.95	Sandy loam

Source: own study based on the results of the Regional Chemical and Agricultural Station in Rzeszów, * farms indicators: A—Lubla$_1$, B—Dobrzechów, C—Lubla$_2$, D—Wiśniowa, E—Markuszowa.

The concentration of assimilable phosphorus ranged from low to high; potassium—from medium to very high; magnesium—from very low to low, with a medium level of copper, manganese, iron, and zinc in soils. The content of humus in the topsoil ranged from 1.1 to 2.78 g kg^{-1}, and the soil pH ranged from acidic to neutral (Table 5). The soils where cereal crops were grown belonged to the agronomic category of medium soils, good rye to defective wheat complex, valuation class IIIa to IVb [16].

Table 5. Physico-chemical properties of soil in the Podkarpackie province (2013–2016) (g kg^{-1}of soil).

Farms *	Macronutrients (mg·100^{-1} of Soil)			$CaCO_3$ (g kg^{-1})	Humus (g kg^{-1})	pH (KCL)	Micronutrients (mg kg^{-1} of Soil)			
	P_2O_5	K_2O	Mg				Cu	Mn	Zn	Fe
A *	5.5–14.1	17.0–28.1	3.8–13.7	0.02	1.79–2.69	5.3–6.7	5.61	177.1	14.4	1581
B	5.1–17.0	11.7–28.3	2.5–15.0	0.03	1.45–1.70	4.9–7.4	5.71	172.9	14.5	1572
C	10.6–16.3	11.4–20.1	2.8–6.4	0.02	1.98–2.75	4.6–5.6	5.61	177.1	14.4	1569
D	5.2–12.4	15.2–25.6	3.1–8.7	0.03	2.13–2.65	4.4–6.1	5.34	169.2	15.1	1610
E	8.5–29.0	13.4–22.0	1.8–7.3	0.02	2.01–2.78	5.1–6.2	5.68	172.3	14.9	1597
Mean	5.1–29.0	11.4–28.3	1.8–15.0	0.02–0.03	1.45–2.78	-	5.59	173.7	14.7	1586

Source: data was compiled on the basis of the results obtained by the District Chemical and Agricultural Station in Rzeszów (2013–2016); * farm indicators: A—Lubla$_1$, B—Dobrzechów, C—Lubla$_2$, D—Wiśniowa, E—Markuszowa.

2.5. Meteorological Conditions

In the vegetation period of winter and spring grains between the years 2013/2014–2015/2016, the weather was changing, which is presented in Table 6. Grain vegetation period 2013/2014 was considered wet, whereas vegetation periods 2014/2015 and 2015/2016 were very wet, which is reflected in the values of the hydrothermal coefficient of Selyaninov (Table 6). However, significant variation of the hydrothermal coefficient was observed between individual months of the vegetation period. In grain vegetation period 2013, September, October, and November were very wet, whereas in 2014, the months from April through November were very dry. In vegetation period 2015/2016, almost all months in 2015 except for June were wet and extremely humid, whereas in 2016, April, May, and June were wet and extremely humid, and July and August were very dry (Table 6).

Table 6. Average monthly air temperature and total rainfalls during grain vegetation between 2013 and 2016 in Dukla.

Year	Month	Rainfall (mm) Decade of the Month 1	2	3	Sum of Rainfall (mm)	Temperature (°C) Decade of the Month 1	2	3	Average Temperature (°C)	Sielianinov Hydrothermal Coefficient
2013/2014	September	35.1	41.4	47.9	124.4	14.3	14.2	10.6	13.0	3.2
	October	21.2	24.6	32.2	78	10.1	8.6	12.5	10.4	2.5
	November	18.1	20.8	16.1	55.0	5.1	5.0	4.6	4.9	3.7
	December	6.2	16.9	8.1	31.2	2.9	3.0	3.1	3.0	-
	January	5.9	4.2	7.5	17.6	2.1	2.0	2.4	2.2	-
	February	6.5	1.5	2.0	10.0	-2.5	-2.8	-2.3	-2.5	-
	March	4.4	7.2	5.8	17.4	1.8	2.2	1.9	2.0	-
	April	7.8	7.2	6.4	21.4	15.1	14.1	17.3	15.5	0.5
	May	19.1	20.2	8.9	48.2	17.6	18.9	19.8	18.8	0.9
	June	12	16.5	13.5	42	18.4	18.9	25.4	20.9	0.7
	July	27.2	20.8	14.5	62.6	20.1	21.6	22.2	21.3	1.0
	August	33.2	45.1	15.3	93.6	20.8	21.9	22.1	21.6	1.4
Total					601.4					
2014/2015	September	11.2	10.1	12.7	34.0	15.6	15.8	16.9	16.1	0.7
	October	11.3	16.8	11.7	39.8	11.6	11.8	12.1	11.8	1.1
	November	2.8	3.0	3.6	9.4	5.9	6.3	6.2	6.1	0.5
	December	13.4	16.1	10.9	40.4	1.2	1.4	1.9	1.5	-
	January	10.1	9.7	11.2	31.0	-1.2	-1.8	-1.7	-1.6	-
	February	3.62	7.45	5.53	16.6	4.3	4.6	4.8	4.6	-
	March	8.0	9.2	8.6	25.8	7.6	7.8	7.1	7.5	-
	April	22.5	23.8	25.1	71.4	10.8	9.8	11.2	10.6	2.2
	May	46.8	75.2	62.1	184.1	13.7	13.8	13.6	13.7	4.5
	June	6.0	7.1	6.5	19.6	20.4	23.1	25.4	23.0	0.3
	July	40.3	45.1	42.7	128.1	19.8	19.3	21.5	20.2	2.1
	August	14.0	21.0	25.0	60.0	20.1	19.6	22.2	20.6	1.0
Total					660.2					
2015/2016	September	22.5	14.5	29	66.0	15.4	15.1	16.1	15.53	1.4
	October	10.1	13.8	18.7	42.6	12.5	12.6	12.8	12.6	1.1
	November	19.2	21.4	17.2	57.8	8.6	7.6	8.7	8.3	2.3
	December	11.4	15.0	18.7	45.1	4.0	5.1	3.6	4.2	-
	January	6.9	2.5	5.2	14.6	1.2	1.8	1.9	1.6	-
	February	12.9	18.2	23.1	54.2	4.6	4.3	4.8	4.6	-
	March	6.8	5.6	4.2	16.6	5.4	7.2	8.0	6.9	-
	April	26.3	19.6	24.1	70.0	9.8	8.7	11.6	10.0	2.3
	May	21.8	47.1	38.2	107.1	11.6	12.6	12.9	12.4	2.9
	June	55.7	49.8	53.1	158.6	16.9	17.3	17.2	17.1	3.1
	July	15.4	9.4	10.2	35.0	20.2	19.6	21.3	20.4	0.6
	August	19.7	11.1	15.2	46.0	21.6	20.6	21.8	21.3	0.7
Total					713.6					

Source: own study according to data from the COBORU meteorological station at SDOO in Dukla. The following ranges of values $p_{0.05}$ for the Selyaninov coefficient were assumed: extremely dry $k \leq 0.4$; very dry $0.4 < k \leq 0.7$; dry $0.7 < k \leq 1.0$; quite dry $1.0 < k \leq 1.3$; optimal $1.3 < k \leq 1.6$; quite damp $1.6 < k \leq 2.0$; wet $2.0 < k \leq 2.5$; very wet $2.5 < k \leq 3.0$; extremely humid $k > 3.0$ [30].

2.6. Statistical Analyses

The research results were analyzed statistically using analysis of variance, by means of statistical software SAS v.9.2 [31]. Statistical analyses were based on a two-way analysis of variance (farm × crop) and Tukey's multiple tests, with the assumed level of relevance $p_{0.05}$. The significance of sources of variance was subject to the F-Snedecor test [32,33].

3. Results

3.1. Number and Air-Dried Mass of Weeds in Fields of Grains

The greatest number of weeds in a field of grain was recorded in spring wheat and spring triticale, and the smallest in winter wheat. Weed infestation in spring wheat and spring triticale was almost twice as high as in winter wheat. The greatest weed infestation in spring wheat was found on farm A, and the smallest on farm D. Winter wheat suffered from the greatest weed infestation on farm C, with rape as forecrop, and the smallest on farm B, with potato as forecrop. In the case of spring triticale, the greatest weed infestation occurred on farm B, and the smallest on farm E, whereas farms A and D turned out homogeneous in terms of this characteristic (Table 7).

Table 7. Number and air-dried mass of weeds in fields of grains (pcs m^{-2}).

Farms *	Number of Weeds ($No\ m^{-2}$)			Air-Dried Mass of Weeds ($g\ m^{-2}$)		
	Winter Wheat	Spring Wheat	Spring Triticale	Winter Wheat	Spring Wheat	Spring Triticale
A	8.0c	76.8a	28.8b	9.1a	17.7a	42.8a
B	2.0d	23.2c	72.0a	0.3e	1.9d	13.6b
C	30.4a	42.0b	21.6bc	3.8c	8.2b	2.4 cd
D	15.6b	6.4d	34.8b	7.5b	2.8d	6.2c
E	13.6b	20.8c	16.8cd	2.3d	5.9c	4.0c
$LSD_{0.05}$	3.55	8.6	8.5	1.2	1.9	3.5

* Farm indicators: A—Lubla$_1$, B—Dobrzechów, C—Lubla$_2$, D—Wiśniowa, E—Markuszowa. Letter indicators at averages determine the so-called homogeneous groups (statistically homogeneous). The occurrence of the same letter pointer at averages (at least one) means that there is no (no) statistically significant difference between them.

The greatest air-dried area of weed in a field of grain was observed in spring triticale crop, and the smallest in winter wheat. In the case of the latter, the greatest weed mass was recorded on farm A, and the smallest on farm B, with corn as forecrop. Weed mass in the field of spring wheat was almost by half lower than weed mass in spring triticale. The greatest air-dried weed mass in spring wheat was found on farm A, and the smallest on farm B. Weed mass on farms B and D did not differ significantly from each other. The greatest air-dried weed mass in spring triticale was noted on farm A, and the smallest on farm E, whereas the weed mass on farms D and E turned out homogeneous (Table 7).

3.2. Floristic Composition of Weeds

Weed infestation of winter wheat grown on all farms included a total of 11 weed species (Table 8). The presence of taxons per a single crop field ranged from two (farm B) to four weed species (farms A, D) and five taxons on farm E. The most frequent were ANTAR, CIRAR, VIOAR [34] from the dicot class, and EQUAR from the *Equisetaceae* family, the fern (*Polypodiop*) class, which were found in three out of five examined fields of winter wheat; CONAR and POLCO were found in a relatively high quantity but lower recurrence per farm. Significant differences in weed infestation in winter wheat between the examined farms were found in such species as ANTAR, APHAR, CIRAR, CONAR, EQUAR, RUMAA, and VIOAR (Table 8). The occurrence of the remaining species was less frequent, and they were not found on all farms. Significant differences in air-dried weed mass were recorded in the case of such species as CIRAR, CONAR, EQUAR. The greatest air-dried weed mass was observed in the case of CONAR, whereas the remaining species produced a significantly lower weed mass, which did not depend on location (Table 8). The greatest soil coverage by segetal plants was recorded on farm C and amounted to 12.3%, with the soil coverage by winter wheat of 87.7%. Dominating weed species in those crops were ANTAR and CIRAR belonging to dicots, and EQUAR belonging to the *Equisetales* order, the *Equisetidae* subclass, the *Polypodiopsida* class [35] (Table 8).

Table 8. The species composition, number and air-dried weed mass and soil coverage in a field of winter wheat per 1 m^2.

No	Species *	Number of Weeds ($No\ m^{-2}$)					$LSD_{0.05}$	Air-Dried Mass of Weeds ($g\ m^{-2}$)					$LSD_{0.05}$
		A	B	C	D	E		A	B	C	D	E	
1	ANTAR	0.0	1.2	8.8	5.2	0.0	0.8	0.0	0.2	1.9	0.4	0.0	0.1
2	APHAR	0.0	0.0	0.0	2.4	0.0	ns	0.0	0.0	0.0	0.5	0.0	ns
3	CIRAR	0.0	0.0	4.0	3.2	3.2	0.5	0.0	0.0	0.5	0.7	0.4	0.1
4	CONAR	1.6	0.0	0.0	4.8	0.0	0.3	4.2	0.0	0.0	5.9	0.0	0.5
5	EQUAR	2.4	0.0	8.0	0.0	3.2	2.7	0.6	0.0	1.3	0.0	0.8	0.1
6	POLLA	0.0	0.0	1.6	0.0	0.0	ns	0.0	0.0	0.0	0.0	0.0	ns
7	POLCO	0.0	0.0	0.0	0.0	3.2	ns	0.0	0.0	0.1	0.0	0.1	ns
8	RUMAA	3.2	0.0	0.0	0.0	0.0	0.6	3.8	0.0	0.0	0.0	0.0	ns
9	VICCR	0.0	0.0	0.0	0.0	0.1	ns	0.0	0.0	0.0	0.0	0.1	ns
10	VICTE	0.0	0.0	8.0	0.0	0.0	ns	0.0	0.0	0.0	0.0	0.0	ns
11	VIOAR	0.8	0.8	0.0	0.0	4.0	0.2	0.5	0.1	0.1	0.0	1.0	ns
Sum of species		4	2	5	4	5	-	4	2	5	4	5	-
	*	97.8	96.5	87.7	92.4	95.1	-	-	-	-	-	-	-
	**	0.0	0.0	0.0	0.9	0.0	-	-	-	-	-	-	-
	***	2.0	3.5	11.2	6.7	4.2	-	-	-	-	-	-	-
	****	0.2	0.0	1.1	0.0	0.7	-	-	-	-	-	-	-

farm indicators: A—Lubla$_1$, B—Dobrzechów, C—Lubla$_2$, D—Wiśniowa, E—Markuszowa; ANTAR—*Anthemis arvensis*; APHAR—*Aphanes arvensis*; CIRAR—*Cirsium arvense*; CONAR—*Convolvulus arvensis*; EQUAR—*Equisetum arvense*; POLCO—*Polygonum convolvulus*; POLLA—*Polygonum lapathifolium*; RUMAA—*Rumex acetosella*; VICCR—*Vicia cracca*; VICTE—*Vicia tetrasperma*; VIOAR—*Viola arvensis*; * soil coverage with crop (%); ** soil coverage with monocot weeds (%); *** soil coverage with dicot weeds (%); **** soil coverage with *Equisetaceae* weeds (%).

During the assessment of weed infestation in fields of spring wheat, 14 weed species were identified, of which the most frequent were dicot weeds: BRANA (52.8%), SINAR (17.0%), and CIRAR (8.8%). The share of these three species in the weed structure amounted to 78.6%. The only monocot weed was APESV, which was only found on farms B, C, and E (Table 9). Significant differences in weed infestation in winter wheat between the examined farms were found in such species as ANTAR, APESV, CIRAR, EQUAR, GAELA, POLAV, POLCO, SINAR, VIOAR. In four out of five farms, CIRAR and EQUAR was recorded, with small numbers of other taxons. The most frequent weed species in fields of spring wheat on farm A was BRANA (Table 9). A very large share of rape in fields of spring wheat was a direct result of the forecrop, which was winter rape, leaving behind numerous fallen seeds of that species. On farms B and E, the dominating species was SINAR, with a 65.5% and 30.8% share in all weeds, respectively. Crops were least affected by weeds on farm D, where only four species were recorded with the smallest quantity of all farms under study. It was observed that the differentiated floristic composition in the examined farms resulted from the forecrop (Table 9). It turned out that the air-dried weed mass depended significantly on the species. A significant influence of that factor was recorded in the case of four species: ANTAR, BRANA, CIRAR, and EQUAR. On average, the highest weed mass was generated by CIRAR. On farm A, the most frequent was BRANA, which was related with the forecrop (Table 9). The greatest soil coverage by segetal plants was recorded on farm C and amounted to 11.7%, with the soil coverage by spring wheat of 88.3%. Dominating weed species in those crops were ANTAR and SINAR belonging to dicots, and EQUAR belonging to the *Equisetaceae* family, the *Equisetales* order [35] (Table 9).

Table 9. Species composition, number, and air-dried weed mass and soil coverage in a field of spring wheat (pcs m^2).

No	Species *	The Number of Weeds ($No\ m^{-2}$)					$LSD_{0.05}$	Air-Dried Mass of Weeds ($g\ m^{-2}$)					$LSD_{0.05}$
		A	B	C	D	E		A	B	C	D	E	
1	ANTAR	0.0	0.0	4.0	0.0	4.8	0.4	0.0	0.0	0.8	1.1	1.0	0.1
2	APESV	0.0	0.8	0.7	0.0	1.6	0.2	0.0	0.1	0.1	0.0	0.1	ns
3	APHAR	0.0	0.0	2.5	0.0	0.0	ns	0.0	0.0	0.5	0.0	0.0	ns
4	BRANA	67.2	0.0	0.0	0.0	0.0	ns	10.8	0.0	0.0	0.0	0.0	0.5
5	CIRAR	6.4	0.0	11.5	0.8	4.0	0.5	6.2	0.0	2.3	0.2	4.3	0.7
6	EQUAR	0.8	0.0	3.5	1.6	2.4	0.4	0.2	0.0	0.7	1.0	0.9	0.2
7	GAELA	0.0	2.4	1.5	0.0	1.6	0.3	0.0	0.3	0.3	0.0	0.2	ns
8	GALAP	0.0	0.8	0.6	0.0	0.0	ns	0.0	0.1	0.1	0.0	0.0	ns
9	POLAV	0.8	0.0	1.1	3.2	0.0	0.3	0.2	0.0	0.2	0.5	0.0	ns
10	POLCO	0.0	4.0	1.0	0.0	0.0	0.3	0.0	0.1	0.2	0.0	0.0	ns
11	POLLA	0.0	0.0	5.0	0.0	0.0	ns	0.0	0.0	1.0	0.0	0.0	ns
12	RUMAA	0.0	0.0	5.6	0.0	0.0	ns	0.0	0.0	1.1	0.0	0.0	ns
13	SINAR	0.0	15.2	3.3	0.0	6.4	1.3	0.0	0.8	0.6	0.0	0.4	ns
14	VIOAR	1.6	0.0	1.7	0.8	0.0	0.2	0.3	0.0	0.0	0.0	0.0	ns
Sum of species		5	5	13	4	6	-	5	5	13	4	6	-
*		93.8	95.6	88.3	96.9	94.5	-	-	-	-	-	-	-
**		0.0	0.1	0.2	0.0	0.2	-	-	-	-	-	-	-
***		6.0	4.3	10.8	2.9	4.0	-	-	-	-	-	-	-
****		0.2	0.0	0.7	0.2	1.3	-	-	-	-	-	-	-

farm indicators: A—Lubla$_1$, B—Dobrzechów, C—Lubla$_2$, D—Wiśniowa, E—Markuszowa; * farm indicators: A—Lubla$_1$, B—Dobrzechów, C—Lubla$_2$, D—Wiśniowa, E—Markuszowa; ANTAR—*Anthemis arvensis;* APESV—*Apera spica-venti;* APHAR—*Aphanes arvensis;* BRANA—*Brassica napus;* CIRAR—*Cirsium arvense;* EQUAR—*Equisetum arvense;* GAELA—*Galeopsis ladanum;* GALAP—*Galium aparine;* POLAV—*Polygonum avicular;* POLCO—*Polygonum convolvulus;* POLLA—*Polygonum lapathifolium;* RUMAA—*Rumex acetosella;* SINAR—*Sinapis arvensis;* VIOAR—*Viola arvensis;* * soil coverage with crop (%); ** soil coverage with monocot weeds (%); *** soil coverage with dicot weeds (%); **** soil coverage with *Equisetaceae* weeds (%).

Fields of spring triticale grown in five farms contained in total 20 weed species, including 13 on farm A, and six in each of the remaining farms (Table 10). Weed species that were found in at least two fields of spring triticale, belonging to the dicot class, included ANTAR, LAMAM, POLLA, and VICCR, the monocot class—APESV, and the *Polypodiopsida* class—EQUAR. The direct forecrop of spring triticale on farm A was rape and winter wheat, which was most probably the reason for higher weed infestation in crops. Significant differences in weed infestation in winter wheat were found in such species as ANTAR, APESV, CENCY, CHEAL, CIRAR, EQUAR, LAMAM, POLLA, POLCO, POLAV, STEME, TAROF, VICCR, VICSA, VIOAR, and VERPE (Table 10). On farm B, the greatest weed infestation was recorded in the case of rape as the forecrop, and the most frequent weeds were VIOAR, VERPE, and TAROF. The lowest quantity of weeds was observed on farm E, and the most frequent weed was VICCR. The great difference in weed infestation between farms did not result from the direct forecrop, which was rape in both cases. After an inquiry, it turned out that it was due to the short usage period of the plot of land on farm B purchased only 2 years earlier. Additionally, high weed infestation was caused by years of set-aside of the purchased land and its location around uncultivated fields, which were a rich source of weed seeds transferred in soil (Table 10).

It turned out that air-dried weed mass significantly depended on the location in the case of such species as BRANA, CIRAR, EQUAR, POLAV, STEME, TAROF, VIOAR, VERPE, and VICCR (Table 10). Among 20 weed species, the largest air-dried weed mass was recorded for BRANA and TAROF found on farm A. Such a large air-dried mass of common dandelion resulted from its late growth stage defined as one of the ripening and fruit coloring stages. Among other species, a significant share in the measured weed mass was noted for VICCR, VIOAR, and VERPE. The share of other species did not influence their frequency in that period. It was expected that a reverse relationship would occur in the case of increasing share of weed species. The greatest air-dried weed mass in a field of spring triticale was observed on farm A, and the smallest on farm C. The air-dried weed mass on farm B was almost three times lower that on farm A (Table 10). The highest soil coverage by segetal plants was recorded on farm A and amounted to 19%, which resulted in an 81% density of the field of spring triticale. The smallest soil coverage by segetal plants was recorded on farm E and amounted to 5.4%, with the soil coverage by spring wheat of 95.4%. The lowest soil coverage in spring triticale crops involved monocot weeds, with their only representative being APESV from the monocot class (Table 10).

Table 10. Species composition, number, and weed mass and soil coverage in a field of spring triticale per 1 m^2.

No	Species	The Number of Weeds ($No\ m^{-2}$)					$LSD_{0.05}$	Air-Dried Mass of Weeds ($g\ m^{-2}$)					$LSD_{0.05}$
		A	B	C	D	E		A	B	C	D	E	
1	ANTAR	0.8	0.6	6.4	0.0	3.2	0.4	0.2	0.1	1.2	0.0	0.5	0.1
2	APESV	2.4	1.1	0.0	2.2	1.6	0.3	0.2	0.2	0.0	0.4	0.2	ns
3	BRANA	4.0	0.0	0.0	0.0	0.0	ns	15.0	0.0	0.0	0.0	0.0	0.8
4	CENCY	0.0	0.0	4.8	2.3	0.0	0.4	0.0	0.0	0.4	0.5	0.0	ns
5	CHEAL	0.0	0.7	2.4	2.1	0.0	0.3	0.0	0.1	0.4	0.4	0.1	ns
6	CIRAR	3.2	5.2	0.0	1.3	0.0	0.5	1.3	1.0	0.0	0.2	0.0	0.1
7	GAELA	0.0	1.1	0.0	0.6	0.0	ns	0.0	0.2	0.0	0.1	0.0	ns
8	GALAP	0.0	0.0	1.6	0.0	0.0	ns	0.0	0.0	0.0	0.0	0.0	ns
9	EQUAR	2.4	9.1	0.0	4.1	2.4	0.8	1.1	1.7	0.4	0.7	1.1	0.3
10	LAMAM	0.8	1.0	0.0	1.3	0.8	0.2	0.3	0.2	0.0	0.2	0.2	ns
11	POLLA	0.8	0.0	4.8	0.0	4.8	0.3	0.0	0.0	0.0	0.0	0.0	ns
12	POLCO	1.6	1.2	0.0	0.7	0.0	0.2	0.3	0.2	0.0	0.1	0.0	ns
13	POLAV	0.0	5.5	0.0	3.1	0.0	0.4	1.6	1.1	0.0	0.6	0.0	0.2
14	STEME	3.2	5.5	0.0	1.2	0.0	0.5	2.0	1.0	0.0	0.1	0.0	0.2
15	TAROF	0.8	10.6	0.0	5.8	0.0	0.9	10.4	2.0	0.0	1.0	0.0	0.7
16	VICCR	4.0	5.8	0.0	2.2	4.0	0.6	1.9	1.1	0.0	0.4	1.9	0.3
17	VICSA	0.0	1.0	1.6	0.7	0.0	0.2	0.3	0.2	0.0	0.1	0.0	ns
18	VIOAR	4.0	12.0	0.0	3.6	0.0	1.0	3.6	2.3	0.0	0.6	0.0	0.3
19	VICTE	0.0	0.8	0.0	0.0	0.0	ns	0.2	0.1	0.0	0.0	0.0	ns
20	VERPE	0.8	10.8	0.0	3.6	0.0	0.8	4.4	2.1	0.0	0.5	0.0	0.4
Sum of species		13	16	6	15	6	-	13	16	6	15	6	-
*		81.0	94.1	89.2	93.3	95.4	-	-	-	-	-	-	-
**		0.9	0.0	0.0	0.0	0.1	-	-	-	-	-	-	-
***		16.1	4.9	11.2	6.2	4.3	-	-	-	-	-	-	-
****		2.0	1.0	0.0	1.1	1.0	-	-	-	-	-	-	-

farm indicators: A—Lubla1, B—Dobrzechów, C—Lubla$_2$, D—Wiśniowa, E—Markuszowa; ANTAR—*Anthemis arvensis;* APESV—*Apera spica-venti;* BRANA—*Brassica napus;* CENCY—*Centaurea cyanus;* CHEAL—*Chenopodium album;* CIRAR—*Cirsium arvense;* GAELA—*Galeopsis ladanum;* GALAP—*Galium aparine;* EQUAR–*Equisetum arvense;* LAMAM—*Lamium* amplexicaule; POLLA—*Polygonum lapathifolium;* POLCO—*Polygonum convolvulus;* POLAV—*Polygonum avicular;* STEME—*Stellaria media;* TAROF—*Taraxacum officinale;* VICCR—*Vicia cracca;* VICSA—*Vicia sativa;* VIOAR—*Viola arvensis;* VICTE—*Vicia tetrasperma;* VERPE—*Veronica persica;* * soil coverage with crop (%); ** soil coverage with monocot weeds (%); *** soil coverage with dicot weeds (%); **** soil coverage with *Equisetaceae* weeds (%).

3.3. Biodiversity Indices

The highest Shannon biodiversity index was recorded in the field of triticale, and the lowest in the field of winter wheat. This means that all species in the field of triticale had the same priority, and the plant community manifested the greatest biodiversity (Table 11). Biodiversity indices reflect the composition of communities in much more detail than just species richness (i.e., number of species present). In fact, they also account for the relative number of various species. A biodiversity index is a mathematical measure of species diversity in a community. Biodiversity indices reflect the composition of communities in much more detail than just species richness (i.e., number of species present). Simpson index (D) determines habitat biodiversity. It also established the probability of randomly picking two specimens of the same species. Index "D" ranges from 0 to 1, where 0 stands for unlimited biodiversity, and 1 no biodiversity. In general, the closer to 1, the poorer the biodiversity. For the sake of intuitiveness and clarity, the end result D was transformed into the so-called Simpson's Index of Diversity: 1—D, and then to the Simpson's Reciprocal Index: 1/D. In the case of such results, the greater the number, the greater the biodiversity. The latter index points to the greatest biodiversity in fields of triticale and the smallest in fields of spring wheat (Table 11).

Table 11. Biodiversity indices in fields of grain.

Species	Farms *					Average
	A	B	C	D	E	
			Shannon index (H′)			
Winter wheat	0.56	0.29	0.64	0.64	0.14	0.45
Spring wheat	0.22	0.45	0.97	0.53	0.72	0.58
Spring triticale	0.92	0.90	0.68	1.00	0.72	0.84
			Simpson index (D)			
Winter wheat	0.20	0.04	0.24	0.24	0.18	0.18
Spring wheat	0.76	0.46	0.09	0.20	0.15	0.33
Spring triticale	0.05	0.10	0.18	0.07	0.16	0.11
			Simpson's Reciprocal Index			
Winter wheat	5.00	25.00	4.17	4.17	5.56	8.78
Spring wheat	1.31	2.17	10.11	5.00	6.66	5.05
Spring triticale	20.00	10.00	5.55	14.28	6.25	11.21

* farm indicators: A—Lubla$_1$, B—Dobrzechów, C—Lubla$_2$, D—Wiśniowa, E—Markuszowa.

4. Discussion

The results of our own research explicitly show that the potential state and biodiversity of weeds in winter and spring wheat and spring triticale was significantly differentiated. Increasing weed infestation in crops is caused by more and more popular grain monoculture and applying the same type of herbicide weeding agents for several years in the same field [7]. Looking for new and efficient solutions to reasonably control weeds in a way that takes into account the eco-demands of protecting floristic biodiversity is the primary goal of modern herbological studies [36]. One of the possibilities, so far rarely used in weed management, is the selection of cultivars based on their natural competitiveness with segetal flora. The degree and condition of weed infestation in fields of grain depends on a number of factors, e.g., the abundance of weed seeds in topsoil, the number of weeding procedures, the type of weed control agent used, the dose of herbicide active substance per surface unit, the time of herbicide application, the growth stage of weeds during chemical weeding procedures, and the weather conditions during and directly after spraying. In the conducted research, weed infestation in grains was related to forecrop and location. Weed infestation in fields can also be significantly influenced by the type and number of mechanical tillage procedures, manure fertilization, crop rotation, crop species, plant density in the field, and plant height [4]. In the experiment, mechanical tillage, chemical weed control, and grain species were the same. Weed infestation was differentiated by soil conditions, quantity of weeds in soil and plot location. According to Kieloch [37], a weed community in a given field is subject to dynamic changes due to various agricultural practices. The size of the weed population and its species composition is shaped by two major factors: crop competition and soil seed bank, which more or less depend on various elements of the agrotechnology, such as crop rotation, soil cultivation, sowing time and density, choice of cultivar. According to Haliniarz et al. [38], the aim of the optimal agrotechnology is to increase crop competitiveness against weeds and reduce the weed seed bank in soil. The conducted studies focused on species diversity, species variety, and species richness. The greatest number of weeds in a field of grain was recorded in spring wheat and spring triticale, and in terms of air-dried weed mass—the greatest mass was found in spring triticale. Based on the analysis of species biodiversity of crops in south-eastern Poland, the richest species composition and the greatest biodiversity according to Shannon index was also noted in spring triticale. In fact, they also account for the relative number of various species. The factors that can modify weed infestation in grains are numerous, with fertilization being the most prominent. In the conducted research, fertilization was adjusted to the fertility of soil and the nutritional needs of grain species. According to a number of authors [3,11,39], it is mainly nitrogen fertilization that strongly affects weed infestation. It is believed that increased nitrogen fertilization results in changes to the species composition of weed communities. It is manifested in the disappearance of oligotrophic species and their replacement with nitrophile species [40,41]. Nitrogen facilitates weed emergence, and during full vegetation—depending on the fertilization type—it can reduce or stimulate their occurrence. In the conducted research, the influence of fertilization on the floristic composition of weed communities in fields of grain was not investigated, but such influence was reported by other authors [39,42,43]. According to Czuba [12], Stępień [41], Kwiatkowski et al. [11], this factor differentiates the population size of weed species. The basic biodiversity means of measurement are based on the evaluation of the species composition in a given area (testing for the occurrence of all species and monitoring selected species) and comparison of the present state of the ecosystem to the past (comparison of the area to protected areas). The most frequently used biodiversity measures include Margalef index (D, R1), simplified Margalef index used by the Ministry of Environment, Shannon diversity index (H), Shannon equitability (species share) index (EH, J), dominance index (of Shannon and Weaver), modified Simpson index (D) [26]. In the conducted research, Shannon diversity index ranged from 0.46 to 0.84, depending on the type of crop field. The highest index value was recorded for fields of spring triticale, and the lowest for winter wheat. According to Zanin et al. [26], Shannon index is the highest when the share of species is even, i.e., all species have the same priority. When the number of species is equal, the community with even species distribution is characterized by higher biodiversity. When the share of species is even (p_i), the area with more species is characterized by greater biodiversity [26]. Based on the results, Shannon–Wiener (H) and Simpson (D) indices of biodiversity were calculated. Using the described methodology, it was established that the differences between indices H and D for the analyzed area and time were minor. According to Kotlarz et al. [44], who estimated the diversity of stands of trees, the investigated area of imaging differentiated after the first and second iteration by means of PCA, and the changes were also minor. The aim of conducting other studies and more thorough analyses, e.g., PCA, and thus complete identification of species in the context of biodiversity, requires further research. Agriculture is one of the key factors causing biodiversity decrease. Beside the loss of biodiversity due to habitat damage resulting from the transformation of natural areas into arable land, increasing agricultural activity has led to a significant decrease in biodiversity of agricultural areas. This threatens not only biodiversity but also the entire ecosystem and ecosystem service that agriculture relies upon. According to Erisman et al. [2], the pressure of feeding the increasing human population around the globe, combined with changing diet including more and more animal protein, has an additional negative effect on available arable land and agricultural areas. An agricultural system based on the full potential (functional agricultural biodiversity) opens up possibilities of establishing a resilient system where both food production and nature can develop [1,43,45,46]. Therefore, we recommend a complex approach in order to boost the development and implementation of agricultural practices which use and support biodiversity and ecosystem service in agricultural areas, seminatural enclaves and ecological grounds. An agricultural system based on the potential of functional agricultural biodiversity opens up possibilities of establishing a resilient system where both food production and nature can develop.

5. Conclusions

The greatest number of weeds in a field of grain was recorded in spring wheat and triticale, and the smallest in winter wheat. Weed infestation in winter wheat grown on all farms included a total of 11 weed species. The most frequent were ANTAR, CIRAR, VIOAR belonging to the dicot class, and EQUAR belonging to the *Polypodiopsida* class. Fields of spring triticale contained in total 20 weed species, including 13 on a single farm (A), and six in each of the remaining farms. During the assessment of weed infestation in fields of spring wheat, 14 weed species were identified, of which the most frequent were dicot weeds: BRANA, SINAR, and CIRAR. The only monocot weed found in that crop was APESV, with specimens reported in three out of five locations. The highest Shannon biodiversity index was recorded in the field of triticale, and the lowest in the field of winter wheat. This means that all species in the field of triticale had the same priority, and the plant community manifested the greatest biodiversity. The biodiversity evaluation of selected fields of grain conducted in several towns in the region enables the assessment of biodiversity on the regional, national, and global level. Some of the approved methods of biodiversity assessment are considered time-consuming and/or subjective, others focus on some parameters only and ignore the rest. However, testing a single index (parameter) does not provide a large picture of the situation, so in examining biodiversity, it is best to use several different methods. The data obtained on the diversity of weeds in the south-eastern part of Poland in the fields of cereal plants will be used in some weed control strategies. Additionally, that will be the added value of this project.

Author Contributions: Conceptualization, B.S.; M.Ć. and B.K.-M.; methodology, B.S.; M.Ć. and B.K.; software, B.S.; validation, B.S. and B.K.-M.; formal analysis, B.S.; and B.K.-M.; investigation, B.K.-M.; resources, B.K.-M. and B.S.; data curation; B.S. and B.K.-M.; writing—original draft preparation, B.S. and B.K.-M.; writing—review and editing, B.K.-M.; B.S.; P.B. and P.P; visualization, B.K.-M.;B.S.; P.B. and P.P.; supervision, B.S. and B.K.-M.; project administration, B.S. and B.K.-M.; funding acquisition, B.K.-M. All authors have read and agreed to the published version of the manuscript.

Funding: This research was funded by the Scholarship Fund of Stanisława Pigoń for students and employees of the State Higher Vocational School in Krosno with the number PDR.SP.0041.6.2020.

Conflicts of Interest: The authors declare no conflict of interest.

Abbreviations

ALOMY	*Alopecurus myosuroides* L.
ANTAR	*Anthemis arvensis* L.
APESV	*Apera spica-venti* L.
APHAR	*Aphanes arvensis* L.
AVRDC	The World Vegetable Center
BBCH	Biologische Bundesanstalt, Bundessortenamt und Chemische Industrie
BNI	Biological Nitrification Inhibition
BRANA	*Brassica napus* L.
CENCY	*Centaurea cyanus* L.
CGIAR	Consultative Group on International Agricultural Research
CHEAL	*Chenopodium album* L.
CIRAR	*Cirsium arvense* L.
CONAR	*Convolvulus arvensis* L.
EQUAR	*Equisetum arvense* L.
GAELA	*Galeopsis ladanum* L.
GALAP	*Galium aparine* L.
LAMAM	*Lamium amplexicaule* L.
POLAV	*Polygonum aviculare* L.
POLCO	*Polygonum convolvulus* L.
POLLA	*Polygonum lapathifolium* L.

RUMAA	*Rumex acetosella* L.
SINAR	*Sinapis arvensis* L.
STEME	*Stellaria media* (L.) Vill./Cyr.
TAROF	Taraxacum officinale Web.
VERPE	Veronica persica Poir.
VICCR	Vicia cracca L.
VICSA	*Vicia sativa* L.
VICTE	*Vicia tetrasperma* L.
VIOAR	Viola arvensis Murr.
WSSA	Weed Science Society of America

References

1. Duer, I.; Feledyn-Szewczyk, B. Species composition and biomass of weeds in winter wheat cultivated under different crop production systems and their contribution to the uptake of mineral components from the soil. *Pamiętnik Puławski* **2003**, *134*, 65–77. (In Polish)
2. Erisman, J.W.; van Eekeren, N.; de Wit, J.; Koopmans, C.H.; Cuijpers, W.; Oerlemans, N.; Koks, B.J. Agriculture and biodiversity: A better balance benefits both. *Agric. Food* **2016**, *1*, 157–174. [CrossRef]
3. Haliniarz, M. The response of selected agrophytocenosis to different doses of biologically active substances of herbicides. In *Monography*; University of Life Science in Lublin: Lublin, Poland, 2019; p. 207.
4. Feledyn-Szewczyk, B.; Smagacz, J.; Kwiatkowski, C.A.; Harasim, E.; Woźniak, A. Weed Flora and Soil Seed Bank Composition as Affected by Tillage System in Three-Year Crop Rotation. *Agriculture* **2020**, *10*, 186. [CrossRef]
5. Moreta, D.E.; Mathur, P.N.; van Zonneveld, M.; Amaya, K.; Arango, J.; Selvaraj, M.G.; Dedicova, B. Current Issues in Cereal Crop Biodiversity. *Adv. Biochem. Eng. Biotechnol.* **2015**, *147*, 1–35. [CrossRef]
6. Feledyn-Szewczyk, B. Biodiversity as an indicator of environmental monitoring. *Stud. Rep. IUNG-PIB* **2016**, *47*, 105–124.
7. Salonen, J.; Ketoja, E. Undersown cover crops have limited weed suppression potential when reducing tillage intensity in organically grown cereals. *Org. Agric.* **2020**, *10*, 107–121. [CrossRef]
8. Gawęda, D.; Cierpiała, R.; Harasim, E.; Haliniarz, M. Effect of tillage systems on yield, weed infestation and seed quality elements of soybean. *Acta Agrophys.* **2016**, *23*, 175–187.
9. Płaza, A.; Ceglarek, F.; Królikowska, A.; Próchnicka, M. The follow-up action of undersown crops and spring barley straw on yielding and structure elements of yield of winter triticale. *Folia Pomer. Univ. Technol. Stetin.* **2010**, *276*, 31–38.
10. Kowalczyk, E.; Patyra, E.; Kwiatek, K. Pyrrolizidine alkaloids as a threat to human and animal health. *Vet. Med.* **2015**, *71*, 602–607.
11. Kwiatkowski, C.A.; Wesołowski, M.; Drabowicz, M.; Misztal-Majewska, B. The effect of adjuvants and reduced rates of crop protection agents on the occurrence of agricultural pests and on winter wheat productivity. *Ann. UMCS* **2012**, *E-67*, 12–21.
12. Czuba, R.; Wróbel, S. Weed competitiveness in nutrient uptake by crops. *Ann. Soil Sci.* **1983**, *34*, 175–184.
13. GUS. Statistical Yearbook of the Republic of Poland. 2019. Available online: https://stat.gov.pl/obszary-tematyczne/roczniki-statystyczne/roczniki-statystyczne/rocznik-statystycznyrzeczypospolitej-polskiej-2019,2,19.html (accessed on 26 October 2020).
14. The EU's Biodiversity Strategy to 2030. Available online: https//ec.europa.eu/environment/nature/biodiversity/strategy/index_en.htm (accessed on 26 October 2020).
15. Mendelowski, S. *The Town and Commune of Strzyżów*; Roksana: Krosno, Poland, 2008.
16. Mocek, A. *Soil Science*; State Scientific Publisher: Warsaw, Poland, 2015.
17. Duer, I.; Fotyma, M.; Madej, A. *Code of Good Agricultural Practice*, 3rd ed.; Foundation for Assistance Programs for Agriculture: Warsaw, Poland, 2004; pp. 64–65. ISBN 83-88010-58-1. (In Polish)
18. Strażyński, P. *Recommendations for the Protection of Agricultural Plants. Vol. I*; Spring Cereals and Corn. Publisher: Poznań, Poland, 2020; ISBN 978-83-64655-53-1. (In Polish)

19. List of Registered Plant Protection Products by the Ministry of Agriculture and Rural Development for 2018/2019. Available online: https://www.gov.pl/web/rolnictwo/rejestr-rodkow-ochrony-roslin (accessed on 12 October 2020).
20. Ryzak, M.; Bartmiński, P.; Bieganowski, A. Methods of determination of granulometric distribution of mineral soils. *Acta Agroph.* **2009**, *175*, 97.
21. PN-R-04020, 1994+AZ1. *Chemical and Agricultural Analysis of Soil*; Polish Committee for Standardization: Warszawa, Poland, 2004.
22. PN-R-04023. *Chemical and Agricultural Analysis of Soil. Determination of Available Phosphorus Content in Mineral Soils*; Polish Committee for Standardization: Warsaw, Poland, 1996.
23. PN-ISO 10390:1997. *Soil Quality—Determination of pH*; Polish Committee for Standardization: Warsaw, Poland, 1997.
24. Nawrocki, S. (Ed.) *Fertilizer Recommendations. Part. I. Limit Numbers for Valuation of Soils in Macro- and Microelements. Part I*; Institute of Soil Science and Plant Cultivation: Puławy, Poland, 1985; p. 44. (In Polish)
25. Hack, V.H.; Bleiholder, H.; Buhr, L.; Meier, U.; Schnock-Fricke, U.; Weber, E.; Witzenberger, A. Einheitliche Codierung der phanologischen Entwicklungsstadien mono- und dikotyler Pflanzen—Erweiterte BBCH-Skala, Allgemein. *Nachrichtenbl. Deut. Pflanzenschutzd* **1992**, *44*, 265–270.
26. Domaradzki, K.; Badowski, M.; Filipiak, K.; Franek, M.; Gołębiowska, H.; Kieloch, R.; Kucharski, M.; Rola, H.; Rola, J.; Sadowski, J. Field experiments. In *Methodology Experiments of Biological Evaluation of Herbicides, Bioregulators and Adjuvants*; Domaradzki, K., Ed.; Institute of Soil Science and Plant Cultivation Publishing House: Puławy, Poland, 2001; Chapter 1; p. 167. (In Polish)
27. Zanin, G.; Mosca, G.; Catizone, P. A profile of the potential flora in maize fields of the Po Valley. *Weed Res.* **1992**, *32*, 407–418. [CrossRef]
28. Simpson, R.L.; Leck, M.A.; Parker, V.T. (Eds.) Seed Banks: General Concepts and Methodological Issues. In *Ecology of Soil Seed Banks Red*; Academic Press Inc.: Cambridge, MA, USA, 1989; pp. 3–8.
29. WRB. World Reference Database for Soil Resources. 2014. Available online: http://www.fao.org/3/a-i3794e.pdf (accessed on 8 June 2020).
30. Skowera, B.; Kopcińska, J.; Kopeć, B. Changes in thermal and precipitation conditions in Poland in 1971–2010. *Ann. Warsaw Univ. Life Sci. SGGW Land Reclam.* **2014**, *46*, 153–162. [CrossRef]
31. SAS. *Users Guide*; SAS Institute Inc.: Cary, NC, USA, 2008.
32. Koronacki, J. *Statistics, for Students of Technical and Natural Sciences*; Scientific and Technical Publishers: Warsaw, Poland, 2009; p. 491.
33. John, H.M. *Handbook of Biological Statistics*, 3rd ed.; Sparky House Publishing: Baltimore, MD, USA, 2014; pp. 148–151.
34. WSSA. *Weed Science*; 1985; pp. 12–15. Available online: https://www.cambridge.org/core/journals/weed-science/article/terms-definitions-and-abbreviations/BA04ABB16C2B5D7D1CDF3A50E3CA8FE7# (accessed on 26 October 2020).
35. PPG. The Pteridophyte Phylogeny Group. A community-derived classification for extant lycophytes and ferns. *J. Syst. Evol.* **2016**, *54*, 563–603. [CrossRef]
36. Twardowski, J.P.; Pastuszko, K. Field margins in winter wheat agrocenosis as reservoirs of beneficial ground beetles (Col., Carabidae). *J. Res. Appl. Agric. Eng.* **2008**, *53*, 123–127.
37. Kieloch, R. Influence of cultivar and agrotechnics on the formation of weed infestation in agricultural crops. *Stud. Rep. JUNG-PIB* **2020**, *61*, 51–63.
38. Haliniarz, M.; Gawęda, D.; Nowakowicz-Dębek, B.; Najda, A.; Chojnacka, S.; Łukasz, J.; Wlazło, Ł.; Różańska-Boczula, M. Evaluation of the Weed Infestation, Grain Health, and Productivity Parameters of Two Spelt Wheat Cultivars Depending on Crop Protection Intensification and Seeding Densities. *Agriculture* **2020**, *10*, 229. [CrossRef]
39. Grundy, A.C.; Mead, A.; Bond, W.; Clark, G.; Burston, S. The impact of herbicide management on long-term changes in the diversity and species composition of weed populations. *Weed Res.* **2010**, *51*, 187–200. [CrossRef]
40. Pawlonka, Z.; Skrzyczyńska, J.; Ługowska, M. Influence of tillage methods and nitrogen fertilisation on weed emergence and development in winter triticale. *Acta Sci. Pol. Agric.* **2007**, *6*, 49–57.
41. Stępień, A. Effect of different fertilisation methods on weed infestation and yielding of spring wheat. *Acta Sci. Pol. Agric.* **2004**, *36*, 45–54.

42. Sawicka, B.; Barbaś, P.; Dąbek-Gad, M. The problem of weed infestation in the conditions of using growth bioregulators and foliar fertilization in potato cultivation. *Sci. Nat. Technol.* **2011**, *5*, 1–12.
43. Bojarszczuk, J.; Podleśny, J.; Nowak, J. The assessment of the diversity of weed flora communities in crops cultivated in selected farms in Lubelskie voivodeship. *Prog. Plant Prot. Postępy Ochr. Roślin* **2018**, *58*, 216–223. [CrossRef]
44. Kotlarz, J.; Kubiak, K.; Kacprzak, M.; Czapski, P. Estimation of tree species diversity of forest stands based on their spectral reflectance. *Sylwan* **2016**, *160*, 1036–1045.
45. Stešević, D.; Jovović, Z. Contribution to the knowledge on the weed flora in potato crop in the vicinity of ikšić (Montenegro). *Herbologia* **2011**, *12*, 1–6.
46. Staniak, M.; Bojarszczuk, J.; Księżak, J. Weed infestation of spring cereals cultivated in pure sowing and undersown with serradella (*Ornithopus sativus* L.) in organic farm. *Water. Environ. Rural. Areas* **2013**, *2*, 121–131.

Article

Effect of Mechanical and Herbicide Treatments on Weed Densities and Biomass in Two Potato Cultivars

Piotr Barbaś [1], Barbara Sawicka [2,*], Barbara Krochmal Marczak [3] and Piotr Pszczółkowski [4]

1 Department Agronomy of Potato, Plant Breeding and Acclimatization Institute—National Research Institute, Branch in Jadwisin, Szaniawskiego 15, Str., 05-140 Serock, Poland; p.barbas@ihar.edu.pl

2 Department of Plant Production Technology and Commodities Science, University of Life Sciences in Lublin, Akademicka 15, Str., 20-950 Lublin, Poland

3 Department of Plant Production and Food Safety, Carpathian State University in Krosno, Dmochowskiego 22 Str., 38-400 Krosno, Poland; barbara.marczak@kpu.krosno.pl

4 Experimental Station for Cultivar Assessment of Central Crop Research Centre, Uhnin, 21-211 Dębowa Kłoda, Poland; p.pszczolkowski.inspektor@coboru.gov.pl

* Correspondence: barbara.sawicka@up.lublin.pl; Tel.: +48-814-456-787

Received: 15 August 2020; Accepted: 30 September 2020; Published: 3 October 2020

Abstract: The effect of potato cultivar and mechanical or herbicide treatments on weed densities and biomass was determined in a research study on a field, conducted from 2007 to 2009 at the Institute of Plant Breeding and Acclimatization. Included in the study were two cultivars and different weed control treatments, including a mechanical method and metribuzin combined with various herbicides and application timings. Chemical methods of controlling weeds were more effective than mechanical methods to reduce weed densities and biomass. The combination of metribuzin with rimsulfuron + SN oil, applied before potato emergence (PRE), was more effective than the other metribuzin combinations. The weed infestation of potato cv. "Irga" was greater than that of cultivar "Fianna" due to differences in the type of growth.

Keywords: potato; cultivars; biodiversity of weeds; mechanical method; chemical method; monocotyledonous weeds; dicotyledonous weeds

1. Introduction

Weeds compete with crops for water, light, and nutrients. Due to the long period from planting to covering inter-rows, potatoes are not as competitive as some other crops. In addition, weeds can reduce air circulation in the potato crop and contribute to favorable conditions for potato infection by pathogens. Effective weed control can eliminate the competitive effects of weeds and should be based on detailed knowledge, including factors impacting performance [1–4]. For example, the same herbicide may have varying efficacy depending on the environmental conditions at and after the application time. Other factors include absorption, translocation, and degradation in the plant [5–7]. Selection of herbicides, herbicide combinations, rates, and application timings should be suited to the degree of weed infestation, the optimal date for using agrochemicals, and the combination of herbicides required to eliminate a wide spectrum of weeds [3–8]. Earlier, Pawlonka [8] used metribuzin pre- and postemergence on potato plants, combined with mechanical treatments, in order to minimize the adverse effect of monoculture on potato yields. However, no publications by other authors concerning the use of metribuzin in mixtures with other herbicides were found. Hence, research was undertaken on the use of metribuzin in combination with other active substances of herbicides, with a different spectrum of action to eliminate both dicotyledonous and monocotyledonous weeds.

The main aim of the research is to develop innovative, effective methods of weed control under the conditions of weed threat, which will provide a larger spectrum of chemical agent action and result in weed control efficiency, costs, rate reductions, and impact on the environment. Hence, the intention of the work is use metribuzin (pre- and post-emergence) and the chosen herbicide mixtures (metribuzin + rimsulfuron + ethoxylated isodecyl alcohol; metribuzin + fluazifop-P-butyl; metribuzin + sulfosulfuron + SN oil) for the control of weeds in the cultivation of two potato cultivars, which will provide a wider spectrum of action of the active substance compared to the mechanical method of combating weeds. This will result in better weed control, a reduction in the dose of herbicides, a reduction in their negative impact on the environment, and a reduction in costs. The selection of these herbicides results from their availability, spectrum of action, and their prices, which is important for farmers, who will make the most economical decision. This research can add value to existing methods. A comparison of chemical weed control methods with mechanical methods that are rarely used today would make it possible to move away from such mechanical methods.

The paper verifies the alternative hypothesis that the use of herbicides and their mixtures, namely, (a) metribuzin—PRE; (b) metribuzin + rimsulfuron + ethoxylated isodecyl alcohol—PRE; (c) metribuzin—POST; (d) metribuzin + rimsulfuron + isodecyl alcohol ethoxylate—PRE; (e) metribuzin + fluazifop-P butyl—POST; (f) metribuzin + sulfosulfuron + SN oil—POST-emergence in potato cultivation will

(a) provide a broader spectrum of action of herbicide chemistry and greater weed control than no weed control and mechanical weed control;
(b) reduce environmental pollution and ensure better effectiveness of chemical treatments, due to the use of lower doses of herbicides, against the null hypothesis that there are no differences between the variants with herbicides and their mixtures and the variant without weed protection and the variant with mechanical weed infestation control in potatoes.

2. Material and Methods

Field trials were carried out for three years, from 2007 to 2009, at the Institute of Plant Breeding and Acclimatization—National Research Institute in Jadwisin (52°29′ N, 21°03′ E). The soil each year was a sandy loam [9]. The experimental design was randomized sub-blocks in a dependent, split-plot system, with three replications. Two factors were examined in the experiment: the first-order factor was potato cultivars ("Irga", an edible, medium-early, stem-type cultivar, and "Fianna", a medium-late, edible, leaf-like cultivar). The second-order factors were weed control methods, including extensive mechanical treatments (every 2 weeks) from planting until row closure and several metribuzin combinations and application timings. Table 1 shows the combinations, rates, and application timings of herbicide treatments. A nontreated control was included for comparison. Herbicides were applied in 400 dm^3 ha^{-1} of water with a backpack sprayer with flat-spray nozzles and a flow rate of 0.35–0.65 $dm.min^{-1}$ and a pressure of 0.1–0.2 MPa. The plot area for treatments was 31.0 m^2, of which 25 m^2 was harvested for tuber yields and quality.

2.1. Agrotechnical Treatments

The previous crops were winter wheat, and white mustard green manure was plowed the fall before potato planting. The type and timing of tillage for field preparation, fertilizer rates and timing, as well as nutrients provided by the white mustard, are shown in Table 2. After harvesting wheat, nitrogen fertilization was applied in an amount of 50 kg N ha^{-1}, followed by stubble cultivation and the sowing of white mustard in an amount of 20 kg·ha^{-1}. White mustard, as a green fertilizer for plowing, brings macro- and micronutrients annually in the amount presented in Table 2.

In addition, in autumn each year, preceding planting, mineral phosphorus–potassium fertilization was applied in the amount of 39.3 kg P·ha^{-1} and 116.2 kg K·ha^{-1} during prewinter plowing. Nitrogen fertilizers were sown in spring, in the amount of 100 kg N ha^{-1}, mixing them with the soil using a tilling set (cultivator + string roller). The fertilizer doses were determined on the basis of the abundance. Potato tubers were planted by hand in the third decade of April, with a spacing of

75 × 33 cm. The propagating material was in Class C/A. Herbicide spraying was done manually using a backpack sprayer. Nursing treatments were performed in accordance with the requirements of good agricultural practice and the methodological assumptions of the experiment. They consisted of covering ridges and ridging. Herbicides in the following combinations were applied to the freshly formed soil, just before the emergence of the potato (metribuzin 1 kg ha^{-1}; metribuzin1 kg ha^{-1} + rimsulfuron 40 g ha^{-1} + ethoxylated isodecyl alcohol 0.1%). After potato emergence, the following preparations were used in reduced doses after prior ridging: metribuzin 0.5 kg ha^{-1}; metribuzin 0.3 kg ha^{-1} + rimsulfuron 30 g ha + ethoxylated isodecyl alcohol 0.1%; metribuzin 0.3 kg ha^{-1} + fluazyfop-P butyl 2 L ha^{-1}; metribuzin 0.3 kg ha^{-1} + sulfosulfuron 26.5 g ha^{-1} + SN oil 1 L ha^{-1} (Table 3). In addition, 400 L ha^{-1} of water was used for spraying with the herbicides.

Potato protection against diseases and insects (Table 4) was used in accordance with the recommendations of the Institute of Plant Protection, National Research Institute, Poland [11].

Table 1. Characteristics of the herbicides and adjuvants used in the experiment.

Trade Names	Common Names	Formulation	Dosage	Utility Forms and Application Date	Grace ***
			Herbicides		
Apyros 75WG	Sulfosulfuron	75%	26.5 g ha^{-1}	granules for water suspension (POST)	Not applicable
Fusilade Forte 150 EC	Fluazyfop-P butyl	150 g in 1 L of measure	2 L ha^{-1}	(POST) concentrate for water suspension	Not applicable
Sencor 70 WG	Metribuzin	70%	0.5 (PRE *) or 0.3 kg ha^{-1} (POST **)	granules for water suspension	42 days
Titus 25 WG	Rimsulfuron	25%	40 g ha^{-1}	granules for water suspension (POST)	Not applicable
			Adjuvants (boosters)		
Atpolan 80 SC	SN oil	76%	1 L ha^{-1}	concentrate for water suspension (POST)	Not applicable
Trend 90 EC	ethoxylated isodecil alcohol	90%	0.1%	concentrate for water suspension (PRE)	Not applicable

Sources: [10,11]; * before potato emergence (PRE); ** after emergence (POST); *** the period from the day of the last treatment to the day of harvesting plants intended for consumption

Table 2. Chemical composition of white mustard as a green fertilizer.

Specification	DM (g kg^{-1})	Ash (g kg^{-1} DM)	N Total (g kg^{-1}DM)	N Miner. (g kg^{-1} DM)	P (g kg^{-1} DM)	K (g kg^{-1} DM)	Mg (g kg^{-1} DM)	Ca (g kg^{-1} DM)	Na (g kg^{-1} DM)	Cu (mg kg^{-1} DM)	Zn (mg kg^{-1} DM)	Mn (mg kg^{-1} DM)	Fe (mg kg^{-1} DM)	Ni (mg kg^{-1} DM)
White mustard biomass	454	55.1	19.2	0.8	3.1	31.6	4.5	1.2	0.8	0.4	35.4	36.0	63.4	7.05

Table 3. The agricultural treatments in the experiment in the years 2006–2009.

Autumn 2006–2008		
Nitrogen fertilization (50 kg ha^{-1})—ammonium nitrate Tillage Sowing and plowing under the forecrop white mustard sowing (20 kg ha^{-1}) disking of white mustard fertilization with phosphorus (39.3 kg P ha^{-1} in the form of granulated superphosphate) and potassium (116.2 kg K ha^{-1} in the form of 60% potassium salt) winter plowing to a depth of about 27 cm		
Spring 2007	Spring 2008	Spring 2009
Tillage and agricultural treatments		
Harrowing fertilization of N (100 kg ha^{-1}—salmag) Cultivation with an aggregate planting of potato seeds—manually Earthing 2 times; in mechanical treatment, earthing and weeding 3 times Herbicide spraying PRE andPOST with atomic knapsack sprayerHarvest with potato elevator digger	Harrowing fertilization of N (100 kg ha^{-1}—salmag) Cultivation with an aggregate planting of potato seeds—manually Earthing 2 times; in mechanical treatment, earthing and weeding 4 times Herbicide spraying PRE and POST with atomic knapsack sprayer Harvest with potato elevator digger	Harrowing fertilization of N (100 kg ha^{-1}—salmag) Cultivation with an aggregate planting of potato seeds—manually Earthing 2 times; in mechanical treatment, earthing and weeding 4 times Herbicide spraying PRE and POST with atomic knapsack sprayer Harvest with potato elevator digger

Source: own research.

Table 4. Potato protection against diseases and insects.

Pesticides	Years		
	2007	2008	2009
Fungicides	chlorothalonil propamocarb hydrochloride SC—2.5 dm^3 ha^{-1} fluazynam—0.4 L ha^{-1}fenamidon + mankozeb—1.25 kg ha^{-1} metalaxyl − M + mancozeb—2.5 kg ha^{-1}	metalaxyl − M + mancozeb—2.5 kg ha^{-1} chlorothalonil (tetrachloroisophthalonitrile)—2 L ha^{-1} propamocarb hydrochloride, fenamidone—2 L ha^{-1}	metalaxyl − M + mancozeb—2.5 kg ha^{-1} mancozeb + cymoxanil/2-cyano-*N*-[(ethylamino) carbonyl]-2-(methoxyimino) acetamide—2 kg ha^{-1} propamocarb hydrochloride, fenamidone—2 L ha^{-1}
Insecticides	thiamethoxam—0.04 kg ha^{-1} acetamiprid—0.08 kg ha^{-1}	thiamethoxam—0.04 kg ha^{-1} thiacloprid—0.75 L ha^{-1} acetamiprid—0.08 kg ha^{-1}	imidacloprid—0.25 L ha^{-1} thiacloprid—0.75 L ha^{-1}

Source: [10,11].

2.2. Assessment of Weeds

The assessment of weed infestation was carried out using the quadrat method, quantitatively and qualitatively, on three randomly selected 1-m^2 areas in each plot, after row closure and just before harvest. Weed species were determined, and before potato harvest, when weed plants entered Stage 97 based on the BBCH scale (BBCH abbreviation comes from the German Biologische Bundesanstalt, Bundessortenamt und CHemische Industrie), fresh biomass from three randomly selected 1.0 m^2 quadrats per plot [11,12] was collected. After measuring fresh weight, samples were dried, and the dry weight was recorded.

A weed species list was established. The number of weeds present in each 1 m^2 quadrat was calculated according to the formula

$$\mathrm{Nw} = \frac{(\mathrm{N1} + \mathrm{N2} + \mathrm{N3})}{(\mathrm{Nm} \times \mathrm{Qa})} \tag{1}$$

where

Nw—number of weeds of a given species growing on an area of 1 m^2,
N1, N2, N3—number of plants in the quadrat in subsequent measurements (pcs),
Nm—number of measurements,
Qa—quadrat of area (m^2) [11].

Plants were uprooted from soil, segregated by species, and the fresh weight of each species was determined. The weight of a given species was averaged across the three randomly selected quadrats. After drying, the dry weight of weeds was weighed for each species separately, and the weighing results were recorded in an evaluation card. Weed samples were placed in a ventilated room until a constant mass was obtained [11].

2.3. Meteorological Conditions

To characterize thermal and humidity conditions in the years of research, data from the meteorological station located at the Institute of Plant Breeding and Acclimatization, National Research Institute, Jadwisin, were used. The conditions of the growing season in 2007–2009 were characterized by varying air temperatures and rainfall (Figure 1, Table 5).

The 2007 and 2008 growing seasons can be described as relatively dry, while 2009 could be characterized as the year with the most favorable humidity and thermal conditions for potato development. Weather data is included in Table 5.

In 2007, the average temperature of the April–September period was 13.7 °C, and this was lower than the long-term average of this period by 0.6 °C. The sum of precipitation was 165.3% of the norm. Weather conditions in May and August of 2008 exceeded the multiyear norm, and, in the remaining months, there was a noticeable water shortage. The average air temperature in the April–September period was 14.2 °C, and this was lower than the multiyear average by 0.3 °C. The meteorological conditions of the 2009 growing season were varied, but their common feature was a drought period at the beginning of the potato-growing season (Figure 1, Table 5).

Table 5. Precipitation and the hydrothermal coefficient of Sielianinov during the potato vegetation period in 2007–2009 according to the meteorological station of the Plant Breeding and Acclimatization Institute, National Research Institute, Jadwisin.

Year	Month	The Sum of Precipitation in the Month (mm)				% of the Long-Term Norm (1971–1995)	Sielianinov Hydrothermal Coefficient	Evaluation of the Month **
		Decade			Month			
		1	2	3				
2007	April	12.0	4.0	0.3	16.3	<25	0.69	Extremely dry
	May	11.6	28.7	38.1	78.4	75–125	1.93	Quite humid
	June	30.4	13.6	65.6	109.6	75–125	2.32	Humid
	July	30.1	6.4	17.6	54.1	50–74	0.99	Dry
	August	43.8	17.2	13.3	74.3	50–74	1.34	Optimum
	September	42.5	49.0	12.2	103.7	75–125	3.20	Extremely moist
	Total	170.4	118.9	146.8	436.4	-	-	-
2008	April	11.4	12.9	5.0	29.3	25–49	1.36	Optimum
	May	36.8	12.8	13.3	62.9	50–74	1.64	Quite humid
	June	0.0	21.0	22.5	43.5	25–49	0.84	Dry
	July	20.2	38.2	10.4	68.8	50–74	1.22	Quite dry
	August	26.8	37.8	16.3	80.9	75–125	1.48	Optimum
	September	21.6	10.3	16.9	48.9	25–49	1.40	Optimum
	Total	116.8	133.0	84.4	334.3	-	-	-
2009	April	0.0	0.0	0.0	0.0	<25	0.0	Extremely dry
	May	8.8	13.3	58.7	80.8	75–125	2.12	Humid
	June	30.2	17.5	24.7	72.4	50–74	1.38	Optimum
	July	36.8	20.5	28.3	85.6	75–125	1.28	Quite dry
	August	12.5	48.5	22.1	83.1	75–125	1.54	Optimum
	September	8.2	4.7	5.9	18.8	<25	0.44	Very dry
	Total	96.5	104.5	139.7	340.7	-	-	-

** Hydrothermal coefficient according to Sielianinov: extremely dry (ed)→$k \leq 0.4$; very dry (vd) → $0.4 < k \leq 0.7$; dry (d) → $0.7 < k \leq 1.0$; quite dry (Qd) → $1.0 < k \leq 1.3$; optimum (o) → $1.3 < k \leq 1.6$; quite humid (pd) → $1.6 < k \leq 2.0$; humid (w) → $2.0 < k \leq 2.5$; very humid (vh) → $2.5 < k \leq 3.0$; extremely moist (em) → $k > 3.0$.

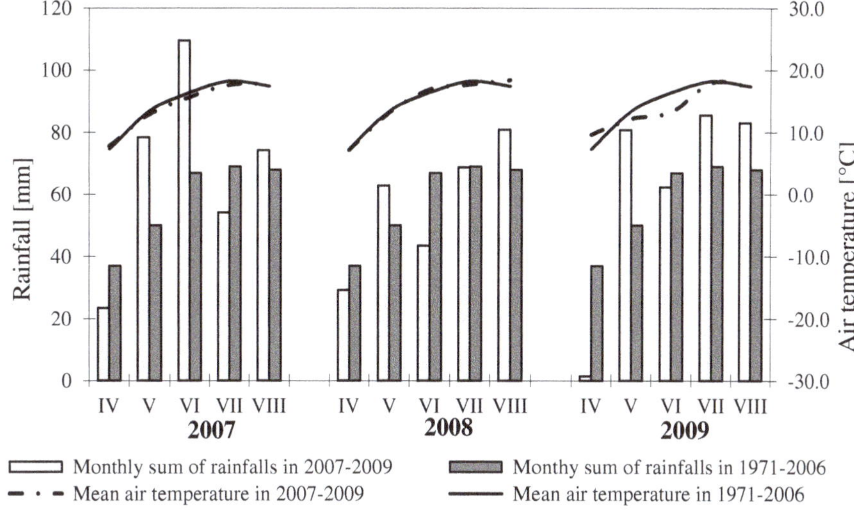

Figure 1. Precipitation and air temperature during the potato-growing season according to the meteorological station of the Institute of Plant Breeding and Seed Production, National Research Institute, Jadwisin (2007–2009), against the background of the long-term average.

2.4. Statistical Analyses

SAS v.9.2 software was used for statistical analyses [13]. The factors for the models were (years × cultivars × weed control treatment), and variance and multiple Tukey's *t*-tests were determined at significance level $p_{0.05}$. The significance of sources of variation was tested using the Fischer–Snedecor F-test [14].

The averaged data are given: for cultivars—means for treatments, years, and repetitions; in the case of care treatments—averages for cultivars, years, and repetitions; in the case of years—averages for treatments, cultivars, and repetitions.

3. Results

3.1. Soil Conditions

Information on the value of soils, on which research with potato are presented in Tables 6 and 7.

Table 6. The granulometric composition of the soil (%).

Years	Percentage of Fraction with Diameter (mm Diameter)			Grain Size Subgroup	Soil–Agricultural Complex
	2.0–0.05	0.05–0.002	<0.002		
2007	72.0	24.0	4.0	Sandy loam	Rye complex
2008	71.0	26.0	3.0	Sandy loam	Rye complex
2009	72.0	24.0	4.0	Sandy loam	Rye complex

Source: Results were determined at the Chemical and Agricultural Station in Wesoła, near Warsaw. The determinations were made according to the applicable methods and standards [9,15,16].

Table 7. Physicochemical properties of soil in Jadwisin (2007–2009).

Years	Content of Available Forms (mg·100 g^{-1} DM of Soil)			pH (1M KCl)	Content of Organic Substance (%)
	P_2O_5	K_2O	Mg		
2007	23.9	22.2	20.0	4.7	0.73
2008	9.9	16.8	15.5	5.4	0.68
2009	3.9	7.3	6.0	5.0	0.70

Source: Results were made at the Chemical and Agricultural Station in Wesoła. The determinations were made according to the applicable methods and standards [15–22].

3.2. Weed Species

In the plantation, before the row closure of plants, only two species of monocotyledonous weeds (*Echinochloa crus-galli*—cockspur grass or barnyard grass (ECHCG) and *Agropyron repens*—quackgrass (*AGRRE*)) and four species of dicotyledonous weeds (*Chenopodium album*—common lambsquarters (CHEAL); *Convolvulus arvensis*—bindweed (*CONAR*); *Erodium cicutarium*—redstem filaree (*EROCI*); *Viola arvensis*—field violet (*VIOAR*)) were recorded. Both the cultivars and methods of weed control, as well as meteorological conditions in the years of research, shaped the floristic composition of weeds that were recorded (Table 8). Weed control methods, cultivars, and years of research differentiated the composition and number of weeds found on the potato plantation. The dominant monocotyledonous species was ECHCG, while the main dicotyledonous species was VIOAR. The least ECHCG was recorded on variants with the application of the mixture of herbicides metribuzin + sulfosulfuron + SN oil POST (Variant 8) and preparations metribuzin + rimsulfuron + ethoxylated isodecyl alcohol, used before crop emergence (Variant 4), compared to the control variant and with the mechanical weed control method (Variant 2). AGRRE was dominant in the variant where the metribuzin and metribuzin + rimsulfuron + ethoxylated isodecyl alcohol mixture was used to control weeds, applied PRE. The least dominance of this weed species was recorded after the application of the mixture of metribuzin + fluazyfop-P butyl (variant 7) and metribuzin + sulfosulfuron + SN oil (Variant 8). CHEAL was best destroyed by a mixture of metribuzin + rimsulfuron + ethoxylated isodecyl alcohol preparations, applied before the potato emerges (Variant 4). Field bindweed was best limited by metribuzin applied after crop emergence. EROCI was best removed when metribuzin + rimsulfuron + ethoxylated isodecyl alcohol was applied PRE and with metribuzin applied after potatoes emergence. VIOAR was the least numerous when applying the mixture of metribuzin + sulfosulfuron + SN oil.

Higher weed infestation before row closure was recorded in the medium-early "Irga" cultivar than the medium-late "Fianna", but only in the case of ECHCG. In the case of EROCI, on the other hand, the "Fianna" cultivar was more weeded. Weed infestation with species such as AGRRE, CHEAL, CONAR, and *VIOAR* did not differentiate the potato cultivars (Table 8).

ECHCG turned out to be the relatively dominant weed species in the years 2007 and 2008; in 2009, the dominating weed was VIOAR, which could possibly be associated with the used forecrop of potato. AGRRE was not present at row closure in the fairly humid conditions of 2008, and CONAR was not present in 2007 and 2008. CHEAL, VIOAR, and CONAR were most frequently recorded in the potato canopy in the quite-dry conditions of 2009. Only the size of *EROCI* did not significantly depend on growing-season conditions (Table 8).

Species composition of weeds before potato harvest was of little variation. The state and structure of weed infestation in variants with chemical plant protection, compared to mechanical cultivation and the control variant, were similar to the weed infestation determined before shorting the rows. In the species composition of weeds, determined before harvesting, monocotyledons taxa species predominated, with *ECHCG* dominating (Table 9).

Table 8. Species composition and the number of monocotyledonous and dicotyledonous weeds before the closure of potato rows, depending on cultivar, weed control method, and year (pcs m^2).

Experiment Factors		Species According WSSA ***					
		ECHCG	AGRRE	CHEAL	CONAR	EROCI	VIOAR
Cultivar	"Irga"	6.6	0.8	1.1	0.4	0.0	3.6
	"Fianna"	5.4	0.7	1.3	0.3	0.5	3.1
	$LSD_{p0.05}$	1.0	ns **	ns	ns	0.2	ns
Weed control methods *	1	14.9	0.9	5.4	1.2	1.8	9.3
	2	6.5	0.7	2.7	0.6	0.2	5.9
	3	5.3	0.7	0.2	0.3	0.1	3.8
	4	3.0	0.7	0.0	0.2	0.0	1.8
	5	5.4	1.0	0.2	0.0	0.0	1.6
	6	5.6	1.1	1.0	0.1	0.1	1.8
	7	4.5	0.5	0.2	0.2	0.2	1.6
	8	3.0	0.6	0.2	0.2	0.2	0.7
	$LSD_{p0.05}$	3.3	ns	1.5	0.8	0.8	2.7
Years	2007	5.1	1.7	0.1	0.0	0.5	1.6
	2008	10.1	0.0	0.6	0.0	0.4	0.4
	2009	2.8	0.5	2.9	0.9	0.0	8.0
	$LSD_{p0.05}$	1.5	0.5	0.7	0.4	ns	1.2
Mean		6.0	0.7	1.2	0.3	0.3	3.3

Source: own research; * (1) Control variant—without chemical protection; (2) extensive mechanical weeding (every 2 weeks) from planting to closure of the rows; (3) metribuzin 1 kg ha^{-1}—PRE; (4) metribuzin 0.3 kg ha^{-1} + rimsulfuron 40 gha^{-1} + ethoxylated isodecyl alcohol 0.1%—PRE; (5) metribuzin 0.5 kg ha^{-1}—POST; (6) metribuzin 0.3 kg ha^{-1} + rimsulfuron 30 g ha^{-1} + ethoxylated isodecyl alcohol 0.1%—PRE; (7) metribuzin 0.3 kg ha^{-1} + fluazifop-P butyl 2 L ha^{-1}—POST; (8) metribuzin 0.3 kg ha^{-1} + ulfosulfuron 26.5 g ha^{-1} + SN oil 1 L ha^{-1}—POST; ns **—not significant at the level $p_{0.05}$; *** AGRRE—*Agropyron repens*; ECHCG—*Echinochloa crus-galli*; CHEAL—*Chenopodium album*; CONAR—*Convolvulus arvensis*; EROCI—*Erodium cicutarium*; VIOAR—*Viola arvensis* [23].

The morphological and physiological features of the studied cultivars differentiated only in the number of ECHCG, VIOAR, and EROCI. The first two were more numerous in the cultivar "Irga" than "Fianna", while, in the last cultivar, it was the opposite. The monocotyledonous weed community was most effectively limited by the following treatments: variant 4 (metribuzin + rimsulfuron + ethoxylated isodecyl alcohol), variant 7 (metribuzin + fluazifop-P butyl), variant 8 (metribuzin + sulfosulfuron + SN oil) (Table 4), limiting their occurrence in comparison with mechanical treatments and the control variant. Dicotyledonous species were eliminated to the greatest extent on Sites 5 (metribuzin) and vriant 8 (metribuzin + sulfosulfuron). Homologous in terms of this feature turned out to be Variant 6 (metribuzin + rimsulfuron + ethoxylated isodecyl alcohol), Variant 7 (metribuzin + fluazifop-P butyl), Variant 5 (Sencor), and Variant 8 (metribuzin + sulfosulfuran + SN oil), on which these herbicides were used after potato emergence. Potato vegetation conditions also significantly differentiated the number and species composition of weeds. Higher numbers of ECHCG were recorded in 2008, with high rainfall in the period of July–August preceding the potato harvest, while there were high numbers of CHEAL, CONAR, and VIOAR in 2009, as it was quite a dry year (Table 9).

The fresh matter of weeds before harvest was 254.1 g on average, whereas their average air-dry matter was 102.8 g·m^{-2}. The weed mass at the end of the growing season was differentiated by cultivars, weed control methods, and years of research (Table 10).

Among the compared cultivars, the medium-early "Irga" was more weeded, the shoots of which dried up earlier and ended the growing season earlier than the medium-late cultivar "Fianna". The best herbicidal effect was achieved using mixtures of the metribuzin + rimsulfuron + ethoxylated isodecyl alcohol (Variant 4) and Sencor + Apyros + Atpolan (Variant 8). Variants 4 and 8 proved to be homologous in terms of this feature. Both fresh and air-dry matter of weeds were the highest in the fairly humid conditions of 2008 and the lowest in 2009, a fairly dry year (Table 10).

Table 9. Species composition and number of mono- and dicotyledonous weeds before potato harvest, depending on cultivar, weed control method, and year (pcs. m^{-2}).

Experiment Factors		ECHCG	AGRRE	CAPBP	CHEAL	CONAR	EROCI	SONAR	VIOAR
Cultivars	"Irga"	8.3	1.4	0.3	2.4	1.6	0.0	0.04	4.7
	"Fianna"	6.4	1.5	0.0	2.0	1.2	0.5	0.01	3.4
	$LSD_{p0.05}$	1.3	ns **	ns	ns	ns	0.1	ns	1.2
Weed control methods *	1	16.2	1.8	1.6	7.4	2.5	1.0	0.10	8.6
	2	8.4	2.0	0.0	4.3	1.8	0.3	0.05	6.6
	3	7.6	1.3	0.0	1.5	1.8	0.2	0.00	6.5
	4	4.8	1.7	0.0	0.6	1.0	0.0	0.00	2.8
	5	6.0	2.1	0.0	0.7	1.0	0.0	0.00	1.5
	6	6.7	1.2	0.0	1.1	1.0	0.2	0.00	2.8
	7	5.7	0.8	0.0	1.3	1.3	0.3	0.00	2.2
	8	3.2	1.0	0.0	0.7	0.8	0.0	0.00	1.4
	$LSD_{p0.05}$	4.0	ns	ns	1.8	ns	0.6	ns	4.2
Years	2007	4.3	3.5	0.0	0.8	0.6	0.4	0.00	0.6
	2008	12.4	0.1	0.0	1.6	0.5	0.4	0.00	0.5
	2009	5.4	0.8	0.6	4.2	3.2	0.0	0.08	11.0
	$LSD_{p0.05}$	1.8	0.8	ns	0.8	0.9	Ns	ns	1.9
Mean		7.3	1.4	0.2	2.2	1.4	0.2	0.02	4.0

Source: own research; * explanations as in Table 8; ** not significant at the level $p_{0.05}$.

Table 10. Fresh and air-dry matter of weeds and their number before potato harvest, depending on cultivar, weed control method, and year.

Experiment Factors		Weed Matter (g·m^{-2})		The Number of Weeds (pcs m^{-2})	
		Fresh Matter	Air-Dry Matter	Monocotyledonous	Dicotyledonous
Cultivars	"Irga"	286.7	116.9	9.7	9.0
	"Fianna"	221.5	88.7	7.9	7.1
	$LSD_{p0.05}$	38.6	16.1	ns **	1.6
Weed control methods *	1	576.9	236.7	18.0	21.2
	2	321.1	142.7	10.4	13.0
	3	231.7	76.7	8.9	10.0
	4	111.4	55.8	6.5	4.4
	5	260.6	90.1	8.1	3.2
	6	227.5	94.5	7.9	5.1
	7	142.5	63.1	6.5	5.1
	8	141.1	62.9	4.2	2.9
	$LSD_{p0.05}$	120.6	50.4	4.3	4.9
Years	2007	276.7	114.0	7.8	2.4
	2008	295.3	132.2	12.5	3.0
	2009	190.3	62.2	6.2	19.0
	$LSD_{p0.05}$	56.7	23.7	2.0	2.3
Mean		254,1	102.8	8.8	8.1

Source: own research; * explanations as in Table 7; ** not significant at the level $p_{0.05}$.

In the case of the fresh mass of weeds assessed before the rows were closed, a significant interaction of cultivars × years was found. The medium-early cultivar "Irga" showed a significantly higher weed weight than the medium-late cultivar "Fianna", but only in 2009, which were characterized by lower humidity in May–June than in 2007. Before the harvest, however, both cultivars showed a similar response to the conditions in the years of research (Table 11).

Table 11. Fresh matter of weeds depending on the cultivar and year (average for weed control methods).

Cultivars	Years		
	2007	2008	2009
	Before row closing		
"Irga"	283.1	310.8	265.0
"Fianna"	270.2	279.7	114.3
$LSD_{p0.05}$		36.6	
	Before tuber harvest		
"Irga"	125.4	142.1	83.2
"Fianna"	102.7	122.3	41.0
$LSD_{p0.05}$		15.3	

In the experiment, the number of monocotyledonous weeds before the potato harvest was 8.8 pcs·m^{-2}; for dicotyledonous weeds, it was 8.1 pcs·m^{-2}. The reduction in the number of monocotyledonous and dicotyledonous weeds was higher in the case of mechanical and chemical care using various herbicides than in mechanical weed control alone. The lowest number of these species, for both classes, was recorded after using the mixture of herbicides Sencor + Apyros + Atpolan (Variant 8) to regulate weed infestation. In the case of the number of monocotyledonous weeds, Variants 4, 7, and 6, and, in the case of dicotyledonous weeds, the combinations of 4, 5, 6, and 7 proved to be homologous in terms of this characteristic. The highest number of monocotyledonous and dicotyledonous weeds was recorded in 2008 and 2009, respectively, while the smallest number was in 2009 and 2007, respectively, as a result of the weather conditions prevailing during the potato growing season (Table 9). In 2009, the cultivar "Irga", with a stem habit, reacted with more weed infestation than the leaf cultivar "Fianna", both before closing the inter-rows and before harvesting the tubers (Table 12).

Table 12. The number weeds per 1 m^2, depending on the cultivar and year (average for weed control methods).

Cultivars	Years		
	2007	2008	2009
	Before row closing		
"Irga"	8.7	11.9	17.0
"Fianna"	9.6	11.7	13.2
$LSD_{p0.05}$		3.5	
	Before tuber harvest		
"Irga"	10.7	15.2	31.0
"Fianna"	10.4	15.7	19.7
$LSD_{p0.05}$		6.8	

4. Discussion

The size of potato yields is largely determined by weeds, and the number, size of the air-dried matter of weeds, and their species composition play a significant role [23–28]. The lowest number of weeds in the tests, regardless of the date of weed infestation measurement timing, was recorded after using the mixture of metribuzin + sulfosulfuran + SN oil, and the highest number of weeds found in the tests was after the application of metribuzin before the emergence of the crop. Mechanical weed control reduced the number of weeds by twice as much as the control variant. These results were confirmed in the studies of Ciesielska and Wysmułek [29]. In their opinion, metribuzin, regardless of the date of use, is characterized by a wide spectrum of weed control, while the addition of rimsulfuron and the ethoxylated isodecyl alcohol adjuvant caused the majority of weeds present in the potato crop to be destroyed by 87–100%. Nowak et al. [30] and Ilić et al. [31] confirmed the improvement of herbicide effectiveness due to the addition of adjuvants to the working liquid. These are chemical substances of organic or inorganic origin, which directly or indirectly affect the herbicidal activity of the herbicide's active substance or change the usable properties of the formulation and application liquid [31–33]. In addition, adjuvants prevent the crystallization of the utilized liquid on the surface of plants and delay its drying, causing an increase in the adhesion and solubility of the herbicide and better hydration of the leaf epidermis [24,31,34–39].

The effect of herbicides is largely dependent on the thermal conditions during the application of the herbicide and a few days after the procedure. This was also confirmed by other authors [3,40,41]. According to Gugała et al. [40], a higher temperature promotes increased herbicide adsorption and translocation. Air humidity is also positively correlated with herbicide effects. In conditions of higher humidity, herbicide uptake increases because the used liquid evaporates more slowly from the leaf surface, and, as a consequence, there is a greater amount of agent that can penetrate plant tissues [1,42]. In addition, in conditions of high humidity, the movement of the herbicide in the plant from its penetration to the site of action is much faster than in the case of low humidity [43–46]. However, as suggested by Lavlesh et al. [5], with excessive humidity, there may be a danger of greater herbicide leaching, faster root uptake, and, therefore, increased plant toxicity. Toxicity and degradation rate of biocides, according to Awasthi et al. [45] and Jezierska and Frąc [47], depend primarily on their dose and the structure of the active substance that is the basic plant protection product, which can also be a cause of the inhibition and growth of microorganisms.

With Zarzecka et al. [48], the smallest number of weeds were observed on variants sprayed with metribuzin, while the number was significantly higher after using flurochloridon and purely mechanical treatments. Additionally, Tomczak et al. [49] demonstrated the high effectiveness of metribuzin and mixtures of prosulfocarb and metribuzin in weed control in potato cultivation. Pszczółkowski and Sawicka [50] obtained the best results of eliminating monocotyledonous weeds after applying linuron, and they limited dicotyledonous weeds after using a mixture of linuron + clomazone. In the study of Wichrowska [51], the most effective chemical in controlling weeds was linuron, reducing the population size by 80.3%, especially of dicotyledonous weeds. The least effective in weed control was promethrin used before potato plant emergence.

ECHCG and AGRRE were the most abundant among the monocotyledonous weed species, while CHEAL was the most abundant of the dicotyledonous species. Similar results were also obtained by Gugała et al. [40] and Baranowska et al. [52]. During the vegetation period, these authors showed the dominance of taxa of the following weed species: AGRRE—19.8%, ECHCG—13.2%, VIOAR—12.1%, and CHEAL—11.2%. A similar tendency also appeared before potato tuber harvesting. AGRRE was the most frequently encountered weed (25.9%), followed by ECHCG (14.7%) and CHEAL (20.3%).

Cultivars were another factor that differentiated potato weed infestation. The difference in weed infestation observed between the two examined potato cultivars may be determined by such features as volume of foliage, plant growth rate, morphological and anatomical type, and resistance to abiotic stress. The highest fresh and dry mass of weeds was found in the medium-early cultivar, with a stalk habit, as well as a shorter vegetation period. A medium-late cultivar turned out to be significantly less weedy, with a leaf habit. These results show that potato cultivars with a strong, erect shoot growth habit (shorter stems, more branching, and a denser and taller canopy in the early stages of plant vegetation) may be less susceptible to weed interference than cultivars with lofty plant habit. Similar observations were made by Zarzecka et al. [50] and Baranowska et al. [52]. The optimal time for weed management and the formation of the physiological age of tubers have not been thoroughly investigated and probably vary by cultivar. These dependencies should be established in future research.

The varied response of the studied cultivars to the meteorological conditions of the growing season, in the case of both the number and fresh mass of weeds, could be explained by a different plant habit, as well as by the length of the vegetation period. The medium-early "Irga" cultivar is characterized by a stem habit that is more favorable to weed infestation, while the middle-late "Fianna" leaf type cultivar shaded the soil more and thus prevented strong weed infestation. Other authors shared similar observations [48–50,53]. In the Polish register of cultivars, and, currently, in the EU register, there is a very large assortment of potato cultivars with different physiological and morphological types and different production possibilities. Therefore, it is necessary to determine not only the potential of the potato cultivars and lines currently grown in Poland but also to test them with biotic and abiotic stresses, including the response to herbicides.

Weed management in the conditions of the changing climate in recent years requires the integration of methods and strategies and a change in the way they are perceived. Weeds in potato crops must be combated in a comprehensive, targeted, and proactive manner. Understanding the interactions between weed control methods and the weed spectrum, as well as managing the crop system to prevent and discourage weeds and keep the level of weed seed in the soil low, is essential for effective weed management in all crop systems [50,53]. Good agricultural and tillage practices, including well-planned crop rotation, planting of cover crops, as well as good sanitary practices [10], optimal row spacing, and early planting dates are important aspects of weed management not only in organic systems but also in other cultivation systems [54,55]. Timely mechanical weed control, before the rows of potato plants close, can eliminate many early weed species. In this regard, the selection of appropriate potato cultivars that have a fast initial growth rate and create a leaf canopy capable of controlling weeds can help to reduce weed infestation and, thereby, increase yield. For this purpose, further research, with the participation of potato breeders, is needed on new creations of cultivars that will allow the reduction of primary weed infestation in potato cultivation.

Thanks to the conducted research, for the first time, a broader spectrum of chemical activity of the active substance metribuzin was obtained, with better weed control than mechanical treatments alone. The use of lower doses of metribuzin will contribute to reducing the pollution of the natural environment. Moreover, it was proven that the deciduous cultivar "Fianna" contributed to less weed infestation of potato stalks by creating a larger canopy of leaves than the stalk cultivar "Irga".

5. Conclusions

The fresh and air-dry matter of weeds were most limited by the combination of the mixture of metribuzin + rimsulfuron and ethoxylated isodecyl alcohol adjuvant used before potato emergence (PRE). The best effect of reducing the number of monocotyledonous and dicotyledonous weeds was obtained in the potato field with the following preparations: metribuzin + sulfosulfuron + SN oil applied POST emergence of the potato plants.

The best way to reduce the number of monocotyledonous weeds in the crop of tested cultivars was to use a mixture of herbicides and an adjuvant (metribuzin + sulfosulfuron + SN oil).

The combination of metribuzin + sulfosulfuron + SN oil eliminated CONAR, VIOAR, and mostly limited the number of CHEAL weeds. In turn, the number of EROCI was best reduced by metribuzin applied after potato emergence and a mixture of herbicides, metribuzin + rimsulfuron + ethoxylated isodecyl alcohol, used before the emergence of the plants, while the numbers of SONAR and CAPBP were successfully eliminated by all herbicides.

The leaf-type cultivar "Fianna" (shorter stems, more branched, and denser and taller leaf canopy) proved to be less susceptible to weed infestation than the stem-type cultivar "Irga".

Selecting cultivars with a fast initial growth rate and a leaf crown capable of controlling weeds can help to reduce weed infestation in potato crops.

Author Contributions: Conceptualization: P.B. and B.S.; methodology: B.S., P.B., and P.P.; data protection: P.B., P.P., and B.S.; data curation—writing: P.B. and B.S.; funding acquisition: P.B.; software: P.P.; validation: P.P.; formal analysis: B.K.M.; investigation: B.K.M.; resources: B.K.M.; project administration: P.B.; writing—original draft preparation: P.B., B.S., B.K.M., and P.P.; writing—review and editing: B.S. and P.B.; visualization: B.S.; supervision: P.B. All authors have read and agreed to the published version of the manuscript.

Funding: This research received no external funding.

Acknowledgments: The authors thank the management of the Institute of Plant Breeding and Seed Production, National Research Institute, for administrative and technical support.

Conflicts of Interest: The authors declare no conflict of interest.

Abbreviation

AGRRE	*Agropyron repens* (L). *P. Beauv.*
ECHCG	*Echinochloa crus-galli* (L.) *P. Beauv*
CAPBP	*Capsella bursa-pastoris* (L.) *Medik.*
CHEAL	*Chenopodium album* (L.)
CONAR	*Convolvulus arvensis* (L.)
EROCI	*Erodium cicutarium* (L.) *L'H R*
SONAR	*Sonchus arvensis* (L.)
VIOAR	*Viola arvenis* Murray
BBCH	Biologische Bundesanstalt, Bundessortenamt und CHemische Industrie
PRE	before potato emergence
POST	after potato emergence

References

1. Riethmuller-Haage, I.; Bastiaans, L.; Kempenaar, C.; Smutny, V.; Kropff, M.; Kropff, M. Are pre-spraying growing conditions a major determinant of herbicide efficacy? *Weed Res.* **2007**, *47*, 415–424. [CrossRef]
2. Azadbakht, A. The effect of chemical and non-chemical control methods on weeds in potato (*Solanum Tuberosum* L.) cultivation in Ardabil province, Iran. *Appl. Ecol. Environ. Res.* **2017**, *15*, 1359–1372. [CrossRef]
3. Barbaś, P.; Sawicka, B. Dependence of potato crop on weed infestation. *Agronom. Sci.* **2020**, *8*, 346–359. [CrossRef]
4. Zarzecka, K.; Gugała, M.; Grzywacz, K.; Sikorska, A. Agricultural and economic effects of the use of biostimulants and herbicides in cultivation of the table potato cultivar Gawin. *Acta Sci. Pol. Agric.* **2020**, *19*, 3–10. [CrossRef]
5. Lavlesh Raghav, M.; Sati, U.C.; Sati, K. Evaluating the manual and chemical methods for weed control in potato (*Solanum tuberosum* L.) under Tarai conditions of Uttarakhand. *Int. Q. J. Life Sci.* **2017**, *12*, 683–686.
6. Mayerová, M.; Madaras, M.; Soukup, J. Effect of chemical weed control on crop yields in different crop rotations in a long-term field trial. *Crop. Prot.* **2018**, *114*, 215–222. [CrossRef]
7. Caldiz, D.; de Lasa, C.; Bisio, P.E. Grass and weed management in potato (*Solanum tuberosum* L.) processing with Clomazone, in Argentine pampas. *Am. J. Plant Sci.* **2016**, *7*, 2339–2348. [CrossRef]
8. Pawlonka, Z. Potato yield in monoculture under differentiated intensity of weed control. *Prog. Plant Prot. Postępy Ochr. Roślin* **2007**, *47*, 229–233.
9. Ryzak, M.; Bartmiński, P.; Bieganowski, A. Methods of determination of granulometric distribution of mineral soils. Theses and monographs. *Acta Agroph.* **2009**, *175*, 1–97.
10. Tomlin, C. *The e-Pesticide Manual*; Version 5.2; British Crop Production Council: Hampshire, UK, 2011.
11. Domaradzki, K.; Badowski, M.; Filipiak, K.; Franek, M.; Gołębiowska, H.; Kieloch, R.; Kucharski, M.; Rola, H.; Rola, J.; Sadowski, J.; et al. Field experiments. In *Methodology Experiments of Biological Evaluation of Herbicides, Bioregulators and Adjuvants*; Domaradzki, K., Ed.; Institute of Soil Science and Plant Cultivation Publishing House: Puławy, Poland, 2001; Chapter 1, p. 167. (In Polish)
12. Roztropowicz, S. (Ed.) *Methodic of Observation, Measurements and Sample Take in Agricultural Experiments with Potato*; Plant Breeding Acclimatization Institute, Section Jadwisin: Wołomin, Poland, 1999; p. 50.
13. SAS. *I.I. SAS/STAT®9.2. Users Guide*; SAS: Cary, NC, USA, 2008.
14. John, H.M. *Handbook of Biological Statistics*, 3rd ed.; Sparky House Publishing: Baltimore, MD, USA, 2014; pp. 148–151.
15. PN–R–0403. *Chemical and Agricultural Analysis of Soil. Sampling*; Polish Committee for Standardization: Warsaw, Poland, 1997.
16. WRB. World Reference Database for Soil Resources. 2014. Available online: http://www.fao.org/3/a-i3794e.pdf (accessed on 8 June 2020).

17. ISO (the International Organization for Standardization). ISO 10390: 2005, *Soil Quality—Determination of pH*; Technical Committee: ISO/TC 190/SC 3; Chemical and Physical Characteristics; ICS: 13.080.10. Chemical Characterization of Soils; ISO (The International Organization for Standardization): Geneva, Switzerland, 2005; Volume 2, pp. 1–7.
18. KQ-07/PO-01/F-02. Version 09/07.04.2014. Regional Chemical And Agricultural Station in Lublin, Poland. 2014. Available online: http://www.oschr.pl/instrukcje/2_GO_dn_2014_01_21.pdf (accessed on 30 August 2020).
19. PN-R-04020,1994+AZ1. *Chemical and Agricultural Analysis of Soil*; Polish Committee for Standardization: Warsaw, Poland, 2004.
20. PN-R-04023. *Chemical and Agricultural Analysis of Soil. Determination of Available Phosphorus Content in Mineral Soils*; Polish Committee for Standardization: Warsaw, Poland, 1996.
21. PN-R-04022,1996+AZ1. *Chemical and Agricultural Analysis of Soil. Determination of Available Potassium Content in Mineral Soils*; Polish Committee for Standardization: Warsaw, Poland, 2002.
22. Nawrocki, S. *Fertilizer Recommendations. Part. I. Limit Numbers for Valuation of Soils in Macro and Microelements*; IUNG: Puławy, Poland, 1990; Volume 44.
23. Rindler, W. Visual horizons in models of the world. *Mon. Not. R. Astron. Soc.* **1956**, *116*, 662–677, reprinted in *Gen. Rel. Grav.* **2002**, *34*, 133–153. [CrossRef]
24. Andreasen, C.; Streibig, C. Evaluation of changes in weed flora in arable fields of Nordic countries based on Danish long-term surveys. *Weed Res.* **2010**, *51*, 214–226. [CrossRef]
25. Grundy, A.C.; Mead, A.; Bond, W.; Clark, G.; Burston, S. The impact of herbicide management on long-term changes in the diversity and species composition of weed populations. *Weed Res.* **2010**, *51*, 187–200. [CrossRef]
26. Stešević, D.; Jovović, Z. Contribution to the knowledge on the weed flora in potato crop in the vicinity of Nikšić (Montenegro). *Herb* **2011**, *12*, 1–6.
27. Mayerová, M.; Mikulka, J.; Kolářová, M.; Soukup, J. Changes in weed community composition in a long-term trial with different crop rotations and herbicide treatments. *Julius-Kühn-Archiv* **2018**, *458*, 58–66.
28. Soren, C. Weed dynamics and yield of potato as influenced by weed management practices. *Int. J. Pure Appl. Biosci.* **2018**, *6*, 398–408. [CrossRef]
29. Ciesielska, A.; Wysmułek, A. The efficacy of tank mixture herbicides Sencor 600 SC + Rimsulfuron in potatoes. *Prog. Plant Protect. Postępy Ochr. Roślin* **2012**, *52*, 885–888. [CrossRef]
30. Nowak, S.; Waindzoch, K.; Świerszcz, S.; Niemczyk, M.; Spałek, K.; Nowak, A. Crop density rather than ruderal plants explains the response of ancient segetal weeds. *Biologia* **2018**, *74*, 351–359. [CrossRef]
31. Ilić, O.; Nikolić, L.; Ilin, Ž.; Mišković, A.; Vujasinović, V.; Kukić, B. Effect of cultural practices on weeds community in function of potato yield. *Acta Sci. Pol. Hortorum Cultus* **2016**, *15*, 31–43.
32. Woźnica, Z.; Idziak, R.; Szewczyk, R.A. New multifunctional adjuvant for glyphosate-based herbicides. *Prog. Plant Protect. Postępy Ochr. Roślin* **2004**, *44*, 536–542.
33. Idziak, R.; Woźnica, Z. Skuteczność chwastobójcza herbicydu Callisto 100 SC stosowanego z adiuwantami i nawozem mineralnym. *Acta Agroph.* **2008**, *11*, 403–410.
34. Pacanoski, Z. Role of adjuvants on herbicide behavior, a review of different experiences. *Herb* **2010**, *11*, 67–79.
35. Villa, P.M.; Rodrigues, A.C.; Márquez, N.; Rodrigues, A.L.; Martins, S.V. Fitosociologíe de malezas después de un cultivo de papa (*Solanum tuberosum* L.) en los andes venezolanos: Un enfoque agroecologico. Phytosociology of weeds after potato cultivation (*Solanum tuberosum* L.) in the Venezuelan Andes: An agroecological approach. *Trop. Subtrop. Agroecosystems* **2017**, *20*, 329–339.
36. Gajbhiye, V.T.; Gajbhiye, V.T. Effect of concentration, moisture and soil type on the dissipation of flufenacet from soil. *Chemosphere* **2002**, *47*, 901–906. [CrossRef]
37. Wang, C.J.; Liu, Z.Q. Foliar uptake of pesticides—Present status and future challenge. *Pestic. Bioch. Physiol.* **2007**, *87*, 1–8. [CrossRef]
38. Stagnari, F.; Chiarini, M.; Pisante, M. Influence of fluorinated surfactants on the efficacy of some post-emergence sulfonylurea herbicides. *J. Pestic. Sci.* **2007**, *32*, 16–23. [CrossRef]
39. Kwiatkowski, C.A.; Wesołowski, M.; Drabowicz, M.; Misztal-Majewska, B. The effect of adjuvants and reduced rates of crop protection agents on the occurrence of agricultural pests and on winter wheat productivity. *Ann. UMCS* **2012**, *E-67*, 12–21.

40. Gugała, M.; Zarzecka, K.; Dołęga, H.; Sikorska, A. Weed infestation and yielding of potato under conditions of varied use of herbicides and bio-stimulants. *J. Ecol. Eng.* **2018**, *19*, 191–196. [CrossRef]
41. Kierzek, R.; Paradowski, A.; Kaczmarek, S. Chemical methods of weed control in maize (*Zea mays* L.) in variable weather conditions. *Acta Sci. Pol. Agric.* **2012**, *11*, 35–52.
42. Johnson, K.; Jacobsen, C.S.; Torsvik, V.; Sørensen, J. Pesticide effects on bacterial diversity in agricultural soils—A review. *Biol. Fertil. Soils* **2001**, *33*, 443–453. [CrossRef]
43. Meena, R.S.; Kumar, S.; Datta, R.; Lal, R.; Vijayakumar, V.; Brtnický, M.; Sharma, S.K.; Yadav, G.S.; Jhariya, M.K.; Jangir, C.K.; et al. Impact of agrochemicals on soil microbiota and management: A review. *Land* **2020**, *9*, 34. [CrossRef]
44. Ramsey, R.J.L.; Stephenson, G.R.; Hall, J.C. Review of the effect of moisture, humectants and surfactant composition on the absorption and effectiveness of highly water-soluble herbicides. *Biochem. Pestic. Phys.* **2005**, *82*, 162–175. [CrossRef]
45. Awasthi, R.; Gaur, P.M.; Turner, N.C.; Vadez, V.; Siddique, K.H.M.; Nayyar, H. Effects of individual and combined heat and drought stress during seed filling on the oxidative metabolism and yield of chickpea (*Cicer arietinum*) genotypes differing in heat and drought tolerance. *Crop. Pasture Sci.* **2017**, *68*, 823. [CrossRef]
46. Spanogiannopoulos, P.; Bess, E.N.; Carmody, R.N.; Turnbaugh, P.J. The microbial pharmacists within us: A metagenomic view of xenobiotic metabolism. *Nat. Rev. Genet.* **2016**, *14*, 273–287.
47. Frąc, M.; Jezierska-Tys, S. Diversity of soil environment microorganisms. *Adv. Microb.* **2010**, *40*, 47–58.
48. Zarzecka, K.; Gugała, M.; Dołęga, H. Regulation of weed infestation degree in potato with the use of herbicides. *Biul. IHAR* **2013**, *267*, 113–120.
49. Tomczak, B.; Bączkowska, E.; Bubniewicz, P.; Górniak, J. Prosulfocarb—Herbicide for monocot and dicot weed control in winter cereals and potatoes. *Prog. Plant Protect. Postępy Ochr. Roślin* **2007**, *47*, 280–284.
50. Pszczółkowski, P.; Sawicka, B. Attempts to reduce weed infestation of potato in cultivation under cover. Part II. Weight, abundance and species composition of weeds. *Biul. IHAR* **2003**, *228*, 261–273.
51. Wichrowska, D. Weed infestation of potato plantation cultivated in Kujawsko-Pomorski Region. *Eko Tech* **2008**, *16*, 91–96.
52. Baranowska, A.; Mystkowska, I.; Zarzecka, K.; Gugała, M. Efficacy of herbicides in potato crop. *J. Ecol. Eng.* **2016**, *17*, 82–88. [CrossRef]
53. Boydston, R.A. Managing weeds in potato rotations without herbicides. *Am. J. Potato Res.* **2010**, *87*, 420–427. [CrossRef]
54. Sawicka, B. Resilient Agricultural Practices. In *Zero Hunger. Encyclopedia of the UN Sustainable Development Goals*; Leal Filho, W., Azul, A., Brandli, L., Özuyar, P., Wall, T., Eds.; Springer: Cham, Switzerland, 2020.
55. De Cicco, A.; Jeanty, J.C. Potato Sector in the EU—Production, Price and Trade Statistics. EUROSTAT. Statistic Explained. Available online: https://ec.europa.eu/eurostat/statistics-explained/index.php/The_EU_potato_sector_-_statistics_on_production,_prices_and_trade (accessed on 14 August 2020).

Article

The Impact of Exotic *Tamarix* Species on Riparian Plant Biodiversity

Kgalalelo Tshimologo Annie Setshedi [1,*] and Solomon Wakshom Newete [1,2]

[1] School of Animal, Plant and Environmental Sciences, University of Witwatersrand, Private Bag X3, Johannesburg 2050, South Africa; solomonnewete@gmail.com

[2] Agricultural Research Council-Soil, Climate and Water (ARC-SCW), Geo-Information Science Division, Private Bag X79, Arcadia, Pretoria 0001, South Africa

* Correspondence: Kgalalelosetshedi@gmail.com

Received: 13 July 2020; Accepted: 10 August 2020; Published: 7 September 2020

Abstract: This study investigated the impact of exotic *Tamarix* species on vascular plant biodiversity in riparian ecosystems in the Western Cape Province, South Africa. Vegetation was sampled, using 5 m wide belt transects, along the Leeu, Swart, and Olifants riparian areas, which had varying invasion intensities. Each transect was split into three zones (Zone 1: 0–15 m; Zone 2: 15–35, and Zone 3: >35 m), which were identified at each site based on species composition across each riparian zone. Woody plant species were identified, counted, and their heights measured within the transects that were laid out from the waterpoint (Zone 1) outwards (Zone 2 and 3). Herbaceous aerial cover (HAC) was determined subjectively and objectified using the Walker aerial cover scale. Leeu River had the highest species richness (Dmg = 2.79), diversity (H′ = 2.17; $-\ln\lambda$ = 1.91; N1 = 8.76 and α = 4.13), and evenness (J′= 0.80). The Swart River had the lowest species richness, which declined from Dmg = 1.96 (Zone 1) to Dmg = 1.82 (Zone 3). Exotic *Tamarix* species ranked in the top three most abundant woody vascular plant species along the Swart and Olifants rivers, where they ranked first and third, respectively. The Jaccard's and Sorenson's coefficients of similarity indicated that species differed greatly between the different sites, $\overline{x} < 27\%$ for both indices. The indices also indicated that the Swart River had the lowest level of species distinctness between zones ($\overline{x} > 80\%$) while the Leeu River had the highest level of species distinctness ($\overline{x} < 50\%$) between the different zones. These findings suggest a possible displacement of herbaceous and woody tree species by exotic *Tamarix* invasion, inter alia, a decrease in ecosystem functions and services associated with the loss in biodiversity, as well as significant bearings on the agricultural ecosystem by reducing the faunal diversity such as crop pollinators, inter alia.

Keywords: exotic *Tamarix*; riparian zone; biodiversity; richness; diversity; evenness; biodiversity indicators

1. Introduction

The major causes of biodiversity loss are believed to be attributed to direct habitat destruction, invasive organisms, pollution, population growth, and over-exploitation/over-harvesting of natural resources [1–3]. All these causes are better known as the 'HIPPO' acronym [4]. Global warming due to climate change has also been identified as one of the factors that cause the loss of biodiversity, consequently endangering multiple species that are not adapted to a wide range of weather conditions [5]. The loss of plant biodiversity is directly linked to a loss in ecosystem functions, maintenance, resilience, and ecosystem services [6]. Ecosystem functions are ecological processes that control energy, nutrient fluxes, and organic matter fluxes, whilst ecosystem services are benefits that are provided to humanity by the interaction between biotic and abiotic components of the environment [7].

The lack of plant conservation directly cascades into the loss of fauna and entomofauna [8]. Exotic *Tamarix* species have been identified as riparian invaders and are well studied in the United States of America (USA) [9]. Several studies conducted in South Africa have identified riparian zones as regions highly invaded by exotic *Tamarix* species [10–13]. Invasive alien plants (IAPs) not only change the plant species composition and distribution patterns but they alter litter quantity and quality. Nutrient cycling regimes change as a result of invasion by alien plants. Furthermore, ground cover which provides surface stability is lost; this eventually leads to an increase in soil erosion [12,14]. Invasive species can alter fire regimes to favour the establishment of alien plants. Catchment hydrology, sediment yield, and geomorphology are also altered upon invasion [12]. Native riparian vegetation and ecosystems have been modified extensively, and there is an urgent need for management and restoration strategies.

Pyšek and Richardson [15] identified key traits associated with the invasive potential of many alien plants that give them a competitive advantage over indigenous plant species by comparing 64 IAPs. Downey and Richardson [16] defined six thresholds along the extinction trajectory and found that, although no plants are extinct in the wild or globally due to alien plant invasions, native plants have crossed other thresholds along the extinction trajectory due to alien plant invasions. The decline in native plant populations and extirpations due to exotic *Tamarix* invasion in South Africa is yet to be quantified. Newete et al. [13] compared the density of exotic *Tamarix* species to the co-occurring plants and found that the *Tamarix* density was greater in seven of the 11 sites investigated.

The plant diversity of riparian communities provides shelter and food to animal populations, thus, promoting integrity and even further growth of overall biodiversity. Maintaining riparian corridors is important for diversity conservation [17]; however, they also provide an introduction pathway for IAPs across landscape patches, which threaten native plant populations. Riparian areas play a critical role with regards to acting as water sources and biodiversity hotspots, and, therefore, need to be managed more-so in South Africa where they have been identified as vulnerable and impacted by anthropogenic activities as well as IAPs [12,17]. Riparian vegetation differs across various biogeographical areas [12]. Therefore, assessments of invasion should be conducted separately for different areas. The determination of biodiversity relies on the assessment of species richness, evenness, and heterogeneity within an ecosystem [7].

Over 10 million ha of land in South Africa is invaded by alien plants [18] and *Tamarix* Ladeb, is one of them. Although the country has an indigenous *Tamarix* species (*T. usneoides*), the two exotic *Tamarix* species (*Tamarix ramosissima* L. and *Tamarix chinensis* L.) have been in the country close to a century after their first introduction reportedly as garden plants for ornamental purposes [19]. The two species, along with their hybrids, are currently threatening many of the riparian ecosystems in the country and they are listed under the National Environmental Management: Biodiversity Act 2014 (NEM: BA) as category 1b invasive weed requiring urgent management intervention [13,20]). Alien *Tamarix* species are also considered as one of the 100 worst invasive species globally, under the International Union for Conservation of Nature (IUCN) list of Global Invasive Species Database [21]. *Tamarix ramosissima* and *T. chinensis* have been found to hybridise between themselves and with the native congeneric, *T. usneoides*, further extending their distribution range across the country [22]. This has accelerated the species' invasive potential, and complicated control of the exotics as the species are cryptic [13,22–25]. Exotic *Tamarix* species are well adapted to survive adverse conditions, i.e., frost, floods, or drought, and they are also able to re-sprout, regenerate, and establish post-natural disasters [20,26].

The Southern African Plant Invaders Atlas (SAPIA) database identified woody species as the most dominant invaders of riparian zones, which prompted this study, to investigate the impact that exotic *Tamarix* species had on vascular plant diversity in riparian ecosystems that had been invaded by the trees. To achieve the aim, species richness, diversity, evenness, and complementarity were determined across sites that had varying *Tamarix* populations in terms of the *Tamarix* species that had been previously identified at those sites. Establishing the species biodiversity would not only fill the knowledge gaps with regards to which ecosystem services and functions are affected as a result of invasion, but this will eventually lead to improved and better-informed mitigation approaches to protect indigenous plant biodiversity.

2. Methods and Materials

2.1. Study Site

The study was conducted at three riparian sites in the Western Cape Province, South Africa, where *Tamarix* invasion had been identified as most prevalent [13]. Both Leeu River (S32,76794 and E21,97958), near the town of Leeu Gamka and the Swart River (S33,16627 and E21,97994) near the town of Prince Albert were in the Central Karoo District of Western Cape. The third site, the Olifants River (S33,50776 and E22,69586), was in the Little Karoo (part of the Cape Floristic Region) about 10–15 km away from the town of De Rust, Western Cape. The World Geodetic System (WGS84) geographical reference system was used to obtain the Global Positioning System (GPS) coordinates. The Leeu River was the only site where the native congeneric, *Tamarix usneoides*, occurred. Both the central Karoo and the Little Karoo are semi-arid regions with dwarf succulent shrubs being a common floral feature [27,28]. Seasonal and periodic droughts are also common in the Karoo area [29].

2.2. Data Collection

Experimental Design

The study was conducted between late October and early November during the flowering season of most Karoo vegetation. Vegetation was sampled, using 5 m wide belt transects, along three riparian areas at Leeu, Swart, and Olifants rivers. Woody plant species were identified and counted before their heights and stem diameters were measured within the transects that were laid out from the waterpoint outwards (transects varied in numbers: 3–6 belt transect, length: 50–75 m, and width: 25–50 m). Herbaceous aerial cover (HAC) was determined subjectively, as a percentage, within three 1 m^2 quadrats laid out systematically within the 25 m^2 quadrats and objectified using the Domin scale [30]. Plant growth forms were later classified into several categories, according to Germishuizen and Meyer [31] and African Plant Database [32]. Each transect was split into three zones (Zone 1: 0–15 m; Zone 2: 15–35; and Zone 3: >35 m), which were identified at each site based on species composition across each riparian zone. GPS coordinates (WGS84 geographical reference system) and altitude were recorded at the beginning of each 25 m^2 quadrat.

2.3. Data Analysis

Determining Species Richness, Plant Diversity, and Evenness

Species accumulation (Sample-based species accumulation) and rarefaction (individual-based species accumulation) curves were plotted for the three sites. The comparisons between both curves provides information on whether the sampled sites are homogeneous or heterogeneous, indicating how plant species differ across the riparian zones. Sample plots were randomised 100 times to compute the mean estimator and expected species richness for each sample plot accumulation level using Estimate S [33].

Species diversity entails richness (the number of species) and evenness (the equality in the number of individuals for every species). The Margalef's index for species richness (D_{mg}), Shannon–Weiner diversity index (H'), the Simpson diversity index ($-\ln\lambda$), Hills diversity number (H1), Fisher's alpha diversity index (α), and the Shannon evenness index (J') were used to determine richness, diversity, and evenness (Magurran, 1988; Williams et al., 2005). Indices were either directly computed by Estimate S or as parameters that were used as substitutes into equations that were obtained from Estimate S [33]. Rank abundance curves (determine the number of individuals for different species and rank species from most to least abundant) were plotted in order to further analyse the patterns of diversity in terms of richness and evenness.

The Shapiro–Wilk's W test was used to examine the normality of the indices. The Fligner–Killeen test was used to test for homoscedasticity. Two-way ANOVAs (R version 4.0.2) were conducted, for the six indices, to determine whether there were significant differences at the zonal and site levels. One-way ANOVAs and Tukey's Honest Significant Difference (HSD) multiple comparison post hoc tests were also conducted in order to establish whether, and to what extent, the zone and site impacted on the indices.

2.4. Complementarity and Similarity

The Jaccard and Sorenson's similarity indices were used to determine how similar or different species composition was between zones sampled at each site. These indices provide a measure of β-diversity (inter-habitat diversity) as well as the species turnover or species composition change along an environmental gradient where environmental dynamics differ. The values were computed on Estimate S and returned as a percentage.

3. Results

3.1. Species Richness, Diversity, and Evenness

Species Identification: Family Name, Binomial Nomenclature, Plant Type, Origin, and Status

A total of 42 vascular plant species from 32 genera within 15 families were identified and sampled across the Swart, Olifants, and the Leeu rivers (Table 1). The most species-rich families were Aizoaceae (seven species), Poaceae (6), Amaranthaceae (4), and Fabaceae (4). Asteraceae and Solanaceae both comprised of three species each, with the remaining nine families having only one species. The *Galenia* L., *Atriplex* L., *Prosopis* L., *Argemone* L., *Lycium* L., and *Tamarix* genera all had more than one species. Species richness varied between the different rivers with the Swart, Olifants, and Leeu rivers having 13, 25, and 18 vascular plants, respectively. A total of nine alien (not endemic or indigenous to South Africa) plants were recorded across all three riparian areas, excluding exotic *Tamarix* species. Six of the alien plants are known invasive species listed either under the Conservation of Agricultural Resources Act, 1983 (Act No. 43 of 1983) [34] legislation or the National Environmental Management: Biodiversity Act, 2004 (Act No. 10 of 2004) [35] legislation. Plants occurring naturally in South Africa were all listed as Least Concern, according to the South African National Biodiversity Institute (SANBI) Redlist plant species list.

Table 1. Plants sampled along the Swart (S), Olifants (O), and Leeu (L) rivers in Western Cape, South Africa, where exotic *Tamarix* invasion was most prevalent. Alien plants have been marked with an * while listed invasive species have been marked with **.

Family	Species	Plant Type	Origin	Site Occurred	Status
Aizoaceae	*Galenia africana* L.	Shrub	South Africa	S, O,	
Aizoaceae	*Galenia pubescens* L.	Dwarf shrub	South Africa	O	
Aizoaceae	*Lampranthus uniflorus* L. Bolus	Succulent-Shrub	South Africa	S	
Aizoaceae	*Mesembryanthemum crystallinum* L.	Succulent	South Africa	S, O, L	
Aizoaceae	*Prenia tetragona* Thunb.	Succulent	South Africa		
Aizoaceae	*Psilocaulon coriarium* Burch. ex N.E. Br.	Succulent-Shrub	South Africa	S	
Aizoaceae	*Tetragonia tetragonioides* Pall.	Shrub	Eastern Asia, Australia, and New Zealand	S, L	**
Amaranthaceae	*Atriplex muelleri* L.	Dwarf shrub	Australia	S, O	*
Amaranthaceae	*Atriplex semibaccata* L.	Dwarf shrub	Australlia	S, O, L	**
Amaranthaceae	*Atriplex vestita* L.	Shrub	South Africa	O, L	
Amaranthaceae	*Bassia salsoloides* Fenzl	Dwarf shrub	South Africa	O	
Amaranthaceae	*Salsola barbata* Aellen	Shrub	South Africa	S	
Anacardiaceae	*Searsia pendulina* Jacq.	Tree	South Africa	L	

Table 1. *Cont.*

Family	Species	Plant Type	Origin	Site Occurred	Status
Asteraceae	*Chrysocoma ciliata* L.	Shrub		O	
Asteraceae	*Oncosiphon piluliferum* L.f.	Herb	South Africa	O	
Asteraceae	*Senecio burchellii* DC.	Shrub/Dwarf shrub	South Africa	O	
Brassicaceae	*Sisymbrium* sp. L.	Herb	Naturalised-introduced in South Africa	O	
Brassicaceae	Unidentified	Herb		O	
Chenopodiaceae	*Chenopodium murale* L.	Herb	Europe and parts of Asia and northern Africa	L	*
Ebenaceae	*Diospyros lycioides* Desf.	Tree/Shrub	South Africa	L	
Fabaceae	*Medicago sativa* L.	Herb	Naturalised-introduced in South Africa	O	
Fabaceae	*Prosopis hybrid* L.	Tree/Shrub	Mexico, Central and northern South America	L	*
Fabaceae	*Prosopis juliflora* L.	Tree/Shrub	Mexico, Central and northern South America	L	**
Fabaceae	*Vachellia karroo* Hayne	Tree/Shrub	South Africa	S, O, L	
Juncaceae	*Juncus kraussii* Hochst.	Herb	South Africa	O	
Papaveraceae	*Argemone albiflora* Hornem.	Forb/Herb	North America	O	*
Papaveraceae	*Argemone mexicana* L.	Forb/Herb	Mexico	O	**
Poaceae	*Cenchrussp.*	Graminoid	South Africa	S	
Poaceae	*Cenchrus ciliaris* L.	Graminoid	South Africa	O, L	
Poaceae	*Cynodon dactylon* L.	Graminoid	South Africa	O, L	
Poaceae	*Enneapogon desvauxii* P.Beauv.	Graminoid	South Africa	O	
Poaceae	*Hordeum murinum* L.	Graminoid	South Africa	O	
Poaceae	*Stipagrostis namaquensis* Nees	Graminoid	South Africa	O, L	
Santalaceae	*Viscum rotundifolium* L.f.	Hemiparasite	South Africa	S	
Scrophulariaceae	*Sutera* sp.Roth	Shrub	South Africa	S	
Solanaceae	*Lycium hirsutum* Dunal	Shrub/Dwarf shrub		S, L	
Solanaceae	*Lycium oxycarpum* Dunal	Shrub/Occasional tree	South Africa	O, L	
Solanaceae	*Solanum tomentosum* L.	Shrub/Dwarf shrub	South Africa	O	
Tamaricaceae	Exotic *Tamarix* (*T. ramossissima* L. and *T. chinensis* L.)	Tree/Shrub	Eurasia	S, O, L	**
Tamaricaceae	*Tamarix usneoides* L.	Tree/Shrub	Eurasia and South Africa	L	
Zygophyllaceae	*Zygophyllum retrofractum* Thunb.	Succulent-Shrub	South Africa	S, L	

3.2. *Species Accumulation and Rarefaction Curves for Woody Vegetation*

The species accumulation and rarefaction curves for the Olifants and Leeu rivers in the first two zones showed heterogeneous species composition, only in the third zone, which was the furthest away from the waterpoint, did the curves suggest a homogenous species composition (Figure 1). Zones were classified as distances across the river bank with similar species composition (i.e., Zone 1: 0–15 m; Zone 2: 15–35; and Zone 3: >35 m). The Swart River species accumulation and rarefaction curves showed that the species composition was homogenous across all the three zones. All the curves reached an asymptote, which suggests that the sampling effort for all three sites was sufficient.

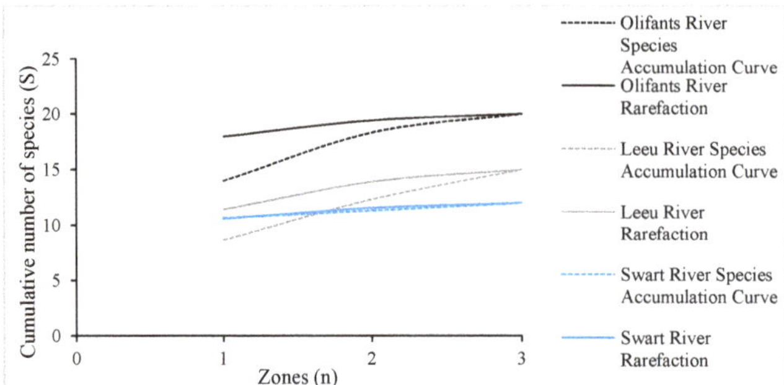

Figure 1. Species accumulation (observed, dotted line) and Rarefaction (expected, solid line) curves for riparian plants at Olifants (Black), Leeu (Grey), and Swart (Blue) river, Western Cape, South Africa. The total number of woody species is 20, 15, and 12, respectively. The curves represent successively pooled and randomly ordered samples between three different zones sampled. Curves were computed using Estimate S [33].

3.3. Species Richness

The Margalef's index (D_{mg}) for species richness showed that the Olifants and Leeu river species richness increased from the waterpoint outwards in contrast to the Swart River species richness, which decreased from the river bank outwards (Figure 2). The species richness indices for all the sites were <3. The Leeu River had the highest species richness index, across all the three zones, compared to the other rivers sampled. Two-way ANOVA tests indicated that there was an overall significant difference ($p < 0.05$) between species richness at the site and zonal level.

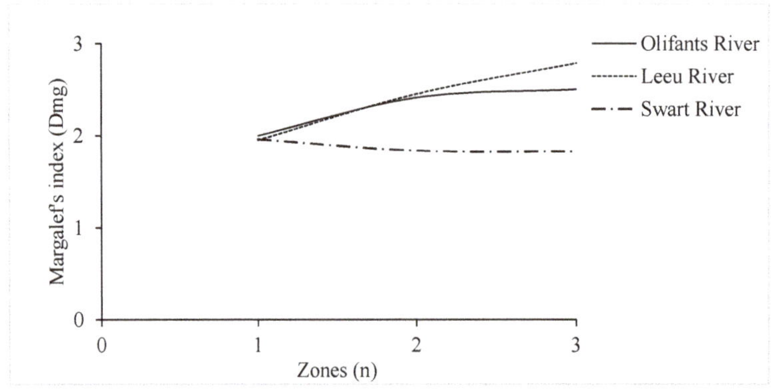

Figure 2. The cumulative species richness curves of Margalef's index (D_{mg}) for plants at Olifants (solid line), Leeu (dotted line), and Swart (solid-dotted line) rivers, Western Cape, South Africa. The curves represent successively pooled and randomly ordered samples at three zones sampled. Curves were computed using Estimate S [33]. The formula for the index is $D_{mg} = (S - 1)/\ln N$.

3.4. Species Diversity

The Shannon diversity index (H′) measures the level of uncertainty in correctly predicting the identity of the next species chosen at random. The Shannon diversity index increased from the waterpoint outwards in all three sites (Figure 3a). The Leeu riparian zones had the highest species richness, i.e., the highest degree of uncertainty in predicting the identity of the next species chosen at random, and, therefore, the highest diversity. The Simpsons diversity index ($-\ln\lambda$) along the Leeu riparian zone increased slightly from the waterpoint outwards, whereas that of the Olifants and Swart riparian zones showed no increase moving outward from the waterpoint (Figure 3b). The zone furthest away from the waterpoint had the highest species diversity. The Olifants riparian zone had the highest dominance (i.e., the vascular plant community was mostly made up of a single species) compared to the Leeu and Swart riparian zones. High dominance (low $-\ln\lambda$ value) is indicative of lower species diversity (Figure 3b). Hill's Diversity Number (N1), which shows diversity in terms of the effective number of species that are present within a sample indicates that the Leeu riparian zone species diversity is likely to increase away from the river bank than the Olifants and Swart riparian zone species diversity, which are more likely to remain low and unchanged (Figure 3c). These findings corroborate those of the Shannon and Simpsons diversity index, which showed that the Leeu riparian zone had the most diverse population. The Leeu riparian zone showed an increasing Fisher's alpha diversity index (α) away from the waterpoint, whereas the α of the Swart riparian zone decreased away from the waterpoint. Furthermore, the Olifants riparian zone α plateaus away from the waterpoint (Figure 3d). Two-way ANOVA tests conducted for the diversity indices indicated that between the three zones and sites species diversity differed significantly ($p < 0.001$). In addition to these findings, the one-way ANOVAs showed that for the Simpson's diversity index, the Olifants River diversity was significantly lower ($p < 0.05$) compared to both the Swart and Leeu river species diversity.

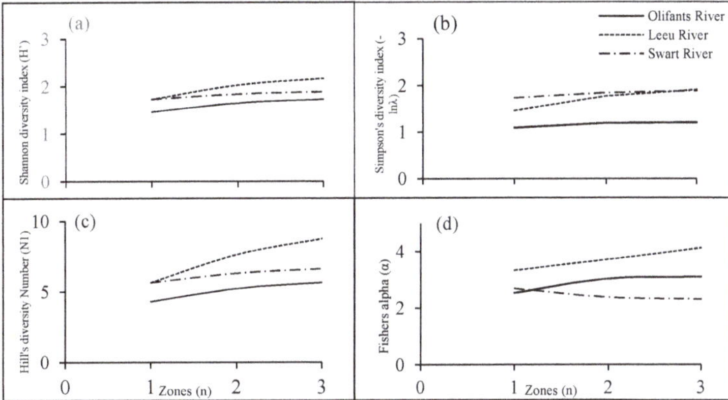

Figure 3. Species diversity curves for Olifants (solid line), Leeu (dotted line) and Swart (solid-dotted line) rivers in Western Cape, South Africa. (**a**) The Shannon diversity index curves with the riparian zone has the highest index, (**b**) the Simpson's diversity index curves with the Leeu River and Swart riparian zones plateau at the same highest value (**c**) Hill's diversity number curves show the Leeu riparian zone to have the highest diversity value and (**d**) the Fisher's alpha index is the highest for the Leeu riparian zone with the Swart River Fisher's alpha value decreasing away from the river bank. Species diversity indices increased towards the zone furthest from the waterpoint.

3.5. Species Evenness

The Leeu riparian zone had the highest Evenness index than the Olifants and Swart riparian zones (Figure 4), and across all sites, Zone 1 diverged from evenness. These results indicate that the Leeu riparian zone species compositions have the lowest dominance of species. The two-way ANOVA tests showed that, overall, the evenness differed significantly ($p < 0.001$) between zones and sites, the one-way ANOVAs corroborated these findings showing that Olifants River species composition was significantly less even than the Swart and Leeu rivers species composition.

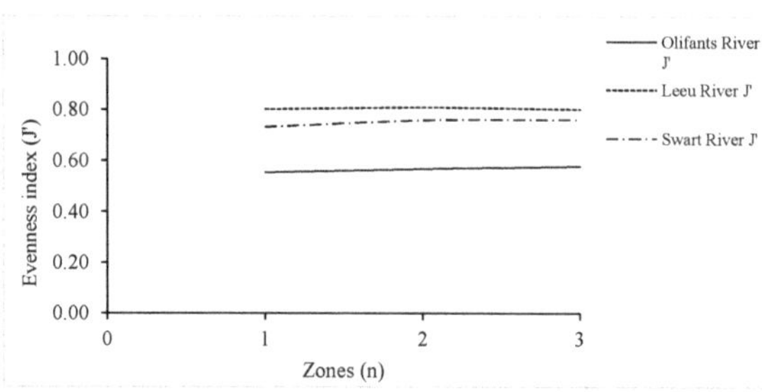

Figure 4. Evenness curves for Olifants (solid line), Leeu (dotted line) and Swart (solid-dotted line) rivers in Western Cape, South Africa. Evenness index was the highest along the Leeu riparian zone.

3.6. Rank Abundance Curves

The rank abundance curves of the three sites sampled were all steep, which is a pattern indicative of uneven plant communities (Figure 5). Only the top three most abundant plant species are displayed in the curves as follows: (a) along the Swart River: (i) Exotic *Tamarix*, (ii) *Lampranthus uniflorus*, and (iii) *Atriplex muelleri*; (b) along the Olifants River: (i) *Oncosiphon piluliferum*, (ii) Brassicaceae, and (iii) Exotic *Tamarix*; (c) along the Leeu River: (i) *Mesembryanthemum crystallinum*, (ii) *Lycium oxycarpum*, and (iii) *Prosopis juliflora*.

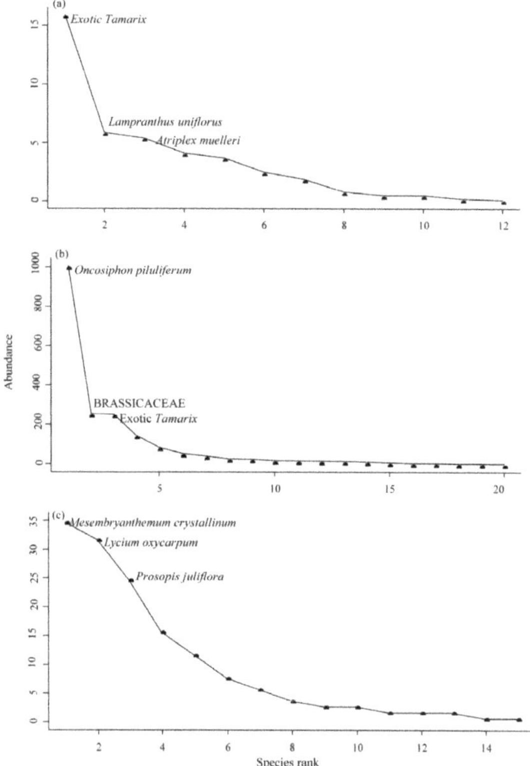

Figure 5. Rank abundance curves showing the top three most abundant species for (**a**) Swart, (**b**) Olifants, and (**c**) Leeu rivers, Western Cape, South Africa. The curves are all steep, and exotic *Tamarix* only occurs in the top three most abundant species along the Swart and Olifants rivers.

3.7. Complementarity and Similarity

The Jaccards and Sorenson's similarity index between the three sampled rivers were all low ($\bar{x} < 27\%$) (Table 2). This indicates that vascular plant species composition varied greatly between the sites that were invaded by exotic *Tamarix* species. Species composition was less similar between the Leeu River (which had the highest species richness, diversity, and evenness) and the Swart River (which had low species richness, diversity, and evenness). Between the three zones sampled along the Leeu, Olifants, and Swart rivers, the Jaccards and Sorenson's similarity index indicated that the Swart riparian zone plant community had the highest species similarity between the zones ($\bar{x} > 80\%$) (Table 3). The Leeu riparian zone plant community had low species similarity between the zones ($\bar{x} < 50\%$) (Table 3). Similarly, the Olifants riparian zone plant community complementarity and similarity indices were indicative of a community, across all three zones, with low degree of similarity ($\bar{x} < 59\%$) (Table 3). Species similarities in all the three sites were relatively low between the first and the third zones, although along the Leeu River, Zone 1 and 3 had the highest species similarities, whereas along both the Olifants and Swart river species, the similarities were highest between Zone 2 and 3.

Table 2. The Jaccard's and Sorenson's coefficients of similarity between the Leeu, Olifants, and Swart riparian zones, expressed as percentages.

Similarity and Complementarity Index	Leeu River	Swart River
Jaccard Index		
Olifants River	21	23
Leeu River		17
Sorenson's index		
Olifants River	34	38
Leeu River		29

Table 3. The Jaccard's and Sorenson's coefficients of similarity between the three zones sampled along the Leeu, Olifants, and Swart rivers, expressed as percentages.

	Leeu River		Olifants River		Swart River	
Similarity and Complementarity Index	**Zone 2**	**Zone 3**	**Zone 2**	**Zone 3**	**Zone 2**	**Zone 3**
Jaccard Index						
Zone 1	36	50	44	41	83	83
Zone 2	-	36	-	70	-	100
Sorenson's index						
Zone 1	53	67	62	52	91	91
Zone 2	-	53	-	82	-	100

4. Discussion

Invasive alien plants are strong drivers of biodiversity change or loss, and exotic *Tamarix* species are such plants, with the ability to modify ecosystems, and this has become evident through research conducted in the Western Cape Province. The Western Cape Province is known to have the highest exotic species (*T. chinensis*—Tc and *T. ramossissima*—Tr) and exotic hybrid species (Tc × Tr) populations in comparison to other South African provinces [36]. This finding is alarming in that Mayonde et al. [22] found hybrid *Tamarix* species (Tc × Tr) to be the dominant (64.7%) invasive genotype. Hybridisation is an evolutionary trait that invasive organisms have developed in order to rapidly increase genetic variation and possibly invasion vigour as well.

Complementarity and similarity assessments of species between the study sites indicated that the Leeu River was highly dissimilar to both the Olifants and Swart rivers with both the Jaccard's and Sorenson's coefficients of similarity being the lowest for the Leeu River (Table 2). A closer look at each individual site showed that the Swart River and the Olifants River had higher species similarity between zones than the Leeu River, which had the lowest zonal species similarity (Table 3).

High complementarity and similarity between zones are suggestive of a riparian community that is made up of a high occurrence of a similar species. The species accumulation and rarefaction curves showed that the Swart River had a homogenous riparian zone plant community (Figure 1). In both, the case of the Olifants River and the Swart River, exotic *Tamarix* was identified as the most abundant tree species. This finding was supported by the rank abundance curves where exotic *Tamarix* ranked first and third along the Swart and Olifants River, respectively (Figure 5).

Newete et al. [13] conducted a study to establish the distribution and abundance of the invasive *Tamarix* genotypes in South Africa and found a strong negative relationship (linear correlation) between the density of the exotic *Tamarix* species and co-occurring tree and shrub species. The Olifants and Swart rivers had the highest exotic *Tamarix* density across the 11 riparian zones sampled [13]. This study developed on these findings to establish the knowledge gaps on how biodiversity, with regards to species richness, diversity, and evenness, was impacted, and which parts of the riparian zone were most vulnerable to invasion by exotic *Tamarix*. It became evident in this study that the Swart and Olifants rivers had lower species richness, diversity, and evenness than the Leeu River, which consistently had the highest biodiversity indices (Figures 2–4). The findings observed at the Swart and Olifants riparian zones were similar to the trends observed in studies conducted in the United States of America where biodiversity decreased in areas where invasion by exotic *Tamarix* increased [26,37].

This study further highlights that Zone 1 across all the three sites had the lowest biodiversity. This is indicated by the gradual increase and peak, at Zone 3, for all biodiversity indices, except for the Fishers alpha index at the Swart River, which decreased from the waterpoint outward (Figures 2–4). The low biodiversity in Zone 1 is an indication of a high abundance of exotic *Tamarix* near the water source. These findings allude to the notion that exotic *Tamarix* is well adapted to displace native plants that naturally take advantage of the phreatic zone as an immediate water source. *Tamarix* shrubs and trees have roots that can extend into alluvial deposits, allowing them access into the groundwater; this further allows them to outcompete native shrubs and trees for water [38–40]. In addition, they form dense stands which are not only less penetrable to animals, but they outcompete native flora for light, this is similar to the observations made by Steenkamp and Chown [38] as well as van Klinken et al. [39] regarding the invasive *Prosopis* species.

The higher abundance of exotic *Tamarix* trees in Zone 1 can most likely result in narrower river banks which can alternatively contribute to the drying out of rivers; such consequences are even more detrimental for areas where the affected rivers are a direct source of water, e.g., the Leeu River feeds into the Gamka Dam, which provides water for the Beaufort West community of >51,000 people and their animals [40]. *Tamarix* species also shed leaves high in salt content, as they are able to accumulate and transport salt from the soil into their leaves [13,26]. The leaf litter acts as fire fuel or excrete salts that remain on the soil surface, inhibiting the germination and growth of native competing flora [41]. In addition, exotic *Tamarix* trees becoming denser towards the water source compete with trees whose growth relies on the catchment, this is a typical characteristic of *V. karroo*, *S. pendulina*, *T. usneoides*, *D. lycioides*, and *L. oxycarpum*, which were sampled at the various sites. We can, thus conclude that exotic *Tamarix* species are vigorous invaders and can act as modifying species in the ecosystems they occur in.

The Leeu River was the only site sampled where the native congeneric to the exotic *Tamarix* species, *Tamarix usneoides*, grew. *Tamarix usneoides* could be acting as a direct competitor to its exotic counterparts because they grow to above-ground heights and root lengths that are similar, and, therefore, they use the same water sources. Native congeneric species are also well adapted to the conditions imposed by their exotic counterparts, e.g., *T. usneoides* is a better salt accumulator than exotic *Tamarix* and their hybrids [36], and, therefore, an increase in soil salt concentrations through invasion does not impact the abundance of *T. usneoides*. It was also the only site were exotic *Tamarix* did not rank among the top three most abundant species. Instead, *Prosopis juliflora*, ranked as the third most abundant tree at the riparian zone. These findings suggest that invasion by exotic *Tamarix* species at a site where they are exposed to other competitors, ideally phreatophytes, halophytes, or xerophytes, have compromised competitive ability. Along the Leeu riparian zone, *Mesembryanthemum crystallinum* and *Lycium oxycarpum* ranked first and second, respectively. This was the only site where native shrub species were more abundant than exotic *Tamarix* trees (Figure 5).

The Swart riparian zone was not only invaded by exotic *Tamarix*, which ranked as the most abundant vascular plant but *Atriplex muelleri* an exotic shrub native to Australia, ranked as the third most abundant plant species (Figure 5a). This finding suggests that at riparian zones where exotic *Tamarix* is the most abundant species, other alien and potentially invasive woody species are most likely to establish. This is even more likely in arid areas like the Little Karroo, which are vulnerable to invasion by exotic plant species [42]. *Lampranthus uniflorus* (succulent shrub) was the only native and perennial species among the top three most abundant species. *Lampranthus uniflorus* is an indicator species for highly degraded habitats and arid environmental conditions [29]. The shrub also provides evidence for impacts of environmental change in an area, giving more information about the biotic or abiotic state of the environment and predicting how diverse other plants species, taxa and communities, are in an area. The high abundance of *Lampranthus uniflorus* is most likely influenced by changed soil characteristics that are a result of exotic *Tamarix* invasion, which is known to increase soil salt levels [43]. This plant is not only alerting of potentially irreparable damage at this site, but it provides a red flag for an urgent need of biodiversity conservation in terms of species richness, diversity, and evenness as well as land rehabilitation.

The majority of the plants that were sampled where exotic *Tamarix* was the most abundant species were indigenous to South Africa and common in the Little Karroo, Western Cape. A matter of concern is that at these sites, the native vascular flora that ranks among the top three most abundant species, particularly along the Olifants riparian zone (Figure 5b), are classified as herbs (Table 1). Both *Oncosiphon piluliferum* and a species from the Brassiaceae have annual life cycles, and, therefore, are at risk of extirpation should natural disasters (e.g., floods and severe droughts) affect environmental and ecological dynamics that allow for the native plants to thrive along the riparian zone. They are both most likely only planted as ornamentals, and, therefore, contribute very little towards the ecosystem functions and services that occur along the river.

Three grass species were identified at the Leeu River; *Cenchrus* sp. (Subfamily: Panicoideae) is a flammable grass genus, and species in the genus are well adapted to dry area conditions and fire alike [44,45]. *Cynodon dactylon* (Subfamily: Cynodonteae), on the other hand, grows well and its seedlings can re-establish in areas prone to seasonal flooding as observed in the Fafan valley of the Jijinga rangelands, Ethiopia [46]. *Stipagrostis namaquensis* (Subfamily: Aristideae) is a typical Nama Karoo riparian grass, which is able to survive maximum disturbance with unpredictable flooding episodes [47].

Grasses adapted to floods, fire, warm, and dry weather conditions dominate areas that have been invaded by exotic *Tamarix*. Along the Olifants River, in addition to the three species described above, *Enneapogon desvauxii* (Subfamily: Pappophoreae) and *Hordeum murinum* (Subfamily: Triticeae) were also sampled. Both grow well in arid environments that were highly degraded [46,48]. However, should disturbance continue for long periods with increasing severity, the grass populations are prone to decline rapidly. This will have direct implications on biodiversity loss, and, therefore, the loss of ecosystem functions and services provided by the grass populations in the study sites.

Indigenous species that were sampled across the three sites were mostly small succulent shrubs or herbs with, generally, lower water consumption than the exotic *Tamarix*. The large native trees, which occurred more often in Zone 2 and Zone 3 of the Leeu riparian zone (i.e., *T. usneoides*, *L. oxycarpum* and *L. hirsutum*) most likely had resource acquisition strategies, which allowed them to compete with exotic *Tamarix*. These strategies could range from occupying a different niche from the exotic *Tamarix* or capitalising on their tolerance to arid and saline conditions through morphological, physiological, or biochemical adaptations.

Morphological adaptions include having deep roots that allow the plants to penetrate deep water sources or having thorns that limit the extent of herbivory, which, if extensive, could result in plant population decline [49,50]. Physiological adaptations include modifications of plants to their C_4/C_3 metabolism pathways of carbon fixation or leaf size and water use efficiency [51,52]. Biochemical adaptations are facilitated by multiple biochemical pathways that facilitate water retention and acquisition, while also protecting chloroplast functions and maintaining ion homeostasis [53].

Unlike the Succulent Karoo, which boasts a wide variety of succulent flora that provide social benefits to the surrounding communities [54], the Little Karoo flora leans more towards provisioning and regulating ecosystem services [55]. This is evident in the host of indigenous trees and shrubs that grew along the riparian areas of the latter region, e.g., *Diospyros lycioides* is a popular forage plant, especially in the Karoo, where forage is scarce, particularly during the dry seasons or extreme drought periods, which the plant is well adapted to [56].

The economic and ecological resources that are provided by the various plants that grew along the sampled riparian zones vary greatly between the different plants (Table 4). Some of the economic resources include grazing, medicinal value, aromatic source, edible food (for humans or animals), fuelwood, fencing, ornamental value, making handicrafts, providing shelter, rope making, and thatching. Ecological resources include sand stabilization, refuge provision, shading, soil fertility, salt tolerance, windbreak, waterpoint stabilisation, and other resources such as bioremediation and pollen supply [57].

Table 4. Summary of the economic and ecological resources provided by some of the trees, shrubs, and herbs sampled along the Swart, Olifants, and Leeu riparian zones [57]. Economic sources include: Grazing (Gr), Medicinal (Md), Aromatic source (Ar), Edible food (Ed), Fuelwood (Fu), Fencing+ Windbreak (Fe), Ornamental (Or), Making handicrafts (Han), and other uses, such as shelter, rope making, and thatching (Ot). Ecological resources include Sand stabilization (Sf), Refuge (Re), Shading (Sh), Soil fertility (Sr), Salt tolerance (St), Wind break (Wb), Weed (We), Waterpoint stabilisation (Rs), and other resources such as bioremediation and pollen supply (Ot).

Plant	Economic Resources	Ecological Resources
Atriplex semibaccata	Gr, Md, Ed	Sf, We
Chenopodium murale	Md, Ed, Ar	Sf
Cynodon dactylon	Gr, Md, Fu	Sf, Re, Sh
Hordeum murinum	Gr, Md	Sf
Juncus kraussii	Gr, Han, Ot	Sf, Re, St, Wb, Ot
Lycium hirsutum	Gr, Md, Ed, Fu, Fe, Ot	Sf, Re, Sh, Sr, Wb, Rs
Lycium oxycarpum	Gr, Md, Ed, Fu, Fe, Ot	Sf, Re, Sh, Sr, Wb, Rs
Medicago sativa	Gr	Sr
Salsola barbata	Gr, Md, Ed, Fu	Ot
Senecio burchellii	Gr, Md, Ed	Sf, We
Sisymbrium sp.	Gr	-
Solanum tomentosum	Md	Sf
Stipagrostis namaquensis	Gr, Md	Sf, Sr
Tamarix usneoides	Gr, Md, Fu, Or, Ot	Sf, Re, Sh, Wb, Rs
Zygophyllum retrofractum	Md, Ot	Sf, St, Wb, Rs

Given the various uses of the flora growing across the riparian zones, losing elements of biodiversity as a result of displacement by the invasion of exotic *Tamarix* species negatively influences the maintenance and resilience of invaded riparian zones. Several plants have been identified as key sources of pollen. The Little Karroo does not boast a wide range of flowering plants, and, therefore, pollen sources are usually scarce yet valuable [55]. This has direct implications on industries that rely on pollinators, such as agriculture and the textile industry. Pollination is a vital process for seed and fruit production and the overall plant life cycle [58]. The site sampled along the Olifants River is adjacent to a farm. Studies conducted by Kremen et al. [59] and Klein et al. [58] showed that farms next to natural habitats attracted more pollinators, and therefore, resulted in higher crop yield. For areas like the Little and Central Karoo, where an arable land is a form of subsistence and gaining revenue, this will have negative impacts on the livelihood of inhabitants in these communities. A loss in biodiversity will have direct implications for food security.

Rutherford and Powrie [29] suggested that severe degradation of land, due to heavy grazing, does, in fact, result in the decrease of species but can ultimately increase the diversity and evenness of indigenous plant species. This was especially true for sites where land was vacant and accessible to plant species such as *Oncosiphon piluliferum*, *Drosanthemum framesii*, and *Galenia sarcophylla*, which are well adapted to overgrazed land [29]. This observation would not be true for sites that have been degraded by IAPs, like exotic *Tamarix*, as they reduce biodiversity by occupying and modifying habitats and ecosystems, making them inhabitable for the bulk of indigenous species. This outcome has become evident in this study, particularly along the Swart and Olifants riparian zones, where biodiversity at both sites decreased and was lower in comparison to the Leeu riparian zone.

Prioritising the rapid removal of exotic *Tamarix* species will gradually allow for the restoration of ecosystem services and functions that are being rapidly lost as the exotic *Tamarix* extends across and along the riparian zones. Not only will ecosystem disturbances become more frequent, it will become easier for other exotic species to establish as they easily take advantage of the degraded and disturbed sites. This is common of smaller herbs and weed plants (e.g., *Argemone* sp.) and other larger woody plant species, which have commonly established easily in degraded sites (e.g., *Prosopis* sp.) [60].

Dwarf succulent shrubs are characteristic of and endemic to the Little and Central Karoo woody plant flora [55]. Indigenous larger trees introduced into riparian zones to stabilise waterpoints amongst other hosts of functions do not typically encroach ranges beyond their introduced ranges. The limited range of the native larger trees grants exotic *Tamarix* species an opportunity to expand into ranges beyond the riparian zone as they are not limited to just the riparian zones; this is due to the plants not being obligate phreatophytes [61,62]. In the long-term, other ecosystems (e.g., wetlands) will most likely become more vulnerable to invasion, and the extirpation, extinction, or a marked local decline of plant populations of other indigenous and endemic vascular plant species will become even more evident.

5. Conclusions

The introduction of exotic *Tamarix* species into South Africa has caused an imbalance in normal ecosystem functions and services. This study further highlighted the importance of thorough consideration of plants selected for phytoremediation purposes. Plants that are not native to novel sites pose a higher threat to biodiversity, while native (hyper) accumulators do not affect ecosystem functions and services negatively. Studies have shown that native congeneric species (e.g., *T. usneoides*) are usually better options for phytoremediation practises, and, therefore, their use should always be prioritised. *Vachellia karroo*, *Lycium oxycarpum*, and *Lampranthus uniflorus*, *Cenchrus* sp., *Cynodon dactylon*, *Stipagrostis namaquensis*, *Enneapogon desvauxii*, and *Hordeum murinum* have been identified as native vascular plants that can be re-planted on previously invaded land since they have been identified as resilient and suited to saline and arid environmental conditions while providing essential roles to the environment.

Successful revegetation projects can only be achieved with constant and consistent post-removal assessments, ensuring that native vascular plant species are establishing and performing the identified ecosystem services and functions. It is the role and responsibility of policy-advisers and policymakers to take into consideration research outputs that serve to conserve biodiversity, and it is the collective role of society at large to call out practises that result in biodiversity loss, especially plant biodiversity.

Author Contributions: Conceptualization, K.T.A.S. and S.W.N.; Data curation, K.T.A.S. and S.W.N.; Formal analysis, K.T.A.S.; Funding acquisition, S.W.N.; Investigation, K.T.A.S.; Methodology, K.T.A.S. and S.W.N.; Supervision, S.W.N.; Writing—original draft, K.T.A.S.; Writing—review & editing, K.T.A.S. and S.W.N. All authors have read and agreed to the published version of the manuscript.

Funding: This research was funded by National Research Foundation (NRF), grant number 114345.

Acknowledgments: The Agricultural Research Council (ARC) and the Postgraduate Merit Award (PMA) at the University of the Witwatersrand for the financial support as well as Prof. Sue Milton for plant identification.

Conflicts of Interest: The authors declare no conflict of interest. The funders had no role in the design of the study; in the collection, analyses, or interpretation of data; in the writing of the manuscript, or in the decision to publish the results.

References

1. Pamela, S.E. *The Root Causes of Biodiversity Loss*; Earthscan: London, UK, 2000.
2. Torrance, A.W. Patent Law, HIPPO, and the Biodiversity Crisis. *John Marshall Rev. Intell. Prop. Law* **2010**, *9*, 1.
3. Viciani, D.; Vidali, M.; Gigante, D.; Bolpagni, R.; Villani, M.; Acosta, A.T.R.; Adorni, M.; Aleffi, M.; Allegrezza, M.; Angiolini, C.; et al. A first checklist of the alien-dominated vegetation in Italy. *Plant Sociol.* **2020**, *57*, 29–54. [CrossRef]
4. Wilson, E.O. *The Future of Life*; Vintage Books: New York, NY, USA, 2002.
5. Intergovernmental Panel on Climate Change (IPCC). *Climate Change 2014—Impacts, Adaptation and Vulnerability: Regional Aspects*; Cambridge University Press: Cambridge, UK, 2014.
6. Lohbeck, M.; Bongers, F.; Martínez-Ramos, M.; Poorter, L. The importance of biodiversity and dominance for multiple ecosystem functions in a human-modified tropical landscape. *Ecology* **2016**, *97*, 2772–2779. [CrossRef]
7. Cardinale, B.J.; Duffy, J.E.; Gonzalez, A.; Hooper, D.U.; Perrings, C.; Venail, P.; Narwani, A.; Mace, G.; Tilman, D.; Wardle, D.A.; et al. Biodiversity loss and its impact on humanity. *Nature* **2012**, *486*, 59–67. [CrossRef] [PubMed]
8. Giam, X.; Bradshaw, C.J.; Tan, H.T.; Sodhi, N.S. Future habitat loss and the conservation of plant biodiversity. *Biol. Conserv.* **2010**, *143*, 1594–1602. [CrossRef]
9. Hultine, K.R.; Bean, D.W.; Dudley, T.L.; Gehring, C.A. Species Introductions and Their Cascading Impacts on Biotic Interactions in desert riparian ecosystems. *Integr. Comp. Biol.* **2015**, *55*, 587–601. [CrossRef]
10. Henderson, L. Invasive alien woody plants of the southern and southwestern Cape region, South Africa. *Bothalia* **1998**, *28*, 91–112. [CrossRef]
11. Foxcroft, L.; Henderson, L.; Nichols, G.; Martin, B. A revised list of alien plants for the Kruger National Park. *Koedoe* **2003**, *46*, 21–44. [CrossRef]
12. Holmes, P.M.; Richardson, D.M.; Esler, K.J.; Witkowski, E.T.F.; Fourie, S. A decision-making framework for restoring riparian zones degraded by invasive alien plants in South Africa. *S. Afr. J. Sci.* **2005**, *101*, 553–564.
13. Newete, S.W.; Mayonde, S.; Byrne, M.J. Distribution and abundance of invasive Tamarix genotypes in South Africa. *Weed Res.* **2019**, *59*, 191–200. [CrossRef]
14. Witkowski, E.T.F. Effects of Invasive Alien Acacias on Nutrient Cycling in the Coastal Lowlands of the Cape Fynbos. *J. Appl. Ecol.* **1991**, *28*, 1. [CrossRef]
15. Pyšek, P.; Richardson, D.M. Traits Associated with Invasiveness in Alien Plants: Where Do we Stand? *Biol. Invasions* **2008**, *193*, 97–125.
16. Downey, P.O.; Richardson, D.M. Alien plant invasions and native plant extinctions: A six-threshold framework. *AoB Plants* **2016**, *8*. [CrossRef] [PubMed]
17. Decamps, H.; Pinay, G.; Naiman, R.J.; Petts, G.E.; McClain, M.E.; Hillbricht-Ilkowska, A.; Hanley, T.A.; Holmes, R.M.; Quinn, J.; Gibert, J.; et al. Riparian zones: Where biogeochemistry meets biodiversity in management practice. *Pol. J. Ecol.* **2004**, *52*, 3–18.
18. Van Wilgen, B.W.; Richardson, D.M.; Le Maitre, D.C.; Marais, C.; Magadlela, D. The economic consequences of alien plant invasions: Examples of impacts and approaches to sustainable management in South Africa. *Environ. Dev. Sustain.* **2001**, *3*, 145–168. [CrossRef]
19. Marlin, D.; Newete, S.W.; Mayonde, S.; Smit, E.R.; Byrne, M.J. Invasive *Tamarix* (Tamaricaceae) in South Africa: Current research and the potential for biological control. *Biol. Invasions* **2017**, *19*, 2971–2992. [CrossRef]
20. Newete, S.W.; Abd Elbasit, M.A.; Araya, T.W. Soil salinity and moisture content under non-native Tamarix species. *Int. J. Phytoremediat.* **2020**, *22*, 1–8. [CrossRef]
21. Lowe, S.; Browne, M.; Boudjelas, S.; De Poorter, M. *100 of the World's Worst Invasive Alien Species: A Selection From the Global Invasive Species Database, 12*; Invasive Species Specialist Group: Auckland, New Zealand, 2000.

22. Mayonde, S.; Cron, G.; Gaskin, J.; Byrne, M.J. Evidence of *Tamarix* hybrids in South Africa, as inferred by nuclear ITS and plastid trnS-trnG DNA sequences. *S. Afr. J. Bot.* **2015**, *96*, 122–131. [CrossRef]
23. Crins, W.J. The Tamaricaceae in the southeastern United States. *J. Arnold Arboretum.* **1989**, *70*, 403–425. [CrossRef]
24. Gaskin, J.F.; Schaal, B.A. Hybrid *Tamarix* widespread in U.S. invasion and undetected in native Asian range. *Proc. Natl. Acad. Sci. USA* **2002**, *99*, 11256–11259. [CrossRef]
25. Mayonde, S.; Cron, G.V.; Gaskin, J.F.; Byrne, M.J. *Tamarix* (Tamaricaceae) hybrids: The dominant invasive genotype in southern Africa. *Biol. Invasions* **2016**, *18*, 3575–3594. [CrossRef]
26. Di Tomaso, J.M. Impact, biology, and ecology of saltcedar (*Tamarix* spp.) in the southwestern United States. *Weed Technol.* **1998**, *12*, 326–336. [CrossRef]
27. Roux, P.W.; Vorster, M.; Zeeman, P.J.L.; Wentzel, D. Stock production in the Karoo region. *Proc. Annu. Congr. Grassl. Soc. S. Afr.* **1981**, *16*, 29–35. [CrossRef]
28. Gallo, J.A.; Pasquini, L.; Reyers, B.; Cowling, R.M. The role of private conservation areas in biodiversity representation and target achievement within the Little Karoo region, South Africa. *Biol. Conserv.* **2009**, *142*, 446–454. [CrossRef]
29. Rutherford, M.C.; Powrie, L.W. Severely degraded rangeland: Implications for plant diversity from a case study in Succulent Karoo, South Africa. *J. Arid. Environ.* **2010**, *74*, 692–701. [CrossRef]
30. Martindale, G.J. Influence of Livestock Grazing on Plant Diversity of Highland Sourveld Grassland in KwaZulu-Natal. Master's Thesis, University of the Witwatersrand, Johannesburg, South Africa, 2007.
31. Germishuizen, G.; Meyer, N.L. *Plants of Southern Africa: An Annotated Checklist*; Strelitzia 14; National Botanical Institute: Pretoria, South Africa, 2003.
32. African Plant Database (version 3.4.0). Conservatoire et Jardin Botaniques (CJB) de la Ville de Genève and South African National Biodiversity Institute, Pretoria. Available online: https://www.ville-ge.ch/musinfo/bd/cjb/africa/recherche.php (accessed on 8 June 2019).
33. Colwell, R.K. EstimateS: Statistical Estimation of Species Richness and Shared Species from Samples. Version 6, User's Guide and Application. Available online: http://viceroy.eeb.uconn.edu/estimates (accessed on 8 June 2019).
34. Department of Agriculture, Forestry and Fisheries (DAFF). *Conservation of Agricultural Resources Act*; Act No. 43 of 1983 (CARA); Department of Agriculture, Forestry and Fisheries: Pretoria, South Africa, 1983.
35. Department of Environmental Affairs (DEA). *National Environmental Management: Biodiversity Act*; Act No. 10 of 2004 (NEM: BA); Department of Environmental Affairs: Pretoria, South Africa, 2004.
36. Newete, S.W.; Allem, S.M.; Venter, N.; Byrne, M.J. *Tamarix* efficiency in salt excretion and physiological tolerance to salt-induced stress in South Africa. *Int. J. Phytoremediat.* **2019**, *22*, 3–9. [CrossRef] [PubMed]
37. Zavaleta, E. Valuing ecosystem services lost to *Tamarix* invasion in the United States. In *Invasive Species in a Changing World*; Island Press: Washington, DC, USA, 2000; pp. 261–300.
38. Steenkamp, H.; Chown, S. Influence of dense stands of an exotic tree, Prosopis glandulosa Benson, on a savanna dung beetle (Coleoptera: Scarabaeinae) assemblage in southern Africa. *Biol. Conserv.* **1996**, *78*, 305–311. [CrossRef]
39. Van Klinken, R.D.; Graham, J.; Flack, L.K. Population Ecology of Hybrid Mesquite (Prosopis Species) in Western Australia: How Does it Differ from Native Range Invasions and What are the Implications for Impacts and Management? *Biol. Invasions* **2006**, *8*, 727–741. [CrossRef]
40. Laldaparsad, S. A geo-referenced census frame of dwellings for the 2011 census of the Republic of South Africa. In Proceedings of the Innovative Methodologies for Censuses in the New Millennium, a Satellite Meeting of the 56 Session of the ISI, Lisbon, Portugal, 31 August–2 September 2007.
41. Busch, D.; Smith, S. Effects of fire on water and salinity relations of riparian woody taxa. *Oecologia* **1993**, *94*, 186–194. [CrossRef]
42. Milton, S.J.; Dean, W.R.J. Plant invasions in arid areas: Special problems and solutions: A South African perspective. *Biol. Invasions* **2010**, *12*, 3935–3948. [CrossRef]
43. Erfanifard, Y.; Khosravi, E. Saltcedar (Tamarix mascatensis) inhibits growth and spatial distribution of eshnan (Seidlitzia rosmarinus) by enrichment of soil salinity in a semi-arid desert. *Plant Soil* **2019**, *440*, 219–231. [CrossRef]

44. Clayton, W.D.; Vorontsova, M.S.; Harman, K.T.; Williamson, H. World Grass Species: Descriptions, Identification, and Information Retrieval. Available online: http://www.kew.org/data/grasses-db.html (accessed on 8 November 2019).
45. Ripley, B.; Visser, V.; Christin, P.A.; Archibald, S.; Martin, T.; Osborne, C. Fire ecology of C3 and C4 grasses depends on evolutionary history and frequency of burning but not photosynthetic type. *Ecology* **2015**, *96*, 2679–2691. [CrossRef] [PubMed]
46. Le Houérou, H.N.; Corra, M. *Some Browse Plants of Ethiopia*; Le Houerou, H.N., Ed.; Browse in Africa. ILCA: Addis Ababa, Ethiopia, 1980; pp. 109–114.
47. Rubin, F.; Palmer, A.R. The physical environment and major plant communities of the Karoo National Park, South Africa. *Koedoe* **1996**, *39*, 25–52. [CrossRef]
48. Davison, A.W. The Ecology of Hordeum Murinum L.: III. Some Effects of Adverse Climate. *J. Ecol.* **1977**, *65*, 523. [CrossRef]
49. Sharif, F.; Khan, A.U. Effect of salinity on tissue nutrient contents of the four dryland tree species of Indus flood plains. *Arid. Land Res. Manag.* **2016**, *30*, 65–78. [CrossRef]
50. El-Keblawy, A.; Abdelfattah, M.A.; Khedr, A.-H. Relationships between landforms, soil characteristics and dominant xerophytes in the hyper-arid northern United Arab Emirates. *J. Arid. Environ.* **2015**, *117*, 28–36. [CrossRef]
51. Mulroy, T.W.; Rundel, P.W. Annual Plants: Adaptations to Desert Environments. *BioScience* **1977**, *27*, 109–114. [CrossRef]
52. Dudley, S. The response to differing delection on plant physiological traits: Evidence for local adaptation. *Evolution* **1996**, *50*, 103. [CrossRef]
53. Muscolo, A.; Sidari, M.; Panuccio, M.R.; Santonoceto, C.; Orsini, F.; De Pascale, S. Plant responses in saline and arid environments: An overview. *Eur. J. Plant Sci. Biotechnol.* **2011**, *5*, 1–11.
54. Lombard, A.; Hilton-Taylor, C.; Rebelo, A.; Pressey, R.L.; Cowling, R.M. Reserve selection in the Succulent Karoo, South Africa: Coping with high compositional turnover. *Plant Ecol.* **1999**, *142*, 35–55. [CrossRef]
55. Vlok, J.; Vlok, A. *Plants of the Klein Karoo*; Umdaus Press: Hatfield, UK, 2015.
56. Dziba, L.; Scogings, P.; Gordon, I.; Raats, J. Effects of season and breed on browse species intake rates and diet selection by goats in the False Thornveld of the Eastern Cape, South Africa. *Small Rumin. Res.* **2003**, *47*, 17–30. [CrossRef]
57. Bidak, L.M.; Kamal, S.A.; Halmy, M.W.A.; Heneidy, S.Z. Goods and services provided by native plants in desert ecosystems: Examples from the northwestern coastal desert of Egypt. *Glob. Ecol. Conserv.* **2015**, *3*, 433–447. [CrossRef]
58. Klein, A.; Vaissière, B.E.; Cane, J.H.; Steffan-Dewenter, I.; Cunningham, S.A.; Kremen, C.; Tscharntke, T. Importance of pollinators in changing landscapes for world crops. *Proc. R. Soc. B Biol. Sci.* **2006**, *274*, 303–313. [CrossRef] [PubMed]
59. Kremen, C.; Williams, N.M.; Thorp, R.W. Crop pollination from native bees at risk from agricultural intensification. *Proc. Natl. Acad. Sci. USA* **2002**, *99*, 16812–16816. [CrossRef] [PubMed]
60. Goel, V.L.; Behl, H.M. Screening of *Prosopis germplasm* for afforestation of degraded soil sites: Performance, leaf nutrient status and influence on soil properties. *J. Sustain. For.* **1998**, *8*, 1–13. [CrossRef]
61. Yuan, G.-F.; Zhang, P.; Shao, M.-A.; Luo, Y.; Zhu, X. Energy and water exchanges over a riparian Tamarix spp. stand in the lower Tarim River basin under a hyper-arid climate. *Agric. For. Meteorol.* **2014**, *194*, 144–154. [CrossRef]
62. Imada, S.; Matsuo, N.; Acharya, K.; Yamanaka, N. Effects of salinity on fine root distribution and whole plant biomass of *Tamarix ramosissima* cuttings. *J. Arid. Environ.* **2015**, *114*, 84–90. [CrossRef]

Article

Biological and Agrotechnical Aspects of Weed Control in the Cultivation of Early Potato Cultivars under Cover

Piotr Pszczółkowski [1,*], Piotr Barbaś [2], Barbara Sawicka [3] and Barbara Krochmal-Marczak [4]

1 Experimental Station for Cultivar Assessment of Central Crop Research Centre, Uhnin, 21-211 Dębowa Kłoda, Poland
2 Department of Potato Agronomy, IHAR-PIB, Branch in Jadwisin, Szaniawskiego 15, Str., 05-140 Serock, Poland; p.barbas@ihar.edu.pl
3 Department of Plant Production Technology and Commodity Science, University of Life Sciences in Lublin Akademicka 15, str., 20-950 Lublin, Poland; barbara.sawicka@up.lublin.pl
4 Department of Plant Production and Food Safety, Carpathian State University in Krosno, Dmochowskiego, 12, 38-400 Krosno, Poland; barbara.marczak@kpu.krosno.pl
* Correspondence: p.pszczolkowski.inspektor@coboru.pl

Received: 30 July 2020; Accepted: 19 August 2020; Published: 21 August 2020

Abstract: Problems with weed infestation under cover were the reason to conduct research on the regulation of weed infestation in potato cultivation for early harvest. The field experiment was carried out in 2015–2017 at the Experimental Station for Cultivar Assessment in Uhnin (51°34′ N, 23°02′ E) using the method of random subblocks, in a dependent system (split-split-plot). The first order factor was edible potato cultivars 'Denar' and 'Lord'. The second order factor was cultivation technologies: (A) traditional technology, (B) technology using polyethylene film cover, (C) technology using polypropylene agrotextile. The third order factor was weed management methods: (1) mechanical, (2) mechanical and chemical method using Afalon Dispersion 450 SC preparation, (3) mechanical and chemical methods using Racer 250 EC herbicide, and (4) mechanical and chemical methods using a mixture of herbicides Afalon Dispersion 450 SC and Command 480 EC. Mechanical and chemical methods proved to be more effective than the mechanical method. The best effectiveness in limiting both fresh and dry weed mass in potato cultivation under cover was achieved using the mechanical and chemical method using a mixture of herbicides, Afalon Dispersion 450 SC and Command 480 EC.

Keywords: early potato production; polypropylene agrotextile; polyethylene sheeting; herbicides; mechanical weed control; chemical weed control; potato cultivars; weed infestation

---

1. Introduction

The cultivation of very early potato cultivars under cover for a very early harvest requires the plants to be provided with appropriate thermal and humidity conditions due to the risk of spring frosts. The unfavorable influence of low temperatures in the initial period of potato growth increases the effectiveness of using foil and agrotextile covers [1–3]. Temperature of the soil under the agrotextile at a depth of 5 cm is 1–2 °C and at a depth of 10 cm it is 2–3 °C higher than that of uncoated soil [4]. Covers are placed flat, immediately after planting, and it is a procedure that accelerates the harvest of tubers and reduces the risk of plant freezing. The most popular covers used in early potato plantations include perforated foil, polypropylene non-woven fabric, and a biofilm that decomposes in contact with the soil [2–4].

Weed infestation is one of the factors determining the potato yield. It is favored by a wide row spacing, a long period from planting to plant emergence, slow initial growth of potato plants, introducing manure combined with intensive mineral fertilization [5–8]. In an integrated cultivation system, chemical weed control is the basic treatment that eliminates segetal vegetation from the field of a cultivated plant [8,9].

Potato yield losses due to weed infestation in Europe are estimated at 10% to 70%. Damage resulting from limiting access to light, water, and nutrients should be added to them, as well as the host role of weeds relating to diseases and pests, harvest difficulties, increase in mechanical damage to tubers, and deterioration of their quality [10–12]. Scalla [13], Ackley et al. [14], Azadbakht et al. [10] proved that during the action of herbicides on weeds, a selection process takes place, due to which the plants that are the least sensitive to the preparation survive, and each weed population is more or less heterogeneous, and the resistance production process may vary in intensity and speed. Physiological and biochemical studies indicate differences between herbicide-resistant and herbicide-sensitive weeds, which are based on the rate of herbicide uptake and decomposition in tissues and different distribution of the root system in the soil (selectivity of uptake) [13,15,16]. Their distribution in cells and tissues also plays an important role. In plants resistant to a given herbicide, they are not displaced, or they migrate in small amounts to the tips of growth, which are very sensitive to toxic substances. The dominant role in the selectivity of action of these compounds is played by the plant's ability to detoxify them [15–17]. In the opinion of Hess and Foy [18] and Starck [16], the phenomenon of weed resistance to herbicides will increase. It is already much more serious than it is supposed, and the survival of weeds in plantations after treatment is most often attributed to inaccurate treatment or low-quality preparation.

Chemical control of weeds should be carried out in their early development stages [7,9,10]. Barbaś [5] and Nowacki et al. [9] recommend three dates of herbicide application: after potato planting (up to 10 days), a few days before the expected emergence of the plants, and after emergence, when the plants develop 2–4 leaves and reach a height of 10–15 cm. It is difficult to implement these recommendations in the cultivation of plants under cover, so the problem of controlling segetal vegetation in this cultivation technology still awaits a solution. Hence, the aim of the research was to develop a technology of growing very early potato cultivars under cover put on "flat" with the use of herbicides or herbicide mixtures, compared to the mechanical method, and the impact of this technology on the degree of weed infestation, species composition, and fresh and air-dried mass of weeds.

It was assumed in the research hypothesis that it is possible to cultivate very early potato cultivars with the use of polyethylene film covers and "flat" agrotextile with comprehensive weed control under cover, against the null hypothesis about the lack of such possibility.

2. Material and Methods

The field experiment was carried out in 2015–2017 at the Experimental Plant for Cultivar Assessment in Uhnin, Lubelskie Voivodeship (51°34′ N, 23°02′ E, H = 155 m.a.s.l.), belonging to the Central Research Center for Cultivar Testing. The experiment was carried out by the method of random sub-blocks, in a dependent design (split-split-plot), in three replications. The first order factor were potato cultivars, 'Denar' and 'Lord'. The second order factor were cultivation technologies: (A) traditional technology as a control object, (B) technology with the use of polyethylene film covers, and (C) technology with the use of agrotextile. Factors of the third order were potato weed management practices: (1) mechanical as a control, (2) mechanical and chemical maintenance with the use of Afalon Dispersive 450 SC in the amount of 2.0 $dm^{3}\cdot ha^{-1}$, (3) mechanical–chemical maintenance with the use of the herbicide Racer 250 EC at a dose of 2.0 $dm^{3}\cdot ha^{-1}$, (4) mechanical–chemical maintenance with the use of a mixture of Afalon Dispersive 450 SC and Command 480 EC herbicides in the amount of 1.0 $dm^{3}\cdot ha^{-1}$ and 0.15 $dm^{3}\cdot ha^{-1}$. The size of the plots to be harvested was 15 m^2.

2.1. Agrotechnical and Plant Protection Treatments

The potato forecrop was winter triticale. After harvesting the forecrop, stubble cultivation was performed. In the fall of each year before planting, winter plowing was carried out to a depth of about 27 cm. In spring, the field was harrowed, then NPK fertilizers were sown and mixed with the soil with a cultivating unit to a depth of 12 cm. Mineral fertilizers—potassium, phosphorus, and sulfur—were applied to the soil in the following amounts: 39.3 kg·P·ha^{-1}, 112.1 kg·K·ha^{-1}, and 15.8 kg·S·ha^{-1}. The amount of mineral fertilization was determined based on the fertility of the soil with these components. Nitrogen fertilizers were sown in spring in the amount of 90 kg·N·ha^{-1} in the form of polifoska (27 kg·ha^{-1}) and urea (63 kg·ha^{-1}). Potato propagating material (EU grade A) was planted annually in spring by hand at the end of April, at a spacing of 67.5 cm × 37 cm. The size of a single plot for harvest was 15 m^2. After the potato tubers had been planted manually, dredging was carefully performed, combined with light harrowing, to remove the top of the ridges, which is eroded by water and wind. Then, the herbicides were sprayed in the given doses, and the covers were applied. The protection of plants against diseases and pests was carried out in accordance with recommendations of the Institute of Plant Protection, National Research Institute, and the principles of Good Agricultural Practice [9,19]. During the growing season, spraying against alternariosis and potato blight was performed in accordance with the warnings of the Institute of Plant Protection, and the Colorado potato beetle was controlled using available preparations during its occurrence (Table 1).

Table 1. The agricultural treatments and plant protection products used in the experiment in the years 2014–2017.

	Autumn 2014–2016	
	Tillage	
	Winter plowing to a depth of about 27 cm	
	Herbicides for forecrop	
	Lentipur Flo 500 SC (chlorotoluron) – 1 dm^3·ha^{-1} (Autumn 2014) Snajper 600 SC (chlorotoluron + diflufenikan) – 1 dm^3·ha^{-1} (Autumn 2014) Glean 75 WG (chlorosulfuron) – 0.01 kg·ha^{-1} (Autumn 2014) Bizon (diflufenikan + florasulam + penoksulam) – 1 dm^3·ha^{-1} (Autumn 2015) Lentipur Flo 500 SC (chlorotoluron) – 1 dm^3·ha^{-1} (Autumn 2016) Snajper 600 SC (chlorotoluron + diflufenikan) – 1 dm^3·ha^{-1} (Autumn 2016) Glean 75 WG (chlorosulfuron) – 0.01 kg·ha^{-1} (Autumn 2016)	
Spring 2015	Spring 2016	Spring 2017
	Tillage and agricultural treatments	
Harrowing Fertilization of NPK Cultivation with an aggregate Planting of potato seeds—manually Earthing Mechanical weed control Harvest with potato elevator digger	Harrowing Fertilization of NPK Cultivation with an aggregate Planting of potato seeds—manually Earthing Mechanical weed control Harvest with potato elevator digger	Harrowing Fertilization of NPK Cultivation with an aggregate Planting of potato seeds—manually Earthing Mechanical weed control Harvest with potato elevator digger
	Fungicides—after removing the covers	
Infinito 867.5 SC (propamocarb hydrochloride + fluopicolide) – 1.6 dm^3·ha^{-1} Ridomil Gold MZ 67.8 (mancozeb + metalaxyl) – 2 kg·ha^{-1} Infinito 867.5 SC (propamocarb hydrochloride + fluopicolide) – 1.6 dm^3·ha^{-1}	Acrobat MZ 69 WG (mancozeb + dimethomorph) – 2 kg·ha^{-1} Infinito 867.5 SC (propamocarb hydrochloride + fluopicolide) – 1.6 dm^3·ha^{-1} Acrobat MZ 69 WG mancozeb + dimethomorph) – 2 kg·ha^{-1}	Acrobat MZ 69 WG mancozeb + dimethomorph) – 2 kg·ha^{-1} Infinito 867.5 SC (propamocarb hydrochloride + fluopicolide) – 1.6 dm·ha^{-1} Acrobat MZ 69 WG mancozeb + dimethomorph) – 2 kg·ha^{-1}
	Insecticides—after removing the covers	
Apacz 50 WG (clothianidin) – 0.04 kg·ha^{-1} Proteus OD 110 (thiacloprid + deltamethrin) – 0.4 dm^3·ha^{-1}	Actara 25 WG (thiamethoxam) – 0.08 kg·ha^{-1} Nuprid 200 SC (imidacloprid) – 0.15 dm^3·ha$^-$	Actara 25 WG (thiamethoxam) – 0.08 kg·ha^{-1} Apacz 50 WG (clothianidin) – 0.04 kg·ha^{-1}

Source: own research.

2.2. *Herbicide Active Substances*

Characteristics of herbicides and adjuvants are given in Table 2.

Table 2. Characteristics of the herbicides and herbicide additives used in the experiment.

Trade Names of the Preparation	Active Substances	Content of Active Substances	Recommended Application Rates per 1 ha	Utility Forms	Grace Period *
Afalon Dispersive 450 SC	Linuron	450 g·dm^{-3}	1.5–2.0·dm^{-3}	granules for water suspension	Not applicable
Racer 250 EC	Flurochloridone	250 g·dm^{-3}	2.0·dm^{-3}	concentrate for making a water emulsion	Not applicable
Command 480 EC	Clomazone	480 g·dm^{-3}	0.2·dm^{-3}	concentrate for making a water emulsion	Not applicable

Sources: Accinelli et al. [12], Caldiz et al. [14], Tomlin [13]; * The period from the last application of the product to the day the crop of potato is harvested.

Afalon Dispersive 450 SC belongs to urea herbicides with toxicity class III. Its biologically active substance is linuron—450 g in 1 kg of the preparation. Linuron [3-(3,4-dichlorocyl)-1-methylurea] is an inhibitor of photosynthesis and electron transport [20,21].

Racer 250 EC is an herbicide of toxicity class IV. The biologically active substance is 250 g of flurochloridone per 1 dm^3 of the preparation—[3-chloro-4-chloromethyl-1-(trifluoro-m-tollyl)-2-pyrrolidone. It is an inhibitor of carotenoid synthesis [19,21].

Command 480 EC is a preparation with the toxicity class III. Its biologically active substance is clomazone—480 g in 1 L of the preparation. Clomazone is also known as oxazolide [2-(2-chlorobenzyl)-4,4-dimethyl-oxazolidine-5-one]. It can be mixed with other herbicides, e.g., with Afalon Dispersive (Table 2) [21,22].

2.3. Soil Sampling and Evaluation

Before establishing the experiment, 20 primary soil samples were collected each year, constituting one aggregate sample weighing of about 0.5 kg [23]. The samples collected in this way were used to determine the basic physicochemical and chemical properties of the soil. Soil samples were taken from the humus level (0–25 cm), after harvesting the crop, from 20 randomly selected places [24].

Chemical and physicochemical properties of the soil were determined in a certified laboratory of the District Chemical and Agricultural Station in Wesoła near Warsaw by the following methods: soil granulometric composition was determined by laser diffraction Ryzak et al. [25,26]; pH, in a suspension of 1 mol·KCl·dm^{-3} and in H_2O suspension by potentiometric method [27]; organic carbon content Corg., the Tiurin method [26]; available magnesium content by the Schachtschabel method [28], and content of available forms of phosphorus and potassium (by the Egner-Riehm method) [29,30]. Cu, Mn, Zn, Fe contents in 1 mole HCl were confirmed by the AAS method (Atomic Absorption Spectrometry) [31–33]; B was determined spectrophotometrically with curcumin [34]. The results of soil analysis were valued according to Mocek and Drzymała [24].

2.4. Soil Analysis

The experiment was carried out on fallow soil with a sandy loam grain composition WRB [35]. In terms of the percentage of sand, silt, and clay fractions, it is a granulometric subgroup—clay sand (light soil). The sand fraction was 67.0%, the dust fraction was 30.6%, and the clay was 2.4%. This proportion of individual fractions corresponds to the composition of clay dust. In terms of agricultural suitability, these soils belong to the weakly acid good rye complex. This soil is classified in the agronomic category as light mineral [36]. The soil was slightly acidic, low humus content, high to very high phosphorus and magnesium contents, and low to medium potassium content. The soil was characterized by an average content of manganese and iron, medium to high content of copper, high content of zinc, and high content of boron [24,36] (Table 3).

Table 3. Physico-chemical characteristics of the soil before establishing the experiment in 2015–2017.

Year of Research	Content Macronutrients [mg·100 g^{-1} soil]			Humus Content [g·kg^{-1}]	pH [KCL]	Micronutrients Content [mg·kg^{-1} soil]				
	P	K	Mg			Cu	Mn	Zn	Fe	B
2015	8.9	10.9	7.8	0.94	5.9	7.51	318	40.1	3760	7.24
2016	8.3	9.1	7.0	1.06	5.8	4.92	337	56.7	3925	5.28
2017	10.6	9.8	6.3	1.03	6.6	8.99	166	41.1	3600	6.04
Mean	9.3	9.9	7.0	1.02		7.02	274	46.0	3762	6.17

Source: own results made at the District Chemical and Agricultural Station in Lublin.

2.5. Assessment of Weed Infestation

To compare the effectiveness of potato weed management, the infestation was determined by the quadrat. The following were determined: fresh and air-dried mass of weeds, number of monocotyledonous and dicotyledonous weeds, and their floristic composition. Weed infestation analysis was performed at three times: before and after row closure and before potato plants ripened [19].

Before counting the weeds, a list of species was established and entered on the analysis sheet. The square was placed randomly in three places of the plot under evaluation. Weed plants were counted in a box covering an area of 0.5 m^2. The number of weeds present on an area of 1 m^2 was calculated according to the formula:

$$\mathrm{Lch} = \frac{(\mathrm{L1} + \mathrm{L2} + \mathrm{L3})}{(\mathrm{lp} * \mathrm{pr})} \tag{1}$$

where:

Lch—number of weeds of a given species growing on an area of 1 m^2,
L_1; L_2; L_3—number of plants in the frame in successive measurements (pcs.),
lp—number of measurements,
pr—frame area (m^2) [19].

The fresh weight of all weed species present on the experimental plot was determined. A floristic list of weeds was established, and their names were entered in the analysis sheet. Then samples of the fresh weight of weeds were taken from the area covered by a randomly placed throw square. Plants were taken out of the soil and placed in plastic containers with an inventory label attached. This operation was repeated three times in each plot. The weeds collected in the sample were sorted, grouping individual weed species separately; fresh weight was determined for each species and the weighing results were recorded in the evaluation sheet. The weights of all weed species were summed and the average of three replicates for the plot was calculated. All results were recorded in the worksheet.

The mean fresh weight of weeds was calculated according to the following formula:

$$\mathrm{Sm} = \frac{(\mathrm{m1} + \mathrm{m2} + \mathrm{m3})}{3} \tag{2}$$

where:

Sm—average weight of the weed species concerned (g),
m_1—weight of the weeds in the first throw of the square (g),
m_2—weight of weeds in the second square projection (g),
m_3—mass of weeds in the third square projection (g).

The determination of the air-dried mass of weeds consisted in manually picking them from the marked places and placing them in a ventilated room until a constant mass was obtained [19].

2.6. Meteorological Conditions

Meteorological conditions in the years of the study were varied. The highest amount of rainfall during the potato vegetation period was recorded in 2015, but their distribution was not favorable for the potato because during the intensive plant growth and accumulation of tuber yield, a significant shortage was observed in June–August. The Sielianinov hydrothermal coefficient defines the months of 2016 as: May, quite dry; June, optimal; July, quite humid; August, dry, and September, extremely dry. In the third year of the research, meteorological conditions were changing. Optimal water supply to potato plants was observed alternately, with a huge shortage of them in the subsequent month of vegetation. The Sielianinov hydrothermal coefficient defines the month of May as optimal, June as very dry, July as humid, August as very dry, and September as quite wet (Table 4).

$$k = \frac{10P}{\sum t} \tag{3}$$

Table 4. Meteorological conditions in the season of potato vegetation in 2015–2017.

Year	Month	Rainfall [mm]				Air Temperature [°C]				Hydrothermal Coefficient of Sielianinov *
		Decade			Month	Decade			Mean	
		1	2	3		1	2	3		
2015	April	14.6	5.9	41.3	61.8	5.4	8.6	12.4	8.8	2.3
	May	23.4	13.9	83.0	120.3	12.6	12.0	13.7	12.8	3.0
	June	5.4	16.5	24.8	46.7	17.7	16.3	16.1	16.7	0.9
	July	10.5	21.6	13.1	45.2	19.6	18.7	19.9	19.4	0.8
	August	0.4	0	5.7	6.1	23.4	20.6	20.3	21.4	0.1
	September	32.4	32.6	65.2	130.2	16.0	17.7	12.8	15.5	2.8
	Total				410.3					
2016	April	11.5	22.2	13.4	47.1	10.9	10.1	9.0	10.0	1.6
	May	4.9	2.8	38.6	46.3	14.4	17.8	12.9	15.3	1.0
	June	10.1	43.2	34.0	87.3	16.6	17.5	23.0	19.1	1.5
	July	22.4	30.8	60.9	114.1	19.5	20.1	21.9	20.5	1.8
	August	22.8	17.7	0.5	41.0	20.7	17.1	20.4	19.5	0.7
	September	7.6	0.1	4.1	11.8	19.5	15.5	11.5	15.5	0.3
	Total				347.6					
2017	April	6.4	7.2	38.2	51.8	10.6	6.8	6.9	8.1	2.1
	May	45.1	1.3	19.1	65.5	10.5	13.0	17.4	13.7	1.5
	June	1.9	9.2	12.0	23.1	16.6	17.7	20.7	18.3	0.4
	July	10.1	80.9	41.0	132.0	17.9	19.0	21.0	19.4	2.2
	August	0.4	24.4	2.2	27.0	22.8	21.3	17.1	20.3	0.4
	September	38.7	35.9	8.7	83.3	16.3	15.3	12.8	14.8	1.9
	Total				382.7					

Source: The meteorological station in Uhnin 2015–2017, * coefficient was calculated according to the formula:

Skowera et al. [37].

where P is the sum of the monthly precipitation in mm, Σt is monthly total air temperature > 0 °C. Ranges of values of this index were classified as follows: extremely dry, 0.0 ≤ k < 0.4; very dry, 0.7 ≤ k < 0.4; dry, 1.0 ≤ k < 0.7; rather dry, 1.3 ≤ k < 1.0; optimal, 1.6 ≤ k < 1.3; rather humid, 2.0 ≤ k < 1.6; wet, 2.5 ≤ k < 2.0; very humid, 3.0 ≤ k < 2.5; extremely humid, 3.0 > k.

2.7. Statistical Calculations

The obtained results were subjected to statistical analysis. The calculations were performed with the SAS/STAT 9.2 software [38]. They were based on a three-factor analysis of the variance model and multiple t-Tukey tests with the significance level $p = 0.05$. Multiple comparison tests allowed for detailed analysis of mean comparisons by distinguishing statistically homogeneous groups of mean (homogeneous groups) and determining so-called the smallest significant mean differences, which in Tukey's tests are denoted as HSD (Tukey's honest significant difference) [39].

3. Results

3.1. Number of Monocotyledonous Weeds

The average population density of monocotyledonous weeds in the potato field was 25.6 pcs. m^{-2}. The highest number of weeds of this class was observed before the rows were closed, and the least was before the tuber ripening. On the other hand, their number after row closure and before potato harvest did not differ significantly (Table 5).

Table 5. The impact of technology, methods of weed management, cultivars, and years of cultivation on the number of monocotyledonous weeds (pcs m^{-2}).

Factors of the Experiment		The Terms of Observation *			Mean
		I	II	III	
Technologies	Traditional	46.8	23.0	18.2	29.3
	PE-Sheeting	34.7	16.4	19.9	23.7
	PP-Sheeting	51.0	16.9	15.9	27.9
	$HSD_{0.05}$		10.3		ns **
Potato weed management practices	Mechanical	44.6	20.1	18.9	27.9
	Afalon	35.5	11.4	8.5	18.5
	Racer	43.0	30.5	29.4	34.3
	Afalon + Command	41.2	11.9	12.7	21.9
	$HSD_{0.05}$		12.5		5.7
Cultivars	'Denar'	39.9	14.4	15.9	23.4
	'Lord'	42.2	22.6	18.8	27.9
	$HSD_{0.05}$		ns		3.1
Years	2015	23.3	33.5	42.7	33.2
	2016	18.6	18.1	5.5	14.1
	2017	80.8	3.9	4.5	29.7
	$HSD_{0.05}$		10.3		5.1
Mean		41.1	18.5	17.4	25.6
$HSD_{0.05}$			5.1		

* I, before shorting rows; II, after closing the rows; III, before potato ripening; ** differences not significant at $p_{0.05}$ level.

Only the cultivation technology under a polyethylene film cover, compared to traditional cultivation, significantly reduced the number of weeds in this group, but only before the potato row closing. Weed control methods with Afalon Dispersive 450 SC and a mixture of herbicides Afalon Dispersive 450 SC and Command 480 EC significantly reduced the number of monocotyledonous weeds compared to mechanical control. The greatest decrease in the number of weeds was observed in mechanical and chemical methods using the herbicide Afalon Dispersive 450 S.C. Only potato methods with Racer 250 EC contributed to an increase in the number of monocotyledonous weeds compared to mechanical method. Such significant increase was observed only before the potato harvest. The potato weed control methods did not differentiate the population density of monocotyledonous weeds before the rows were closed, while after the rows closing and before the tuber ripening, development of this group of weeds was most limited by the treatment with Afalon Dispersive 450 S.C. The objects where 'Denar' cv. was grown were characterized by a smaller population density of monocotyledonous weeds than those with 'Lord' cv., which results from greater foliage and better shading of the soil by plants of the former cultivar. Regardless of the experimental factors, the randomness factor played an

3.4. Air-Dried Mass of Weeds

Phases of plant development were the factor determining the amount of air-dried mass of weeds in a potato field. The air-dried mass of weeds was the lowest before the rows were closed, and the highest was before potato harvesting (Table 8).

Table 8. Impact of cultivation technology, methods of weed management, cultivars, and years of cultivation on the air-dried mass of weeds (g. m^{-2}).

Factors of the Experiment		Observation Dates *			Mean
		I	II	III	
Technologies	Traditional	8	11	82	34
	PE-Sheeting	5	13	89	36
	PP-Sheeting	6	9	107	41
	$HSD_{0.05}$		19		ns **
Potato weed management practices	Mechanical	4	9	89	34
	Afalon	5	9	82	32
	Racer	7	9	112	43
	Afalon + Command	4	15	83	34
	$HSD_{0.05}$		19		9
Cultivars	'Denar'	5	10	92	36
	'Lord'	5	11	91	36
	$HSD_{0.05}$		ns		ns
Years	2015	2	11	101	38
	2016	1	2	31	11
	2017	12	18	143	58
	$HSD_{0.05}$		ns		7
Mean		6	10	92	36
$HSD_{0.05}$			7		

* I, before shorting rows; II, after closing the rows; III, before potato ripening; ** differences not significant at $p_{0.05}$ level.

Technologies of potato cultivation and varietal properties did not significantly differentiate this feature. However, a significant interaction of potato cultivation technology and dates of weed infestation observation was found. In the cultivation under cover of polypropylene agrotextile, a significant increase in the air-dried mass of weeds was observed, compared to the traditional technology, but only before the potato was mature. Among the methods of mechanical and chemical weed control used in the experiment, the use of the Racer 250 EC preparation contributed to a significant increase in the dry mass of weeds, compared to both the standard object and other methods of potato cultivation. The latter turned out to be homogeneous in terms of the value of this feature. However, this negative effect of method with Racer 250 EC was noticed only in the period before tuber ripening, i.e., regarding secondary weed infestation. Meteorological conditions during the growing season had a decisive influence on the value of this feature. The lowest air-dried mass of weeds was recorded in 2016, which was the most optimal, both in terms of rainfall and air temperature, and the highest value of this characteristic was recorded in 2017, when there were periodical excesses or deficiencies of rainfall during the potato growing season (Table 8).

3.5. Weed Species Composition

In the potato field, seven species of monocotyledonous weeds and 20 species of dicotyledonous weeds were recorded. Among monocotyledonous weeds, the most numerous species were: *Echinochloa crus-galli*, *Setaria glauca*, and *Setaria viridis*, and only occasionally, *Avena fatua* and *Poa annua*. The highest number of *Echinochloa crus-galli* was recorded before the row closing, and the lowest one was before the potato was ripened. In turn, the highest numbers of *Setaria viridis* and *Setaria glauca* were observed before the potato ripening, which was related to the development biology of these species. Cultivation technologies used significantly modified the number of *Echinochloa crus-galli* and *Setaria glauca* species. The agrotextile cover, for *Echinochloa crus-galli*, contributed to the increase in the number of this species, compared to the foil cover, but did not differ significantly from the abundance of this species in traditional technology. It follows that not only crops, but also weeds, and especially thermophilic species, have better growth and development conditions under the agrotextile cover. The number of *Echinochloa crus-galli*, the most onerous species of the monocotyledonous species, was most effectively limited by mechanical and chemical treatment with Afalon Dispersive 450 SC, compared to mechanical treatment. The use of the Afalon Dispersive 450 SC and Command 480 EC herbicide mixture significantly reduced the number of *Echinochloa crus-galli* per unit area, while the use of the Racer 250 EC herbicide did not reduce the weed infestation with this species, and there was even a tendency to increase its number under the influence of this herbicide compared to the standard facility. In the objects with 'Denar' cv., the species *Echinochloa crus-galli* in the field was less numerous than in the combinations with the 'Lord' cv. It should be assumed that this was related to the faster growth rate of 'Denar' cv. and more leafy conformation of these plants, which probably contributed to faster and better soil protection by the cultivated plant. The most significant variation in the number and species composition of weeds occurred in the years of the study. This was due to the meteorological conditions of the growing season, differences in the years of the study, different abundances of available nutrients in the soil—especially potassium, phosphorus, calcium, and magnesium—and the weed seed abundance that varied between the years. In 2016, there were only three species of monocotyledonous weeds, while in the warm and alternately severe and dry 2017, there were as many as seven species, including *Echinochloa crus-galli*, in large numbers. While an abundant species composition of monocotyledonous weeds was observed in the primary weed infestation, only *Echinochloa crus-galli*, *Setaria glauca*, and *Setaria viridis* occurred in the secondary weed infestation. The most abundant species of dicotyledonous weeds turned out to be *Anthemis arvensis* and *Polygonum convolvulus*, and less abundantly in the soil were *Viola arvensis*, *Veronica hederaefolia*, *Cirsium arvense*, and *Stellaria media*, while the remaining species occurred sporadically. Mechanical weed control only limited the incidence of *Anthemis arvensis*. The application of mechanical and chemical treatment with the herbicide Afalon Dispersive 450 SC reduced the number of *Stellaria media*, *Galium aparine*, *Centaurea cyanus*, and *Anagalis arvensis*, and it was ineffective against *Anthemis arvensis*, *Polygonum convolvulus*, and *Veronica hedereafolia*. In turn, weed management methods with Racer 250 EC eliminated *Polygonum lapathifolium* and *Galium aparine*, and it limited *Polygonum convolvulus* and *Viola arvensis*. Mechanical and chemical treatment with the use of a mixture of Afalon Dispersive 450 SC and Command 480 EC preparations limited the occurrence of the following species: *Chenopodium album*, *Cirsium arvense*, *Spergula arvensis*, and *Vicia tetrasperma*. Variable meteorological and soil conditions in the years of research also modified the number of dicotyledonous weeds. In 2015—which was poor in rainfall, but warm and sunny—the largest number of the following species was recorded: *Anthemis arvensis*, *Stellaria media*, *Veronica hedereafolia*, *Vicia tetrasperma*, and *Viola arvensis*. In 2016, the most optimal in terms of temperature and rainfall were the most numerous species requiring thermal and wetness conditions—*Cirsium arvense*, *Polygonum lapathifolium*, *Polygonum aviculare*, *Spergula arvensis*, *Raphanus raphanistrum*, and *Rumex acetosella*—while in 2017, when a periodic excess or a shortage of precipitation was observed, the most numerous were *Chenopodium album*, *Polygonum convolvulus*, *Galeopsis tetrahit*, and *Vicia hirsute*. The species composition and the number of dicotyledonous weeds also varied over time. Before the rows of the crop plant were closed, the most numerous were *Chenopodium album*, *Raphanus raphanistrum*, and *Polygonum convolvulus*. After row closure, *Cirsium arvense* was the most abundant species, and before the potato ripening, it was *Anthemis arvensis*, *Spergula arvensis*, *Stellaria media*, *Viola arvensis*, *Veronica hedereafolia* and *Myosotis arvensis*, which results from the phenology of weeds (Table 9). It should be noted that there were no weeds from other botanical groups in the species composition.

Table 9. Species composition and number of mono- and dicotyledonous weeds (pcs m^{-2}).

Factors of the Experiment		*Agropyron repens*	*Apera spica-venti*	*Avena fatua*	*Echinochloa crus-galli*	*Poa annua*	*Setaria glauca*	*Setaria viridis*	*Anagalis arvensis*	*Anthemis arvensis*	*Centaurea cyanus*	*Chenopodium album*	*Cirsium arvense*	*Galium aparine*	*Galeopsis tetrahit*	*Myosotis arvensis*	*Poligonum convolvulus*	*Poligonum lapathifolium*	*Poligonum aviculare*	*Stellaria media*	*Spergula arvensis*	*Veronica hedereafolia*	*Vicia hirsuta*	*Vicia cracca*	*Vicia tetrasperma*	*Viola arvensis*	*Raphanus raphanistrum*	*Rumex acetosella*
Technologies	Traditional	0.6	0.3	0.2	24.4	0.2	2.6	1.1	0.1	5.1	0.9	4.5	0.5	0.0	0.0	0.4	1.5	0.4	0.1	0.5	0.7	1.0	0.1	0.1	0.2	1.3	2.9	0.3
	PE-Sheeting	0.6	0.2	0.1	19.8	0.2	1.8	1.1	0.1	5.9	0.9	4.5	1.2	0.0	0.0	0.3	1.7	0.3	0.3	0.6	0.8	0.7	0.1	0.0	0.3	0.9	3.2	0.1
	PP-Sheeting	0.8	0.1	0.0	26.3	0.0	0.2	1.2	0.2	5.8	0.6	4.4	0.7	0.1	0.1	0.2	1.7	0.1	0.2	0.5	0.4	0.6	0.1	0.0	0.3	0.5	5.5	0.1
	$HSD_{0.05}$	ns *	ns	ns	5.6	ns	0.4	ns	0.1	ns	ns	ns	0.6	ns	ns	0.2	ns	0.2	0.2	ns	0.7	0.5	ns	ns	ns	0.5	2.1	ns
Potato weed management practices	Mechanical	0.7	0.1	0.1	24.4	0.2	1.5	1.4	0.1	3.0	1.0	5.3	0.6	0.0	0.2	0.2	1.9	0.1	0.1	0.6	0.4	0.4	0.1	0.0	0.3	0.6	4.4	0.1
	Afalon	0.5	0.1	0.0	16.5	0.0	0.6	0.7	0.1	8.8	0.6	4.0	0.6	0.0	0.1	0.4	2.4	0.3	0.2	0.4	0.6	1.4	0.1	0.1	0.6	1.1	2.9	0.3
	Racer	0.5	0.3	0.1	29.8	0.1	1.9	1.6	0.2	6.7	0.8	9.0	1.2	0.0	0.0	0.4	1.0	0.0	0.1	0.6	0.9	0.6	0.0	0.0	0.3	0.6	2.9	0.2
	Afalon + Command	0.5	0.2	0.1	18.9	0.0	0.8	1.0	0.1	3.9	0.9	1.9	0.5	0.1	0.0	0.2	1.9	0.4	0.2	0.9	0.1	0.3	0.0	0.1	0.1	1.3	2.6	0.1
	$HSD_{0.05}$	0.5	ns	ns	5.6	ns	0.4	0.7	0.1	2.3	ns	3.1	0.6	0.1	0.1	0.2	1.1	0.2	n	0.4	0.7	0.5	ns	ns	0.3	0.5	ns	ns
Cultivars	'Denar'	0.7	0.2	0.1	20.6	0.1	1.1	1.0	0.1	5.4	0.7	4.7	0.7	0.0	0.0	0.4	1.9	0.3	0.1	0.7	0.4	0.5	0.1	0.1	0.4	0.8	3.2	0.1
	'Lord'	0.8	0.1	0.1	24.2	0.1	1.3	1.3	0.1	5.7	0.9	5.4	0.7	0.0	0.1	0.2	1.7	0.1	0.2	0.6	0.6	0.9	0.1	0.0	0.3	1.0	3.2	0.2
	$HSD_{0.05}$	ns	ns	ns	3.0	ns	ns	ns	ns	ns	0.2	ns	ns	ns	0.1	0.1	ns	0.1	ns	ns	ns	0.3	ns	ns	ns	ns	ns	ns
Years	2015	0.3	0.5	0.3	25.8	0.3	3.1	2.9	0.2	13.5	1.2	3.4	0.1	0.0	0.0	0.6	1.1	0.0	0.0	1.2	0.3	2.0	0.0	0.0	1.0	1.3	0.0	0.0
	2016	1.6	0.0	0.0	12.5	0.0	0.6	0.0	0.0	0.0	0.0	2.7	2.1	0.1	0.0	0.1	0.2	0.5	0.5	0.0	1.1	0.0	0.0	0.1	0.0	0.2	8.7	0.5
	2017	0.3	0.0	0.0	29.0	0.0	0.0	0.5	0.1	0.0	1.0	9.1	0.0	0.0	0.2	0.2	4.1	0.1	0.0	0.6	0.1	0.0	0.2	0.0	0.0	1.2	0.9	0.0
	$HSD_{0.05}$	0.4	0.2	0.2	4.4	0.2	0.3	0.5	0.1	1.8	0.4	2.5	0.4	0.1	0.1	0.2	0.9	0.2	0.1	0.3	0.5	0.4	0.1	0.1	0.2	0.4	1.7	0.3
Term of Observations *	I	0.9	0.1	0.0	39.9	0.0	0.0	0.0	0.1	4.1	0.7	7.8	0.1	0.0	0.1	0.1	3.2	0.2	0.1	0.4	0.1	0.3	0.2	0.0	0.4	1.1	6.1	0.2
	II	0.6	0.1	0.1	17.1	0.3	0.0	0.8	0.0	6.3	0.8	3.8	1.3	0.0	0.0	0.0	0.5	0.1	0.3	0.2	0.1	0.8	0.0	0.1	0.2	0.4	2.9	0.0
	III	0.7	0.3	0.2	10.2	0.1	3.7	2.7	0.2	6.4	0.9	3.5	0.8	0.1	0.0	0.7	1.7	0.2	0.2	1.3	1.4	0.9	0.0	0.0	0.4	1.2	0.6	0.3
	$HSD_{0.05}$	ns	0.2	0.2	4.4	0.2	0.3	0.5	0.1	1.8	ns	2.5	0.4	0.1	ns	0.2	0.9	ns	0.1	0.3	0.5	0.4	0.1	0.1	ns	0.4	1.7	ns
Mean		0.7	0.2	0.1	22.4	0.1	1.2	1.2	0.1	5.6	0.8	5.0	0.7	0.0	0.0	0.3	1.8	0.2	0.2	0.6	0.5	0.7	0.1	0.0	0.3	0.9	3.2	0.2

* insignificant differences at $p_{0.05}$ level, before row closing; after shorting the rows; before ripening potato.

4. Discussion

The conducted research proved that the degree of soil coverage with weeds, both monocotyledonous and dicotyledonous plants in covered objects, was not significantly higher than in traditional cultivation without cover. These results are confirmed by Pszczółkowski and Sawicka [3]. Lutomirska and Roztropowicz [40], Lutomirska [4], Wadas [2], Krzysztofik [1], and Sawicka [41] are of different opinion. These authors prove that the better development of weeds under cover results from the fact that the temperature of the soil under cover, in comparison with the open area, is 1–2 °C higher; moreover, air saturation with water vapor is much higher than in the open area. This difference can be explained by the fact that conditions under the covers favor the germination and growth of weeds, and high saturation with water vapor under cover promotes the formation of the so-called herbicide filter on the surface of ridge, which eliminates germinating zoospores, and the fact that strict field research was carried out in the years with air and soil temperatures significantly exceeding the long-term average for the research point.

Own research confirmed previous reports by Gugała et al. [7], Azadbakht et al. [10], and Baranowska et al. [11] that the number of weeds increases until the potato is fully flowering, and in the later stages it decreases, while their air-dried mass increases until the end of the potato vegetation. Technologies of cultivation with the use of covers did not significantly differentiate this feature, but it was modified by the methods of weed management. The highest fresh and dry mass of weeds was observed during the full ripeness period of the potato, which was caused by secondary weed infestation, which is determined, as reported by Zarzecka et al. [12], by the length of the period of vegetation and the aboveground potato mass. Sawicka and Pszczółkowski [42] proved that the highest tolerated weed infestation, expressed in dry mass of weeds, in the cultivation of very early potato cultivars on light soils at the end of potato vegetation, for the total yield it amounts to 98 g, in mechanical and chemical treatment with the use of herbicide Afalon 50 WP it was 64 g, in the method with Racer 25 EC it was 135 g, and in the method with the Afalon 50 WP and Command 480 EC herbicide mixture it was 204 $g{\cdot}m^{-2}$. Cultivation with the mixture of herbicides Afalon Dispersive 450 SC and Command 480 EC gave a similar effect in destroying the fresh weed mass as their mechanical control. The research by Gugała et al. [7] shows that the mechanical and chemical method reduces the weight of weeds by about 59%. Ciesielska and Wysmułek [43] confirm this view as well.

In cultivation under cover, mechanical method treatments can be performed only after planting, before placing the covers, thus supporting this method of weed control with chemical treatments is the only way to reduce the weed infestation in the potato field [1,42]. The use of herbicides in potato cultivation immediately after planting in combination with previous mechanical method did not result in complete elimination of weeds. Weed species that survive after the treatment may constitute, in the opinion of Sawicka et al. [44], Barbaś [5], Gugała et al. [7], and Zarzecka [12], a great threat to the cultivated plant, as it may contribute to significant losses of tuber yield. Hess and Foy [18] report the interaction of surfactants with plant skin and possible damage to it.

The total number of weeds per unit area in the field of cultivated potato cultivars exceeded the weed harmfulness threshold adopted by Pszczółkowski and Sawicka [45]. Their number was most severely limited employing a mixture of Afalon Dispersive 450 SC and Command 480 EC.

According to Accinelli et al. [20], Grundy et al. [46], and Haliniarz [47], the complete elimination of weeds from crops is impossible, as they are an inseparable component of agrocenoses. In the opinion of Zarzecka and Gugała [48], each crop may contain a certain number of weeds, which do not reduce its yield. According to Zarzecka et al. [12], the complete elimination of weeds from crops is not possible, because they are an inseparable component of agrocenoses, but their occurrence can be limited to a level that does not exceed the harmfulness threshold, without reducing the yield. In the opinion of Gugała et al. [7], the limit of harmless weed infestation should be defined as "tolerated weed infestation" and, according to Barralis [49], "critical density".

Results of the author's own research stating that the weather conditions, varied in particular years of research, significantly affect the population density of weeds, and the amount of fresh and air-dried weed mass are consistent with the reports of many authors [2,7,12,42,50], as in recent years, abiotic stresses have had a large impact on the growth and productivity of crops. Sawicka et al. [51] and Wadas and Dziugieł [52] proved that in regions with extensive plant production, there are more and more frequent periods of high temperature and drought, which disrupt photosynthesis and limits the potential of potato yielding. Krzyżewska et al. [53] proved that in Poland, the most unfavorable bioclimatic conditions occur in the south-east of Poland, where forceful heat stress occurs with a frequency of 10%. This phenomenon is associated with a drastic reduction in yields, especially of root crops, as well as with the development of drought-resistant segetal species. The highest number of cases with strong and very strong heat stress was recorded in 2015, and heat waves were observed in the first half of August. This is an important period for the accumulation of yield from root crops. The authors recorded an increase in the number of days with strong and forceful heat stress and their highest frequency in July. Inhibition of potato plant growth is observed at the temperature of 29 °C [41], but most dicotyledonous and monocotyledonous species are resistant to high air temperatures and survive this period, and after a drought period, they increase their weight and density, taking advantage of the inhibition of crop growth. Simultaneously, the increased number of weeds uses the same water resources as the cultivated plants and limits the access of light to them [47].

Diversity of the number and species composition of weeds was also determined by the course of weather conditions in the years of the study. Their higher numbers and more abundant species composition were observed in warm and dry year than in cooler and wetter years. These results are consistent with the observations of Gugała et al. [7], Baranowska et al. [11], and Barbaś and Sawicka [54].

Our research proved that the morphological properties of tested cultivars and length of their growing season determined only the number of monocotyledonous weeds, while Gugała et al. [55,56] found that the physiological and morphological properties of cultivars, i.e., conformation of bushes, their foliage, and position of leaves relating to the light, are factors that most strongly modify the weed mass.

The species composition of weeds in the field of the studied potato cultivars was quite abundant, even though the experiment was conducted on light soil (seven monocotyledon species and 20 dicotyledonous species). Stešević and Jovović [57], in the Lublin region Haliniarz [47], made similar observations. They observed 36–48 species of weeds in potato cultivation, but only five to eight species were dominant. The most abundant of the monocotyledons was *Echinochloa crus-galli*, and of the dicotyledons, it was *Chenopodium album*, *Anthemis arvensis*, and *Raphanus raphanistrum*. Haliniarz [47] came to similar conclusions.

The species composition of weeds was significantly modified by the applied treatments. Afalon Dispersive 450 SC used in the mechanical and chemical control of weeds showed high efficiency in reducing *Echinochloa crus-galli*, *Setaria glauca*, *Setaria viridis*, *Centaurea cyanus*, *Stellaria media*, *Anagalis arvensis*, and no or weak action against *Anthemis arvensis*, *Polygonum convolvulus*, *Vicia tetrasperma*, and *Veronica hederaefolia*. Accinelli et al. [20], Andreasen and Streibig [58], and Gugała et al. [55] recorded similar efficiency of Afalon 50 WP herbicide in other formulation. According to the research of Haliniarz [47], the following weed species are particularly high in nutrient uptake from the soil: *Stellaria media*, *Amaranthus retroflexus*, *Echinochloa crus-galli*, and *Chenopodium album*. Considering their common occurrence, these species should be considered particularly competitive for potato cultivars grown under cover; moreover, the uptake of minerals from the soil by weeds exceeds the uptake by cultivated plants several times and reaches over 80% of macroelements and over 90% of their total uptake by weeds and the plant being grown.

Herbicide Racer 250 EC used in the care of potatoes turned out to be effective against *Polygonum convolvulus*, *Polygonum lapathifolium*, *Galium aparine*, and *Viola arvensis* and ineffective against *Echinochloa crus-galli*, *Setaria glauca*, *Setaria viridis*, *Chemopodium album*, *Cirsium arvensis*, *Spergula arvensis*, and *Mysostis arvensis*. However, it did not limit the fresh and dry mass of weeds, which mainly applies to secondary weed infestation. Similar results were obtained by Sawicka and Pszczółkowski [42], which is consistent with the reports of Gupta and Gajbhiye [59].

Our own research confirms the reports of Barbaś [5] and Zarzecka et al. [48,60], with high efficiency of herbicide mixtures used in potato cultivation. The use of a mixture of Afalon Dispersive 450 SC and Command 480 EC in potato cultivation showed high efficiency in reducing such species as *Chenopodium album, Raphanus raphanistrum, Vicia tetrasperma, Cirsium arvense, Galeopsis tetrahit, Spergula arvensis, Veronica hederaefolia*, and *Myosotis arvensis*, while it showed weak action towards *Polygonum lapathifolium, Galium aparine, Viola arvensis*, and *Stellaria media*. Mystkowska et al. [61] and Mayerová et al. [62] showed that both the application of mechanical and chemical treatments, with the use of single herbicides and their mixtures, significantly reduced the weed infestation in potato crops. Research by Riethmuller-Haage et al. [63] demonstrated the importance of the sensitivity of herbicide combinations and species and proved that pre-spray weather information is crucial in developing the dose reduction recommendations.

In the conducted studies, the phytotoxic effect of the herbicides used was observed not only on the weeds to be controlled, but also on the cultivated plant. To provide potato plants with the best possible conditions for growth and development, according to Gugała et al. [7], along with the application of herbicides, biostimulators should also be used, which, in their opinion, reduces the phytotoxic effect of herbicides on potato plants and significantly reduces the fresh mass of weeds, compared to the standard object. According to Eshel et al. [64], cultivation under cover (CC) has direct benefits to a society by reducing the cost of agricultural damage and restoring the infrastructure, and by protecting the environmental benefits provided by agriculture.

According to Wang and Liu [65], Haliniarz [47], Barbaś and Sawicka [54], and Zarzecka et al. [12], the occurrence or significant delay in the resistance of weeds to herbicides is prevented, among others, by crop rotation; alternating the use of agents from various chemical groups, as well as the use of mixtures of herbicides with different mechanisms of action; the application of herbicides to weeds in the period of their greatest sensitivity; the application of herbicides in doses guaranteeing complete destruction of weeds; the addition of adjuvants to the spray liquid in case of dose reduction, including mechanical treatments in the weed control system, and the use of non-selective herbicides before the emergence of the crop.

Chemical methods of weeding with the use of the tested preparations reduced the population of weeds to a varying degree, depending on the herbicide used. Afalon Dispersive 450 SC showed high efficiency in reducing *Echinochloa crus-galli, Setaria glauca, Setaria viridis, Centaurea cyanus, Stellaria media*, and *Anagalis arvensis* and showed no or weak activity against *Anthemis arvensis, Polygonum convolvulus, Vicia tetraspderaolia*, and *Vicia tetraspderaolia*. Racer 250 EC herbicide proved to be effective in reducing *Polygonum convolvulus, Polygonum lapathifolium, Galium aparine, and Viola arvensis* and was ineffective against *Echinochloa crus-galli, Setaria glauca, Setaria viridis, Chemopodium album, Cirsium arvensis, Spergula arvensis*, and *Mysostis arvensis*. The use of a mixture of Afalon Dispersive 450 SC and Command 480 EC has shown high efficiency in reducing species, such as *Chenopodium album, Raphanus raphanistrum, Vicia tetrasperma, Cirsium arvense, Galeopsis tetrahit, Spergula arvensis, Veronica hederaefolia*, and *Myosotis arvensis*, while the activity against *Polygoneum lapathifolium, Galium aparine, Viola arvensis*, and *Stellaria media* is weak.

5. Conclusions

The use of covers in the cultivation of very early potato cultivars increased the number and mass of weeds and enriched the species composition of segetal vegetation.

Mechanical and chemical methods proved to be more effective than the mechanical method. The best effectiveness in limiting both fresh and dry weed mass in potato cultivation under cover was achieved using the mechanical and chemical method using a mixture of herbicides Afalon Dispersion 450 SC and Command 480 EC.

Variable courses of atmospheric conditions in the years of the study differentiated both the number of monocotyledonous and dicotyledonous weeds, their fresh and dry mass, and the species composition of segetal vegetation.

Author Contributions: Conceptualization, P.P., B.S.; methodology, P.P., B.S., P.B.; software, B.K.-M.; validation, B.S., P.B. and B.K.-M.; formal analysis, P.P.; investigation, P.B., P.P.; data curation, P.P.; writing—original draft preparation P.P., B.S.; writing—review and editing, B.S.; visualization, B.K.-M.; supervision, B.S.; project administration, P.B., P.P.; funding acquisition, P.P. All authors have read and agreed to the published version of the manuscript.

Funding: This research received no external funding—own financing.

Conflicts of Interest: The authors declare no conflict of interest.

References

1. Krzysztofik, B. Effects of early potatoes cultivation under cover. *IR* **2013**, *3*, 169–177.
2. Wadas, W. Production of potatoes for agrotextile. *Ziemn. Pol.* **2012**, *1*, 13–17.
3. Pszczółkowski, P.; Sawicka, B. Productivity of very early potato varieties cultivated under coverage. Part I. Yield and its structure. *Acta Sci. Pol. Agric.* **2003**, *2*, 61–72.
4. Lutomirska, B. *Production for the Earliest Harvest. [w:] Potato Production and Market*; Chotkowski, J., Ed.; Wieś Jutra: Warsaw, Poland, 2012; pp. 222–237.
5. Barbaś, P. Protection of potato plantations against weed infestation. *Ziemniak Polski* **2015**, *1*, 11–17.
6. Barbaś, P.; Sawicka, B. Changes of weed infestation in potato under and integrated ecological and production systems. *Prog. Plant Prot.* **2010**, *50*, 1161–1165.
7. Gugała, M.; Zarzecka, K.; Sikorska, A.; Mystkowska, I.; Dołęga, H. Effect of herbicides and growth biostimulants on weed reduction and yield of edible potato. *Fragm. Agron.* **2017**, *34*, 59–66.
8. Soren, C.; Chowdary, K.A.; Sathish, G.; Patra, B.C. Weed dynamics and yield of potato as influenced by weed management practices. *Int. J. Pure App. Biosci.* **2018**, *6*, 398–408. [CrossRef]
9. Nowacki, W.; Czerko, Z.; Goliszewski, W.; Jankowska, J.; Lutomirska, B.; Trawczyński, C.; Zarzyńska, K. *Methodology of Integrated Potato Production*; State Plant Health and Seed Institute, Main Inspectorate: Warsaw, Poland, 2014; p. 79.
10. Azadbakht, A.; Akbar, M.T.; Ghavidel, A. Effect of chemical and non-chemical methods of weed control in potato (Solanum tuberosum L.) cultivation in Ardabil Province, Iran. *AEER* **2017**, *15*, 1359–1372. [CrossRef]
11. Baranowska, A.; Mystkowska, I.; Zarzecka, K.; Gugała, M. Efficacy of herbicides in potato crop. *J. Ecol. Eng.* **2016**, *17*, 82–88. [CrossRef]
12. Zarzecka, K.; Gugała, M.; Grzywacz, K.; Sikorska, A. Agricultural and economic effects of the use of biostimulants and herbicides in cultivation of the table potato cultivar Gawin. *Acta Sci. Pol. Agric.* **2020**, *19*, 3–10. [CrossRef]
13. Scalla, R. *Les Herbicides. Mode D'action et Principes D'utilisation*; INRA: Paris, France, 1990.
14. Ackley, J.A.; Wilson, H.P.; Hines, T.E. Weed management programs in potato (Solanum tuberosum) with rimsulfuron. *Weed Technol.* **1996**, *10*, 354–358. [CrossRef]
15. Starck, Z.; Chołuj, D.; Niemyska, B. *The physiological Reaction of Plants to Difficult Environmental Conditions*; Starck, Z., Ed.; Publishing House SGGW: Warsaw, Poland, 1995; p. 55.
16. Strack, Z. Plant physiology: What was yesterday, what today and what will tomorrow bring? *Kosmos Probl. Nauk. Biol.* **2014**, *63*, 569–589.
17. Kacperska, A. *Plant Reactions to Stress Factors [w:] Fundamentals of Plant Physiology*; Kopcewicz, J., Lewak, S., Eds.; Polish Scientific Publishers: Warsaw, Poland, 1998.
18. Hess, F.D.; Foy, C.L. Interaction of surfactants with plant cuticles. *Weed Technol.* **2000**, *14*, 807–813. [CrossRef]
19. Korbas, M. (Ed.) Potato Protection Program. 2017. Available online: https://www.ior.poznan.pl/plik,2913, program-ochrony-ziemniaka.pdf (accessed on 20 June 2020).
20. Accinelli, C.; Screpanti, C.; Vicari, A. Influence of flooding on the degradation of linuron, isoproturon and metolachlor in soil. *Agron. Sustain. Dev.* **2005**, *25*, 401–406. [CrossRef]
21. Tomlin, C. *The e-Pesticide Manual Version 5.2.*; British Crop Protection Council: Berkshire, UK, 2011.
22. Caldiz, D.; de Lasa, C.; Bisio, P.E. Grass and weed management in potato (Solanum tuberosum L.) processing with clomazone, in Argentine pampas. *AJPS* **2016**, *07*, 2339–2348. [CrossRef]

23. PN-ISO 10381-2: 2007. *Soil Quality—Sampling—Part 2: Rules for Collection Techniques PN-ISO 10381-2*; Polish Committee for Standardization: Warsaw, Poland, 2007.
24. Mocek, A.; Drzymała, S. *Genesis, Analysis and Classification of Soils*; Publishing House of the University of Life Sciences: Poznań, Poland, 2010.
25. Ryzak, M.; Bartmiński, P.; Bieganowski, A. Methods of determination of granulometric distribution of mineral soils. *Acta Agroph. Theses Monogr.* **2009**, *175*, 97.
26. KQ/PB-34 version 04 from 07.10.2011. 2011.
27. PN-R-BS 0403: 1997. *Chemical and Agricultural Analysis of Soil*; Sampling. Polish Committee for Standardization: Warsaw, Poland, 1997.
28. PN-R-04020 1994 + AZ1: 2004. *Chemical and Agricultural Analysis of Soil*; Polish Committee for Standardization: Warsaw, Poland, 1994.
29. PN-R-04023: 1996. *Chemical and Agricultural Analysis of Soil*; Determination of Available Phosphorus Content in Mineral Soils; Polish Committee for Standardization: Warsaw, Poland, 1996.
30. PN-R-04022: 1996 + AZ1: 2002. *Chemical and Agricultural Analysis of Soil*; Determination of Available Potassium Content in Mineral Soils; Polish Committee for Standardization: Warsaw, Poland, 2002.
31. PN-R-04017: 1992. Content of Copper Soluble in HCl. In *Mineral Soil*; Polish Committee for Standardization: Warsaw, Poland, 1992.
32. PN-R-04019: 1993. *Content of Manganese Soluble in HCl in Mineral Soil*; Polish Committee for Standardization: Warsaw, Poland, 1993.
33. PN-R-04021: 1994. *The Content of Iron Soluble in HCl in Mineral Soil*; Polish Committee for Standardization: Warsaw, Poland, 1994.
34. PN-R-04018: 1993. *The Content of Boron Soluble in HCl in Mineral Soil*; Polish Committee for Standardization: Warsaw, Poland, 1993.
35. WRB. World Reference Database for Soil Resources. 2014. Available online: http://www.fao.org/3/a-i3794e.pdf (accessed on 8 June 2020).
36. Mocek, A. *Soil Science*; State Scientific Publisher: Warsaw, Poland, 2015.
37. Skowera, B.; Kopcińska, J.; Kopeć, B. Changes in thermal and precipitation conditions in Poland in 1971–2010. *Ann. Wars. Univ. Life Sci. SGGW Land Reclam.* **2014**, *46*, 153–162. [CrossRef]
38. SAS Institute Inc. *SAS/STAT®9.2 User's Guide*; SAS Institute Inc.: Cary, NC, USA, 2008.
39. Koronacki, J. *Statistics, for Students of Technical and Natural Sciences*; WNT: Warsaw, Poland, 2009; p. 491.
40. Roztropowicz, S.; Lutomirska, B. Potato Production Technology for Early Harvest. In *Potato Production. Technology—Economics—Marketing*; Chotkowski, J., Ed.; Plant Breeding and Acclimatization Institute: Bonin, Poland, 1997; pp. 82–98.
41. Sawicka, B. *Potato [in:] Growing Plants, Vol. II, Part VII. Root Crops Red*; Kotecki, A., Ed.; University of Life Sciences: Wroclaw, Poland, 2020; pp. 395–520.
42. Sawicka, B.; Pszczółkowski, P. Attempts to control infestation of potato canopy cultivated under shields. Part III. The effect of weed overgrowth in the stand upon the total and market yield of tubers. *Biul. IHAR* **2003**, *228*, 233–247.
43. Ciesielska, A.; Wysmułek, A. The efficacy of tank mixture herbicides Sencor 600 SC + Titus 25 WG in potatoes. *Prog. Plant Prot.* **2012**, *52*, 885–888.
44. Sawicka, B.; Barbaś, P.; Dąbek-Gad, M. The problem of weed infestation in the conditions of using growth bioregulators and foliar fertilization in potato cultivation. *Sci. Nat. Technol.* **2011**, *9*, 1–12.
45. Pszczółkowski, P.; Sawicka, B. Attempts to reduce weed infestation of potato in cultivation under cover. Part II. Weight, abundance and species composition of weeds. *Biul. IHAR* **2003**, *228*, 261–273.
46. Grundy, A.C.; Mead, A.; Bond, W.; Clark, G.; Burston, S. The impact of herbicide management on long-term changes in the diversity and species composition of weed populations. *Weed Res.* **2010**, *51*, 187–200. [CrossRef]
47. Haliniarz, M. The response of selected agrophytocoenoses to different doses of biologically active substances of herbicides. In *Monography*; University of Life Science in Lublin: Lublin, Poland, 2019; p. 207, ISSN 1899-2374.
48. Zarzecka, K.; Gugała, M. Comparison of various methods of weeding potato plantations. *Pam. Puł* **2006**, *142*, 607–615.

49. Barralis, G. Senils de nuisibilité des mauvaises herbes. IV. Determination et utilisation des senils de nuisibilité. *Phytoma* **1977**, *29*, 290.
50. Wadas, W.; Dziugieł, T. Quality of New Potatoes (Solanum tuberosum L.) in Response to Plant Biostimulants Application. *Agriculture* **2020**, *10*, 265. [CrossRef]
51. Sawicka, B.; Michałek, W.; Pszczółkowski, P. The relationship of potato tubers chemical composition with selected physiological indicators. *Zemdirb. Agric.* **2015**, *102*, 41–50. [CrossRef]
52. Wadas, W.; Dziugieł, T. Changes in Assimilation Area and Chlorophyll Content of Very Early Potato (Solanum tuberosum L.) Cultivars as Influenced by Biostimulants. *Agronomy* **2020**, *10*, 387. [CrossRef]
53. Krzyżewska, A.; Wereski, S.; Dobek, M. Summer variability of UTCI in Poland in the 21st century. *Int. J. Biometeorol.* **2020**. [CrossRef]
54. Barbaś, P.; Sawicka, B. Dependence of potato crop on weed infestation. *Agrom. Sci.* **2020**, *8*, 346–359. [CrossRef]
55. Gugała, M.; Zarzecka, K.; Sikorska, A. The weed infestation and yielding of potato depending on the mechanical and chemical treatments. *Fragm. Agron.* **2014**, *31*, 50–57.
56. Gugała, M.; Zarzecka, K.; Dołęga, H.; Sikorska, A. Weed Infestation and Yielding of Potato Under Conditions of Varied Use of Herbicides and Bio-Stimulants. *J. Ecol. Eng.* **2018**, *19*, 191–196. [CrossRef]
57. Stešević, D.; Jovović, Z. Contribution to the knowledge on the weed flora in potato crop in the vicinity of Nikšić (Montenegro). *Herbology* **2011**, *12*, 1–6.
58. Andreasen, C.; Streibig, C. Evaluation of changes in weed flora in arable fields of Nordic countries—Based on Danish long-term surveys. *Weed Res.* **2010**, *51*, 214–226. [CrossRef]
59. Gupta, S.; Gajbhiye, V.T. Effect of concentration, moisture and soil type on the dissipation of flufenacet from soil. *Chemosphere* **2002**, *47*, 901–906. [CrossRef]
60. Zarzecka, K.; Gugała, M.; Dołęga, H. Regulation of weed infestation degree in potato with the use of herbicides. *Biul IHAR* **2013**, *267*, 113–120.
61. Mystkowska, I.; Zarzecka, K.; Baranowska, A.; Gugała, M. Weed infestation of potato cultivars depending on weed control methods and weather conditions. *Acta Agroph.* **2017**, *24*, 111–121.
62. Mayerová, M.; Mikulka, J.; Kolářová, M.; Soukup, J. Changes in weed community composition in a long-term trial with different crop rotations and herbicide treatments. *Julius-Kühn-Archiv* **2018**, *458*, 58–66.
63. Riethmuller-Haage, I.; Bastiaans, L.; Kempenaar, C.; Smutny, V.; Kropff, M.J. Are pre-spraying growing conditions a major determinant of herbicide efficacy? *Weed Res.* **2007**, *47*, 415–424. [CrossRef]
64. Eshel, G.; Egozi, R.; Goldwasser, Y.; Designi, D.; Kashtic, Y.; Fine, P.; Hayutb, E.; Kazukro, H.; Rubin, B.; Dar, Z.; et al. The benefits of growing potatoes under cover in Mediterranean climate. *Agric. Ecosyst. Environ.* **2015**, *211*, 1–9. [CrossRef]
65. Wang, C.J.; Liu, Z.Q. Foliar uptake of pesticides—Present status and future challenge. *Pestic. Biochem. Physiol.* **2007**, *87*, 1–8. [CrossRef]

Article

Glyphosate Resistance in *Amaranthus viridis* in Brazilian Citrus Orchards

Ricardo Alcántara-de la Cruz [1,*], Gabriel da Silva Amaral [1], Guilherme Moraes de Oliveira [2], Luiz Renato Rufino [3], Fernando Alves de Azevedo [4], Leonardo Bianco de Carvalho [5] and Maria Fátima das Graças Fernandes da Silva [1]

1 Departamento de Química, Universidade Federal de São Carlos, São Carlos 13565-905, Brazil; amaral91gabriel@gmail.com (G.d.S.A.); dmfs@ufscar.br (M.F.d.G.F.d.S.)

2 Centro de Ciências Biológicas e da Saúde, Universidade Federal de São Carlos, São Carlos 13565-905, Brazil; gui.moraesoliveira95@gmail.com

3 Centro de Ciências Agrárias, Universidade Federal de São Carlos—Campus Araras, Araras 13604-900, Brazil; luizrufino@gmail.com

4 Centro de Citricultura "Sylvio Moreira", Instituto Agronômico de Campinas (IAC), Cordeirópolis 13490-970, Brazil; fernando@ccsm.br

5 School of Agricultural and Veterinary Sciences, São Paulo State University (UNESP), Jaboticabal 14884-900, Brazil; leonardo.carvalho@unesp.br

* Correspondence: ricardo.cruz@ufscar.br; Tel.: +55-(16)-3351-8293

Received: 30 June 2020; Accepted: 15 July 2020; Published: 20 July 2020

Abstract: Glyphosate is the main tool for weed management in Brazilian citrus orchards, where weeds, such as *Conyza bonariensis* and *Digitaria insularis*, have been found with resistance to this herbicide. Field prospections have allowed the identification of a possible new case of glyphosate resistance. In this work, the susceptibility levels to glyphosate on three *Amaranthus viridis* L. populations, with suspected resistance (R1, R2, and R-IAC), collected in citrus orchards from the São Paulo State, Brazil, as well as their accumulation rates of shikimic acid, were determined. The fresh weight of the susceptible population (S) was reduced by 50% (GR_{50}) with ~30 g ea ha^{-1} glyphosate, while the GR_{50} values of the R populations were between 5.4 and 11.3 times higher than that for S population. The LD_{50} (herbicide dose to kill 50% of individuals of a weed population) values of the S population were ≤150 g ea ha^{-1} glyphosate, while the LD_{50} of the R populations ranged from 600 to 920 g ea ha^{-1}. Based on the reduction of fresh weight and the survival rate, the R1 population showed the highest level of glyphosate resistance, which had GR_{50} and LD_{50} values of 248 and 918 g ea ha^{-1} glyphosate, respectively. The S population accumulated 240 μg shikimic acid at 1000 μM glyphosate, while the R1, R2, and R-IAC populations accumulated only 16, 43, and 33 μg shikimic acid, respectively (between 5.6 to 15 times less than the S population). Enzyme activity assays suggested that at least one target site-type mechanism was involved in resistance. This result revealed the first report of glyphosate resistance in *A. viridis* reported in the world.

Keywords: dose-response; enzyme activity; herbicide resistance; shikimic acid; slender amaranth

1. Introduction

Brazil is the world's largest producer and exporter of citrus [1]. The State of São Paulo (SP) accounts for ~77% of the national production of orange (*Citrus sinensis* (L.) Osbeck), being the main exporter of concentrated orange juice [2]. However, the yield of citrus fruit is not the best in the world, mainly due to improper management of the plantations by growers [3]. The average orange yield is 27.6 tons ha^{-1}, occupying the 13th world rankings [4]. Weed presence can be directly or indirectly responsible for up to 30–52% of yield losses in citrus orchards of young and growing trees [3,5].

In mature citrus orchards in full production, weeds do not have a great impact on yield, but they can hinder agricultural operations [6,7].

Small-scale cultural and ecological practices, such as the use of dead covers of *Brachiaria* shoot residues, are implemented for weed control in Brazilian citrus [3]; however, a chemical method based on the use of herbicides is the most employed by growers [8]. Glyphosate is the most used herbicide and, in some cases, the only weed control tool, which is applied up to four times a year in high doses (≥720 g ae ha^{-1}) [3]. In susceptible plants, this herbicide inhibits the activity of 5-enolpyruvylshikimate-3-phosphate synthase (EPSPS), causing the accumulation of shikimic acid [9,10]. Extensive and excessive use of glyphosate to control weeds has led to select at least 45 species with resistance to this herbicide in the world [9,11]. Specifically, in Brazilian citrus orchards, *Conyza bonariensis*, *Conyza canadensis*, and *Digitaria insularis* [12,13] have been reported as being resistant to glyphosate. However, species of the genera *Amaranthus*, *Bidens*, *Chloris*, *Conyza*, *Eleusine*, *Lolium*, among others, which present high occurrence rates in citrus orchards [8], have a great risk of evolving resistance not only to glyphosate but also to other herbicides [14].

Growth rate, reproductive capacity, genetic variability, and stress tolerance are all high in *Amaranthus* species [15,16]. These traits give them a large capacity to evolve resistance to herbicides [15], which makes it difficult to control these species. Slender amaranth (*Amaranthus viridis* L.) is widely spread over tropical and subtropical regions in more than 80 countries [17]. These species, as well as other *Amaranthus* species, are the most abundant weeds in the Brazilian south and center-west regions [18]. In 2010, this weed was reported with multiple resistance to the acetolactate synthase (ALS) and photosystem II inhibiting herbicides in cotton in the states of Bahia and Mato Grosso, Brazil [19], and in 2015, *Amaranthus palmeri* was found with resistance to glyphosate and ALS-inhibiting herbicides in soybean in the state of Mato Grosso, Brazil [20].

This study hypothesized that slender amaranth might have evolved glyphosate resistance due to the high survival rates in the field after herbicide applications observed in the last cropping seasons. The objectives were to evaluate glyphosate resistance, monitor the accumulation rates of shikimic acid, and check the activity of the EPSPS in three slender amaranth populations with suspected resistance, which were harvested in lemon and sweet orange orchards of Mogi-Mirim and Cordeirópolis municipalities, SP, aiming to confirm a new unique case of herbicide resistance.

2. Materials and Methods

2.1. Biological Material and Seedling Propagation in Greenhouse

Seeds of three slender amaranth populations with suspected resistance to glyphosate were collected from at least 20 plants that survived the last application of glyphosate (≥720 g ea ha^{-1}) in December 2018. The populations R1 and R2 were collected in sweet orange orchards that were 500-m apart in the municipality of Mogi-Mirim, SP, Brazil (22°25′ S, 47°09′ W). The R-IAC population was collected in the citrus experimental field of the Agronomic Institute (IAC) in Cordeirópolis, SP, Brazil (22°32′ S, 47°27′ W). Seeds of a susceptible population (S-Lim) used as control were collected in an organic orchard (without the use of herbicides) of Tahiti acid lime, also in Mogi Mirim [3].

Seeds of each population were germinated in plastic containers (10 × 20 × 8 cm) containing substrate (Insumax, Nova Europa, SP, Brazil) and sand (1:2, *v*/*v*) moistened to field capacity. Some of these seeds were deposited and covered with vermiculite (2-mm), and the containers were hermetically sealed. Germinated seedlings were individualized in 250 mL pots filled with substrate, sand, and vermiculite (2:2:1, *v*/*v*/*v*). Two days after transplanting, seedlings were fertilized with ~100 mg (5–6 granules) of 14-14-14 (Forth Cote, Osmocote, Froth Jardim Ltd., Cerquinho, SP. Brazil) and irrigated as needed until use. Seedlings with 3–5 true leaves were used in all experiments, which were kept under greenhouse conditions (25–32 °C, 60 ± 10% relative humidity, and 16-h photoperiod) from germination to evaluations.

2.2. Glyphosate Dose-Response Assays

Plants from the R and S populations were treated with the following doses of glyphosate (Roundup Original DI, 370 g acid equivalent (ae) L^{-1}, Monsanto do Brasil Ltd.): 0, 45.25, 92.5, 185, 370, 740 (reference field dose), 1480, and 2960 g ae ha^{-1}. The glyphosate was sprayed using a pneumatic backpack sprayer equipped with an LBD-110015E nozzle (KGF Bicos para Pulverização, Vinhedo, SP) calibrated to deliver 200 L of ha^{-1} herbicide mixture at 30 psi (pressure measured with a glycerin manometer (Model GCN, Cotergavi Instrumentos de Medição Ltd., São Paulo, Brazil) coupled to the spray bar). Experimental design (completely randomized) included six replicates per glyphosate dose, and they were repeated. When the dose-response experiments were repeated, the 2960 g ae ha^{-1} glyphosate dose was deleted, but a dose of 22.63 g ae ha^{-1} was included. In both experiments, the fresh weight by cutting the plants at ground level and the plant mortality rates were determined 21 days after the treatment. Data were expressed as a percentage in relation to the untreated control, and the effective mean doses that reduce the fresh weigh (GR_{50}) and cause plant mortality (LD_{50}) by 50% were determined by non-linear regression analysis.

2.3. Shikimic Acid Accumulation

Shikimic acid accumulation was determined following two approaches. In the first one, the methodology of Cromartie and Polge [21] was followed. A set of plants of each slender amaranth population was treated with 370 g ae ha^{-1} glyphosate, and another set of untreated plants was reserved as a control to build the calibration curve with known concentrations of shikimic acid (0, 0.01, 0.05, 0.1, and 0.2). The first and second leaf of treated and untreated plants were cut in small segments 96 h after treatment (HAT). Samples of 50 mg of fresh leaf tissue were placed in tubes containing 1 mL of HCl 0.25 N, frozen in liquid nitrogen, and stored at −40 °C for further analysis. Samples were thawed at room temperature and then incubated for 45 min at 37 °C. Aliquots of 50 µL were transferred to new tubes containing 200 µL of periodic acid 0.25% (*w*/*v*) + sodium m-periodate 0.25% (*w*/*v*) (Solution 1). Samples were incubated at 37 °C for 30 min, following which, 200 µL of 0.6 N sodium hydroxide + 0.22 N sodium sulfite (Solution 2) was added and, finally, they were homogenized. Volumes of 300 µL were transferred to spectrophotometric cuvettes containing 600 µL of distilled water. Absorbance was measured at 380 nm wavelength in a diode-array spectrophotometer (HP 8425A, Palo Alto, CA, USA). The shikimic acid accumulation was determined from the difference between treated and untreated plants, and the results were expressed as µg of shikimic acid g^{-1} fresh tissue. Five samples with three technical replicates were analyzed per population in a completely randomized design.

In the second approach, shikimic acid was quantified in vivo, according to Dayan et al. [22], with modifications. Different glyphosate concentrations (0, 10, 50, 100, 200, 500, and 1000 µM) were prepared in 10-mM ammonium phosphate monobasic solution (pH adjusted to 4.4 with 0.1 HCl). Samples of 50 mg young leaf segments (4 × 4 mm) were placed in tubes containing 1 mL of the corresponding glyphosate concentration. Tubes were incubated for 24 h in a BOD (biochemical oxygen demand) chamber at 25 °C under fluorescent lights (150 µM m^{-2} s^{-1}). After incubation, tubes were rapidly frozen in liquid nitrogen, thawed at room temperature, and incubated at 60 °C for 30 min. Samples received 250 µL of 1.25 N HCl, and they were incubated again at 60 °C for 15 min. Volumes of 75 µL of the samples were transferred to spectrophotometric cuvettes containing 300 µL of solution 1 and incubated at 25 °C for 90 min in a BOD chamber. Finally, samples received 300 µL of solution 2, and the absorbance was measured at 380 nm, as described above. The experiment had a completely random design with three replications (three technical replicates each one) per glyphosate concentration. Results were expressed as µg shikimate per mL HCl solution (µg mL^{-1}).

2.4. EPSPS Enzyme Activity Assays

Five grams of fresh leaf tissue from each *A. viridis* population was harvested, frozen immediately in liquid N_2, and stored at −80 °C. EPSPS enzyme extraction was performed following the protocol described by Dayan et al. [22]. The total soluble protein (TPS) in the extract (EPSPS basal activity in the absence of glyphosate) was determined by the Bradford assay [23]. The specific EPSPS activity was assayed in the presence of glyphosate (0, 1, 10, 100, 1000 μM) using the EnzChek Phosphate Assay Kit (Invitrogen, Carlsbad, CA, USA) to determine the amount of inorganic phosphate (μmol) released by μg^{-1} TSP min^{-1} in comparison to the control (basal activity). Three replications per glyphosate concentration were assayed. The results were expressed as the inhibition rate of the EPSPS by 50% (I_{50}).

2.5. Statistical Analysis

The GR_{50} and LD_{50} values were calculated using a log-logistic model of four-parameters $Y = c + \{(d - c)/[1 + (x/g)^b]\}$ [24], where: Y = response by 50%; d and c are the upper and lower limits of the curve; b is the slope of the line; x is the herbicide dose, and g is the herbicide rate at the point of an inflection curve. SigmaPlot 10.0 (Systat Sofware Inc. San Jose, CA, USA) was used to obtain those parameters. The resistance factors (RF = R/S) were computed as R-to-S GR_{50} or LD_{50} ratios. Because the estimated LD_{50} and GR_{50} values for population S differed between repetitions of the dose-response experiments, the results were shown separately for each repetition.

One-way ANOVA was performed for the analysis of shikimic acid accumulation data. Differences of $p < 0.05$ were considered significant, and Tukey's test at $\alpha = 0.05$ probability was conducted for means comparison. Statistical analyses were performed using the Statistics 9.0 software (Analytical Software, Tallahassee, FL, USA).

3. Results

3.1. Fresh Weight Reduction and Plant Survival

The fresh weight reduction was higher in individuals of the S-Lim slender amaranth population than the R populations. In the first experiment, the estimated GR_{50} for population S was 22 g ea ha^{-1} (Table 1). Thus, the RF of the R populations ranged were 7.5, 8.7, and 11.3 for the R2, R-IAC, and R1 populations, respectively. After repetition of dose-response experiments, more representative value of GR_{50} was estimated for the population S-Lim (38 g ea ha^{-1}) due to the smaller dose of glyphosate included; while the GR_{50} values of the R populations were similar (according to the confidential intervals) to those estimated in the first experiment. Because the GR_{50} of the S-Lim population was higher in the second experiment, the ratio of glyphosate resistance level in relation to the R populations decreased and ranged from 5.4 to 7.9 (Figure 1).

Table 1. Parameters of the sigmoidal equations [1] used to estimate the effective mean dose (g ea ha^{-1}) of glyphosate required to reduce the fresh weight by 50% (GR_{50}) and to cause plant mortality (LD_{50}) by 50% in resistant (R1, R2 and R-IAC) and susceptible (S-Lim) slender amaranth (*Amaranthus viridis* L.) populations (Pop.) collected in citrus orchards from São Paulo State, Brazil.

Pop.	Fresh Weight Reduction					Plant Mortality				
	c	d	b	$GR_{50} \pm$ CI	RF [2]	c	d	b	$LD_{50} \pm$ CI	RF [2]
					Experiment I					
S-Lim	0.3	99.9	0.9	22.0 ± 7.4	-	1.9	102.2	2.8	149.5 ± 11.8	-
R1	1.8	100.2	2.2	248.4 ± 17.7	11.3	1.1	100.7	3.4	918.9 ± 31.4	6.1
R2	0.6	100.1	1.9	165.9 ± 16.1	7.5	2.2	100.3	3.4	607.8 ± 15.1	4.1
R-IAC	2.9	101.3	1.9	192.2 ± 17.4	8.7	4.1	101.1	2.3	768.3 ± 32.6	5.1
					Experiment II					
S-Lim	1.7	100.0	1.6	38.1 ± 3.8	-	0.4	101.4	3.4	113.1 ± 8.1	-
R1	0.4	98.4	2.1	301.8 ± 39.4	7.9	0	99.4	3.4	706.3 ± 45.4	6.2
R2	5.5	98.4	1.4	205.5 ± 42.5	5.4	4.6	96.7	3.6	639.7 ± 75.1	5.7
R-IAC	3.7	97.9	1.5	259.5 ± 52.4	6.8	4.9	97.0	2.7	726.7 ± 59.2	6.4

[1] $Y = c + \{(d - c)/[1 + (x/g)b]\}$, where Y = fresh weight reduction or plant mortality by 50% with respect to the control, c = lower limit, d = upper limit, b = slope of the curve, g = herbicide dose at the inflection point (i.e., GR_{50} or LD_{50}), and x = herbicide dose. CI values are the 95% limits of the confidence intervals (n = 6). [2] Resistance factors are the R-to-S GR_{50} or LD_{50} ratios.

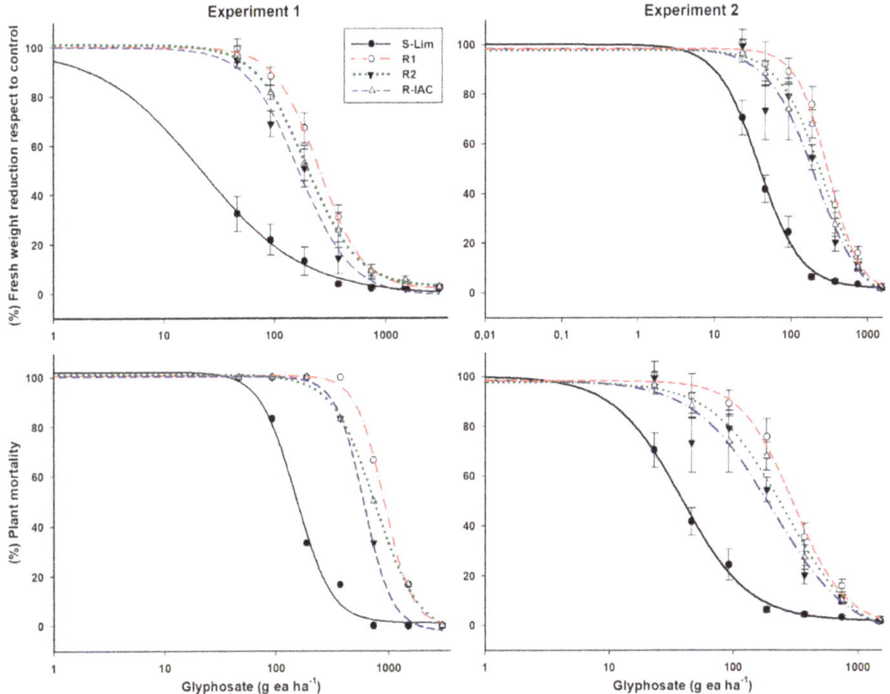

Figure 1. Glyphosate dose-response curves relative to percentages of fresh weight reduction and plant survival of susceptible and resistant slender amaranth (*Amaranthus viridis* L.) populations collected in citrus orchards from São Paulo State, Brazil. Vertical bars of the fresh weight reduction plots represent the standard error of the mean ($n = 6$).

Based on plant survival, the S-Lim population had LD_{50} values of 149 and 113 g ea ha^{-1} in the experiments I and II, respectively. However, individuals of this population did not survive glyphosate doses greater than 370 g ea ha^{-1}. In the resistant populations, although their individuals had a large weight loss in relation to their respective untreated controls, around 50% of their individuals survived to the field dose of 720 g ea ha^{-1}. In addition, plants of the R1 and R-IAC populations survived at 1480 g ea ha^{-1}. Thus, the R populations were between 4.1 and 6.4 times more resistant than the S-Lim population. In both experiments, the R1 population was the most resistant based on its lower weight reduction or the highest survival rate of its individuals (Figure 1, Table 1).

3.2. Shikimic Acid Accumulation

Monitoring the shikimic acid at 96 HAT, the S-Lim population accumulated ~26 μg g^{-1} fresh weight. This amount of shikimic acid found in the S population was between 6 and 9 times higher than that accumulated in R populations, which accumulated less than 4 μg g^{-1} fresh weight (Figure 2A). In the in vivo assays, the highest amount of shikimic acid was also found in the S-Lim population that began to accumulate it from the lowest glyphosate concentration (6.4 μg mL^{-1} at 10-μM glyphosate). The accumulation increased as glyphosate concentrations increased, reaching up ~240-μg shikimic acid mL^{-1} at 1000 μM glyphosate. R populations began to accumulate shikimic acid only from 50 μM glyphosate; however, such accumulation was much lower than in population S-Lim. At 1000 μM glyphosate, the R1, R2, and R-IAC populations accumulated 16, 43, and 33 μg shikimic acid mL^{-1}, respectively (between 5.6 and 15 times less than the S-Lim population) (Figure 2B).

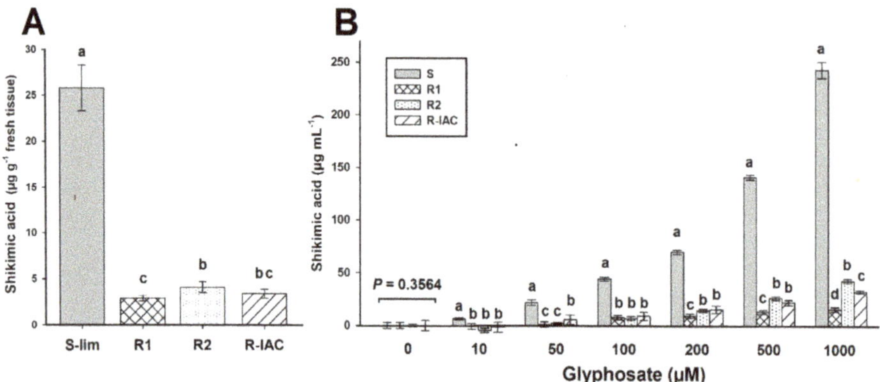

Figure 2. Shikimic acid accumulation in glyphosate-susceptible and -resistant slender amaranth (*Amaranthus viridis* L.) populations collected in citrus orchards from São Paulo State, Brazil. (**A**) Shikimic acid accumulation in sprayed plants with 360 g ae ha^{-1} at 96 h after treatment. (**B**) In vivo shikimic acid accumulation at different glyphosate concentrations. Groups of bars with the same letter above them are not different using the Tukey test at 95%. Vertical bars represent the standard error of the mean (n = 9 technical replicates).

3.3. EPSPS Activity

Resistant *A. viridis* populations presented higher EPSPS basal activity than the S population (0.22 µmol Pi µg^{-1} TSP min^{-1}), but there were differences between them (ranged from 0.38 to 0.52 µmol Pi µg^{-1} TSP min^{-1}) (Figure 3A). To inhibit the EPSPS activity in the S-Lim population, only 4 µM glyphosate was necessary. The R1 population was 20 times more resistant than to the S-Lim population, and the R2 and R-IAC populations were 9 and 12 times more resistant, respectively (Figure 3B, Table 2).

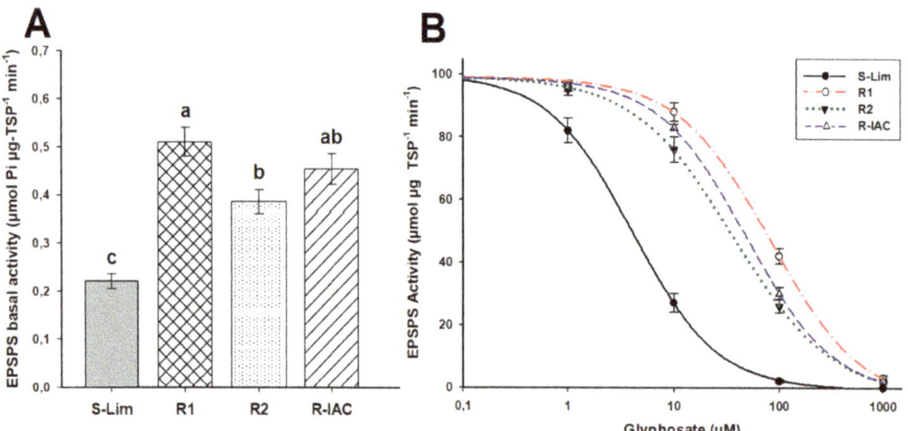

Figure 3. EPSPS (5-enolpyruvylshikimate-3-phosphate synthase) activity in glyphosate-susceptible and -resistant slender amaranth (*Amaranthus viridis* L.) populations collected in citrus orchards from São Paulo State, Brazil a. (**A**) Basal EPSPS activity (absence of glyphosate). Same letter above bars are not different using the Tukey test at 95%. (**B**) Dose-response curves of the EPSPS enzyme activity, expressed as a percentage of the untreated control, exposed to different glyphosate concentrations (µM). Histograms represent the means, and vertical bars the standard error (n = 3).

Table 2. Parameters of the sigmoidal equations [1] used to estimate the glyphosate concentration (μM) required to inhibit the EPSPS by 50% (I_{50}) in glyphosate-resistant and -susceptible slender amaranth (*Amaranthus viridis* L.) populations (Pop.) collected in citrus orchards from São Paulo State, Brazil.

Pop.	c	d	b	I_{50} ± CI	RF [2]
S-Lim	0.5	99.9	1.1	4.0 ± 0.3	-
R1	1.1	99.2	0.9	82.4 ± 6.3	20.6
R2	2.6	99.4	0.9	36.4 ± 3.6	9.1
R-IAC	2.1	98.9	1.0	48.4 ± 4.4	12.1

[1] $Y = c + \{(d - c)/[1 + (x/g)b]\}$, where Y = EPSPS inhibition by 50% with respect to the control, c = lower limit, d = upper limit, b = slope of the curve, g = herbicide concentration at the inflection point (i.e., I_{50}), and x = herbicide dose. CI values are the 95% limits of the confidence intervals (n = 3). [2] Resistance factors are the R-to-S GR_{50} or LD_{50} ratios.

4. Discussion

The GR_{50} values found for the S-Lim population of slender amaranth were lower than 100 g ae ha^{-1} and showed the great susceptibility to glyphosate of this species. Susceptible populations of *Amaranthus hybridus* [16], *A. palmeri* [25], and *Amaranthus tuberculatus* [26] also showed GR_{50} values less than 100 g ae ha^{-1} (17, 89, and 61 g ae ha^{-1}, respectively). In general, *Amaranthus* species are very sensitive to glyphosate, so the recommended label doses by manufacturers range from 370 (initial growth stage) to 740 (adult stage) g ae ha^{-1}, while for other weed species, such as *Panicum maximum*, *Richardia brasiliensis*, and *Sida rhombifolia*, the minimum dose is 1850 g ae ha^{-1} [27]. Taking into account the phenological stage of the plants used in our experiments (3–5 true leaves), the reference dose must be 370 g ae ha^{-1}. However, a dose of 740 g ae ha^{-1} was considered as the field reference dose because it is the minimum glyphosate dose sprayed by Brazilian citrus growers [3].

One of the main criteria to consider a weed as a new case of herbicide resistance is that individuals have been survived to a dose normally lethal to individuals of a wild (susceptible) population of the same species, and these individuals are able to reproduce sexually, i.e., the resistance must be inheritable [28,29]. The estimated LD_{50} values for the R slender amaranth populations were close to 740 g ae ha^{-1}, with RFs ranging from 4.1 to 6.4 in relation to the S-Lim population ($LD_{50} \leq$ 150 g ae ha^{-1}). These results demonstrated that the field reference dose did not satisfactorily control ~50% of individuals of R populations. Although not arbitrary, the herbicide amount enough to achieve an acceptable control level in a resistant weed population often requires at least twice its LD_{50} [30]. However, increasing the herbicide dose is not recommended because the selection pressure also increases, as well as the resistance level, depending on the resistance mechanism involved [31].

Glyphosate is a potent inhibitor of the EPSPS, enzyme responsible for the biosynthesis of the chorismate from shikimic acid [9]. Therefore, glyphosate effect can be measured by monitoring the accumulation of shikimic acid [22]. In both experiments, the R plants of slender amaranth accumulated less shikimic acid than the S-Lim plants, showing the low sensitivity of the R populations to this herbicide. The lower accumulation of shikimic acid in resistant plants is due to glyphosate that does not reach the EPSPS in sufficient amount to inhibit this enzyme [32]. The R1 population had a shikimate accumulation pattern different from those observed in the R2 and R-IAC, suggesting that the mechanisms endowing glyphosate resistance among R populations were different from each other.

Enzymatic activity tests did not directly reveal the mechanism that is involved in resistance but could guide whether it is a target-site or non-target-site mechanism. In this case, the three R populations showed high levels of basal enzyme activity than the susceptible population, which might suggest that an EPSPS overexpression participated in the resistance. This mechanism has been the most common resistance mechanism found in glyphosate-resistant *Amaranthus* sp., such as *A. palmeri*, *A. tuberculatus*, and *Amaranthus spinosus*, which have presented different EPSPS gene copy numbers, ranging from 5 to more than 160 [33–36]. This mechanism may confer an unpredictable resistance level to glyphosate [37]; however, the higher I_{50} of the R1 population suggests that another mechanism is also involved in resistance. Single mutations at 106 position in the EPSPS gene confer low levels of glyphosate resistance (2 to 4 times the recommended dose); reduced glyphosate translocation and vacuole sequestration endow moderate resistance levels (4 to 8 times), and double or triple mutations at 102, 103, and 106 positions confer high resistance levels (>10 times) [16,37,38]. Our results did not allow us to confirm the mechanism(s) that endows glyphosate resistance in the R populations of slender amaranth; therefore, biochemical and molecular studies are necessary to unravel them.

5. Conclusions

These results confirmed that slender amaranth (*A. viridis*) was selected for glyphosate resistance in Brazilian citrus orchards, being the first case in the world reported for this species. At least one target site-type mechanism participated in this glyphosate resistance; however, further experiments are required to elucidate the resistance mechanisms involved.

Author Contributions: Conceptualization, R.A.-d.l.C. and M.F.d.G.F.d.S.; methodology, R.A.-d.l.C., G.d.S.A., G.M.d.O., and L.B.d.C.; investigation, R.A.-d.l.C., G.d.S.A., G.M.d.O., and L.R.R.; resources, R.A.-d.l.C., F.A.d.A., and M.F.d.G.F.d.S.; writing—original draft preparation, R.A.-d.l.C., G.M.d.O., and G.d.S.A.; writing—review and editing, all authors; funding acquisition, R.A.-d.l.C. and M.F.d.G.F.d.S.; Supervision, M.F.d.G.F.d.S. All authors have read and agreed to the published version of the manuscript.

Funding: The RAC, GMO and MFGFS thank the "Fundação de Amparo à Pesquisa do Estado de São Paulo" for the financial support (main-grant: 2014/50918-7, sub-grants: 2018/15910-6 and 2019/15527-0).

Acknowledgments: Authors thank Pamela Freitas dos Santos for her technical assistance in the greenhouse experiments.

Conflicts of Interest: No conflicts of interest have been declared.

References

1. U.S. Department of Agriculture [USDA]. Citrus: World Markets and Trade. Available online: https://apps.fas.usda.gov/psdonline/circulars/citrus.pdf (accessed on 17 August 2019).
2. Instituto Brasileiro de Geografia e Estatística [IBGE]. Produção Agrícola Municipal. 2017. Available online: http://www.sidra.ibge.gov.br/bda/tabela/protabl.asp?c=1613&z=p&o=18&i=p (accessed on 14 August 2019).
3. Martinelli, R.; Monquero, P.A.; Fontanetti, A.; Conceição, P.M.; Azevedo, F.A. Ecological mowing: An option for sustainable weed management in young citrus orchards. *Weed Technol.* **2017**, *31*, 260–268. [CrossRef]
4. Food and Agriculture Organization [FAO]. FAOSTAT: Statistical Database. Available online: http://faostat.fao.org/site/567/default.aspx (accessed on 17 August 2019).
5. Singh, M.; Sharma, S.D. Benefits of triazine herbicides and other weed control technology in citrus management. In *The Triazine Herbicides—50 Years Revolutionizing Agriculture*, 1st ed.; Lebaron, H.M., McFarland, J.E., Burnside, O.C., Eds.; Elsevier: San Diego, CA, USA, 2008; pp. 199–209.
6. La Cruz, R.A.-D.; Domínguez-Martínez, P.A.; Da Silveira, H.M.; Cruz-Hipólito, H.E.; Palma-Bautista, C.; Vazquez-Garcia, J.G.; Domínguez-Valenzuela, J.A.; De Prado, R. Management of glyphosate-resistant weeds in Mexican citrus groves: Chemical alternatives and economic viability. *Plants* **2019**, *8*, 325. [CrossRef]
7. Bracamonte, E.; Da Silveira, H.M.; La Cruz, R.A.-D.; Domínguez-Valenzuela, J.A.; Cruz-Hipolito, H.E.; De Prado, R. From tolerance to resistance: Mechanisms governing the differential response to glyphosate in *Chloris barbata*. *Pest Manag. Sci.* **2018**, *74*, 1118–1124. [CrossRef]
8. Caetano, R.S.X.; Christoffoleti, P.J.; Victoria-Filho, R. Weed seed bank of a 'Pera' citrus orchard. *Sci. Agric.* **2001**, *58*, 509–517. [CrossRef]
9. Duke, S.O. The history and current status of glyphosate. *Pest Manag. Sci.* **2018**, *74*, 1027–1034. [CrossRef]
10. Vázquez-García, J.G.; Castro, P.; Torra, J.; Alcántara-de la Cruz, R.; De Prado, R. Resistance evolution to EPSPS inhibiting herbicides in false barley (*Hordeum murinum*) harvested in Southern Spain. *Agronomy* **2020**, *10*, 992. [CrossRef]
11. Heap, I. The International Herbicide-Resistant Weed Database. Available online: http://www.weedscience.org (accessed on 30 June 2020).
12. Moreira, M.S.; Nicolai, M.; Carvalho, S.J.P.; Christoffoleti, P.J. Glyphosate-resistance in *Conyza canadensis* and *C. Bonariensis*. *Planta Daninha* **2008**, *25*, 157–164. [CrossRef]

13. De Carvalho, L.B.; Cruz-Hipólito, H.; González-Torralva, F.; Alves, P.; Christoffoleti, P.; De Prado, R. Detection of sourgrass (*Digitaria insularis*) biotypes resistant to glyphosate in Brazil. *Weed Sci.* **2011**, *59*, 171–176. [CrossRef]
14. Alcántara-de la Cruz, R.; Oliveira, G.M.; Carvalho, L.B.; Silva, M.F.G.F. Herbicide resistance in Brazil: Status, impacts, and future challenges. In *Herbicides-Current Research and Case Studies in Use*, 2nd ed.; Ferreira, K.M., Ed.; IntechOpen: Londod, UK, 2020. [CrossRef]
15. Sellers, B.A.; Smeda, R.J.; Johnson, W.G.; Kendig, J.A.; Ellersick, M.R. Comparative growth of six *Amaranthus* species in Missouri. *Weed Sci.* **2003**, *51*, 329–333. [CrossRef]
16. García, M.J.; Palma-Bautista, C.; Rojano-Delgado, A.M.; Bracamonte, E.; Portugal, J.; La Cruz, R.A.-D.; De Prado, R. The triple amino acid substitution TAP-IVS in the EPSPS gene confers high glyphosate resistance to the superweed *Amaranthus hybridus*. *Int. J. Mol. Sci.* **2019**, *20*, 2396. [CrossRef]
17. Thomas, W.E.; Burke, I.C.; Spears, J.F.; Wilcut, J.W. Influence of environmental factors on slender amaranth (*Amaranthus viridis*) germination. *Weed Sci.* **2006**, *54*, 316–320. [CrossRef]
18. Raimondi, M.; Oliveira, J.R.; Constantin, J.; Rios, F.; Gemelli, A.; Raimondi, R. Dose-response curve to soil applied herbicides and susceptibility evaluation of different *Amaranthus* species using model identity. *Planta Daninha* **2015**, *33*, 157–164. [CrossRef]
19. Francischini, A.C.; Constantin, J.; Oliveira, R.S., Jr.; Santos, G.; Braz, G.B.P.; Dan, H.A. First report of *Amaranthus viridis* resistence to herbicides. *Planta Daninha* **2014**, *32*, 571–578. [CrossRef]
20. Küpper, A.; Borgato, E.A.; Patterson, E.; Netto, A.G.; Nicolai, M.; De Carvalho, S.J.P.; Nissen, S.J.; Gaines, T.A.; Christoffoleti, P. Multiple resistance to glyphosate and acetolactate synthase inhibitors in Palmer amaranth (*Amaranthus palmeri*) identified in Brazil. *Weed Sci.* **2017**, *65*, 317–326. [CrossRef]
21. Cromartie, T.H.; Polge, N. Method of Detecting Shikimic Acid. U.S. Patent 006482654B1, 8 February 2002. Available online: https://patents.google.com/patent/US6482654B1/en (accessed on 30 June 2020).
22. Dayan, F.; Owens, D.K.; Corniani, N.; Silva, F.M.L.; Watson, S.B.; Howell, J.; Shaner, D.L. Biochemical markers and enzyme assays for herbicide mode of action and resistance studies. *Weed Sci.* **2015**, *63*, 23–63. [CrossRef]
23. Bradford, M.M. A rapid and sensitive method for the quantitation of microgram quantities of protein utilizing the principle of protein-dye binding. *Anal. Biochem.* **1976**, *72*, 248–254. [CrossRef]
24. Ritz, C.; Baty, F.; Streibig, J.C.; Gerhard, D. Dose-response analysis using R. *PLoS ONE* **2015**, *10*, e0146021. [CrossRef]
25. Fernández-Escalada, M.; Gil-Monreal, M.; Zabalza, A.; Royuela, M. Characterization of the *Amaranthus palmeri* physiological response to glyphosate in susceptible and resistant populations. *J. Agric. Food Chem.* **2016**, *64*, 95–106. [CrossRef]
26. Lorentz, L.; Gaines, T.A.; Nissen, S.J.; Westra, P.; Strek, H.J.; Dehne, H.W.; Ruiz-Santaella, J.P.; Beffa, R. Characterization of glyphosate resistance in *Amaranthus tuberculatus* populations. *J. Agric. Food Chem.* **2016**, *62*, 8134–8142. [CrossRef]
27. Agência de Defesa Agropecuária do Paraná [ADAPAR]. Trade Label of Roundup Original DI. Available online: http://www.adapar.pr.gov.br/arquivos/File/defis/DFI/Bulas/Herbicidas/rounduporiginaldi190118.pdf (accessed on 30 June 2020).
28. Heap, I. Criteria for Confirmation of Herbicide-Resistant Weeds—With Specific Emphasis on Confirming Low Level Resistance. Available online: http://www.weedscience.org/Documents/ResistanceCriterion.pdf (accessed on 30 June 2020).
29. Takano, H.K.; Oliveira, R.S., Jr.; Constantin, J.; Braz, G.B.P.; Gheno, E.A. Goosegrass resistant to glyphosate in Brazil. *Planta Daninha* **2017**, *35*, e017163071. [CrossRef]
30. Alcántara-de la Cruz, R.; Fernández-Moreno, P.T.; Ozuna, C.V. Target and non-target site mechanisms developed by glyphosate-resistant hairy beggarticks (*Bidens pilosa* L.). *Front. Plant Sci.* **2016**, *7*, 1492. [CrossRef] [PubMed]
31. Salas, R.A.; Scott, R.C.; Dayan, F.E.; Burgos, N.R. EPSPS Gene amplification in glyphosate-resistant Italian ryegrass (*Lolium perenne* ssp. *multiflorum*) populations from Arkansas (United States). *J. Agric. Food Chem.* **2015**, *63*, 5885–5893. [PubMed]
32. Shaner, D.L.; Nadler-Hassar, T.; Henry, W.B.; Koger, C.H. A rapid in vivo shikimate accumulation assay with excised leaf discs. *Weed Sci.* **2005**, *53*, 769–774. [CrossRef]

33. Gaines, T.A.; Zhang, W.; Wang, D.; Bukun, B.; Chisholm, S.T.; Shaner, D.L.; Nissen, S.J.; Patzoldt, W.L.; Tranel, P.J.; Culpepper, A.S.; et al. Gene amplification confers glyphosate resistance in *Amaranthus palmeri*. *Proc. Natl. Acad. Sci. USA* **2010**, *107*, 1029–1034. [CrossRef] [PubMed]
34. Nandula, V.K.; Ray, J.D.; Ribeiro, D.N.; Pan, Z.; Reddy, K.N. Glyphosate resistance in tall waterhemp (*Amaranthus tuberculatus*) from Mississippi is due to both altered target-site and nontarget-site mechanisms. *Weed Sci.* **2013**, *61*, 374–383. [CrossRef]
35. Riggins, C.W.; Peng, Y.H.; Stewart, C.N.; Tranel, P.J. Characterization of de novo transcriptome for waterhemp (*Amaranthus tuberculatus*) using GS-FLX 454 pyrosequencing and its application for studies of herbicide target-site genes. *Pest Manag. Sci.* **2010**, *66*, 1042–1052. [CrossRef] [PubMed]
36. Gaines, T.A.; Patterson, E.L.; Neve, P. Molecular mechanisms of adaptive evolution revealed by global selection for glyphosate resistance. *New Phytol.* **2019**, *223*, 1770–1775. [CrossRef]
37. Giacomini, D.A.; Westra, P.; Ward, S.M. Variable inheritance of amplified EPSPS gene copies in glyphosate-resistant Palmer Amaranth (*Amaranthus palmeri*). *Weed Sci.* **2019**, *67*, 176–182. [CrossRef]
38. Yu, Q.; Jalaludin, A.; Han, H.; Chen, M.; Sammons, R.D.; Powles, S.B. Evolution of a double amino acid substitution in the 5-enolpyruvylshikimate-3-phosphate synthase in *Eleusine indica* conferring high-level glyphosate resistance. *Plant Physiol.* **2015**, *167*, 1440–1447. [CrossRef]

Article

Phytotoxic Effect of Herbicides on Various Camelina [*Camelina sativa* (L.) Crantz] Genotypes and Plant Chlorophyll Fluorescence

Łukasz Sobiech [1,*], Monika Grzanka [1], Danuta Kurasiak-Popowska [2] and Dominika Radzikowska [1]

1 Agronomy Department, Faculty of Agronomy and Bioengineering, Poznan University of Life Sciences, 60-632 Poznan, Poland; monika.grzanka@up.poznan.pl (M.G.); dominika.radzikowska@up.poznan.pl (D.R.)

2 Genetics and Plant Breeding Department, Faculty of Agronomy and Bioengineering, Poznan University of Life Sciences, 60-632 Poznan, Poland; danuta.kurasiak-popowska@up.poznan.pl

* Correspondence: lukasz.sobiech@up.poznan.pl; Tel.: +48-61-848-7599

Received: 20 April 2020; Accepted: 21 May 2020; Published: 23 May 2020

Abstract: Camelina is an oil plant classified as a minor crop. The small acreage is the main cause of the small amount of plant protection products that are registered for use on camelina plantations. This contributes to difficulties in the protection of this plant. In the conducted experiment, the genetic similarity of genotypes of camelina was compared. The effect of selected herbicides (propaquizafop at rate 70 g a.i. ha^{-1}, quizalofop-p-ethyl at rate 50 g a.i. ha^{-1}, clopyralid at rate 90 g a.i. ha^{-1}, and picloram at rate 24 g a.i. ha^{-1} applied in the three-four-leaves growth stage of camelina) on six individual genotypes of the plant and plant chlorophyll fluorescence after the use of these substances was also determined. The Przybrodzka variety showed the lowest level of damage in the assessment carried out 42 days after herbicide application and the damages of plants after quizalofop-p-ethyl and propaquizafop was completely gone. The variety Przybrodzka had the lowest genetic similarity to all analyzed genotypes. In other cases, genetic similarity of analyzed genotypes could not be linked to herbicide-related damage. Picloram contributed to the greatest damage to test plants and had the greatest impact on the operation of photosystem II (PSII). However, the level of plant chlorophyll fluorescence parameter values indicates small PSII damage for all substances and the possibility of subsequent plant regeneration. The results of the presented research indicate that it is worth referring to several plant varieties in phytotoxicity studies of herbicides towards arable crops.

Keywords: camelina; herbicides; genetic similarity; phytotoxicity; formulation; plant chlorophyll fluorescence

1. Introduction

One of the features of modern agriculture is the simplification of crop rotation [1]. Diversity in agriculture plays an important role, both in terms of preserving biodiversity and the agricultural environment and protecting crops against pests, diseases, and weeds [2,3]. Therefore, it is worth looking for plants that can diversify the species composition of contemporary farming plants. Camelina is a self-pollinating annual plant belonging to the *Brassicaceae* family. The vegetation period of spring forms in the climatic conditions of Poland is about 100–130 days, while winter forms are 280–300 days. The first of these forms is mainly grown in the world. Plant height in the case of spring camelina varies in the range of 50–90 cm; for winter forms it is 80–120 cm. The diameter of the flowers is 5–7 mm. The fruit of the plant in question is the silicule, in which there are 8–15 seeds reaching a length of about 2 mm. The mass of one thousand seeds is about 0.8–1.8 g [4].

Drought is becoming a growing problem in agricultural production in the world [5,6]. Camelina is considered a plant with low water requirements [7]. The production of phytalexins exhibiting antimicrobial activity and the limited area of cultivation of camelina contribute to the fact that this plant is characterized by high resistance to diseases and pests [8]. In the cultivation of camelina, the main problem is the presence of weeds, whose quantity and fresh mass are sometimes greater than on plantations of other plants belonging to the *Brassicaceae* family [9].

Camelina was an important oilseed plant in Europe until the mid-20th century, especially in the northern and central parts of the continent. Later, its production level decreased [10]. In recent years, interest in camelina has been noted due to the unique composition of oil obtained from camelina. High oil content (28–40%) [11] makes it a suitable oil source for biodiesel production [12]. Camelina oil is rich in oleic (18:1, 14–16%), linoleic (18:2, 15–23%), linolenic (18:3, 31–40%), and eicosenoic (20:1, 12–15%) acid [13].

The most attention is paid to the use of camelina oil for the production of biodiesel, renewable diesel, and renewable jet fuel [13,14]. The first tests of hydroprocessed renewable jet fuel (a mixture of the branched aromatics and the C7-C18 branched alkanes) were carried out by the US Air Force in 2009 [15]. Since then, numerous fighter jets (Thunderbirds) and jets (KLM Royal Dutch and Japan Airlines) have been successfully tested on a blend of JP-8 (typical jet fuel) and camelina-derived jet fuel [13].

Camelina oil is also used in the food industry as functional food and in the cosmetics industry [16]. Camelina oil can be used in the petrochemical-based polymer industry, because epoxidized camelina oil has the properties of peel adhesion and, after further formulation, the potential to be used for preparation of pressure-sensitive adhesives, coating, or resin [17].

Camelina meal, or ground camelina oil cake or press cake, which is a by-product of the oil extraction process can be used as animal feed [18,19]. The US Food and Drug Administration (USDA) approved the use of camelina meal in feed rations given to beef cattle and broiler chickens, provided that the percentage is not higher than 10% [20].

The possibility of full use of *Camelina sativa* is highly economically advantageous. The cost of producing fuel for *Camelina sativa* grown in Canada was USD 0.83/L, with approximately 88% of the total production cost resulting from the cost of feedstock [21].

Due to the fact that this plant is grown in a small area, the availability of plant protection products registered for use in this crop is an important problem. In the European Union, the availability of plant protection products registered for use in this crop is a limitation. However, it closes within the limits of several active substances that can be used [22].

Winter oilseed rape is the most important oilseed plant in Europe [23]. Herbicides registered for use in this include propaquizafop, quizalofop-p-ethyl, clopyralid, and picloram [24]. Camelina and winter oilseed rape belong to the same family; hence, it is likely that these herbicides could also be used in minor crops such as camelina. However, it is important to determine the phytotoxicity of these herbicides to camelina. Determining the impact of herbicides on crop development is also important in the context of different varieties, which is associated with their intraspecific variability [25].

Propaquizafop and quizalofop-p-ethyl are herbicides that belong to the group of aryloxyphenoxypropionates [26,27]. Aryloxyphenoxypropionates are classified as inhibitors of acetyl coenzyme A carboxylase (ACCase inhibitors) [28]. This enzyme is involved in the synthesis of fatty acids [29]. It catalyzes the reaction of attaching carbon dioxide to acetyl-CoA, resulting in the formation of malonyl-CoA. There are two forms of this enzyme in plant cells—prokaryotic and eukaryotic. The latter of these forms is sensitive to the effects of these herbicides [30]. Most higher plants have both forms of CoA acetyl carboxylase. The exception are plants from the family *Gramineae*, in which only the eukaryotic form is found [31]. Weeds belonging to this family are therefore sensitive to the action of herbicides belonging to the inhibitors of acetyl coenzyme A carboxylase [32].

Clopyralid and picloram are substances classified as pyridine carboxylates. They belong to the group of synthetic auxins [33]. These herbicides mimic the action of excessive amounts of indole-3-acetic acid (IAA), which is the main natural phytohormone found in higher plants [34]. The reason for the selectivity of these preparations for monocots is not yet known [35].

Chlorophyll fluorescence is becoming a very powerful tool in agricultural, environmental, and ecological research [36]; along with increasing the sensitivity of fluorimeters measurement, these methods become also useful in ecotoxicological studies [37]. The chlorophyll fluorescence signal can be used as a probe for photosynthetic activity [38]. As is known, plant stress, both biotic and abiotic, negatively affects the physiological state of plants and their photosynthetic activity. Plant protection treatments, including herbicidal ones, cause short-term stress in crops, whereas they are toxic to the undesirable plants (weed). In many cases, herbicidal preparations directly reduce photosynthetic activity or indirectly cause damage to plants, which in turn reduces the efficiency of photosynthesis. Herbicides that directly affect PS-II usually work by inhibiting the transport of electrons in PS-II [39]. Analysis of chlorophyll fluorescence can yield important information on how herbicides interfere with photosystem I and II mechanisms [40]. The assessment of the physiological state of the photosynthetic apparatus is performed on the basis of analyses of several groups of measured and calculated parameters based on induction and general identification of two different states of the photosynthetic apparatus: photochemically inactive (after adaptation to the dark)—electron transport is completely stopped, the H^+ gradient across thylakoid membranes is minimal, the concentration of NADPH and ATP is minimal—and photochemically active (after adaptation of the sample to light)—NADPH and ATP synthesis and CO_2 binding occur. In addition, the technique for measuring chlorophyll fluorescence allows the recording of changes occurring in PSII under the influence of stress factors before visual symptoms are seen, which makes this technique the perfect complement to the visual assessment of phytotoxicity. Chlorophyll fluorescence measurement is a non-destructive tool that allows scientists to obtain information on the photosynthesis process without destroying the test sample [41].

The aim of the study is to assess the effect of selected herbicides on different camelina genotypes, compare the sensitivity of different genotypes with the genetic distance between them, and determine the effect of herbicides on plant chlorophyll fluorescence.

2. Materials and Methods

2.1. Genetic Distance Testing

2.1.1. Plant material

The spring variety Omega and winter varieties Luna and Przybrodzka of camelina were obtained from the Polish National Plant Breeders' Rights (PBR). The winter genotype Lenka has been protected by temporary rights in Poland since 2017. The spring 57L3 genotype is stable breeding material. All of the above genotypes were bred at the Poznań University of Life Sciences (Poznań, Poland) [42]. The spring genotype Hoga (PI 650150) collected in Denmark came from the US National Plant Germplasm System (NPGS, Copenhagen, Denmark).

2.1.2. DNA Extraction

DNA was isolated from the leaves of 10-day-old seedlings with the use of an extraction kit (Genomic Mini AX Plant, A&A Biotechnology, Gdynia, Poland) by means of the column-based method.

2.1.3. Random Amplified Polymorphic DNA Assays

The RAPD-PCR reaction was carried out in a 12.5 μL mixture consisting of: water, 1 M Tris HCl (pH 8.3), 2 mM $MgCl_2$, 2 mM dNTP, 5 pmol/μL primer, 5 U/μL Taq polymerase, 25 ng/μL DNA extract. The temperature regime was set to an initial denaturation of 60 s at 94 °C, followed by 34 cycles (60 s at 94 °C, 45 s at primer annealing at the optimal temperature, 30 s at 72 °C), and finishing with an extension step of 5 min at 72 °C. The PCR was carried out in a T Professional Basic Gradient

Thermocycler. A set of 20 10-mer RAPD-PCR primers was used for DNA amplification (Supplementary Table S1).

2.1.4. Microsatellite Markers of DNA Assays

A set of 7 primer pairs (P3H4, P4B3, P4C2, P4H3, P6C2, P6E4, P7D4) was used for DNA amplification according to the method developed by Manca et al. [43]. The polymerase chain reaction (PCR) was conducted in a mixture composed of water, 5 μL; DreamTaqTMGreen PCR Master Mix, 6.25 μL; primers, 2 × 0.25 μL (final concentration was 20 μM); and DNA matrix, 1 μL. The amplification took place under the following conditions: initial denaturation for 5 min at 94 °C, 40 cycles (denaturation at 94 °C for 30 s, primer annealing at the optimal temperature for 30 s, synthesis 72 °C for 30 s), final synthesis 72 °C for 30 min. Two slightly different touchdown PCR protocols called TD1 and TD2 were used for selected primers (P4H3, P6C2, and P7D4). The TD1 profile consisted of initial denaturation for 5 min at 94 °C, followed by 10 cycles of 30 s at 94 °C, annealing for 30 s at a temperature of 65 °C reduced by 1 °C every cycle, and for 30 s at 72 °C. This was followed by 30 cycles of 30 s at 94 °C, 30 s at 55 °C, 30 s at 72 °C, and one final extension step at 72 °C for 30 min. The TD2 program differed from the TD1 profile in the annealing temperature and number of cycles. The annealing temperature was reduced in steps of 0.5 °C in each cycle (8 cycles) from 54 °C to 50 °C, followed by 32 cycles at 50 °C.

2.1.5. Electrophoresis Conditions

The PCR products were separated by electrophoresis lasting 90 min at 100 V in 1.5% agarose gels containing TBE buffer. Next, they were visualized under UV light after being stained with ethidium bromide. To visualize PCR products, a Molecular Imager Gel Doc^{TM} XR UV transilluminator was used with the ImageLab^{TM}Software (Bio-Rad, Hercules, CA, USA).

2.1.6. Statistical Analysis

The coefficients of genetic similarity (S) between the cultivars were calculated using the formula developed by Nei and Li [44]:

$$\text{Sij} = \frac{2\text{Nij}}{\text{Ni} + \text{Nj}} \tag{1}$$

where:

Nij is the number of alleles present in the ith and jth cultivars,

Ni is the number of alleles present in the ith cultivar,

Nj is the number of alleles present in the jth cultivar, and

i, j = 1, 2, . . . , 20.

The cultivars were grouped hierarchically according to the coefficients, using the unweighted pair group method of arithmetic means (UPGMA). The relationships between genotypes were presented in a dendrogram.

2.2. Greenhouse Experiment

Camelina were grown for 76 days in a greenhouse on photoperiod 16 h day/8 h night. Greenhouse temperature was maintained at 25 ± 2 °C during the day and at 20 ± 2 °C during the night. Camelina were planted in the greenhouse in plastic pots (1.0 L, 15 cm diameter) containing a mixture of peat and soil at a 1:1 ratio. Soil moisture was systematically measured by ML3 ThetaProbe Soil Moisture Sensor (ThetaProbe, Eijkelkamp, The Netherlands). By regular replenishment, the appropriate weight of pots was maintained at 65–75% of soil water capacity. Relative air humidity was around 50–80%. Natural sunlight with an intensity of 600 μE m^{-2} s^{-1} was supplemented with sodium lamps (HPS) with a capacity of 400 W (Elektro-Valo Oy. Netafim. Avi: 13473, Uusikaupunki, Finland). Two weeks after emergence, camelina plants were thinned to 6 uniform seedlings per pot.

Herbicides were applied at doses recommended for use in winter rapeseed cultivation. Tested herbicides were: propaquizafop at rate 70 g a.i. (active ingredient)* ha^{-1} (Agil S 100 EC, Adama Agan Ltd, Airport city, Israel), quizalofop-p-ethyl at rate 50 g a.i.* ha^{-1} (Targa Super 05 EC, Nissan Chemical Europe, Saint-Didier-au-Mont-d'Or, France), clopyralid at rate 90 g a.i. ha^{-1} (Major 300 SL, Innvigo Sp. z o.o., Warsaw, Poland), and picloram at rate 24 g a.i.* ha^{-1} (Zorro 300 SL, Innvigo Sp. z o.o., Warsaw, Poland). Herbicides have been applied in the three-four-leaves growth stage of camelina.

Treatments were applied using a spray chamber with Tee Jet 1102 (TeeJet Technologies GmbH, Schorndorf, Germany) nozzles delivering 200 L* ha^{-1} at 0.2 MPa. Water used for herbicide application contained (mg/L): 114 Ca^{2+}, 7.4 Mg^{2+}, 0 Na^{+}, 0 K^{+}, <1 Fe^{3+}, 356 $CaCO_3$, and pH 7.3. Camelina plants were assessed 21 and 42 days after treatment (DAT) by estimating the herbicidal phytotoxicity compared to untreated control. The data was calculated using the Henderson–Tilton formula [45]. Visual evaluation of the herbicidal phytotoxicity was based on the comparison of the condition of camelina plants from objects treated with herbicides with plants from the untreated control. Visual phytotoxicity was shown using a scale ranging from 0% (untreated) to 100% (completely destroyed plant). The greenhouse trial was designed as a randomized complete block with four replications. Four herbicide variants and untreated controls for six camelina genotypes were tested in each series, resulting in 30 combinations (120 replicates) in both series. Assessments were made for two series of tests. They were made in accordance with EPPO Standards PP 1/49 (3) [46]. At the end of the experiments, the weight of the plants was tested (data not shown). Statistical analysis was conducted using Statistica software (Version 12, StatSoft Inc., Tulsa, OK, USA). Data were subjected to ANOVA followed by Tukey's protected LSD test at the 0.05 probability level.

2.3. Plant Chlorophyll Fluorescence

Chlorophyll fluorescence was measured at the first young fully mature healthy leaf with Multi-Mode Chlorophyll Fluorometer (OS5p, Opti-Sciences, Inc., Hudson, USA) with PAR Clip. The Fv/Fm protocol was selected [47]. Fluorescence measurements were done in triplicate for each replication, which gives 12 results for combination for one series of test. Measurements were made 14 and 21 days after herbicide application. Before the measurements, the leaves were dark-adapted for 30 minutes using white clips to silence photosynthesis. Measurements were made at the same time of day, at midday. Settings fluorometer protocols were selected according to the OS5p User's Guide, the standard in Plant Stress Measurement, Opti-Sciences, as follow: Modulation Source: Red, Modulation Intensity: 15, Detector Gain: 05, Saturation Flash Intensity: 29 (non-nominated units). Before measurement, settings parameters were adjusted to an appropriate level. Modulation intensity and detector gain values were adjusted so that the set Ft (fluorescence signal) was stable and in the range of 150–250 counts, which means that modulated illumination is not driving photosynthesis. Similarly, based on the values of the measured parameters, the saturation flash intensity was determined to get the saturation point. The following parameters were measured: F0—minimum fluorescence, Fm—maximum fluorescence, Fv—variable fluorescence. The following parameters were calculated: Fv/Fm—Maximum Photochemical Efficiency of Photosystem II; Fv/Fm = (Fm − F0)/Fm. Statistical analysis of average results for all tested genotypes was conducted using Statistica software (Version 12, StatSoft Inc., Tulsa, OK, USA). Data were subjected to ANOVA followed by Tukey's protected LSD test at the 0.05 probability level.

3. Results

3.1. Genetic Distance Testing

To assess genetic variation of *Camelina sativa* genotypes, RAPD (random amplified polymorphic DNA) and SSR (simple-sequence repeats also known as microsatellites) analyses were carried out. Finally, 6 polymorphic PCR products were obtained from SSR analysis (P4C2 and P6C2 were monomorphic) and 69 for RAPD analysis.

The RAPD and SSR data were used for grouping genotypes by the UPGMA method. The relationship among genotypes is presented in the form of a dendrogram (Figure 1). Genotypes (spring and winter) from Polish breeding of the last 20 years had the greatest genetic similarity: Omega (registered in PBR in 2013), Luna (registered in PBR in 2012), Lenka (in registration process), and the stabilized breeding line 57L3. The old Polish variety Przybrodzka was characterized by the lowest genetic similarity to all analyzed genotypes. The Przybrodzka variety was donated by the Institute of Soil Science and Plant Cultivation (Poznań, Poland) to the US National Plant Germplasm System (NPGS) in 1966.

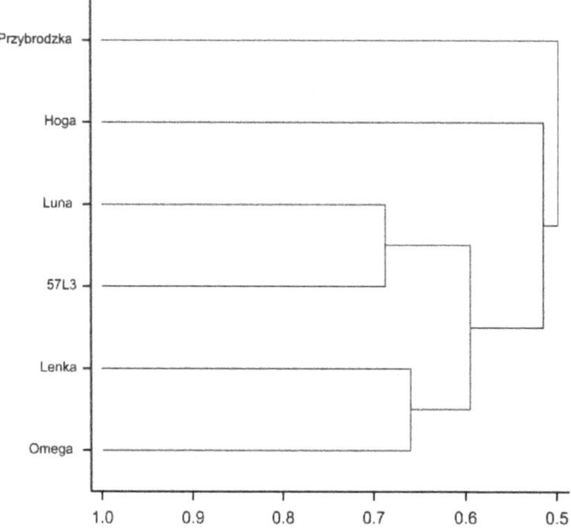

Figure 1. A dendrogram of *Camelina sativa* genotypes. The cultivars were grouped hierarchically using the unweighted pair group method of arithmetic means (UPGMA).

3.2. Greenhouse Experiment

Quizalofop-p-ethyl and propaquizafop caused the smallest damage to all camelina varieties and genotypes. At 21 days after the use of propaquizafop, the largest damage (15%) was observed on the Omega variety (Table 1). In the assessment at 42 DAT (Table 2), no statistically significant camelina damage was found after the use of these herbicides. The greatest damage occurred after using herbicides from the group of synthetic auxin herbicides (clopyralid, picloram). Picloram caused the most damage to camelina and Luna at 21 DAT. Unlike graminicides, damage after clopyralid and picloram was not completely transient, and their levels were still high. In the assessment carried out 42 DAT, the largest damages (30%, 30%, 25%, respectively) were observed on the Hoga genotype and on the Omega and Luna varieties. The smallest damage after using picloram was found on the variety Przybrodzka and genotype Lenka and after the application of clopyralid on the Przybrodzka variety and genotypes Lenka and 57 L3. In the experiment, the mass of plants was analyzed. However, the results were not presented in the results and discussions. The reason is a problem with drawing reliable conclusions for this parameter.

Table 1. Visual assessment of phytotoxicity (%) 21 days after treatment (21 DAT).

Treatment	Varieties/Genotypes						
	Przybrodzka	**Lenka**	**Luna**	**Omega**	**Hoga**	**57 L3**	**LSD (0.05)**
Untreated	0 C	0 D	0 E	0 E	0 C	0 C	-
Propaquizafop 70 g a.i.* ha^{-1}	10b B	0c D	10b C	15a C	10b B	7.5b B	2.9
Quizalofop-p-ethyl 50 g a.i.* ha^{-1}	0c C	7.5ab C	5b D	10a D	0c C	0c C	3.1
Clopyralid 90 g a.i.* ha^{-1}	10d B	22.5c A	20c B	35ab B	40a A	32.5b A	5.7
Picloram 24 g a.i.* ha^{-1}	20d A	20d B	55a A	45b A	40bc A	35c A	7.2
LSD (0.05)	2.8	2.4	0.2	4.8	1.6	3.7	-

0% (untreated) to 100% (completely destroyed plant); a–e and A–E different letters indicate statistically different mean LSD ($p < 0.05$); capital letters—vertical, e.g. A; lower-case letters—horizontal, e.g., a.

Table 2. Visual assessment of phytotoxicity (%) 42 days after treatment (42 DAT).

Treatment	Varieties/Genotypes						
	Przybrodzka	**Lenka**	**Luna**	**Omega**	**Hoga**	**57 L3**	**LSD (0.05)**
Untreated	0 C	0 C	0 C	0 D	0 D	0 D	-
Propaquizafop 70 g a.i.* ha^{-1}	0a C	0a C	0a C	10a C	10a C	5a C	n.s.
Quizalofop-p-ethyl 50 g a.i.* ha^{-1}	0a C	0a C	0a C	10a C	10a C	0a D	n.s.
Clopyralid 90 g a.i.* ha^{-1}	7.5c B	7.5c B	15b B	20a B	20a B	10c B	3.2
Picloram 24 g a.i.* ha^{-1}	15cd A	10d A	25ab A	30a A	30a A	20bc A	5.3
LSD (0.05)	1.9	1.9	3.9	1.4	2.6	3.2	-

n. s.—no significant differences; 0% (untreated) to 100% (completely destroyed plant); a–e and A–E different letters indicate statistically different mean LSD ($p < 0.05$); capital letters—vertical, e.g., A; lower-case letters—horizontal, e.g., a.

3.3. Plant Chlorophyll Fluorescence

Chlorophyll fluorescence was measured 14 and 21 days after herbicide application. Averages of results of herbicide effects on plant chlorophyll fluorescence for all genotypes are shown. The measurement at the first date showed no significant differences in the parameters: Fv (Figure 2) and Fm (Figure 3). However, significant differences were noted in the Fv/Fm parameter (Figure 4). They were observed in all combinations where herbicides were used, relative to the values obtained for the unsprayed control. The highest decrease in Maximum Photochemical Efficiency of Photosystem II was found in plants sprayed with picloram. During the second, subsequent measurement, significant differences were found between the preparations used in all three parameters of chlorophyll fluorescence measurement: Fm, Fv, and Fv/Fm. Similar to the case of changes in the Fv/Fm parameter value, a significant decrease in Fm and Fv parameters was observed in all sprayed objects relative to the control, to the greatest extent in plants treated with picloram and to the smallest those treated with quizalofop-p-ethyl and propaquizafop. The highest decrease in Fm and Fv parameters in relation to the unsprayed control was recorded for objects treated with picloram. The measurements showed no significant differences in the parameter F0 (data not shown).

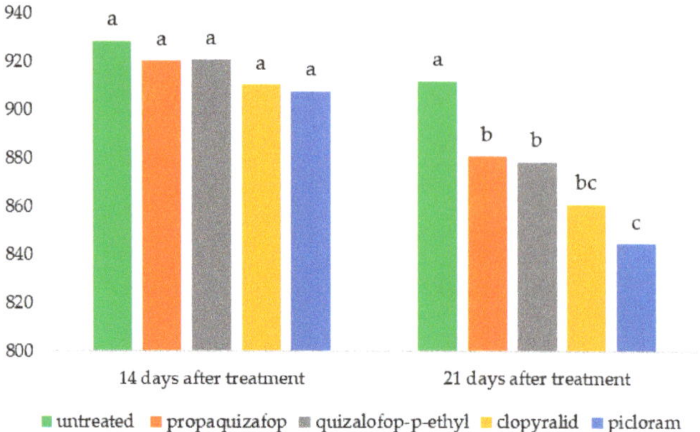

Figure 2. The effect of herbicides on Fv—variable fluorescence (non-nominated units). Different letters ([a–c]) indicate statistically different mean LSD ($p < 0.05$) 14 DAA = 19.52; LSD ($p < 0.05$) 21 DAA = 25.91; DAA—days after herbicides application.

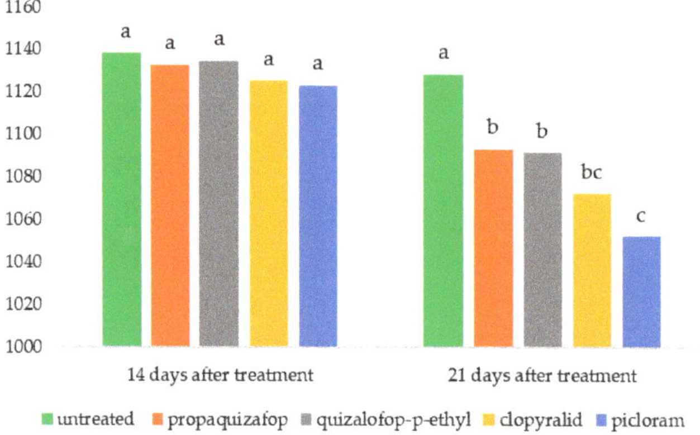

Figure 3. The effect of herbicides on Fm—maximum fluorescence (non-nominated units). Different letters ([a–c]) indicate statistically different mean LSD ($p < 0.05$) 14 DAA = 23.80; LSD ($p < 0.05$) 21 DAA = 32.17; DAA—days after herbicides aplication.

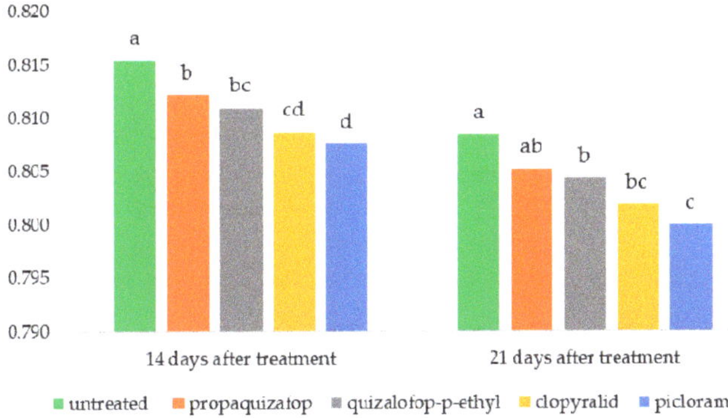

Figure 4. The effect of herbicides on Fv/Fm—maximum Photochemical Efficiency of Photosystem II (non-nominated units). Different letters ($^{a–c}$) indicate statistically different mean LSD ($p < 0.05$) 14 DAA = 0.029; LSD ($p < 0.05$) 21 DAA = 0.0036; DAA—days after herbicides aplication.

4. Discussion

The phytotoxicity of herbicides towards a crop means exceeding the maximum protective capacity of this plant related to the mechanisms of selectivity or the possibility of detoxification of these substances [48]. Selective herbicides can therefore cause transient stress in crop plants [49]. In some cases, they can damage plants and reduce yields [50].

In the conducted experiment, propaquizafop and quizalofop-P-ethyl contributed to the inhibition of plant growth. For most varieties and genotypes, however, these were transient symptoms. Quizalofop-P-ethyl is a registered substance for use on camelina plantations in different parts of the world [13,22]. This substance is therefore widely used in practice to control monocotyledonous weeds in this crop. According to data available in the literature, propaquizafop contributed little to the phytotoxicity effect on Indian mustard [51]. Damage was also observed in the case of the application of other substances classified as acetyl coenzyme A carboxylase inhibitors in canola cultivation [52]. These plants, like camelina, belong to the *Brassicaceae* family. The results of the conducted research indicate that substances belonging to synthetic auxin herbicides have contributed to greater damage to plants than graminicides. The application of these preparations resulted in twisting of camelina leaves. Picloram has the highest degree of phytotoxicity. Slight damage caused by picloram and aminopyralid (also included in synthetic auxin herbicides) in relation to some species of plants belonging to *Brassicaceae* were also noted in other papers [53]. The Przybrodzka variety was characterized by the smallest genetic similarity in relation to all tested varieties and genotypes. At the same time, it was characterized by the greatest tolerance to the effects of applied substances. In the case of other varieties and genotypes, no relationship between genetic similarity and sensitivity to the herbicides used was observed.

The herbicides used in the study included synthetic auxins. Substances belonging to this group may cause calluses of plants, which affect their mass [54]. It is estimated that for studies using synthetic auxins, visual assessment of phytotoxicity is more desirable than testing the weight of plants [55].

One of the biggest problems associated with camelina breeding is the low diversity within this species. In addition, only spring camelina genotypes were collected in gene banks. Due to the low intraspecific variability, breeders look for variability of related species and genera, or induce variation through mutagenesis or due to genetic engineering [56]. Analysis of the genetic similarity of spring genotypes (currently grown in Poland and Ukraine) and winter varieties of *Camelina sativa* (which are traditionally grown only in Western Europe) was performed by Kurasiak-Popowska et al. (2018) [56]. Apart from the 'Przybrodzka' cultivar, the Polish cultivars of spring and winter camelina were included in one similarity group. The plant material for this publication was selected based on the results of molecular analyses, yielding, and the content of chemical compounds in camelina genotypes collected at the University of Life Sciences in Poznań [56,57].

An important factor influencing the herbicidal effectiveness of herbicides is the formulation of the preparations [58]. The type of formulation also plays an important role in relation to herbicide phytotoxicity to the crop [59]. In the experiment, quizalofop-p-ethyl and propaquizafop were used in the EC (emulisfiable concentrate) formulation. This formulation is a mixture of pesticides, adjuvants, and emulsifiers that are dissolved in oil. Together with the water in the sprayer tank, they form a diluted, stable emulsion. One of the disadvantages of this formulation, however, is the possibility of damage to the leaves of crop plants [60]. The oil-based adjuvants contained in the herbicide formulation may contribute to the dissolution of the wax layer covering the surface of the crop [61]. Oil adjuvants work in a similar way and are included in the composition of the spray liquid. In some cases, they increase the risk of phytotoxicity but at the same time contribute to improving the herbicidal herbicide effectiveness [62]. Most of the graminicides available on the market occur in the EC formulation, which due to the oil content can cause damage to some plants. In our research, herbicides from the aryloxyphenoxypropionates group that control monocotyledonous weeds were phytotoxic for some genotypes of camelina because they were in EC formulation, injurious for cuticular wax.

The Fv/Fm parameter is considered to be a reliable indicator of the photochemical activity of the photosynthetic apparatus, and its reduction indicates the occurrence of stress in plants, the consequence of which is damage to the PSII function [63]. During both the first and second measurements, a significant decrease in the Fv/Fm parameter (maximum photochemical efficiency of PSII) was observed. However, according to studies of other authors [63,64], the obtained values do not indicate strong damage, and it can be concluded that they did not affect the health of plants. The Fv/Fm value for unstressed control of most plant species is close to 0.83 [65,66]. The decrease in the Fm parameter—maximum fluorescence—indicates the occurrence of stress in plants, the effect of which means that not all electron acceptors in PSII can be completely reduced [67]. In turn, a decrease in the Fv parameter—variable fluorescence—may indicate a decrease in PSII activity and scattering of excitation energy in the form of heat [63]. However, the level of the obtained parameters indicates slight damage to the PSII. The measurements were to show the trend of plant stress caused by individual herbicides, but the level of this reaction, due to the selected mechanisms of herbicide action, did not indicate high stress of the plants; hence, the average results for genotypes were shown. Larger differences in the response of individual genotypes were found in the visual assessment; therefore, the results of visual observation are presented separately for individual genotypes. However, the trend in plant reactions was the same for both parameters—for plant chlorophyll fluorescence and visual assessment: the highest plant damage was observed for picloram and the lowest for ACCase inhibitors. Dayan et al. (2012) [40], studying the mechanisms of herbicide action using chlorophyll fluorescence, showed that the effect on photosynthesis caused by herbicides from the group of aryloxyphenoxypropionate inhibitors of ACCase was very modest on cucumber cotyledons, which may be due to the grass selectivity of this herbicide. In the same research, some herbicides acting as superauxins did not affect photosynthesis in the cucumber cotyledon assay. It can therefore be concluded that the decrease in fluorescence parameters as a consequence of spraying is secondary and results from leaf damage caused by herbicide injury, and the photosynthetic activity carried out is proportional to the surface of the undamaged leaf.

5. Conclusions

The small number of herbicides registered for use on camelina plantations makes weeds a difficult problem. The conducted research determined the genetic distance between different varieties and genotypes of camelina (spring and winter forms). The effect of individual herbicides on chlorophyll fluorescence was tested. The Przybrodzka variety was characterized by the greatest genetic distance in relation to other varieties and genotypes. At the same time, it was highly resistant to phytotoxic effects of herbicides. For the other varieties and genotypes, no similar relationship was observed. Damage to camelina after herbicide application was largely transient. Among the substances used, picloram contributed most to the occurrence of phytotoxicity. Plant fluorescence measurements indicate that PSII damage was not large and plants have a chance of regeneration. Further research is needed on the possibility of using various herbicides in camelina and other minor crops. It should be carried out taking into account the effect of the substance on individual plant varieties.

Supplementary Materials: The following are available online at http://www.mdpi.com/2077-0472/10/5/185/s1, Table S1. Nucleic sequences of random amplified polymorphic DNA (RAPD-PCR) primers.

Author Contributions: Conceptualization, Ł.S.; methodology, Ł.S. and D.K.-P.; data curation, Ł.S., M.G., D.K.-P.; writing—original draft preparation, Ł.S., M.G., D.K.-P. and D.R.; writing—review and editing, Ł.S., M.G., D.K.-P. and D.R.; visualization, Ł.S. and D.K.-P.; supervision, Ł.S. All authors have read and agreed to the published version of the manuscript.

Funding: The publication was co-financed within the framework of Ministry of Science and Higher Education programme as "Regional Initiative Excellence" in years 2019–2022, Project No. 005/RID/2018/19.

Conflicts of Interest: The authors declare no conflict of interest.

References

1. Barbieri, P.; Pellerin, S.; Nesme, T. Comparing crop rotations between organic and conventional farming. *Sci. Rep.* **2017**, *7*, 1–10. [CrossRef] [PubMed]
2. Kremen, C.; Iles, A.; Bacon, C. Diversified Farming Systems: An Agroecological, Systems-based Alternative to Modern Industrial Agriculture. *Ecol. Soc.* **2012**, *17*, 44. [CrossRef]
3. Sawinska, Z.; Świtek, S.; Głowicka-Wołoszyn, R.; Kowalczewski, P.Ł. Agricultural Practice in Poland Before and after Mandatory IPM Implementation by the European Union. *Sustainability* **2020**, *12*, 1107. [CrossRef]
4. Kurasiak-Popowska, D. Lnianka siewna—Roślina historyczna czy perspektywiczna? *Fragm. Agron.* **2019**, *36*, 42–54.
5. Bandurska, H.; Niedziela, J.; Pietrowska-Borek, M.; Nuc, K.; Chadzinikolau, T.; Radzikowska, D. Regulation of proline biosynthesis and resistance to drought stress in two barley (*Hordeum vulgare* L.) genotypes of different origin. *Plant Physiol. Biochem.* **2017**, *118*, 427–437. [CrossRef]
6. Vogel, E.; Donat, M.G.; Alexander, L.V.; Meinshausen, M.; Ray, D.K.; Karoly, D.; Meinshausen, N.; Frieler, K. The effects of climate extremes on global agricultural yields. *Environ. Res. Lett.* **2019**, *14*, 1–12. [CrossRef]
7. Iskandarov, U.; Kim, H.J.; Cahoon, E.B. Camelina: An Emerging Oilseed Platform for Advanced Biofuels and Bio-Based Materials. In *Plants and bioenergy*; Carpita, N., McCann, M., Buckeridge, M.S., Eds.; Springer Science & Business Media: New York, NY, USA, 2014; pp. 131–140.
8. Vollmann, J.; Steinkellner, S.; Glauninger, J. Variation in resistance of camelina (*Camelina sativa* (L.) Crtz.) to downy mildew (Peronospora camelinae Gäum.). *J. Phytopathol.* **2001**, *149*, 129–133. [CrossRef]
9. Lenssen, A.W.; Iversen, W.M.; Sainju, U.M.; Caesar-TonThat, T.C.; Blodgett, S.L.; Allen, B.L.; Evans, R.G. Yield, Pests, and Water Use of Durum and Selected Crucifer Oilseeds in Two-Year Rotations. *Agron. J.* **2012**, *104*, 1295–1304. [CrossRef]
10. Román-Figueroa, C.; Padilla, R.; Uribe, J.M.; Paneque, M. Land Suitability Assessment for Camelina (*Camelina sativa* L.) Development in Chile. *Sustainability* **2017**, *9*, 154. [CrossRef]
11. Nguyen, H.T.; Silva, J.E.; Podicheti, R.; Macrander, J.; Yang, W.; Nazarenus, T.J.; Nam, J.-W.; Jaworski, J.G.; Lu, C.; Scheffler, B.E.; et al. Camelina seed transcriptome: A tool for meal and oil improvement and translational research. *Plant Biotechnol. J.* **2013**, *11*, 759–769. [CrossRef]
12. Lu, C.; Napier, J.A.; Clemente, T.E.; Cahoon, E.B. New frontiers in oilseed biotechnology: Meeting the global demand for vegetable oils for food, feed, biofuel, and industrial applications. *Curr. Opin. Biotechnol.* **2011**, *22*, 252–259. [CrossRef] [PubMed]
13. Berti, M.; Gesch, R.; Eynck, C.; Anderson, J.; Cermak, S. Camelina uses, genetics, genomics, production, and management. *Ind. Crops Prod.* **2016**, *94*, 690–710. [CrossRef]
14. Goímez-Monedero, B.; Bimbela, F.; Arauzo, J.S.; Faria, J.; Ruiz, M.P. Pyrolysisof red eucalyptus, camelina straw, and wheat straw in an ablative reactor. *Energy Fuels* **2015**, *29*, 1766–1775. [CrossRef]
15. Eynck, C.; Falk, K.C. Camelina (*Camelina sativa*). In *CAB International 2013. Biofuel Crops: Production, Physiology and Genetics*; Singh, B.P., Ed.; Fort Valley State University: Fort Valley, GA, USA, 2013; pp. 369–391.

16. Kurasiak-Popowska, D.; Stuper-Szablewska, K.; Nawracała, J. Olej rydzowy jako naturalne źródło karotenoidów dla przemysłu kosmetycznego. *Przem. Chem.* **2017**, *96*, 2077–2080. [CrossRef]
17. Kim, N.; Li, Y.; Sun, X.S. Epoxidation of Camelina sativa oil and peel adhesion properties. *Ind. Crops. Prod.* **2015**, *64*, 1–8. [CrossRef]
18. Hixson, S.M.; Parrish, C.C.; Anderson, D.M. Full substitution of fish oil with camelina (*Camelina sativa*) oil, with partial substitution of fish meal with camelina meal, in diets for farmed Atlantic salmon (*Salmo salar*) and its effecton tissue lipids and sensory quality. *Food Chem.* **2014**, *157*, 51–61. [CrossRef]
19. Woyengo, T.A.; Patterson, R.; Slominski, B.A.; Beltranena, E.; Zijlstra, R.T. Nutritive value of cold-pressed camelina cake with or without supplementation of multi-enzyme in broiler chicken. *Poultry Sci.* **2016**, *95*, 2314–2321. [CrossRef]
20. Moser, B.R. Camelina (*Camelina sativa* L.) oils as a biofuels feedstock: Golden opportunity or false hope? *Lipid Technol.* **2010**, *22*, 270–273. [CrossRef]
21. Chu, P.L.; Vanderghem, C.; MacLean, H.L.; Saville, B.A. Financial analysis and risk assessment of hydroprocessed renewable jet fuel production from camelina, carinata and used cooking oil. *Appl. Energy* **2017**, *198*, 401–409. [CrossRef]
22. Matyjaszczyk, E. Protection possibilities of agricultural minor crops in the European Union: A case study of soybean, lupin and camelina. *J. Plant. Dis. Prot.* **2020**, *127*, 55–61. [CrossRef]
23. Pullens, J.W.M.; Sharif, B.; Trnka, M.; Balek, J.; Semenov, M.A.; Olesen, J.E. Risk factors for European winter oilseed rape production under climate change. *Agric. For. Meteorol.* **2019**, *272*, 30–39. [CrossRef]
24. Available online: www.gov.pl/web/rolnictwo/wyszukiwarka-srodkow-ochrony-roslin (accessed on 14 April 2020).
25. Bountin, C.; Aya, K.L.; Carpenter, D.; Thomas, P.J.; Rowland, O. Phytotoxicity testing for herbicide regulation: Shortcomings in relation to biodiversity and ecosystem services in agrarian systems. *Sci. Total Environ.* **2012**, *415*, 79–92. [CrossRef] [PubMed]
26. Singh, S.P.; Rawal, S.; Dua, V.K.; Roy, S.; Sadaworti, M.J.; Sharma, S.K. Evaluation of propaquizafop: A new molecule as post emergence herbicide in potato. *Int. J. Chem. Stud.* **2017**, *5*, 1216–1220.
27. Bradley, K.W.; Wu, J.; Hatzios, K.K.; Hagood, E.S. The mechanism of resistance to aryloxyphenoxypropionate and cyclohexanedione herbicides in a johnsongrass biotype. *Weed Sci.* **2001**, *49*, 477–484. [CrossRef]
28. Délye, C.; Calmès, É.; Matéjicek, A. SNP markers for black-grass (Alopecurus myosuroides Huds.) genotypes resistant to acetyl CoA-carboxylase inhibiting herbicides. *Theor. Appl. Genet.* **2002**, *104*, 1114–1120. [CrossRef] [PubMed]
29. Ohlrogge, J.; Browse, J. Lipid Biosynthesis. *Plant Cell* **1995**, *7*, 957–970.
30. Ali, K.; Tyagi, A. Recent advances in acetyl CoA carboxylase: A key enzyme of fatty acidbiosynthesis in plants. *Int. J. Appl. Biol. Pharm.* **2016**, *7*, 264–277.
31. Konishi, T.; Shinohara, K.; Yamada, K.; Sasaki, Y. Acetyl-CoA Carboxylase in Higher Plants: Most Plants Other Than Gramineae Have Both the Prokaryotic and the Eukaryotic Forms of This Enzyme. *Plant Cell Physiol.* **1996**, *37*, 117–122. [CrossRef]
32. Konishi, T.; Sasaki, Y. Compartmentalization of two forms of acetyl-CoA carboxylase in plants and the origin of their tolerance toward herbicides. *Proc. Natl. Acad. Sci. USA* **1994**, *91*, 3598–3601. [CrossRef]
33. Busi, R.; Goggin, D.E.; Heap, I.M.; Horak, M.J.; Jugulam, M.; Masters, R.A.; Napier, R.M.; Riar, D.S.; Satchivi, N.M.; Torra, J.; et al. Weed resistance to synthetic auxin herbicides. *Pest Manag. Sci.* **2018**, *74*, 2265–2276. [CrossRef]
34. Grossmann, K. Mediation of Herbicide Effects by Hormone Interactions. *J. Plant Growth Regul.* **2003**, *22*, 109–122. [CrossRef]
35. McSteen, P. Auxin and Monocot Development. *Cold Spring Harb. Perspect. Biol.* **2010**, *2*, a001479. [CrossRef] [PubMed]
36. Gottardini, E.; Cristofori, A.; Cristofolini, F.; Nali, C.; Pellegrini, E.; Bussotti, F.; Ferretti, M. Chlorophyll-related indicators are linked to visible ozone symptoms: Evidence from a field study on native *Viburnum lantana* L. plants in northern Italy. *Ecol. Ind.* **2014**, *39*, 65–74. [CrossRef]
37. Juneau, P.; Qiu, B.; Deblois, C. Use of chlorophyll fluorescence as a tool for determination of herbicide toxic effect: Review. *Toxicol. Environ. Chem.* **2007**, *89*, 609–625. [CrossRef]

38. Kalaji, H.M.; Schansker, G.; Brestic, M.; Bussotti, F.; Calatayud, A.; Ferroni, L.; Goltsev, V.; Guidi, L.; Jajoo, A.; Li, P.; et al. Frequently asked questions about chlorophyll fluorescence, the sequel. *Photosyn. Res.* **2017**, *132*, 13–66. [CrossRef]
39. Kaiser, Y.I.; Menegat, A.; Gerhards, R. Chlorophyll fluorescence imaging: A new method for rapid detection of herbicide resistance in Alopecurus myosuroides. *Weed Res.* **2013**, *53*, 399–406. [CrossRef]
40. Dayan, F.E.; de Zaccaro, M.L. Chlorophyll fluorescence as a marker for herbicide mechanisms of action. *Pestic. Biochem. Phys.* **2012**, *102*, 189–197. [CrossRef]
41. Kalaji, H.M.; Jajoo, A.; Oukarroum, A.; Brestic, M.; Zivcak, M.; Samborska, I.A.; Cetner, M.D.; Łukasik, I.; Goltsev, V.; Ladle, R.J. Chlorophyll a fluorescence as a tool to monitor physiological status of plants under abiotic stress conditions. *Acta Physiol. Plant.* **2016**, *38*, 102. [CrossRef]
42. Kurasiak-Popowska, D.; Ryńska, B.; Stuper-Szablewska, K. Analysis of distribution of selected bioactive compounds in Camelina sativa from seeds to pomace and oil. *Agronomy* **2019**, *9*, 168. [CrossRef]
43. Manca, A.; Pecchia, P.; Mapelli, S.; Masella, P.; Galasso, I. Evaluation of genetic diversity in a *Camelina sativa* (L.) Crantz collection using microsatellite markers and biochemical traits. *Genet. Resour. Crop Evol.* **2013**, *60*, 1223–1236. [CrossRef]
44. Nei, M.; Li, W.H. Mathematical model for studying genetic variation in terms of restriction endonucleases. *Proc. Natl. Acad. Sci. USA* **1979**, *76*, 5269–5273. [CrossRef] [PubMed]
45. Smith, J.; Wherley, B.; Reynolds, C.; White, R.; Senseman, S.; Falk, S. Weed control spectrum and turfgrass tolerance to bioherbicide Phoma macrostoma. *Int. J. Pest. Manag.* **2015**, *61*, 91–98. [CrossRef]
46. Available online: https://www.eppo.int/RESOURCES/eppo_standards/pp1_list (accessed on 2 May 2020).
47. Kowalczewski, P.Ł.; Radzikowska, D.; Ivanišová, E.; Szwengiel, A.; Kačániová, M.; Sawinska, Z. Influence of Abiotic Stress Factors on the Antioxidant Properties and Polyphenols Profile Composition of Green Barley (*Hordeum vulgare* L.). *Int. J. Mol. Sci.* **2020**, *21*, 397. [CrossRef] [PubMed]
48. de Carvalho, S.J.P.; Nicolai, M.; Ferreira, R.R.; de Oliveira Figueira, A.V.; Christoffoleti, P.J. Herbicide selectivity by differential metabolism: Considerations for reducing crop damages. *Sci. Agric.* **2009**, *66*, 136–142. [CrossRef]
49. Dyer, W.E.; Weller, S.C. Plant response to herbicides. In *Plant Abiotic Stress*; Jenks, M., Hasegawa, P.M., Eds.; Blackwell Publishing Ltd.: Oxford, UK; Ames, IA, USA, 2005; pp. 171–214.
50. Young, B.G.; Young, J.M.; Matthews, J.L.; Owen, M.D.K.; Zelaya, I.A.; Hartzler, R.G.; Wax, L.M.; Rorem, K.W.; Bollero, G.A. Soybean Development and Yield as Affected by Three Postemergence Herbicides. *Agron. J.* **2003**, *95*, 1152–1156. [CrossRef]
51. Gupta, S.; Sharma, P.K.; Kumar, S.; Singh, S.; Singh, S.; Yadav, H.L. Effect of herbicides and fertilizers on the phytotoxicity, growth and yield of Indian mustard [*Brassica juncea* (L.) Czern and Coss.]. *J. Pharmacogn. Phytochem.* **2019**, *8*, 2451–2454.
52. Mohamed, I.A. Efficiency of Some Post Emergence Acetyl Coenzyme a Carboxylase-Inhibitor Herbicides Against Certain Grassy Weeds in Canola Fields. *Egypt. Acad. J. Biolog. Sci.* **2017**, *8*, 49–58. [CrossRef]
53. Harris, B.M.; Bath, S.; Degenhardt, R. Safening Aminopyralid Compositions in Brassica Species with Clopyralid and Methods of Use Thereof. U.S. Patent Application No 15/611,956, 14 December 2017.
54. Roskamp, J.M.; Chahal, G.S.; Johnson, W.G. The Effect of Cations and Ammonium Sulfate on the Efficacy of Dicamba and 2,4-D. *Weed Technol.* **2013**, *27*, 72–77. [CrossRef]
55. Dowler, C.C. A cucumber bioassay test for the soil residues of certain herbicides. *Weed Sci.* **1969**, *17*, 309–310. [CrossRef]
56. Kurasiak-Popowska, D.; Tomkowiak, A.; Czołpińska, M.; Bocianowski, J.; Weigt, D.; Nawracała, J. Analysis of yield and genetic similarity of Polish and Ukrainian Camelina sativa genotypes. *Ind. Crop. Prod.* **2018**, *123*, 667–675. [CrossRef]
57. Kurasiak-Popowska, D.; Stuper-Szablewska, K. The phytochemical quality of Camelina sativa seed and oil. *Acta Agric. Scand. Sect. B-Soil Plant Sci.* **2020**, *70*, 39–47. [CrossRef]
58. Kudsk, P. Optimising herbicide dose: A straightforward approach to reduce the risk of side effects of herbicides. *Environmentalist* **2008**, *28*, 49–55. [CrossRef]
59. Mashayamombe, B.K.; Mazarura, U.; Chiteka, A. The Effect of Two Formulations of Sulfentrazone on Soil and Leaf Residues and Phytotoxicity in Tobacco (*Nicotiana Tabacum* L.). *Asian J. Agric. Dev.* **2013**, *3*, 135–140.
60. Hazra, D.K.; Purkait, A. Role of pesticide formulations for sustainable crop protection and environment management: A review. *J. Pharmacogn. Phytochem.* **2019**, *8*, 686–693.

61. Hazen, J.L. Adjuvants—Terminology, Classification, and Chemistry. *Weed Technol.* **2000**, *14*, 773–784. [CrossRef]
62. Zhang, J.; Jaeck, O.; Menegat, A.; Zhang, Z.; Gerhards, R.; Ni, H. The Mechanism of Methylated Seed Oil on Enhancing Biological Efficacy of Topramezone on Weeds. *PLoS ONE* **2013**, *8*, e74280. [CrossRef]
63. Kalaji, M.H.; Łoboda, T. *Fluorescencja Chlorofilu w Badanach Stanu Fizjologicznego Roślin*; SGGW Publisher: Warsaw, Poland, 2010; p. 116.
64. Cetner, M.D.; Dąbrowski, P.; Samborska, I.A.; Łukasik, I.; Swoczyna, T.; Pietkiewicz, S.; Bąba, W.; Kalaji, H.M. Zastosowanie pomiarów fluorescencji chlorofilu w badaniach środowiskowych. *Kosmos* **2016**, *65*, 197–205.
65. Björkman, O.; Demmig, B. Photon yield of O_2 evolution and chlorophyll fuorescence characteristics at 77 k among vascular plants of diverse origins. *Planta* **1987**, *170*, 489–504. [CrossRef]
66. Radzikowska, D.; Grzanka, M.; Kowalczewski, P.-Ł.; Głowicka-Wołoszyn, R.; Blecharczyk, A.; Nowicki, M.; Sawinska, Z. Influence of SDHI Seed Treatment on the Physiological Conditions of Spring Barley Seedlings under Drought Stress. *Agronomy* **2020**, *10*, 731. [CrossRef]
67. Sayed, O.H. Chlorophyll Fluorescence as a Tool in Cereal Crop Research. *Photosynthetica* **2003**, *41*, 321–330. [CrossRef]

Article

Weed Infestation and Health of Organically Grown Chamomile (*Chamomilla recutita* (L.) Rausch.) Depending on Selected Foliar Sprays and Row Spacing

Cezary A. Kwiatkowski, Małgorzata Haliniarz and Elżbieta Harasim *

Department of Herbology and Plant Cultivation Techniques, University of Life Sciences, Akademicka 13, 20-950 Lublin, Poland; czarkw@poczta.onet.pl (C.A.K.); malgorzata.haliniarz@up.lublin.pl (M.H.)
* Correspondence: elzbieta.harasim@up.lublin.pl; Tel.: +48-81-4456820

Received: 22 April 2020; Accepted: 12 May 2020; Published: 13 May 2020

Abstract: Chamomile is a herbal plant of very high economic importance worldwide. Its organically grown raw material is particularly valuable. Under organic farming conditions, weeds and fungal diseases are an important problem in a chamomile plantation. Seeking agronomic solutions designed to eliminate the occurrence of these pathogens in chamomile crops is constantly valid. The aim of the present study was to evaluate the effect of some foliar sprays (enhancing the condition of the crop plant and its competition against pathogens) and different row spacing of two chamomile cultivars on weed infestation and health of a chamomile plantation. The study results presented in this paper were collected from field experiments carried out in the organic system in the village of Dys (the central Lublin region, Poland) over the period 2014–2016. Experiments were conducted on podzolic soil (class III) as a split-block design in 3 replicates in plots with an area of 525 m^2 (6.25 m^2 a single plot). This study included two chamomile cultivars ("Złoty Łan", "Mastar"). The second experimental factor was single or double foliar application of three bioproducts (Herbagreen Basic, Bio-algeen, Effective Microorganisms—EM Farming). The other experimental factor was a different row spacing of chamomile (40 cm and 30 cm). The obtained study results show that 10–16 annual weed species and 1–3 perennial species occurred in both chamomile cultivars. Foliar application of the bioproducts contributed to a reduction in the total number of weeds in the crop, but at the same time to greater weed species diversity. In the control treatments (without the bioproducts), the dominance of several weed species (*Viola arvensis, Galeopsis tetrahit, Spergula arvensis, Juncus bufonius, Scleranthus annuus*) and lower biodiversity of the weed flora were observed. The largest reduction (by about 20%) in the number of annual weeds was found under the influence of the bioproducts Herbagreen Basic and Bio-algeen applied once. Bio-algeen and Effective Microorganisms (EM), in turn, had a significant effect on decreasing the weed weight. A narrower (30 cm) row spacing of chamomile had a significant impact on reducing the weight of weeds in chamomile crops compared to the wider spacing, which was 40 cm. It should be concluded that infection of the chamomile plantation with fungal diseases was overall at a low level. Significantly higher infection with fungal diseases was found in the case of the cultivar "Mastar", regardless of the experimental factors. A statistically proven decrease in infection of chamomile plants with fungal diseases was determined under lower crop density conditions (a row spacing of 40 cm). Chamomile plants were found to exhibit better health under the influence of double application of the biofertilizers Herbagreen Basic and Bio-algeen.

Keywords: chamomile; organic system; bioproducts; seeding density; quantitative weed infestation indicators; weed species; fungal diseases

1. Introduction

Chamomile (*Chamomilla recutita* (L.) Rausch.) is one of the most popular herbal plants grown in different countries across the world. The great economic importance of this herbal plant is due to the valuable chemical composition of chamomile raw material (flower heads) and its application in medicine, pharmacy, and cuisine [1,2]. Chamomile raw material obtained from organic plantations is particularly valuable. In organic cultivation where, by definition, no crop protection chemicals (herbicides, fungicides, insecticides) or synthetic mineral fertilizers are used, chamomile plants are more susceptible—than in conventional cropping—to weed pressure and infection with fungal diseases [3,4]. The negative effects of weed infestation on yield quantity and quality of crops are due to the high competitiveness of weeds for all environmental resources (light, water, nutrients, space) as well as their initial faster rate of growth and more effective use of CO_2 [5–7]. Differences in the number and weight of weeds and in their biological diversity between the organic and conventional farming systems predominantly result from crop agronomy, which involves crop rotation, tillage and cropping technology, and weed control methods [8–10]. By depleting nutrients from the soil, weeds increase the susceptibility of crops to fungal diseases [11].

Seeking environmentally friendly agronomic solutions that enhance the resistance and competitiveness of herbal plants against weeds and pathogenic fungi is currently very popular. A proper selection of a chamomile variety as well as an appropriate seeding rate and density (the width of interrows in herbal crops), which will allow for natural competition with weeds, can be of major importance in this respect [3,12]. In science and agricultural practice, special hopes are associated with bioproducts (such as growth biostimulators, organic foliar fertilizers, Effective Microorganisms) which increase crop plant resistance to adverse habitat conditions and pathogens [4,13,14].

This study hypothesized that foliar application of biological products in chamomile crops would contribute to better competitiveness of this crop plant against weeds and its higher resistance to fungal pathogens compared to the control treatment (without foliar application). An assumption was also made that the study would allow us to determine a species-specific optimal row spacing in the context of crop competition with pathogens. It was also hypothesized that the cultivar factor would not have a significant effect on weed infestation and health of chamomile.

The aim of the present study was to evaluate the effect of some foliar sprays: biostimulator, foliar fertilizer, Effective Microorganisms (enhancing the condition of the crop plant and its competition against pathogens), and different row spacing (30 and 40 cm) of two chamomile cultivars ("Złoty Łan" and "Mastar") on weed infestation and health of a chamomile plantation.

2. Methods

Field experiments in growing chamomile (*Chamomilla recutita* (L.) Rausch.) were conducted in the village of Dys located in the Municipality of Niemce (51°18′57″ N 22°35′06″ E), Lubelskie Voivodeship, Poland, over the period 2014–2016. The experiment was set up as a split-block design with 3 replicates. The total experimental area was 525 m^2. Treatments with the two chamomile cultivars were the experimental blocks. These blocks comprised alternately arranged plots with 2 different row spacings of chamomile. Seven foliar spray fertilization treatments were randomly assigned to these 2 chamomile cultivars and 2 row spacings of this herbal plant. The experiment included a total of 84 plots with a single plot area of 6.25 m^2 (each plot was in the shape of a 2.5 m × 2.5 m square). The characteristics analyzed in this study were determined in each of the 84 experimental plots. The specific design of this field experiment is shown in Figure 1. Chamomile was grown on a podzolic soil classified as very good rye soil complex (soil class II). Crops were grown in the experimental field in the organic farming system (without using synthetic mineral NPK fertilizers and without application of crop protection chemicals—herbicides, fungicides, and insecticides).

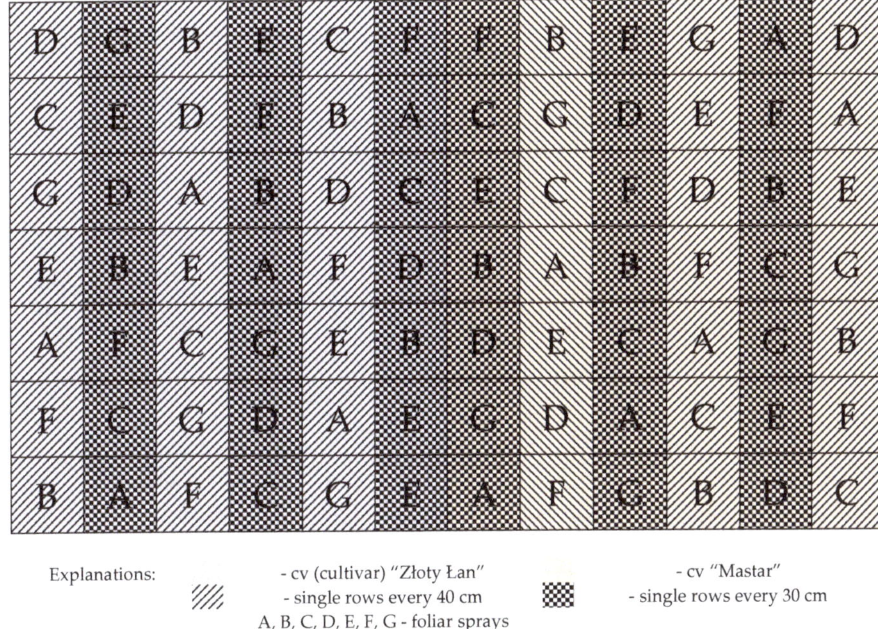

Figure 1. Design of the field experiment on the cultivation of chamomile.

In the field where the experiment was conducted, organic farming had been carried out since 2009 and thus for 5 years before the establishment of the experiment in question. During the period 2009–2013, the following crops were organically grown in this field: oats, potato, spring barley and white mustard. Throughout this entire period (2009–2016), this field was (and still is) under supervision of a certification body (Polish Society of Organic Farming "Eco-guarantee") and it has an Organic Farming Certificate. The experiment conducted in the organic system was surrounded by a 200-m "buffer zone" (with organically grown lacy phacelia, red clover, and oats). The distance between the experimental plots and the nearest traffic artery was 700 m.

During the study period, the soil was characterized by slightly acidic pH (in 1 M KCl = 6.2–6.4) and a medium content of available macronutrients (P = 78.1–78.6; K = 84.9–86.1; Mg = 31.6–32.6 mg kg^{-1}). The soil humus content was 1.39–1.43%.

In both experiments with the above-mentioned chamomile cultivars, the following factors were included:

I. Chamomile cultivars

a. "Złoty Łan"—a commonly grown cultivar;

b. "Mastar"—a new and less popular cultivar.

II. Foliar sprays:

A—Without application of foliar sprays (control treatment);

B—Foliar spray Herbagreen Basic—(10 g in 1.0 L of water);

C—Foliar spray Bio-algeen S90—(4.0 mL in 1.0 L of water);

D—Foliar spray Effective Microorganisms—EM Farming—(60.0 mL in 1.0 L of water);

E—Double application of the foliar spray Herbagreen Basic—(2 × 5 g in 1.0 L of water);

F—Double application of the foliar spray Bio-algeen S90—(2 × 2.0 mL in 1.0 L of water);

G—Double application of the foliar spray Effective Microorganisms—EM Farming—(2 × 30.0 mL in 1.0 L of water).

III. Row spacing and seeding density:

1. Single rows every 40 cm;
2. Single rows every 30 cm.

In each year of the study, white mustard grown for green manure was the previous crop for chamomile. Pre-sowing soil fertilization included a mineral fertilizer approved for use in organic farming—Humac Agro (Table 1.).

Table 1. Chemical composition of Humac Agro fertilizer.

Nutrient Content on A Dry Weight Basis *								
Humic acid %	K $g\ kg^{-1}$	Ca $g\ kg^{-1}$	Na $g\ kg^{-1}$	Fe $g\ kg^{-1}$	Zn $mg\ kg^{-1}$	Br $mg\ kg^{-1}$	Cu $mg\ kg^{-1}$	Se $mg\ kg^{-1}$
62.0	1.18	16.80	12.80	14.50	64.0	77.0	19.0	6.0

* Moisture content = 20%.

Chamomile seeds were sown directly into the soil in the third decade of April (using a hand seeder with a press wheel). The seeding rate was 2.0 kg ha^{-1} (a row spacing of 40 cm)–2.5 kg ha^{-1} (a row spacing of 30 cm). Weed control involved mechanical weed removal (with a weeder) at the 3–5 leaf stage of chamomile. The foliar sprays (treatments B-G) were applied using a field sprayer under a pressure of 0.25 MPa. Double application was done at the 2–3 leaf stage of chamomile and at the 5–7 leaf stage (after mechanical weed removal). Single application of the sprays was carried out at the 5–7 leaf stage of chamomile. The specific composition of the bioproducts used in this experiment is shown in Table 2.

Table 2. Components of the sprays used in the experiment.

Name of Spray	Spray Composition
Bio-algeen S90	An extract from sea algae; the spray contains 90 groups of chemical compounds, including amino acids, vitamins, alginic acid, and other unidentified active ingredients of seaweeds; the major elements include the following: nitrogen—0.02%, phosphorus—0.006%, potassium—0.096%, calcium—0.31%,magnesium—0.021%, as well as boron—16 mg kg^{-1}, iron—6.3 mg kg^{-1}, copper—0.2 mg kg^{-1}, manganese—0.6 mg kg^{-1}; zinc—1.0 mg kg^{-1}; moreover, the spray contains molybdenum and selenium.
Herbagreen Basic	Calcium oxide (CaO)—36.7%, silicon dioxide (SiO_2)—17.0%, iron trioxide (Fe_2O_3)—3.4%, magnesium oxide (MgO)—2.4%, titanium dioxide (TiO_2)—0.5%, potassium oxide (K_2O)—0.5%, sodium oxide (Na_2O)—0.5%, sulfur trioxide (SO_3)—0.4%, phosphorus pentoxide (P_2O_5)—0.5%, manganese oxide (MnO)—0.1%; and trace amounts of boron (1), cobalt (13), copper (26), zinc (34) (mg kg^{-1} DM).
EM Farming	Anaerobic organisms which release free, chemically uncombined oxygen into the environment during metabolic processes (photosynthetic bacteria, actinobacteria, lactic acid bacteria, fermentation fungi, yeasts)—the percentage contributions of particular microorganism strains in the spray is the manufacturer's secret (patent) and this information is not included in any available data sheets.

Before harvesting the chamomile crop, the following activities were performed:

- Evaluation of weed infestation of the crop using the agro-phytosociological method at the 2–3 leaf stage of chamomile (moving around the 6.25 m^2 plots according to the scheme in Figure 2). Weed infestation intensity was determined using the Braun-Blanquet scale [15], where: - = no individuals; r = 1–3 individuals; + means less than 1% of ground cover; 1 = 1–5% of ground cover; 2 = 5–25% of ground cover; 3 = 25–50% of ground cover; 4 = 50–75% of ground cover; 5 = 75–100% of ground cover.
- Determination of health of chamomile plants according to a five-point scale carried out 15 days before harvest, based on 30 randomly selected plants per 6.25 m^2 plot (Table 3).
- Over a period of 10 days before harvest of chamomile, evaluation of weed infestation of the plantation was performed by the dry-weight-rank method (number and air dry weight of weeds as well as their botanical composition) using a 0.25 × 1.0 m quadrat frame (Figure 3) in 4 randomly selected places in each 6.25 m^2 plot. The frame was placed in the randomly selected places in

each plot (Figure 4), taking care not to damage chamomile plants. Within the frame, weeds were cut down just above ground and subsequently the species composition and number of weeds were determined. Next, weeds collected from 4 replicates in a given plot were combined into a composite sample and placed in paper bags with labels (plot number). The collected weed samples were dried in a plant house, bringing them slowly to air-dry weight.

Table 3. The scale used to determine health of chamomile plants.

Degree of Health	% of Plants Infected
I	1–10
II	11–25
III	26–50
IV	51–75
V	76–100

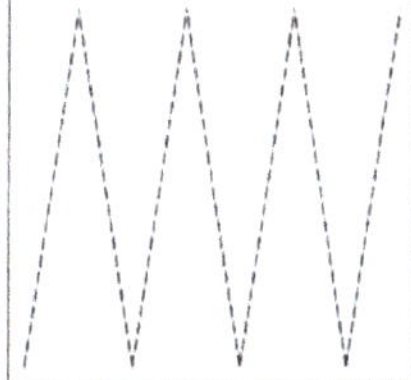

Figure 2. Agro-phytosociological method scheme—in each 6.25 m^2 plot.

Figure 3. Frame (0.25 m × 1.0 m) used in this research.

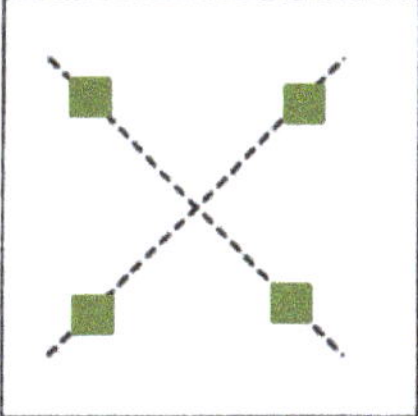

Figure 4. Weed infestation evaluation scheme—in each 6.25 m^2 plot.

3. Statistical Analysis

Analysis of variance (ANOVA) was used to statistically analyze the results by employing Statistica PL 13.3, while Tukey's test was applied to determine HSD (Honest Significant Difference) values at $p < 0.05$. The results tables show the mean for the study period because the year-to-year differences between the characteristics analyzed were statistically not significant. No significant interactions were found between the main experimental factors: cultivars, foliar sprays and row spacing. The statistical calculations also included the standard deviation (SD) and the coefficient of variation (CV). Correlation coefficients (r) between the quantitative weed infestation indicators for the chamomile crop and the level of health of chamomile plants were also calculated.

4. Results

4.1. Weed Infestation of the Chamomile Crop

The first stage of evaluation of the chamomile plantation was to analyze in-crop weed infestation at the 2–3 leaf stage of this herbal plant. The evaluation performed using the agro-physiological method allowed the intensity of occurrence of dominant weed species to be determined. The data included in Tables 4 and 5 show that 7–11 annual weed species and 1–2 perennial species occurred in crops of the cultivars "Złoty Łan" and "Mastar", depending on the experimental plots included in the experimental design. *Galeopsis tetrahit* and *Viola arvensis* were predominant in the weed flora, *Scleranthus annuus, Spergula arvensis,* and *Juncus bufonius* were found at a medium level, while the other weed species occurred at the lowest level (r) or were found not to occur at all in some plots. It should be indicated that at the 2–3 leaf stage of chamomile the effect of different row spacings on weed infestation was not yet observed and that the foliar sprays had not been applied yet, either.

Number of weeds in the chamomile crop determined several days before its harvest was independent of the cultivar and almost identical values were found for the "Złoty Łan" and "Mastar" crops (Table 6). However, number of weeds as affected by row spacing exhibited significant relationships. Denser sowing of chamomile (at a row spacing of 30 cm) promoted a reduction in the number of weeds by about 20% compared to a row spacing of 40 cm. Some of the foliar sprays had a statistically proven effect on decreasing the number of weeds in chamomile crops—single spraying with the bioproducts Herbagreen Basic and Bio-algeen as well as double spraying with EM Farming promoted a reduction in the number of weeds by about 23% compared to the control treatment.

Table 4. Number and species composition of weeds found in the crop of the chamomile cultivar "Złoty Łan" at the 2–3 leaf stage of this herbal plant.

Species	Row Spacing—40 cm							Row Spacing—30 cm						
	A *	B	C	D	E	F	G	A	B	C	D	E	F	G
Annual														
Galeopsis tetrahit L.	1 **	+	+	+	1	+	+	+	+	+	+	+	+	+
Viola arvensis Murray	+	1	1	+	+	+	1	1	+	+	+	+	+	+
Juncus bufonius L.	+	r	r	+	+	r	+	+	r	r	r	r	+	r
Spergula arvensis L.	+	r	r	r	+	+	+	+	+	r	r	r	r	r
Scleranthus annuus L.	+	+	+	+	+	+	+	+	+	+	+	+	+	+
Geranium pusillum L.	+	+	r	r	+	-	r	r	-	r	-	+	+	+
Erigeron canadensis (L.) Cronquist)	r	r	+	-	+	r	r	+	r	-	+	r	r	r
Tripleurospermum maritimum (L.) W. D. J. Koch	r	r	r	r	-	-	-	r	r	r	-	-	r	r
Polygonum convolvulus L.	r	-	-	-	-	-	r	r	r	r	-	-	-	-
Chenopodium album L.	-	-	-	-	r	r	-	r	-	r	r	-	-	-
Veronica arvensis L.	-	-	-	-	r	-	r	-	r	-	-	-	-	-
Erodium cicutarium (L.) L'Her.	r	-	r	r	-	-	r	-	r	r	r	-	-	-
Capsella bursa-pastoris (L.) Medik	-	-	-	r	r	r	-	-	r	-	-	-	-	-
Number of annual species	10	8	9	9	10	8	10	10	11	10	8	7	8	8
Perennial														
Elymus repens (L.) Gould	r	r	r	-	r	r	r	r	r	-	r	r	r	r
Convolvulus arvensis L.	-	r	-	-	-	-	-	-	-	-	-	r	-	-
Number of perennial species	1	2	1	0	1	1	1	1	1	0	1	2	1	1

* A—without application of foliar sprays (control treatment), B—single application of Herbagreen Basic, C—single application of Bio-algeen S90, D—single application of EM Farming, E—double application of Herbagreen Basic, F—double application of Bio-algeen S90, G—double application of EM Farming; ** r, +, -, 1—explanation in the Methods section.

Table 5. Number and species composition of weeds found in the crop of the chamomile cultivar "Mastar" at the 2–3 leaf stage of this herbal plant.

Species	Row Spacing—40 cm							Row Spacing—30 cm						
	A *	B	C	D	E	F	G	A	B	C	D	E	F	G
Annual														
Galeopsis tetrahit L.	1 **	1	+	1	+	1	+	1	+	1	+	+	+	+
Viola arvensis Murray	+	+	1	1	1	1	1	1	1	+	1	1	+	+
Juncus bufonius L.	+	+	r	+	r	+	+	+	+	+	r	r	+	r
Spergula arvensis L.	+	+	+	r	r	r	r	+	+	r	r	r	r	r
Scleranthus annuus L.	+	r	r	+	r	+	+	+	r	r	+	r	+	+
Geranium pusillum L.	+	+	-	r	+	-	-	r	-	r	-	+	+	+
Erigeron canadensis (L.) Cronquist)	r	+	+	-	+	r	r	+	r	-	+	r	r	r
Tripleurospermum maritimum (L.) W. D. J. Koch	r	r	-	r	-	r	-	r	-	r	-	-	r	r
Polygonum convolvulus L.	r	r	-	-	r	-	-	r	-	r	-	r	-	-
Chenopodium album L.	r	r	-	-	r	-	-	r	r	-	r	-	-	-
Veronica arvensis L.	-	-	-	r	r	-	r	-	r	-	-	-	-	-
Erodium cicutarium (L.) L'Her.	r	-	r	-	-	r	r	-	r	-	r	-	r	-
Capsella bursa-pastoris (L.) Medik	-	r	r	-	-	-	-	-	-	-	-	-	-	-
Number of annual species	11	11	8	8	10	8	8	10	9	8	8	8	9	8
Perennial														
Elymus repens (L.) Gould	r	r	r	r	r	r	r	r	r	-	r	r	-	r
Convolvulus arvensis L.	r	-	-	-	r	-	-	-	r	-	-	r	-	-
Equisetum arvense L.	-	-	-	r	-	r	-	-	-	r	-	-	-	-
Number of perennial species	2	1	1	2	2	2	1	1	2	1	1	2	0	1

* A—without application of foliar sprays (control treatment), B—single application of Herbagreen Basic, C—single application of Bio-algeen S90, D—single application of EM Farming, E—double application of Herbagreen Basic, F—double application of Bio-algeen S90, G—double application of EM Farming; ** r, +, -, 1—explanation in the Methods section.

Table 6. Number of weeds in the chamomile crop (plants per m^{-2})—evaluation before harvest of the herbal crop.

Treatment	Złoty Łan			Mastar			Mean for Spacing		
	40 cm	30 cm	Mean	40 cm	30 cm	Mean	40 cm	30 cm	Mean
A *	21.6 ± 1.9 **	19.4 ± 1.8	20.5	23.8 ± 1.9	19.6 ± 1.2	21.7	22.7	19.5	21.1
B	17.1 ± 1.7	13.2 ± 1.4	15.2	19.9 ± 1.7	15.1 ± 0.9	17.5	18.5	14.2	16.3
C	17.1 ± 1.6	17.4 ± 1.6	17.2	19.1 ± 1.1	12.9 ± 0.8	16.0	18.0	15.2	16.6
D	21.2 ± 1.1	18.6 ± 1.7	20.0	22.9 ± 1.5	14.8 ± 0.9	18.3	22.2	16.7	19.4
E	21.4 ± 1.8	15.8 ± 1.3	18.5	22.8 ± 0.9	18.2 ± 1.3	20.5	22.0	17.0	19.5
F	20.1 ± 1.5	18.2 ± 1.8	19.2	19.7 ± 1.0	17.0 ± 1.0	18.3	19.8	17.7	18.8
G	19.7 ± 1.6	13.3 ± 1.4	16.5	17.6 ± 1.5	15.1 ± 0.6	16.3	18.7	14.2	16.4
CV (%) ***	6.2	5.6		6.7	5.9				
Mean	19.7	16.6	18.1	20.8	16.1	18.5	20.3	16.3	-
$HSD_{(0.05)}$	for years—n.s. ****; for cultivars—n.s.; for row spacing—2.5; for foliar sprays—4.3; for interaction: cultivar × row spacing—n.s.; for interaction: cultivar × foliar spray—n.s.; for interaction: row spacing × foliar spray—n.s.; for interaction: cultivar × row spacing × foliar sprays—n.s.								

* A—without application of foliar sprays (control treatment), B—single application of Herbagreen Basic, C—single application of Bio-algeen S90, D—single application of EM Farming, E—double application of Herbagreen Basic, F—double application of Bio-algeen S90, G—double application of EM Farming; ** SD—standard deviation; *** CV—coefficient of variation; **** n.s.—not significant differences.

The cultivar factor did not also have a major influence on the dry weight of weeds in the chamomile crop (Table 7). The weed infestation indicator in question was however significantly dependent on row spacing—the wider row spacing (40 cm) contributed to a higher weight of weeds in the crop by about 35% relative to a spacing of 30 cm. Among the foliar sprays used, Bio-algeen (applied once and twice) and EM Farming (applied twice) had a statistically significant effect on decreasing the dry weight of weeds in the chamomile crop. In the case of the other bioproduct application treatment combinations, the bioproducts were found to tend to decrease the weed weight.

Table 7. Dry weight of weeds in the chamomile crop (g m^{-2})—evaluation before harvest of the herbal crop.

Treatment	Złoty Łan			Mastar			Mean for Row Spacing		
	40 cm	30 cm	Mean	40 cm	30 cm	Mean	40 cm	30 cm	Mean
A *	15.01 ± 1.22 **	9.23 ± 0.87	12.12	15.41 ± 1.90	10.73 ± 0.92	13.07	15.21	9.98	12.59
B	11.77 ± 0.74	6.15 ± 0.93	8.96	14.78 ± 0.99	7.43 ± 0.58	11.10	13.27	6.79	10.03
C	7.28 ± 0.71	6.81 ± 0.66	7.04	15.95 ± 1.33	4.68 ± 0.42	10.31	11.61	5.74	8.67
D	11.52 ± 0.86	7.45 ± 0.91	9.48	10.67 ± 0.83	7.74 ± 0.61	9.20	11.09	7.59	9.34
E	12.43 ± 1.04	9.74 ± 0.65	11.08	11.31 ± 0.92	11.15 ± 0.79	11.23	11.87	10.52	11.19
F	8.91 ± 0.83	8.80 ± 0.52	8.85	8.66 ± 0.49	6.23 ± 0.47	7.44	8.78	7.51	8.14
G	11.33 ± 0.92	5.32 ± 0.39	8.32	9.15 ± 0.59	6.67 ± 0.56	7.91	10.24	5.99	8.11
CV (%) ***	4.8	5.2		5.1	6.1				
Mean	11.17	7.64	9.41	12.27	7.80	10.03	11.72	7.72	-
$HSD_{(0.05)}$	for years—n.s. **; for cultivars—n.s.; for row spacing—2.79; for foliar sprays—3.52; for interaction: cultivar × row spacing—n.s.; for interaction: cultivar × foliar spray—n.s.; for interaction: row spacing × foliar spray—n.s.; for interaction: cultivar × row spacing × foliar spray—n.s.								

* A—without application of foliar sprays (control treatment), B—single application of Herbagreen Basic, C—single application of Bio-algeen S90, D—single application of EM Farming, E—double application of Herbagreen Basic, F—double application of Bio-algeen S90, G—double application of EM Farming; ** SD—standard deviation; *** CV—coefficient of variation; n.s.—not significant differences.

Table 8 presents the number of dominant weed species in crops of the chamomile cultivars "Złoty Łan" and "Mastar" relative to the other experimental factors (foliar sprays, row spacing). The data contained in this table show that 10–16 annual weed species and 1–3 perennial species occurred in crops of both chamomile cultivars, and hence 5–6 species more than the numbers found during the initial plantation development period (2–3 leaf stage of chamomile). The following species were dominant in terms of numbers: *Viola arvensis*, *Galeopsis tetrahit*, *Spergula arvensis*, *Juncus bufonius*, and *Scleranthus annuus*, and thus the species that had been observed to occur with greater intensity already during the first (agro-phytosociological) evaluation of weed infestation. When analyzing the effect of row spacing in chamomile crops on weed infestation, we note that it had no impact on the number of annual or perennial weed species, regardless of the foliar sprays, but the total number of annual weeds was found to be lower (by 21%) under narrow row spacing conditions (a spacing of 30 cm) compared to a spacing of 40 cm. The foliar sprays used in the experiment had a different impact on the number of annual weeds and the number of annual weed species, but did not modify the perennial flora almost at all. Overall, it can be stated that the foliar sprays had an indirect effect on decreasing the number of annual weeds in comparison with the control plots. The largest reduction (by about 19-20%) in the number of annual weeds was recorded under the influence of the bioproducts Herbagreen Basic and Bio-algeen applied once (treatments B and C) and under the influence of Effective Microorganisms (treatment G). The cultivar factor had no impact on the number of weeds and their species composition (Table 8).

Table 8. Number and species composition of weeds in the chamomile crop depending on the experimental factors—evaluation before harvest of the herbal crop.

Species	Foliar Spray Application Treatments							Row Spacing		Cultivar	
	A *	B	C	D	E	F	G	40 cm	30 cm	Złoty Łan	Mastar
Annual											
Viola arvensis Murray	4.3	3.0	3.3	3.5	2.7	2.9	2.8	3.9	2.7	3.5	3.1
Spergula arvensis L.	3.9	2.4	3.1	2.4	2.2	2.1	2.4	3.1	2.1	2.5	2.7
Galeopsis tetrahit L.	3.7	3.1	2.9	2.5	3.4	2.8	2.9	3.4	2.4	2.7	3.1
Juncus bufonius L.	1.9	1.4	1.5	1.5	1.2	2.6	2.1	2.2	1.5	1.5	2.1
Scleranthus annuus L.	1.1	1.9	1.4	1.2	2.0	1.5	1.2	1.5	1.6	1.5	1.2
Erigeron canadensis (L.) Cronquist)	0.8	0.7	0.3	1.1	0.7	0.9	0.5	0.7	0.6	0.8	0.9
Tripleurospermum maritimum (L.) W. D. J. Koch	0.7	0.4	0.1	0.4	0.3	0.7	0.9	0.7	0.4	0.5	0.6
Echinochloa crus-galli (L.) P. Beauv.	0.7	0.2	0.4	0.7	1.2	1.1	0.7	0.6	0.7	0.9	0.4
Geranium pusillum L.	0.6	0.5	0.2	1.5	1.0	0.5	0.3	0.6	0.8	0.7	0.8
Setaria pumila (Poir.) Roem.	0.5	0.4	-	0.4	0.2	0.4	0.2	0.3	0.1	0.3	0.2
Polygonum convolvulus L.	0.5	0.3	0.6	0.3	0.5	0.8	0.4	0.4	0.5	0.6	0.5
Hypericum humifusum L.	0.5	0.5	0.8	0.6	0.7	0.4	0.3	0,5	0.7	0.7	0.3
Chenopodium album L.	0.3	0.2	-	0.4	0.4	-	0.6	0.4	0.2	0.1	0.5
Capsella bursa-pastoris (L.) Medik	0.3	0.1	0.1	0.5	0.5	0.6	0.3	0.2	0.4	0.2	0.3
Veronica arvensis L.	0.1	-	0.2	0.3	0.3	0.4	0.2	0.3	0.1	0.2	0.2
Erodium cicutarium (L.) L'Her.	-	0.1	0.2	0.2	0.5	-	0.2	0.1	0.2	0.2	0.1
Galinsoga parviflora Cav.	-	-	0.1	0.5	0.2	0.3	-	0.1	0.2	0.1	0.2
Number of annual weeds	19.6	15.1	15.2	18.1	17.8	17.7	15.8	19.1	15.1	17.0	17.1
Number of annual species	15	15	15	17	17	15	16	17	17	17	17
Perennial											
Elymus repens (L.) Gould	1.1	0.7	1.1	0.9	1.3	0.9	0.6	1.0	0.9	0.7	1.0
Equisetum arvense L.	0.3	0.3	0.2	0.1	-	0.2	-	0.1	0.2	0.1	0.1
Convolvulus arvensis L.	0.1	0.3	0.1	0.4	0.4	-	-	0.1	0.1	0.2	0.3
Number of perennial weeds	1.5	1.2	1.4	1.3	1.7	1.1	0.6	1.2	1.2	1.1	1.4
Number of perennial species	3	3	3	3	2	2	1	3	3	3	3

* A—without application of foliar sprays (control treatment), B—single application of Herbagreen Basic, C—single application of Bio-algeen S90, D—single application of EM Farming, E—double application of Herbagreen Basic, F—double application of Bio-algeen S90, G—double application of EM Farming.

4.2. Health of Chamomile Plants

Infection of chamomile plants with fungal pathogens was at a low level (ranging on average 2.7–10.1%). Nonetheless, the experimental factors were found to significantly affect the analyzed trait (Table 9). The chamomile cultivar "Mastar" was characterized by almost twice higher susceptibility to infection with fungal diseases than cv. "Złoty Łan", which is probably attributable to its poor adaptation to local habitat conditions. The foliar sprays contributed to a significant reduction in the occurrence of fungal diseases in relation to the level of occurrence of fungal pathogens in the control plots (without application of the bioproducts). Herbagreen Basic and Bio-algeen applied twice during the growing season of this herbal plant had a particularly beneficial effect on the health of chamomile plants. Regardless of the foliar sprays, chamomile plants sown at a row spacing of 40 cm were found to be significantly healthier in comparison with the narrower spacing (30 cm). The greater infection of chamomile plants with fungal diseases in the control treatments (without application of the bioproducts) and in the case of the narrower row spacing was positively correlated with the higher number of weeds in these treatments, as evidenced by the calculated correlation coefficients Table 10.

Table 9. Health of chamomile plants (%)—evaluation before harvest of the herbal crop.

Treatment	Złoty Łan			Mastar			Mean for Row Spacing		
	40 cm	30 cm	Mean	40 cm	30 cm	Mean	40 cm	30 cm	Mean
A *	6.2 ± 0.9 **	7.9 ± 1.1	7.0	11.5 ± 1.3	15.4 ± 1.7	13.4	8.8	11.6	10.2
B	3.4 ± 0.6	4.5 ± 0.6	3.9	7.5 ± 1.0	8.0 ± 0.9	7.7	5.4	6.2	5.8
C	3.2 ± 0.6	3.8 ± 0.7	3.5	6.1 ± 0.7	6.5 ± 0.8	6.3	4.6	5.1	4.8
D	4.0 ± 0.7	5.1 ± 0.8	4.5	8.4 ± 0.5	9.7 ± 0.9	9.0	6.2	7.4	6.8
E	2.0 ± 0.3	2.4 ± 0.5	2.2	3.1 ± 0.3	3.4 ± 0.4	3.2	2.5	2.9	2.7
F	2.2 ± 0.5	2.8 ± 0.6	2.5	3.7 ± 0.4	4.0 ± 0.5	3.8	2.9	3.4	3.1
G	3.4 ± 0.8	4.1 ± 0.9	3.7	5.5 ± 0.6	6.2 ± 0.7	5.8	4.4	5.1	4.7
CV (%) ***	8.4	7.3		12.5	13.2				
Mean	3.4	4.4	3.9	6.5	7.6	7.0	5.0	6.0	-
$HSD_{(0.05)}$	for years—n.s. ****; for cultivars—0.97; for row spacing—0.94; for foliar sprays—1.11; for interaction: cultivar × row spacing—n.s.; for interaction: cultivar × foliar spray—n.s.; for interaction: row spacing × foliar spray—n.s.; for interaction: cultivar × row spacing × foliar spray—n.s.								

* A—without application of foliar sprays (control treatment), B—single application of Herbagreen Basic, C—single application of Bio-algeen S90, D—single application of EM Farming, E—double application of Herbagreen Basic, F—double application of Bio-algeen S90, G—double application of EM Farming; ** SD—standard deviation; *** CV—coefficient of variation; **** n.s.—not significant differences.

Table 10. Correlation coefficients (r) between the number and weight of weeds in the crop and infection of chamomile plants with fungal diseases.

Treatment	Row Spacing (cm)	Number of Weeds		Weight of Weeds	
		Złoty Łan	Mastar	Złoty Łan	Mastar
A *	30	0.71 **	0.89 **	0.36	0.55 **
	40	0.58 **	0.62 **	0.28	0.33
B	30	0.18	0.23	0.11	0.15
	40	0.09	0.16	0.07	0.06
C	30	0.20	0.22	0.09	0.13
	40	0.08	0.10	0.06	0.09
D	30	0.27	0.33	0.21	0.26
	40	0.22	0.28	0.14	0.18
E	30	0.22	0.21	0.20	0.17
	40	0.28	0.32	0.31	0.26
F	30	0.18	0.19	0.22	0.28
	40	0.13	0.11	0.12	0.15
G	30	0.13	0.15	0.11	0.23
	40	0.09	0.12	0.09	0.17

* A—without application of foliar sprays (control treatment), B—single application of Herbagreen Basic, C—single application of Bio-algeen S90, D—single application of EM Farming, E—double application of Herbagreen Basic, F—double application of Bio-algeen S90, G—double application of EM Farming; ** significant correlation $coefficient_{(0.05)}$.

5. Discussion

There are few scientific studies regarding direct and indirect effects of foliar applied bioproducts on weed infestation and health of organically grown chamomile. Therefore, the discussion of these study results is based primarily on other researchers' reports dealing generally with the problem of weeds and fungal diseases in crop plantations grown in the organic system.

Segetal flora occurring in crop fields is one of the more important elements influencing the functioning of the entire agroecosystem. Its competitiveness against a crop and also its reproductive potential significantly impact agricultural production [16–18]. The farming system, which includes crop rotation, selection of varieties, fertilization method and level, and weed control methods, is of key significance for the preservation of the biodiversity of segetal flora in an agroecosystem [19–22].

The present study confirmed the research hypothesis and demonstrated lower weed infestation of chamomile crops fertilized with the foliar sprays. At the same time, greater richness of the segetal flora was observed in the plots where the foliar sprays were used. Foliar biological products (biostimulators, micronutrient fertilizers, Effective Microorganisms) contribute to greater resistance of a crop to unfavorable habitat conditions (including pressure from agricultural pests) through, among others, a faster generative growth rate. Due to this, the crop competes more effectively for light and nutrients with weeds [3,13]. In conventional agriculture, when crops are fertilized, for example, with soil applied mineral nitrogen, the result is that weeds, taking up N from the soil substrate, show an even greater growth rate than the crop [23]. In organic farming, weeds are perceived as an integral part of an agroecosystem [24]. If weeds occur in low numbers, but exhibit high species variation (greater biodiversity), they are even perceived as a positive aspect of an ecosystem [25,26]. The present study confirms this thesis because in the treatments with application of the foliar sprays the number and weight of weeds were observed to be lower, but at the same time exhibiting greater biodiversity in chamomile crops. Furthermore, this research reveals that the weed species composition determined before harvest of chamomile was slightly greater (by only 5–6 species) than that found at the 2–3 leaf stage of chamomile. This situation is explained by other authors' studies which reveal that in organic crops changes in weed species composition occur more slowly than changes in weed numbers [27]. This evidences relatively high stability of weed flora in such crops [28].

Another method to reduce weed infestation in organic plantations is to select an appropriate spacing of crop rows. If a narrower row spacing is used, weeds occurring in interrows of a crop have limited light access and competition with the crop is more difficult [29]. The results of the present study confirm this view because chamomile sown at the narrower row spacing (30 cm) exhibited a lower degree of weed infestation than at the wider spacing (40 cm). Kraska et al. [30] also recorded a lower number and weight of weeds in organically grown lentil crops in the case of the narrowest row spacing, which was 20 cm. Kwiatkowski et al. [31], in turn, observed significantly lower weed infestation of a winter oilseed rape crop using a row spacing of 18 cm in comparison with a spacing of 30 cm. Similar relationships (a lower degree of weed infestation of crops with narrowing row spacing) were also found by Lins and Boerboom [32] on the example of soybean as well as by Primot and Valantin-Morison [33] on the example of winter oilseed rape. Lutman et al. [34] also report on the importance of a higher crop density as regards competition with weeds in a study investigating the competitiveness of oilseed rape against the dominant species *Stellaria media* in oilseed rape crops.

The cultivar factor can also be of significance in controlling weed infestation of crops. This research did not find significant differences in the quantitative weed infestation indicators between the chamomile cultivars "Złoty Łan" and "Mastar". Feledyn-Szewczyk [35] also found a lower effect of the crop variety on the level and state of weed infestation on the example of winter wheat.

The results of this study show infection of chamomile with fungal diseases to be significantly lower compared to the control plots (without application of the bioproducts), regardless of the bioproduct used (foliar fertilizer, biostimulator, Effective Microorganisms) and the number of its applications (single or double). Bioproducts are not directly fungicides [36]. Their indirect role in controlling plant infection with fungal diseases consists in "feeding plants" and enhancing their resistance to environmental stresses as well as to fungal and bacterial pathogen activity [37,38]. Numerous scientific studies confirm the positive effects of various biological products on the biometric characteristics, nutritional composition, and health of herbal, horticultural, cereal, and root plants [39–44].

While being effective in reducing weed infestation in a field (as evidenced by this research), a dense crop (a smaller interrow width) poses at the same time a greater risk of development of fungal diseases, which is promoted by, among others, higher humidity and temperature in a dense crop [45,46]. As shown by the present study, infection of a crop with fungal diseases is due not only to the crop density, but also results from a higher number and weight of weeds in the crop, which are hosts for many disease-causing fungi [47,48]. Resistance to fungal diseases can also be driven by the cultivar factor [45,49]. This is confirmed by the results of this study because infection of the chamomile cultivar "Złoty Łan" (better adapted to habitat conditions of the central Lublin region, Poland) was found to be significantly lower than in the case of cv. "Mastar".

This research demonstrates that in the case of the bioproducts Herbagreen Basic and Bio-algeen single application of these sprays proved to be sufficient in controlling agricultural pests, with its effect being similar to that of double application. The studies by Kwiatkowski [3] on chamomile and Kwiatkowski et al. [4] on common thyme also confirm the high efficacy of single application of bioproducts.

6. Conclusions

Weed species composition was quite a stable characteristic (the number of weed species in the chamomile crop during the period from the 2–3 leaf stage until its harvest increased only slightly). The experimental factors modified the number and species composition of the weed flora to a small degree. Sixteen annual weed species and 3 perennial species were found in chamomile crops. *Viola arvensis, Galeopsis tetrahit, Spergula arvensis, Juncus bufonius,* and *Scleranthus annus* dominated in terms of numbers.

Among the foliar sprays tested in this experiment, the bioproducts Bio-algeen and Effective Microorganisms (number and weight of weeds) as well as Herbagreen Basic (number of weeds) showed the most beneficial effect on reducing the quantitative weed infestation indicators. All the foliar sprays used in this experiment significantly improved the health of chamomile plants.

From the point of view of plant health, the wider spacing (40 cm) undoubtedly proved to be more beneficial for the cultivation of chamomile. The narrower row spacing (30 cm), on the other hand, reduced weed infestation of the chamomile crop more effectively.

The chamomile cultivar "Złoty Łan" exhibited better plant health when grown organically than the cultivar "Mastar". Nevertheless, both chamomile cultivars were characterized by a similar degree of weed infestation of the plantation, regardless of the other experimental factors.

Author Contributions: C.A.K., M.H. conceptualization; C.A.K., M.H., E.H. formal analysis; C.A.K., M.H. investigation and methodology; C.A.K., E.H., M.H. resources, C.A.K., E.H. writing-original draft. All authors have read and agreed to the published version of the manuscript.

Acknowledgments: This research was supported by the Ministry of Agriculture and Rural Development in Poland.

Conflicts of Interest: The authors declare no conflict of interest.

References

1. Srivastava, J.K.; Shankar, E.; Gupta, S. Chamomile. A herbal medicine of the past with bright future. *Mol. Med. Rep.* **2010**, *3*, 895–901. [CrossRef]
2. Cavalieri, E.; Rigo, A.; Bonifacio, M.; Carcereri de Prati, A.; Guardalben, E.; Bergamini, C.; Fato, R.; Pizzolo, G.; Suzuki, H.; Vinante, F. Pro-apoptotic activity of α-bisabolol in preclinical models of primary human acute leukemia cells. *J. Transl. Med.* **2011**, *9*, 45. [CrossRef]
3. Kwiatkowski, C.A. Yield and quality of chamomile (*Chamomila recutita* (L.) Rausch.) raw material depending on selected foliar sprays and plant spacing. *Acta Sci. Pol. Hortorum Cultus* **2015**, *14*, 143–156.
4. Kwiatkowski, C.A.; Haliniarz, M.; Harasim, E.; Kołodziej, B.; Yakimovich, A. Foliar applied biopreparations as a natural method to increase the productivity of garden thyme (*Thymus vulgaris* L.) and to improve the quality of herbal raw material. *Acta Sci. Pol. Hortorum Cultus* **2020**, *19*, 107–118. [CrossRef]
5. Radosevich, S.; Holt, J.; Ghersa, C. *Ecology of Weeds and Invasive Plants: Relationship to Agriculture and Natural Resources Management*, 3rd ed.; John Wiley&Sons, Inc.: Hoboken, NJ, USA, 2007.
6. Rao, V.S. Principles of Weed Science. In *Weed Biology and Ecology*, 2nd ed.; Science Publishers, Inc.: Enfield, NH, USA, 2000. [CrossRef]

7. Milberg, P.; Hallgren, E. Yield loess due to weeds in cereals and its large-scale variability in Sweden. *Field Crops Res.* **2004**, *86*, 199–209. [CrossRef]
8. Demjanová, E.; Macák, M.; Ćaloviü, I.; Majernik, F.; Týr, S.; Smatana, J. Effects of tillage systems and crop rotation on weed density, weed species composition and weed biomass in maize. *Agron. Res.* **2009**, *7*, 785–792.
9. Santin-Montanyáa, M.I.; Martin-Lammerding, D.; Walter, I.; Zembrana, E.; Tenorio, J.L. Effects of tillage, crop systems and fertilization of weed abundance and diversity in 4-year dry land winter wheat. *Eur. J. Agron.* **2013**, *48*, 43–49. [CrossRef]
10. Dolijanovic, Z.; Kovacevic, D.; Momirovic, Ä.; Oljaca, S.; Jovovic, Z. Effect of crop rotations on weed infestation in winter wheat. *Bulg. J. Agric. Sci.* **2014**, *20*, 416–420.
11. Datnoff, L.E.; Elmer, W.H.; Huber, D.M. (Eds.) . *Mineral Nutrition and Plant Disease;* APS Press: St. Paul, MN, USA, 2007.
12. Kwiatkowski, C.A.; Harasim, E.; Yakimovich, A.; Kołodziej, B.; Tomczyńska-Mleko, M. Evaluation of spent mushroom substrate, mineral NPK fertilization and manure fertilization on chamomile (*Chamomilla recutita* L. Rausch) yield and raw material quality. *Acta Sci. Pol. Hortorum Cultus* **2018**, *17*, 3–16. [CrossRef]
13. Javaid, A. Foliar application of effective microorganisms on pea an alternative fertilizer. *Agron. Sustain. Dev.* **2006**, *26*, 257–262. [CrossRef]
14. Dobromilska, R.; Mikiciuk, M.; Gubarewicz, K. Evaluation of cherry tomato yielding and fruit mineral composition after using of Bio-algeen S-90 preparation. *J. Elem.* **2008**, *13*, 491–499.
15. Braun-Blanquet, J. *Pflanzensoziologie;* Springer-Verlag: Wien, Austria, 1951.
16. Menalled, F. Weed Seedbank Dynamics & Integrated Management of Agricultural Weeds. Bozeman: Extension Publication, Montana State University. Available online: https://www.mssoy.org/uploads/files/weed-seed-bank-mont (accessed on 12 May 2020).
17. Swanton, C.; Nkoa, R.; Blackshaw, R.E. Experimental methods for crop-weed competition studies. *Weed Sci.* **2015**, *63*, 2–11. [CrossRef]
18. Keller, M.; Böhringer, Ä.; Möhring, J.; Rueda-Ayala, V.; Gutjar, C.; Gerhards, R. Changes in weed communities, herbicides, yield levels and effect of weeds on yield in winter cereals based on three decades of field experiments in south-western Germany. *Gesunde Pflanzen* **2015**, *67*, 11–20. [CrossRef]
19. Lunkvist, A.; Salomonsson, L.; Karlsson, L.; Gustavsson, A.D. Effects of organic farming on weed flora composition in a long term perspective. *Eur. J Agron.* **2008**, *28*, 570–578. [CrossRef]
20. Edesi, L.; Järvan, M.; Adamson, A.; Lauringson, E.; Kuht, J. Weed species diversity and community composition in conventional and organic farming: A five-year experiment. *Zemdirb. Agric.* **2012**, *99*, 339–346.
21. Kolářová, M.; Tyšer, L.; Soukup, J. Weed species diversity in the Czech Republic under different farming and site conditions. *Acta Univ. Agric. Silvic. Mendel. Brun.* **2015**, *63*, 741–749. [CrossRef]
22. Petit, S.; Munier-Jolain, Ä.; Bretagnolle, V.; Bockstaller, C.; Gaba, S.; Cordeau, S.; Lechenet, M.; Mézière, D.; Colbach, N. Ecological intensification through pesticide reduction: Weed Control, Weed Biodiversity and Sustainability in Arable Farming. *Environ. Manag.* **2015**, *56*, 1078–1090. [CrossRef]
23. Blackshaw, R.E.; Brandt, R.N.; Janzen, H.H.; Entz, T.; Grant, C.A.; Derksen, D.A. Differential response of weed species to added nitrogen. *Weed Biol. Ecol.* **2003**, *51*, 532–539. [CrossRef]
24. Marshall, E.J.P.; Brown, V.K.; Boatman, N.D.; Lutman, P.J.W.; Squire, G.R.; Ward, L.K. The role of weeds in supporting biological diversity within crop fields. *Weed Res.* **2003**, *43*, 77–89. [CrossRef]
25. Baessler, C.; Klotz, S. Effects of changes in agricultural land-use on landscape structure and arable weed vegetation over the last 50 years. *Agric. Ecosyst. Environ.* **2006**, *115*, 3–50. [CrossRef]
26. Fried, G.; Petit, S.; Dessaint, F.; Reboud, X. Arable weed decline in Northern France: Crop edges as refugia for weed conservation? *Biol. Conserv.* **2009**, *142*, 238–243. [CrossRef]
27. Odum, E.P.; Park, T.Y.; Hutcheson, K. Comparison on the weedy vegetation in old-fields and crop fields on the same site reveals the fallowing crop fields does not result in seedbank buildup of agricultural weeds. *Agric. Ecosyst. Environ.* **1994**, *49*, 247–252. [CrossRef]
28. Zanin, G.; Otto, S.; Riello, L.; Borin, M. Ecological interpretation of weed flora dynamics under different tillage systems. *Agric. Ecosyst. Environ.* **1997**, *66*, 177–188. [CrossRef]
29. Alnenhofen, L.M.; Dekker, J. Complex regulation of *Chenopodium album* seed germination. *Appl. Ecol. Environ. Sci.* **2013**, *1*, 133–142. [CrossRef]

30. Kraska, P.; Andruszczak, S.; Kwiecińska-Poppe, E.; Staniak, M.; Różyło, K.; Rusecki, H. Supporting crop and different row spacing as factors influencing weed infestation in lentil crop and seed yield under organic farming conditions. *Agronomy* **2020**, *10*, 9. [CrossRef]
31. Kwiatkowski, C.A.; Gawęda, D.; Drabowicz, M.; Haliniarz, M. Effect of diverse fertilization, row spacing and sowing rate on weed infestation and yield of winter oilseed rape. *Acta Sci. Pol. Agric.* **2012**, *11*, 53–63.
32. Lins, R.D.; Boerboom, C.M. Effect of Soybean Row Spacing on Weed Competition. In Proceedings of the 2002 Wisconsin Fertilizer Aglime, and Pest Management Conference, Madison, WI, USA, 15–17 January 2002.
33. Primot, S.; Valantin-Morison, M.; Makowski, D. Predicting the risk of weed infestation in winter oilseed rape crops. *Weed Res.* **2006**, *46*, 22–33. [CrossRef]
34. Lutman, P.J.W.; Bowerman, P.; Palmer, G.M.; Whytock, G.P. Prediction of competition between oilseed rape and Stellaria media. *Weed Res.* **2000**, *40*, 255–269. [CrossRef]
35. Feledyn-Szewczyk, B. Comparison of the competitiveness of modern and old winter wheat cultivars in relation to weeds. *J. Res. Appl. Agric. Eng.* **2009**, *54*, 60–67.
36. Trewavas, A. Urban myths of organic farming. *Nature* **2001**, *410*, 409–410. [CrossRef]
37. De Ponti, T.; Rijk, B.; Van Ittersum, M.K. The crop yield gap between organic and conventional agriculture. *Agric Syst.* **2012**, *108*, 1–9. [CrossRef]
38. Du Jardin, P. Plant biostimulants: Definition, concept, main categories and regulation. *Sci. Hortic.* **2015**, *196*, 3–14. [CrossRef]
39. Elad, Y. Biological control of foliar pathogens by means of *Trichoderma harzianum* and potential modes of action. *Crop Prot.* **2000**, *19*, 709–714. [CrossRef]
40. Prokkola, S.; Kivijärvi, P. Effect of biological sprays on the incidence of grey mould, fruit yield and fruit quality in organic strawberry production. *Agric. Food Sci.* **2007**, *16*, 25–33. [CrossRef]
41. Jamiołkowska, A. The influence of bio-preparation Biosept 33 SL on fungi colonizing of sweet pepper plants (*Capsicum annuum* L.) cultivated in the field. *Electron. J. Polish Agric. Univ. Ser. Hortic.* **2009**, *12*, 117.
42. Das, K.; Tiwari, R.K.S.; Shrivastava, D.K. Techniques for evaluation of medicinal plant products as antimicrobial agent: Current methods and future trends. *J. Med. Plants Res.* **2010**, *4*, 104–111. [CrossRef]
43. Escobar, Ä.; Solarte, V. Microbial diversity associated with organic fertilizer obtained by composting of agricultural waste. *Int. J. Biosci. Biochem. Bioinform.* **2015**, *5*, 70–79. [CrossRef]
44. De Pascale, S.; Rouphael, Y.; Colla, G. Plant biostimulants: Innovative tool for enhancing plant nutrition in organic farming. *Eur. J. Hortic. Sci.* **2017**, *82*, 277–285. [CrossRef]
45. Ballaré, C.L.; Mazza, C.A.; Austin, A.T.; Pierik, R. Canopy light and plant health. *Plant Physiol.* **2012**, *160*, 145–155. [CrossRef]
46. Lal, M.; Kanwar, H.S.; Kanwar, R.; Lal, C. Effect of planting density and training on plant health and seed quality of bell pepper (*Capsicum annuum* L.) under protected conditions. *J. Appl. Nat. Sci.* **2016**, *8*, 1219–1222. [CrossRef]
47. Vila, M.; Williamson, M.; Lonsdale, M. Competition experiments on alien weeds with crops: Lessons for measuring plant invasion impact? *Biol. Invasions* **2004**, *6*, 59–69. [CrossRef]
48. Tracy, B.F.; Sanderson, M.A. Forage productivity, species evenness and weed invasion in pasture communities. *Agric. Ecosyst. Environ.* **2004**, *102*, 175–183. [CrossRef]
49. French, R.J.; Smart, W.L.; McCarthy, K. Optimum plant population densities for lupin (*Lupinus angustifolius* L.) in the Western Australian wheat belt. *Aust. J. Exp. Agric.* **1994**, *34*, 491–497. [CrossRef]

MDPI
St. Alban-Anlage 66
4052 Basel
Switzerland
Tel. +41 61 683 77 34
Fax +41 61 302 89 18
www.mdpi.com

Agriculture Editorial Office
E-mail: agriculture@mdpi.com
www.mdpi.com/journal/agriculture

* 9 7 8 3 0